Electricity-Electronics Fundamentals:

A Text-Lab Manual

Second Edition

Paul B. Zbar
Joseph G. Sloop

GREGG DIVISION
McGRAW-HILL BOOK COMPANY
New York St. Louis Dallas San Francisco Auckland Bogotá Düsseldorf Johannesburg London Madrid Mexico Montreal New Delhi Panama Paris São Paulo Singapore Sydney Tokyo Toronto

Library of Congress Cataloging in Publication Data

Zbar, Paul B (date)
Electricity-Electronics Fundamentals.

1. Electric engineering. 2. Electronics. 3. Electric engineering –Laboratory manuals. 4. Electronics–Laboratory manuals. I. Sloop, Joseph G, joint author. II. Title
TK146.Z35 1977 621.3 76–49501
ISBN 0-07-072748-1

Other Books by Paul B. Zbar

Basic Electricity: A Text-Lab Manual, Fourth Edition
Basic Electronics: A Text-Lab Manual, Fourth Edition
Basic Radio: Theory and Servicing, A Text-Lab Manual, Third Edition
Basic Television: Theory and Servicing, A Text-Lab Manual, Second Edition (with P. Orne)
Electronic Instruments and Measurements: A Text-Lab Manual
Industrial Electronics: A Text-Lab Manual, Second Edition

ELECTRICITY-ELECTRONICS FUNDAMENTALS: A Text-Lab Manual, Second Edition

3 4 5 6 7 8 9 0 MUMU 7 8 3 2 1 0 9 8

The editors for this book were Gordon Rockmaker and Alice V. Manning, the designer was Marsha Cohen, the art supervisor was George T. Resch, and the production supervisor was Regina R. Malone. It was set in Times Roman by Kingsport Press, Inc.

CONTENTS

NOTE ON EXPERIMENT CONTENT

Each of the experiments described below is set up in the following manner:

OBJECTIVES. The behavioral objectives are clearly stated.
INTRODUCTORY INFORMATION. The theory and basic principles involved in the experiment are set forth.
SUMMARY. A summary of the salient points in the introductory information is given.
SELF-TEST. A self-test, based on the material included in introductory information, helps students evaluate their understanding of the principles covered prior to the experimental procedure. The self-test should be completed before the experiment is undertaken. *Answers to the self-test questions are given at the end of each experiment.*
MATERIALS REQUIRED. All the materials required to do the experiment, including test equipment and components, are enumerated.
PROCEDURE. A detailed step-by-step procedure is given for performing the experiment.
QUESTIONS. The conclusions reached by the students are brought out by a series of pertinent questions based on the experimental data.

SECTION II—ELECTRONICS

SERIES PREFACE

Faced with a need for increasing numbers of technicians for an expanding industry, as well as faced with a rapidly advancing technology, the Consumer Electronics Group Service Committee of the Electronic Industries Association (EIA), in association with various publishers, has long been active in developing and constantly revising educational materials to meet these challenges. In recent years, the pressing need for training programs to serve students of various backgrounds and abilities has also become evident.

These reasons, coupled with an increasingly critical shortage of trained technicians, especially for home-entertainment electronics servicing, have induced EIA to sponsor the preparation of a wide range of materials in three specific categories, tailored to meet a variety of requirements.

The following paragraphs explain these instructional materials and suggest their proper use to achieve the desired results.

The text-laboratory manuals in the Basic Electricity-Electronics Series provide in-depth, detailed technical materials, including a closely coordinated program of experiments, each preceded by a comprehensive discussion of the objectives, theory, and underlying principles. *Electricity-Electronics Fundamentals* provides introductory electrical-electronic theory and experiments especially suitable for the basic preparation of television and audio service technicians or for other broad-based programs of instruction. *Basic Electricity* and *Basic Electronics* are planned as 280-hour courses, one to follow the other, providing a more thorough background for all levels of technician training. A related instructor's guide is available.

THE BASIC ELECTRICITY-ELECTRONICS SERIES

Title	*Author*	*Publisher*
Electricity-Electronics Fundamentals	Zbar and Sloop	McGraw-Hill Book Company
Basic Electricity	Zbar	McGraw-Hill Book Company
Basic Electronics	Zbar	McGraw-Hill Book Company

The Radio-Television Servicing Series includes materials in two categories: those designed to prepare apprentice technicians for performing in-home servicing and other apprenticeship functions, and those designed to prepare technicians for performing more sophisticated and complicated job skills such as bench-type servicing in the shop.

The apprenticeship servicing group of materials, entitled *Television Sympton Diagnosis,* includes a text, student response manual (workbook) and related film materials. This group is used for a semester course where trouble diagnosis skills are developed through

THE RADIO-TELEVISION SERVICING SERIES

Title	*Author*	*Publisher*
Television Symptom Diagnosis—An Entry into TV Servicing	Tinnell	Howard W. Sams & Co. Inc.
Television Symptom Diagnosis (*film loops or slides*)	Tinnell	Howard W. Sams & Co. Inc. & Hickok Teaching Systems, Inc.
Basic Radio: Theory and Servicing	Zbar	McGraw-Hill Book Company
Basic Television: Theory and Servicing	Zbar and Orne	McGraw-Hill Book Company

audio and video symptom recognition. The television receiver is presented on a functional block diagram basis, and related malfunctions are demonstrated. The student in turn is trained to recognize a symptom and relate it to a given functional block, for the purpose of effective servicing. A complete instructor's guide for this group of materials is provided.

The related film materials for *Television Symptom Diagnosis* come in two forms, a set of color-sound motion picture film loops or a set of color slides for use with any standard slide projector. The film medium is especially suitable for students who may have reading and, in turn, learning difficulties. The film materials can be used, along with the student response manual, as an integrated learning system to teach color-television adjustment and set-up procedures, trouble symptom diagnosis, and, in turn, the ability to isolate troubles to a given stage. Or the film materials can be used independently to supplement any color television servicing course.

Since only a minimum of theory is presented in *Television Symptom Diagnosis,* apprentices completing this program can be expected to progress to the following more comprehensive text-lab manuals for the purpose of deepening their understanding of electronics theory and procedures for the bench servicing of television and audio-electronics products.

The text-lab manuals for bench servicing are *Basic Radio: Theory and Servicing* and *Basic Television: Theory and Servicing.* These books provide a series of experiments, with preparatory theory, designed to teach the in-depth theory and procedures required for servicing all types of consumer electronics products. An instructor's guide for these books is also available.

THE INDUSTRIAL ELECTRONICS SERIES

Title	*Author*	*Publisher*
Industrial Electronics	Zbar	McGraw-Hill Book Company
Electronic Instruments and Measurements	Zbar	McGraw-Hill Book Company

Materials for basic laboratory courses in industrial control and computer circuits and laboratory standard measuring equipment are provided by the Industrial Electronics Series and their related instructor's guides. *Industrial Electronics* is concerned with the fundamental building blocks in industrial electronics technology, giving the student an understanding of the basic control circuits and their applications. *Electronic Instruments and Measurements* fills the need for basic training in the complex field of industrial instrumentation. Prerequisites for both courses are *Basic Electricity* and *Basic Electronics.*

The foreword to the first edition of the EIA co-sponsored basic series said: "The aim of this basic instructional series is to supply schools with a well-integrated, standardized training program, fashioned to produce a technician tailored to industry's needs." This statement is still the objective of the varied training program that has been developed through joint industry educator-publisher cooperation.

V. J. Adduci, President
Electronic Industries Association

AUTHORS' PREFACE

In response to the suggestions of educators using the first edition of *Electricity-Electronics Fundamentals*, the second edition is an updated, self-pacing text-laboratory manual which presents *single concept* experiments. Employing modern educational methods, it features behavioral objectives and self-learning techniques. It is intended for use in vocational schools, in industrial and military training programs, and in industrial arts programs.

This book provides an industry-recommended, tested laboratory program in dc and ac circuits; in transistors, ICs, and other solid-state devices and circuits; in vacuum tubes; and in digital circuits. *Electricity-Electronics Fundamentals is* admirably suited as an *introductory course* for *radio and television technicians*. The 103 experiments provide students with "live" experiences in the use of those test instruments they will find in industry. A down-to-earth, *practical* approach characterizes the book throughout.

There are significant changes in the second edition. The first is related to experiment content—each experiment deals with a single concept. The student's total effort is concentrated on mastering just that one concept before going on to the next. This approach improves understanding and reduces the complexity that tends to inhibit learning. To facilitate the single-concept approach, related experiments are grouped in a single chapter. For example, chapter 9 deals with troubleshooting resistive circuits. It consists of three single-concept experiments: Experiment 9.1, Troubleshooting a Series Circuit; Experiment 9.2, Troubleshooting a Parallel Circuit; Experiment 9.3, Troubleshooting a Series-Parallel Circuit. The slower student may complete the first of these experiments, while the fastest student may do all the experiments in this group. Self-paced learning is therefore encouraged by this arrangement.

In a second major change, the experiments' objectives are formulated in measurable behavioral terms. At the conclusion of each experiment, student and teacher can evaluate the results of that learning experience in terms of the stated objectives.

In a third major change, a detailed summary follows the introductory information, in which the principles of the experiment are discussed. Immediately after the Summary comes a self-test (questions with answers) to help the students determine how well they have understood the principles of the experiment. The results of the self-test will indicate to the students their readiness to undertake the experimental procedure which follows, or they may suggest a need to review the basic principles. The presentation of objectives, introductory information, summary, self-test, and answers to self-test will therefore pace the students' progress and prepare them for a meaningful hands-on experience in the procedural steps of the experiment.

In a fourth change, new solid-state devices have been further emphasized, and vacuum tubes de-emphasized. The electronics experiments in the second edition are approximately 85 percent solid state and 15 percent vacuum tube.

New topics featuring late developments in the state of the art are included. These are: complementary-symmetry push-pull amplifiers, field-effect transistors and amplifiers, digital ICs, AND/OR/NOT gates, NOR and NAND logic, binary addition and binary adders, light-emitting diodes (LEDs), light-dependent resistors (LDRs), and magnetism and electromagnetism. Experiments on measurement of power and rms and peak-to-peak values are also included.

The second edition provides greater emphasis on troubleshooting. There are troubleshooting experiments for: (1) resistive circuits, (2) power-supply circuits, (3) transistor amplifiers, (4) audio amplifiers, (5) vacuum-tube amplifiers, and (6) IC linear amplifiers.

As in the first edition, the introductory information contains basic discussion, presented clearly and simply, of circuits and of testing and measuring techniques. Solutions of sample problems are included, as are graphs and interpretations of these graphs. Stimulating questions at the end of each experiment oblige the student to evaluate the experimental data and draw the necessary conclusions.

The authors wish to thank for their guidance and help the members of the Service Education Subcommittee of the Consumer Electronics Group of the Electronic Industries Association:

Ed Mueller (Quasar), Chairman
Ray Guichard (Magnavox), Vice Chairman
*Jack Berquist (GTE Sylvania)
Jim Farrell (Dynascan)
Howard Gerrish (CEG Consultant)
Joe Groves (Howard Sams)
Frank Hadrick (Zenith)
†Bob Hannum (GE)
Ed Milbourn (RCA)
*Jack Moore (Hickok)
Andy Murnick (Admiral)
Irv Rebeschini (Simpson)
*Gordon Rockmaker (McGraw-Hill)
Don Sabatini (Panasonic)
Stan Stamatis (Simpson Electric)
Frank Steckel (CEG Consultant)
Chuck Vollmer (Zenith)
Marlin Westra (Sencore)
Marv Whittenberg (Admiral)
Jack Zupko (Panasonic)

† indicates the chairman and * the members of the Textbook Advisory Review Committee.

Thanks also to Ed Bedford (Brodhead-Garrett) for his review of the manuscript.

Acknowledgment is also made for permission to use equipment and components to: B & K Dynascan Corp., Fairchild Semiconductor Co., General Electric Co., and Hickok Teaching Systems, Inc. Experiments were performed on equipment furnished by Hickok Teaching Systems, Inc.

And last, but not least, the authors wish to thank their wives, May and Andrea, for their encouragement, suggestions, and many hours of proofreading.

Paul B. Zbar
Joseph G. Sloop

SAFETY

Electronics technicians work with electricity, electronic devices, motors, and other rotating machinery. They are often required to use hand and power tools in constructing prototypes of new devices or in setting up experiments. They use test instruments to measure the electrical characteristics of components, devices, and electronic systems. They are involved in any of a dozen different tasks.

These tasks are interesting and challenging, but they may also involve certain hazards if technicians are careless in their work habits. It is therefore essential that the student technicians learn the principles of safety at the very start of their career and that they practice these principles.

Safe work requires a careful and deliberate approach to each task. Before undertaking a job, technicians must understand what to do and how to do it. They must plan the job, setting out on the work bench in a neat and orderly fashion tools, equipment, and instruments. Extraneous items should be removed, and cables should be secured as far as possible.

When working on or near rotating machinery, loose clothing should be anchored, ties firmly tucked away.

Line (power) voltages should be isolated from ground by means of an isolation transformer. Power-line voltages can kill, so these should *not* be contacted with the hands or body. Line cords should be checked before use. If the insulation on line cords is brittle or cracked, these cords must *not* be used. TO THE STUDENT: Avoid direct contact with any voltage source. Measure voltages with one hand in your pocket. Wear rubbersoled shoes or stand on a rubber mat when working at your experiment bench. Be certain that your hands are dry and that you are not standing on a wet floor when making tests and measurements in a live circuit. Shut off power before connecting test instruments in a live circuit.

Be certain that line cords of power tools and non-isolated equipment use safety plugs (polarized 3-post plugs). Do not defeat the safety feature of these plugs by using ungrounded adapters. Do not defeat any safety device, such as fuse or circuit breaker, by shorting across it or by using a higher amperage fuse than that specified by the manufacturer. Safety devices are intended to protect you and your equipment.

Handle tools properly and with care. Don't indulge in horseplay or play practical jokes in the laboratory. When using power tools, secure your work in a vise or jig. Wear gloves and goggles when required.

Exercise good judgment and common sense and your life in the laboratory will be safe, interesting, and rewarding.

FIRST AID

If an accident should occur, shut off power immediately. Report the accident at once to your instructor. It may be necessary for you to render first aid before a physician can come, so you should know the principles of first aid. A proper knowledge of these may be acquired by attendance at a Red Cross first-aid course.

Some first-aid suggestions are set forth here as a simple guide.

Injured people should be kept lying down until medical help arrives. They should be kept warm to prevent shock. Do not attempt to give them water or other liquids if they are unconscious. Be sure nothing is done to cause further injury. Keep them comfortable and cheerful until medical help arrives.

ARTIFICIAL RESPIRATION

Severe electrical shock may cause stoppage of breathing. Be prepared to start artificial respiration at once if breathing has stopped. The two recommended techniques are:

1 Mouth-to-mouth breathing, considered the most effective
2 Schaeffer method

These techniques are described in first-aid books. You should master one or the other so that if the need arises you will be able to save a life by applying artificial respiration.

These instructions are not intended to frighten you but to make you aware that there are hazards attendant upon the work of an electronics technician. But then there are hazards in every job. Therefore you must exercise common sense, good judgment, and safe work habits in this, as in every job.

LETTER SYMBOLS

As noted in the Author's Preface, primary emphasis in this manual has been placed on semiconductor (solid-state) devices and circuits. However, vacuum tubes and their associated circuits are also treated, making it desirable to use letter symbols that have the same meaning throughout the text for both solid-state and vacuum-tube circuits. Accordingly, the IEEE (Institute of Electrical and Electronics Engineers) Letter Symbols for Semiconductor Devices (IEEE Standard #255) were used, with modifications for vacuum tubes.

The following summary of symbols for electrical quantities is intended to clarify their use throughout the text.

Quantity Symbols

1 Instantaneous values of current, voltage, and power, that vary with time, are represented by the lowercase letter of the proper symbol.

Examples: i, v, p

2 Maximum (peak), average (direct current), and root-mean-square values of current, voltage, and power are represented by the uppercase letter of the appropriate symbol.

Examples: I, V, P

Subscripts for Quantity Symbols

1 Direct-current values and instantaneous total values are indicated by uppercase subscripts.

Examples: i_C, I_C, v_{EB}, V_{EB}, p_C, P_C

2 Alternating-component values are indicated by lowercase subscripts.

Examples: i_c, I_c, v_{eb}, V_{eb}, p_c, P_c

3 Symbols to be used as subscripts:

- E, e emitter terminal
- B, b base terminal
- C, c collector terminal
- A, a anode terminal
- K, k cathode terminal
- G, g grid terminal
- P, p plate terminal
- M, m maximum value
- Min, min minimum value

Examples:

- I_E emitter direct-current (no alternating current component)
- I_e rms value of alternating component of emitter current
- i_e instantaneous value of alternating component of emitter current

4 Supply voltages may be indicated by repeating the terminal subscript.

Examples: V_{EE}, V_{CC}, V_{BB}, V_{PP}, V_{GG}

The one exception to this system is the occasional use of $V+$ for the plate supply voltage of a tube. Note that $V+$ replaces the more usual $B+$.

5 The first subscript designates the terminal at which current or voltage is measured with respect to the reference terminal, which is designated by the second subscript.

CHAPTER 1 USING THE MULTIMETER

EXPERIMENT 1.1. Electronic Voltmeter (EVM)—Familiarization

OBJECTIVES

1. To identify the operating controls and write down the function of each control of an electronic voltmeter (EVM)
2. To identify each of the leads of a VTVM/TVM (vacuum-tube voltmeter/transistor voltmeter) or DVM (digital voltmeter)
3. To set the zero and ohms adjust on the meter

INTRODUCTORY INFORMATION

Electronic Voltmeter

Voltage measurements are made primarily with an instrument called a *voltmeter*. There are other instruments which may be used to measure voltage, for example, the oscilloscope, but voltage measurement is not the primary purpose of these devices.

Two types of voltmeter are used today, (1) the analog and (2) the digital. A pointer on the analog meter indicates the voltage on a calibrated scale. The voltage measured by a digital meter appears as a number on a numerical (digital) display, so that any one who reads numbers can read the voltage.

The reading errors associated with analog meters (these are discussed in the experiment) are eliminated by the digital voltmeter (DVM). Moreover, DVMs are more accurate than analog meters. Despite the fact that digital meters are more costly than analog meters, their advantages are leading to their widespread adoption.

Another way of identifying voltmeters is to specify them as (1) electronic voltmeters or (2) electromechanical meters. The vacuum-tube voltmeter (VTVM), the transistorized voltmeter (TVM), and the digital voltmeter (DVM) are examples of electronic voltmeters (EVMs). The volt-ohm-milliammeter (VOM) is an example of an electromechanical multimeter which is used to measure voltage.

Meters may be single-function devices, such as the voltmeter, which can measure only voltage. Or meters may be multipurpose devices such as the multimeter, which is used to measure a variety of electrical quantities, such as voltage, current, and resistance. The VTVM, the TVM, the DMM (digital multimeter) or the digital volt-ohm-milliammeter (DVOM), and the VOM are examples of multimeters.

In this experiment we shall be concerned with electronic multimeters which we shall designate simply as EVMs. When an EVM is specified in this experiment or in those which follow, any one of the three types of electronic multimeter (TVM, DVOM, or VTVM) may be used.

Prior to the invention of transistors, the VTVM (vacuum-tube voltmeter) was the electronic meter used to measure dc and ac voltage and resistance. On rare occasions it also included a current-measuring function. VTVMs are still in use in industry and in school laboratories. However, they are being replaced by the TVM and the DVM. The TVM and DVM include all the measuring functions of a VTVM plus current-measuring capability.

Electronic voltmeters sometimes have other specialized scales such as the decibel (dB) scale for sound-level measurement.

It is evident that the transistor part of the name TVM stems from the use of transistors and other solid-state devices, while in the VTVM, vacuum tubes are employed to perform similar functions. The electromechanical VOM uses neither vacuum tubes nor transistors. The general-purpose VTVM is usually a line-operated instrument—it is plugged into an electrical outlet and derives its operating power from the line. TVMs and DVMs are either battery or line operated, and frequently they contain facilities for both types of power operation.

Operating Controls (VTVM)

Figure 1.1-1*a* is an illustration of a typical vacuum-tube voltmeter. Note that the operating controls and meter scales are on the front panel. Of the four controls, two are labeled "Zero" and "Ohms." Though the other two are not identified by name, their functions are apparent. The lower left-hand knob might

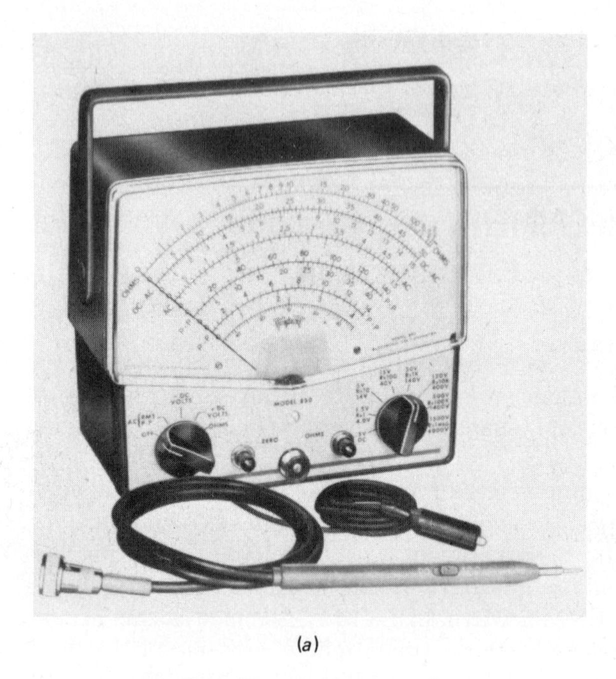

(a)

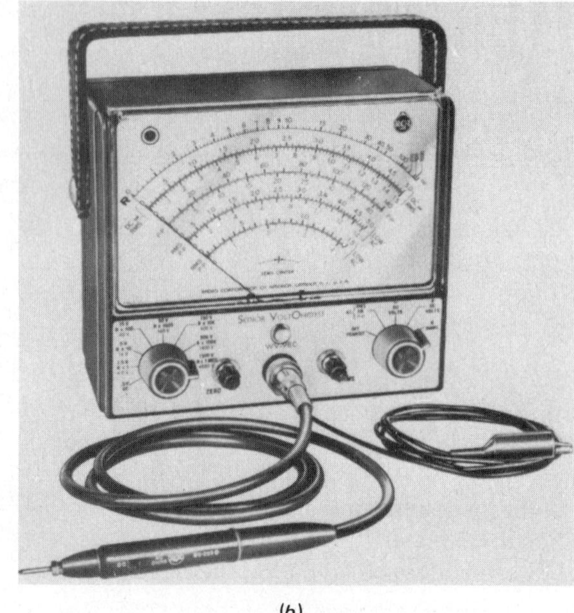

(b)

Fig. 1.1-1. (a) VTVM (*The Triplett Electrical Instrument Co.*); (b) VTVM. (*Radio Corp. of America*)

well be called the function switch. The instrument is turned ON when it is switched from OFF to any of its four other settings. The "–Volts" and "+Volts" positions are for measurement of positive and negative dc volts. The "AC Volts" position is for measuring ac voltage. The "Ohms" position is for measuring resistance.

The lower right-hand control is a range switch. It works in conjunction with the function switch to select the proper operating range for a specific measurement. For example, if technicians wish to measure positive dc volts and know that the voltage they will be measuring lies between 5 and 15 V, they will turn the function switch to +Volts, and the range switch to 15. Note that the range number represents the maximum voltage which can be read on that range. The meter in Fig. 1.1-1*a* has eight voltage ranges, 0.5 to 1500 V.

The range switch shown in Fig. 1.1-1*a* acts also as a resistance-scale multiplier. Thus, the 15-V position is also the $R \times 100$ position. When resistance is being measured, the function switch is set on ohms. Resistance readings on the R scale must be multiplied by 100 on the $R \times 100$ position. On the $R \times 1000$ position, they must be multiplied by 1000, etc.

The purpose of the Zero control is to set the meter pointer on zero (at the left side) of the meter scales. When the meter is first turned to any voltage position, the pointer may go off zero during the warmup period. After the meter has been ON for about five minutes, it should stabilize and the pointer should return to its zero setting. If it does not, it can be brought to zero by the Zero adj.

When the function switch is in the Ohms position, the meter pointer will move all the way to the right, on the R scale. The pointer should line up on the last calibration mark on the R (top) scale. If it does not, the Ohms adj. will bring the pointer into position. It should be noted that the Zero control must be properly set, *before* Ohms control is adjusted.

There are many variations among manufacturers both as to appearance of the instrument and as to operating controls. In the VTVM of Fig. 1.1-1*b*, for example, the function switch is on the right, and the range selector on the left. The two controls labeled "Zero" and "Ohms" are comparable to those on the previously discussed VTVM.

Operating Controls (TVM)

The operating controls on a TVM are very similar to those on a VTVM. In Fig. 1.1-2, for example, a function switch is on the lower left, a range switch is on the lower right, and there are a Zero ("0") control and an Ohms (Ω) control. The Zero and Ohms controls are adjusted like the same controls on the VTVM, and the range switch is similar in operation to that of a VTVM. It should be noted, however, that the range switch has 12 positions, including a low 0.05 dc volts position and a 0.005 ac volts position, as contrasted with the 8 positions on the VTVMs

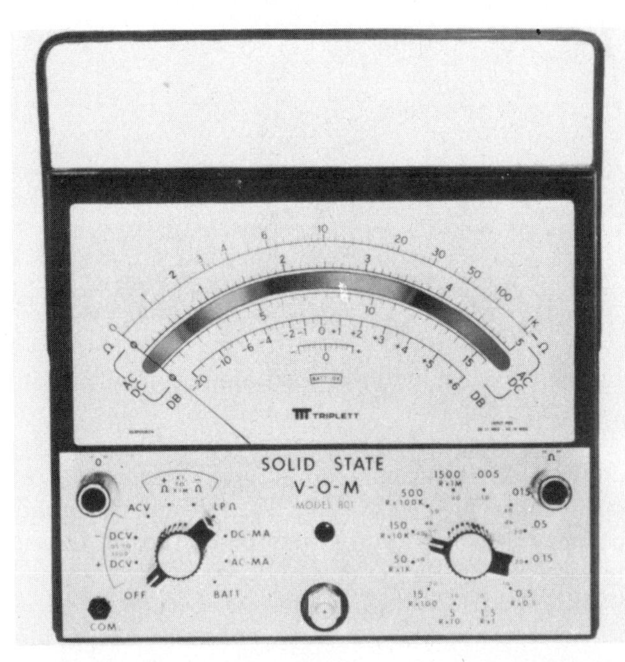

Fig. 1.1-2. TVM: solid state VOM. (*The Triplett Electrical Instrument Co.*)

in Fig. 1.1-1. The function switch is also more complex, reflecting the greater versatility of the TVM. Thus, in addition to the OFF, + and − DCV, ACV, and conventional ohms (Ω) positions, there are LPΩ (low-power ohms), DC-MA, AC-MA, and Batt. positions. The low-power ohms function is used for resistance measurements in solid-state circuits. DC and AC-MA positions are for measuring, respectively, dc and ac current, and the Batt. position is used to test the battery which powers this instrument. This is *not* a line-operated TVM.

In TVMs, also, there are variations among manufacturers both as to appearance of the instrument and operating controls.

Operating Controls (DVOM)

There are two operating controls on the DVOM in Fig. 1.1-3. On the right is a six-position function switch for selecting any one of the following measurement functions: (1) alternating current, (2) direct current, (3) resistance, (4) dc volts, or (5) ac volts. The fourth position on the switch labeled "Off" shuts power off. The left-hand control is a six-position range switch. The ranges are common to each of the measurement functions. The control positions shown in Fig. 1.1-3 indicate that the meter is set to measure resistance and that the maximum resistance which can be measured on the range shown is 20,000 Ω.

An unusual feature of this digital instrument is an analog scale which uses an edgewise analog meter in addition to its digital capabilities. The analog readings are useful in certain electronic applications.

Meter Scales

In general, the voltage scales on a TVM and VTVM are linear, the ohms scale is nonlinear.

The scale on a ruler is the most familiar example of a linear scale. On a ruler equal distances are represented by equally spaced calibrations. Thus, a 1-in distance is represented by the same interval whether it is between 1 and 2 in or between 9 and 10 in. This same relationship holds true for any linear scale; that is, the distance between consecutive corresponding calibration marks is everywhere the same on the scale.

The voltmeter scale (Fig. 1.1-4) is another example of a linear scale. Here the meter pointer travels equal distances along the scale *arc* for equal voltage changes. Thus, the *arc* distance between 0 and 10 V is the same as between 10 and 20 V or between 90 and 100 V.

In reading the scale in Fig. 1.1-4 it is necessary to

Fig. 1.1-3. DVOM: solid-state digital VOM. (*Simpson Electric Co.*)

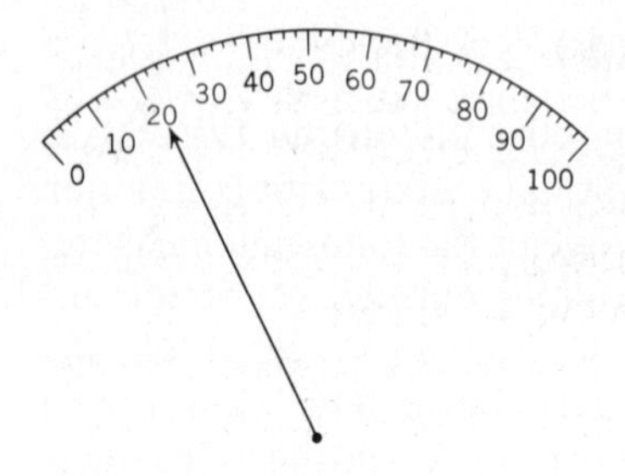

Fig. 1.1-4. Linear meter scale.

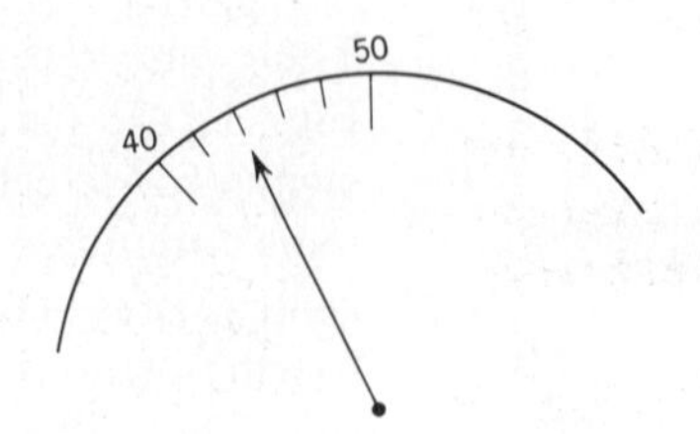

Fig. 1.1-5. Meter pointer reads 44.

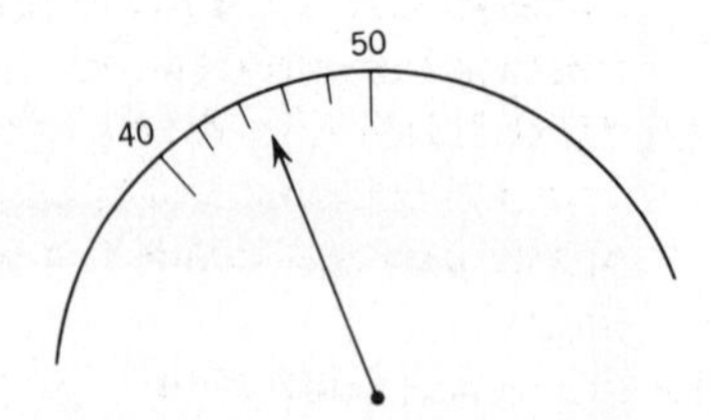

Fig. 1.1-6. Meter pointer reads 45.

supply numbers for the unnumbered calibrations. This is easily accomplished. For example, there are five equally spaced intervals between numbered voltage calibrations. Each of these distances represents one-fifth of the voltage difference between the numbered calibrations. In the case of Fig. 1.1-4, every calibration marker would therefore represent a 2-V interval. Thus, in Fig. 1.1-5, the meter pointer reads 44 V, that is, 40 V plus 2 intervals × 2 V per interval.

When the pointer falls between scale markers, the reading is approximated. For example, in Fig. 1.1-6 the reading is 45 V. As you can see, reading linear meter scales is like reading the speedometer on your car. When the needle pointer of the speedometer points halfway between 40 and 50, the speed of the car is 45 miles per hour. Keep this in mind as you read the scales of the multimeter.

The voltage scales of a voltmeter are not always calibrated in the same units as the range setting. For example, the scale of Fig. 1.1-4 would serve admirably for a 0- to-100-V range. However, it could serve as well for a 0- to-10-V range, or a 0- to-500-V range. In the case of a 10-V range, every scale reading must be divided by 10. For the 500-V range, every scale reading must be multiplied by 5.

EVM Leads

A TVM, a DVM, and a VTVM usually require two or three leads. The three-lead arrangement normally consists of a flexible common ground (black), a shielded coaxial lead for direct current, and a shielded coaxial lead for alternating current and ohms. The ground lead is terminated in an alligator clip which is clipped onto the common return or ground of the circuit. The two coaxial leads are terminated in probes, completely insulated except for the metallic tip which extends from the end of the probe. The technician holds the insulated probe in measuring and does not come into contact with the conductive tip which is applied to the point of measurement. Since there are other variations of the three-lead arrangement, technicians should consult the instruction manual for the meter in use until they are entirely familiar with the instrument.

The two-lead arrangement shown in Fig. 1.1-1 includes a flexible common "ground" lead and a coaxial cable. The coaxial cable is terminated in a probe. There is a two-position switch on the probe. In one position the cable serves as a dc voltage lead. In the other it is an ac voltage or ohms lead.

The meter end of the common ground lead is terminated in a pin tip or banana plug, which plugs into an appropriate jack on the meter. The coaxial cable is connected to the meter by means of a microphone-type connector. Leads should be left on the meter when it is not in use.

SUMMARY

1. The TVM (transistorized voltmeter), DVM (digital voltmeter), and VTVM (vacuum-tube voltmeter) are classified as electronic voltmeters, EVMs.
2. Electronic voltmeters are used to measure dc and ac voltages and resistance. The TVM and DVM also have current-measuring ranges.
3. Electronic voltmeters are line-powered or battery-powered.
4. The meter operating controls include: (*a*) function switch, (*b*) range switch, (*c*) zero adjust (*d*) ohms adjust. The DVM does not have zero or ohms adjusts.
5. The function switch selects the desired measurement function: voltage, resistance, or current.
6. The range switch selects the proper operating range for a particular measurement.
7. The zero adjust sets the meter pointer at the zero calibration marker on the left side of the meter scales.
8. The ohms adjust is used to line up the meter pointer with the last calibration marker (infinity) on the ohms scale.
9. The voltage and current scales of an electronic voltmeter are linear.
10. The ohms scale is nonlinear.

11. The electronic voltmeter uses two or three leads. In both cases a flexible black lead serves as the common or ground lead.
12. In a three-lead meter, the other two leads are coaxial (shielded) leads, one for dc measurements and the other for ac and resistance readings.
13. In a two-lead meter, the single coaxial lead is terminated in a probe containing a two-position switch. One switch position is for dc measurements; the other position for ac and resistance measurements.

SELF-TEST

Check your understanding by answering these questions.

1. The two types of instruments used for measurement of electrical quantities are the electronic voltmeter and the ________.
2. The meter which employs transistors in its circuitry is called a ________.
3. VTVMs are normally ________-powered devices.
4. Electronic voltmeters usually have four operating controls. These are: (*a*) ________; (*b*) ________; (*c*) ________; and (*d*) ________.
5. The purpose of each control listed in the answer to question 4 is: (*a*) ________; (*b*) ________; (*c*) ________; and (*d*) ________.
6. An EVM has two or three leads. These are for: (*a*) ________; (*b*) ________; (*c*) ________ and ________.
7. The voltage and current scales of an electronic voltmeter are linear. ________ (true/false)
8. Refer to Fig. 1.1-4. The range switch is set to the 10-V range. What is the meter reading? ________ V

MATERIALS REQUIRED

- Equipment: EVM

PROCEDURE

1. Examine the meter assigned to you. Draw a panel view of the meter, showing operating controls and the functions and ranges associated with the switches. Draw also the voltage scales and the ohm scale.
2. Read the instruction manual and learn how to operate the meter.
3. Turn the EVM **on.** Set the function switch on +dc volts. Permit the meter to warm up if required. Then vary the Zero adj. control. Observe its effect on the pointer. Set the pointer on zero. Now turn the range switch through every setting and note if the meter pointer remains on zero.
4. Check the zero setting of the meter on each range of −dc volts, and on each range of ac volts.

NOTE: A well-designed, properly adjusted meter should maintain its zero setting on every voltage function and range.

5. Set the function switch on ohms and range switch on $R \times 1$. Be sure the lead tips do not touch each other or a metal object. Vary Ohms adj. and observe its effect on the pointer. Set this control so that the pointer rests on the maximum calibration on the resistance scale. In this position the resistance circuit is open and the meter registers infinite resistance. Check the setting of the pointer on every position of the range switch ($R \times 10$, $R \times 100$, etc.).
6. Short the ohms and ground leads (that is, touch the metal tips of these leads). The pointer should swing to zero. If it does not come to rest on zero, with the leads shorted, set it on zero with the *Zero adj.* Now open the leads and check the position of the pointer. If it is no longer on the maximum-resistance marker (infinity), set it there with the Ohms adj.
7. Recheck the zero and infinity positions.

NOTE: Do not leave meter leads shorted together on Ohms for any length of time. In the Ohms position, an internal battery is connected in the circuit. The battery voltage will be greatly reduced in a short period of time if the leads remain shorted together.

QUESTIONS

1. List the controls on the panel of your EVM and state the purpose of each.
2. Is it possible to use the same scale on the 3-V range as on the 300-V range of your meter? Explain.
3. Draw a linear scale with number calibrations 0, 1, 2, etc., through 10. Set off each major subdivision into ten minor subdivisions. Show where *8.7* would be on your scale.
4. Is the ohms scale on your meter linear or nonlinear? Justify your answer by referring specifically to resistance calibrations.
5. Explain in detail how you would zero the meter for dc voltage readings. Identify the control or controls you would use.

6. Explain in detail how you would check to determine if the meter is properly zeroed on the $R \times 1$ range. Identify the controls you would use.
7. If in question 6 you find the meter is not properly zeroed, explain in detail how you would zero it.

Answers to Self-Test

1. electromechanical VOM
2. TVM or solid-state VOM
3. line
4. (*a*) function; (*b*) range; (*c*) zero; (*d*) zero ohms
5. (*a*) Select the operating function; voltage, current or resistance; (*b*) select the appropriate range of measurement; (*c*) set the pointer at zero on the left-hand side of the scales; (*d*) set the pointer at infinity on the right-hand side of the ohms scale.
6. (*a*) common or ground; (*b*) direct current; (*c*) alternating current and resistance
7. true
8. 2

EXPERIMENT 1.2. Measuring Resistance

OBJECTIVES

1. To calibrate (to "zero") the ohmmeter
2. To measure resistance of different value resistors

INTRODUCTORY INFORMATION

The Ohms Scale

The service technician must be able to measure resistance accurately. This measurement is probably the most often used "check" in electronic servicing to find bad components and circuits.

The *ohm* is the unit of resistance, and the symbol for ohm is Ω (Greek letter *omega*).

The measuring of resistance is one of the functions of an electronic voltmeter (EVM) or volt-ohm-milliammeter (VOM). Each manufacturer provides operating instructions for the use of each particular instrument. It will therefore be necessary to refer to the instruction manual before using any electronic voltmeter. The student should be thoroughly familiar with the operation of the ohmmeter function before attempting to use it in this experiment.

One fact common to all EVMs is that they contain a basic resistance scale from which readings are made in ohms directly on the $R \times 1$ range of the meter. Figure 1.2-1 shows an ohmmeter scale. Notice that the scale is not linear. That is, the arc distance between numbers is not equal. Also, note that not all scale calibrations are numbered, and so the technician has to supply numbers for the unnumbered markers. For example, if the pointer is on the second mark to the right of 3, between 3 and 4 in Fig. 1.2-1, the corresponding value on the $R \times 1$ range is 3.4 Ω.

To read resistance values greater than the maximum

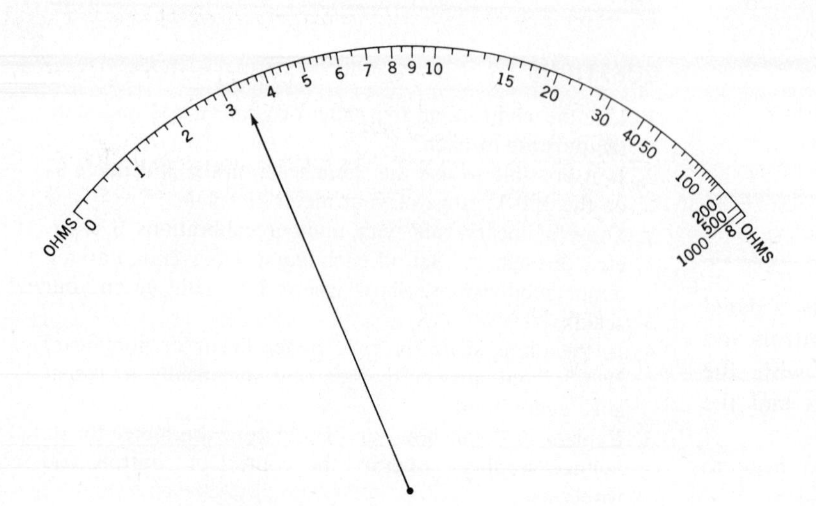

Fig. 1.2-1. Ohmmeter scale. On $R \times 1$ range, pointer shows 3.4 Ω.

value shown on the basic scale, a higher range must be selected. So, in addition to the $R \times 1$ range there will usually be found $R \times 100$, $R \times 10{,}000$, $R \times 100{,}000$, etc., ranges on the meter. In the $R \times 100$ range, any reading made on the basic scale must be multiplied by 100. In the $R \times 10{,}000$ range, any reading made must be multiplied by 10,000, etc.

If the EVM to be used has a digital or numerical readout rather than the conventional meter scale with needle pointer, the range setting usually indicates the maximum resistance which can be measured with the meter when it is set to this particular range. For instance, the range switch may indicate R 1000 Ω, meaning that the maximum resistance which could be measured at this setting is 1000 Ω.

Zeroing the Meter

All nondigital ohmmeters and some digital ohmmeters have a zero ohms control. This control is used to calibrate or adjust the meter to zero. This is similar to setting a clock to the correct time. It is necessary to make sure the meter is zeroed before each measurement and each time ranges are changed.

SUMMARY

1. The ohmmeter measures resistance.
2. Resistance is measured in ohms.
3. In reading resistance, it is sometimes necessary to multiply the scale reading by a number on the range switch.
4. It is necessary to zero the ohmmeter before each measurement when using a nondigital ohmmeter.
5. It is necessary to zero the ohmmeter when the range is changed.

SELF-TEST

Check your understanding by answering these questions.

1. The measuring of resistance is one function of a(n) EVM.
2. The OHMMETER is used to measure resistance.
3. One type of ohmmeter may have a DIGITAL readout rather than the conventional scale and needle pointer.
4. The ohmmeter must be ZEROED before measuring resistance.
5. In order to read resistance values greater than the maximum value shown on the basic scale, a HIGHER range must be selected.

MATERIALS REQUIRED

- Equipment: EVM
- Resistors: Five, different values

PROCEDURE

1. Secure the resistors to be used for this experiment.
2. Refer to the instruction manual of the EVM for procedure in zeroing the meter.
3. Zero the meter.
4. Refer to the instruction manual of the EVM for the procedure to measure resistance.
5. Measure each resistor with the ohmmeter.

NOTE: Do not touch both leads of the resistor or the metal parts of the test probes while measuring resistance as this will result in a reading error, especially on the higher-resistance ranges.

6. Fill in the information required in Table 1.2-1.

TABLE 1.2-1. Measuring Resistors with an Ohmmeter

Resistor Number	*1*	*2*	*3*	*4*	*5*
Measured value	27 Ω	2 Ω	985 Ω	677 KΩ	840 Ω

QUESTIONS

1. What resistance is in the center of your ohmmeter scale, $R \times 1$ range?
2. At which end of the scale are resistance readings more accurate, the crowded or uncrowded end?
3. What is the symbol for ohms?
4. What happens to the ohmmeter reading when both leads of the resistor being measured are held in your fingers as the reading is being made?

Answers to Self-Test

1. EVM
2. ohmmeter
3. digital
4. zeroed
5. higher

CHAPTER 2 RESISTANCE: SERIES AND PARALLEL CIRCUITS

EXPERIMENT 2.1. Resistor Color Code

OBJECTIVES

1. To learn the Electronic Industries Association (EIA) color code
2. To gain practice in the use of the ohmmeter

INTRODUCTORY INFORMATION

Color Code

Resistance values are indicated by a standard color code adopted by manufacturers. This code involves the use of color bands or dots on the body of the resistor. The colors and the numerical values are given in the Resistor Color Chart, Table 2.1-1.

The basic resistor types are shown in Fig. 2.1-1. Type A uses color bands. The color of the first band tells the first significant figure of the resistor. The color of the second band tells the second significant figure. The color of the third band tells the multiplier (number of zeros to be added or the placement of the decimal point). A fourth color band is used for tolerance designation. The absence of the fourth color band means that the resistor has a 20 percent tolerance.

In Fig. 2.1-1 resistor type A is color-coded red, red, black, gold. Its value would be 22 Ω at 5 percent tolerance. Resistor types B and C are now obsolete, but may be found in old electronic devices.

In type B, the color of the body is the first figure, the color band at one end of the resistor is the second figure, the dot is the multiplier. Tolerance is shown by a gold or silver band, or a band of no color at the other end. The system employed in type C is the same as type B except that a color band is used as the multiplier rather than a dot.

Resistor type B is coded red body, yellow band at one end, orange dot, and silver band at the other end. Its value is, therefore, 24,000 Ω at 10 percent tolerance.

TABLE 2.1-1. Resistor Color Chart

Resistors, EIA and MIL		*Significant figures*	*Color*
Multiplier	*Tolerance, %*		
1	–	0	Black
10	–	1	Brown
100	–	2	Red
1000	–	3	Orange
10,000	–	4	Yellow
100,000	–	5	Green
10^6	–	6	Blue
10^7	–	7	Violet
10^8	–	8	Gray
10^9	–	9	White
0.1	5	–	Gold
0.01	10	–	Silver
–	20	–	No color

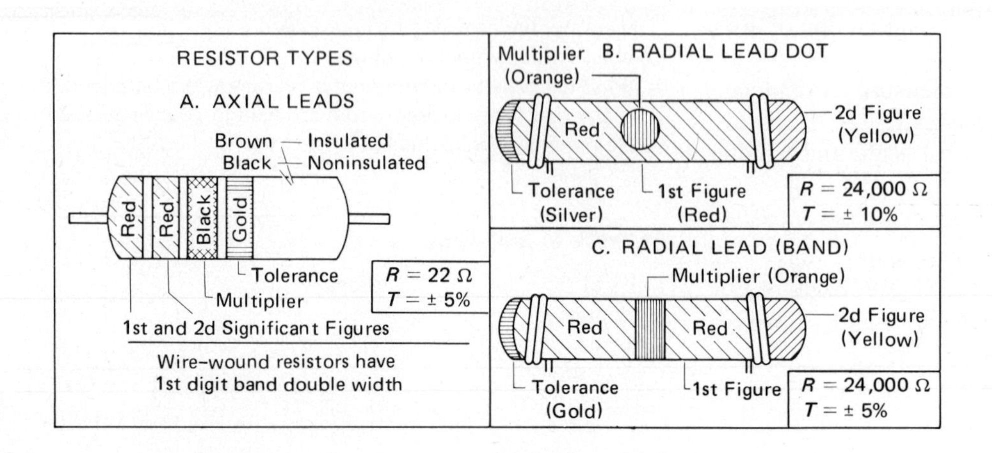

Fig. 2.1-1. Basic resistor types. (*Centralab, a Division of Globe-Union, Inc.*)

Resistor type C is coded red body, yellow band at one end, orange band in center, gold band at the other end. Its value is 24,000 Ω at 5 percent tolerance.

In case of a resistor whose value is less than 1 Ω, the multiplier is silver (band or dot). In the case of a resistor whose value is greater than 1 Ω but less than 10 Ω, the multiplier is gold.

Wire-wound, high-wattage resistors usually are not color-coded, but have the resistance value printed on the body of the resistor.

Resistors with brown body color are insulated; those with black body color are not insulated.

In writing the values of resistors, the following designations are employed:

k, a multiplier which stands for 1000

M, a multiplier which stands for 1,000,000

For example, 33 kilohms (33 kΩ) stands for 33,000 ohms; 1.2 megohms (1.2 MΩ) stands for 1,200,000 ohms.

SUMMARY

1. The unit of resistance is the ohm.
2. The body of a fixed carbon resistor is color-coded to specify its ohmic value and tolerance.
3. The first three bands on the resistor designate its resistance.
4. The fourth band on the resistor designates its tolerance.
5. High-wattage, wire-wound resistors are not color-coded but have the resistance and wattage values printed on their bodies.

SELF-TEST

Check your understanding by answering these questions.

1. What is the value of each of type A (Fig. 2.1-1) radial resistor color-coded as follows:
   a. red, violet, silver? 0.27 Ω
   b. red, red, gold? 2.2 Ω
   c. green, blue, brown? 560 Ω
2. What is the color code for each of the resistors?
   a. 39 Ω ORANGE WHITE BLACK
   b. 68,000 Ω BLUE GRAY ORANGE
   c. 4700 Ω YELLOW VIOLET RED
3. The fourth band on a type A resistor is silver. The tolerance of this resistor is 10 percent.
4. A type A resistor has only three bands. Its tolerance is 20 percent.
5. A resistor with a black body color is UN-INSULATED
6. Two points in a circuit are said to be "shorted" if the resistance between these points is 0 Ω.

MATERIALS REQUIRED

- Equipment: EVM
- Resistors: Five assorted values
- Miscellaneous: A piece of hookup wire and hand tools

PROCEDURE

1. Determine the value of each resistor supplied from its color code. Fill in the information required in Table 2.1-2.
2. Zero the ohmmeter. Measure each resistor with the ohmmeter. Fill in the results in the row titled "Measured value." The measured value and the coded value should agree within the tolerance range of the resistor.
3. Measure and record the resistance of a small piece of uninsulated hookup wire. .3 Ω.
4. **a.** Measure and record the resistance of one of the resistors supplied. 9670 Ω.
   b. Connect the uninsulated wire across this resistor, as in Fig. 2.1-2. We say that the resistor has been "shorted" or "short-circuited."
   c. Measure and record the resistance of this combination. 0 Ω.

TABLE 2.1-2. Using the Resistor Color Code

		Resistor Number				
	Example	*1*	*2*	*3*	*4*	*5*
1st color	(red) 2	RED 2	BLU 6	RED 2	BWN 1	BWN 1
2d color	(red) 2	VIOLET 7	GRA 8	OR 3	RED 2	BLK 0
3d color	(red) 00	BLK	YEL 0000	OR 000	RED 00	OR 000
4th color	(silver) 10%	SILV 10%	GOLD 5%	GOLD 5%	GOLD 5%	GOLD 5%
Coded value, Ω	2200	27	680K	23K	1.2K	10K
Tolerance, Ω	220	2.7	34K	1150	60	500
Measured value, Ω	2160	27	676K	23K	1.18K	10.03K

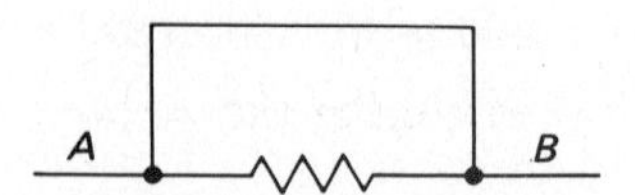

Fig. 2.1-2. Resistor short-circuited.

5. Remove the wire from the resistor. Cut the wire in half. Place the two pieces of wire near each other, but not touching. This is an *open* circuit. Measure and record the resistance between the two pieces. ________ Ω.

QUESTIONS

1. What is the resistance of a short circuit?
2. What is the resistance of an open circuit?
3. If a resistor reads *infinite ohms* on the ohmmeter, the resistor is ________ circuited.
4. A resistor color-coded red, red, orange, silver, reads 20,000 Ω on the ohmmeter. Is it within tolerance?

Answers to Self-Test

1. (*a*) 0.27 (*b*) 2.2 (*c*) 560
2. (*a*) orange, white, black (*b*) blue, gray, orange (*c*) yellow, violet, red
3. 10
4. 20
5. not insulated
6. zero

EXPERIMENT 2.2. Variable Resistors

OBJECTIVE

To measure resistance between the variable (center) terminal and the terminals on either side of it as the shaft of a potentiometer is turned from its minimum to maximum position

INTRODUCTORY INFORMATION

Variable Resistors

In addition to fixed-value resistors, variable resistors are used extensively in electronics. There are two types of variable resistors, the rheostat and the potentiometer. Volume controls used in radio, and the contrast and brightness controls of television receivers, are familiar examples of potentiometers.

A rheostat is a two-terminal device whose circuit symbol is shown in Fig. 2.2-1. Points *A* and *B* connect into the circuit. A rheostat has a maximum value of resistance, specified by the manufacturer, and a minimum value, usually 0 Ω. The arrowhead in Fig. 2.2-1 indicates a mechanical means of adjusting the rheostat so that the resistance, measured between the two terminals, can be set to any value between the minimum and maximum values.

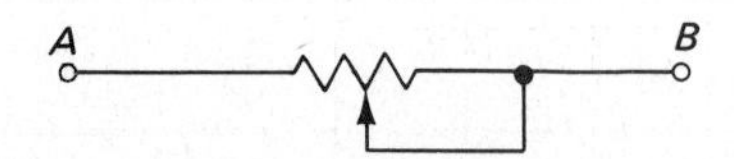

Fig. 2.2-1. A rheostat is a variable resistor with two terminals.

The circuit symbol for a potentiometer (Fig. 2.2-2) shows that this is a three-terminal device. The resistance between points *A* and *B* is fixed. Point *C* is the variable arm of the potentiometer. The arm is a metal contact which moves along the surface of the resistance element, selecting different lengths of resistive surface. The longer the surface between points *A* and *C*, the greater the resistance in ohms between these two points. Similarly, the resistance between points *B* and *C* varies as the length of element included between points *B* and *C* varies.

The total resistance of the potentiometer can be measured between the two outside terminals (*A* and *B*). The resistance from the variable arm (terminal *C*) to one of the outside terminals *plus* the resistance from the variable arm to the other outside terminal equals

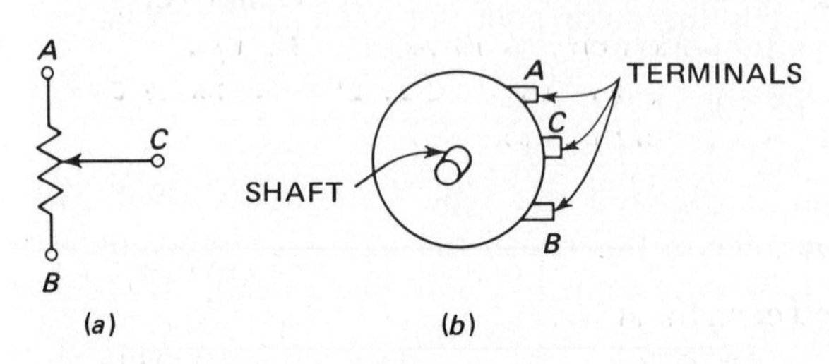

Fig. 2.2-2. A potentiometer is a three-terminal device. (*a*) Circuit symbol; (*b*) end view showing shaft and terminals.

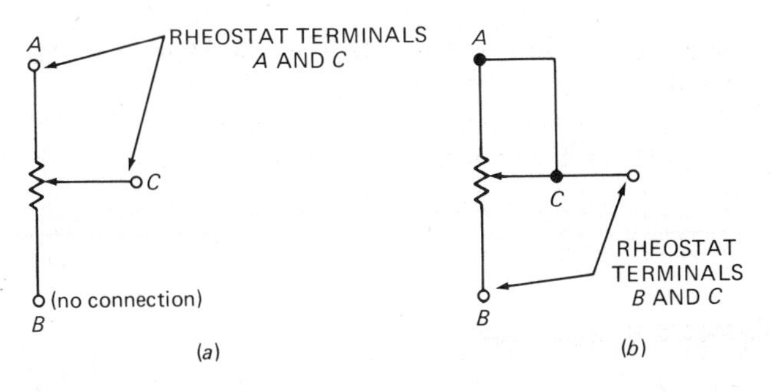

Fig. 2.2-3. Potentiometer connected as a rheostat.

the total resistance. The action of the arm, then, is to increase the resistance between the variable arm (terminal *C*) and one of the end terminals, and at the same time to decrease the resistance between the variable arm (terminal *C*) and the other end terminal, while the sum of the two resistances remain constant.

Converting a Potentiometer to a Rheostat

A potentiometer may be used as a rheostat if the center arm and one of the end terminals are connected into the circuit and the other end terminal is left disconnected, as in Fig. 2.2-3*a*. Another way to convert a potentiometer to a rheostat is to connect a piece of hookup wire between the arm and one of the end terminals, for example, *C* connected to *A*, as in Fig. 2.2-3*b*. Points *B* and *C* now serve as the terminals of the rheostat. (When two points in a circuit are connected by hookup wire, we say that these points are *shorted* together.)

SUMMARY

1. Variable resistors are of two types, the rheostat and the potentiometer.
2. A rheostat is a two-terminal device.
3. A potentiometer is a three-terminal device. The center terminal is the variable.
4. The resistance between the two outside terminals of a potentiometer is fixed, say 10 kΩ, between *A* and *B* (R_{AB}) of Fig. 2.2-2. This device is called a 10-kΩ potentiometer.
5. The two variable resistances in the potentiometer, Fig. 2.2-2, are R_{AC} and R_{BC}.
6. The sum of R_{AC} and R_{BC} equals the fixed resistance between *A* and *B*, that is $R_{AC} + R_{BC} = R_{AB}$.
7. A potentiometer may be used as a rheostat by connecting it as in Fig. 2.2-3*a* or *b*.

SELF-TEST

Check your understanding by answering these questions.

1. Figure 2.2-2 is a 100-kΩ potentiometer. The resistance between terminals *A* and *B* is 100 K Ω.
2. In the potentiometer in question 1, $R_{AC} = 70$ kΩ; then $R_{BC} =$ 30 K Ω.
3. A 5-kΩ potentiometer is connected as a rheostat, as in Fig. 2.2-3*b*. The maximum resistance of the rheostat is 5 K Ω.

MATERIALS REQUIRED

- Equipment: EVM
- Miscellaneous: 10,000-Ω potentiometer

PROCEDURE

1. Examine the potentiometer assigned to you. Place it so that the shaft points toward you. Call the terminals of the potentiometer *A*, *B*, and *C* as in Fig. 2.2-2*b*. Measure and record in Table 2.2-1 the value of the potentiometer between the two outside (*A* and *B*) terminals.
2. Turn the shaft to any position (1) and measure the resistance between the left terminal (terminal *A*) and the center terminal (terminal *C*). Record this reading R_{AC} in Table 2.2-1.
3. Without moving the shaft, measure the resistance between the right terminal (terminal *B*) and the center terminal (terminal *C*). Record this reading R_{BC} in Table 2.2-1.
4. Add the readings in steps 2 and 3 and record the sum in Table 2.2-1. The sum should equal the reading of the total resistance in step 1.
5. Repeat steps 2, 3, and 4 for another setting (2) of the potentiometer shaft. Record the results in Table 2.2-1.

TABLE 2.2-1. Measuring Variable Resistors

Step	*Potentiometer Shaft Setting*	R_{AB} Ω	R_{AC} Ω	R_{BC} Ω	$R_{AC} + R_{BC}$
1	Any	10.31k	X	X	X
2, 3, 4	Position 1	X	3.53k	7.20K	10.73 KΩ
5	Position 2	X	640	9.95k	10.59 KΩ
6	CW	X	0	10.32k	10.32 KΩ
7	CCW	X	10.32K	0	10.32 KΩ

6. Turn the shaft completely clockwise (CW). Measure and record R_{AC} and R_{BC}; compute and record the sum of $R_{AC} + R_{BC}$.
7. Turn the shaft completely counterclockwise (CCW). Measure and record R_{AC} and R_{BC}; compute and record $R_{AC} + R_{BC}$.

QUESTIONS

1. In a potentiometer what is (*a*) fixed in value? (*b*) variable?
2. (*a*) In the potentiometer in Fig. 2.2-2, what is the relationship between R_{AC}, R_{BC}, and R_{AB}? (*b*) Do your measurements confirm this relationship?
3. With the potentiometer held as in Fig. 2.2-2*b*, in what position of the shaft is the resistance between *A* and *C* zero?
4. In what position of the shaft is the resistance between *A* and *B* maximum?

Answers to Self-Test

1. 100 kΩ
2. 30 kΩ
3. 5 kΩ

EXPERIMENT 2.3. Resistance of Series-Connected Resistors

OBJECTIVE

To determine by experiment the total resistance of resistors connected in series

INTRODUCTORY INFORMATION

Series Circuit Connection

An electronic circuit is a closed path for current flow. Resistance in a circuit opposes current flow.

In electronic circuits there may be one or more resistors in a series arrangement. In a circuit with a voltage source and series-connected resistors, there is only *one* path for current, which must pass through each of the resistors in the circuit. If any one resistor is removed without the circuit being reconnected, no current will flow and the circuit is said to be "open" (recall Exp. 2.1).

Figure 2.3-1 shows a voltage source *V* connected across a resistor R_1. The resistor is in series with the power source because current would stop if the resistor were lifted from the circuit.

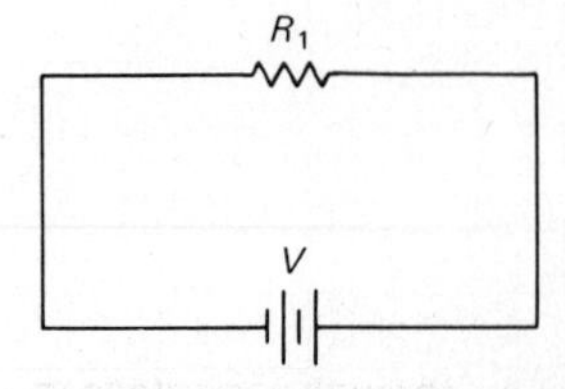

Fig. 2.3-1. Single resistor in series with a voltage source.

Figure 2.3-2 shows a voltage source *V* applied across resistors R_1, R_2, and R_3. These resistors are in series since removal of any of the resistors would open the circuit and current would stop.

It may help you to understand this idea better if you draw Fig. 2.3-2 on paper with a pencil. Now erase one of the resistors and note that the path (wires and resistors) for electric current is broken and no current can pass.

Total Resistance R_T of Series-Connected Resistors

Since the electric current flowing in a series circuit must pass through each resistance in its path, it would appear that two series resistors would offer more opposition to current than any one of the resistors individually. Three series-connected resistors would offer more resistance to current than any combination of two resistors, and so on. This fact is true because the total resistance R_T of a series circuit is equal to the sum of all the resistors in the circuit. The mathematical formula for this is:

$$R_T = R_1 + R_2 + R_3 + \cdots \qquad (2.3\text{-}1)$$

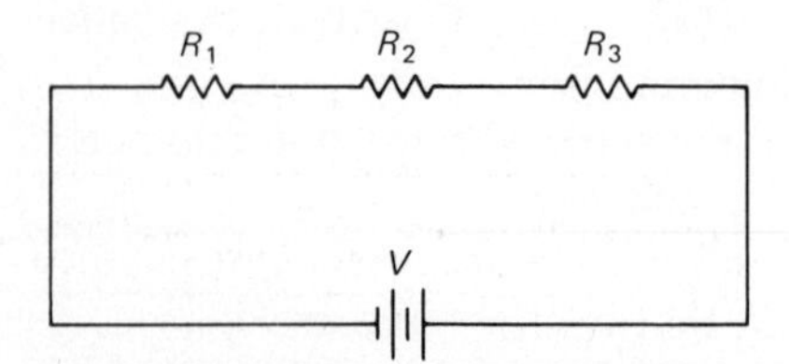

Fig. 2.3-2. Three resistors in series with a voltage source.

For example, for three 500-Ω resistors connected in series, $R_T = 1500\ \Omega$. Another example: if $R_1 = 220\ \Omega$ and $R_2 = 330\ \Omega$, then $R_T = 550\ \Omega$ for these series-connected resistors.

In radio and television, series circuits are frequently used. The technician must therefore understand series circuit operation in order to predict what will happen in the circuit under normal operation. This will help in locating defective components in troubleshooting a circuit. In radio and television receivers the signal follows a series path, moving from one amplifier to the next, Fig. 2.3-3. Therefore, just as with resistors, if one in the series of amplifiers is defective, the signal flow either stops or is changed in some way, creating trouble in receiver operation.

SUMMARY

1. A series circuit has only one path for electric current flow.
2. Removing a series component "opens" the circuit and stops current flow.
3. The total resistance R_T of a circuit is equal to the sum of the values of the individual resistors.
4. In radio and television there are many series components and circuits.
5. A circuit is a closed path for electric current flow.

SELF-TEST

Check your understanding by answering these questions.

1. Two 100-Ω resistors connected in series would have a total resistance of 200 Ω.
2. Five resistors wired in series have a total resistance that is greater than that of only four of the same resistors. TRUE. (true/false)
3. A circuit which has only one path for electric current flow is called a SERIES circuit.
4. A RESISTOR opposes the flow of electric current.
5. The mathematical formula for total resistance of series-connected resistors R_1, R_2, R_3, etc., is $R_T =$ $R_1 + R_2 + R_3$ . . .

MATERIALS REQUIRED

- Equipment: EVM
- Resistors: ½-W 330-, 470-, 1200-, 2200-, 3300-, and 4700-Ω

PROCEDURE

1. Measure the resistance of each resistor supplied and record its value in Table 2.3-1 beneath its color-coded value.
2. Connect series arrangement 1 shown in Fig. 2.3-4 by connecting resistors R_1 and R_4 in series. From Table 2.3-1, record the measured values of R_1 and R_4 in the spaces provided in Table 2.3-2.
3. Using the measured values of the individual resistors, compute the total resistance of this series combination and write the value in the column labeled "Computed Value R_T" in Table 2.3-2.
4. Measure the total resistance of this combination of series-connected resistances from points *A* to *B* and record that value in the column labeled "Measured Value R_T."
5. Repeat steps 2, 3, and 4 for resistor combinations 2 and 3 in Fig. 2.3-4.

TABLE 2.3-1. Color-coded and Measured Resistor Values

Color-coded Value	*R_1: 330 Ω*	*R_2: 470 Ω*	*R_3: 1200 Ω*	*R_4: 2200 Ω*	*R_5: 3300 Ω*	*R_6: 4700 Ω*
Measured value	330Ω	470Ω	1170Ω	2300Ω	3250Ω	4690Ω

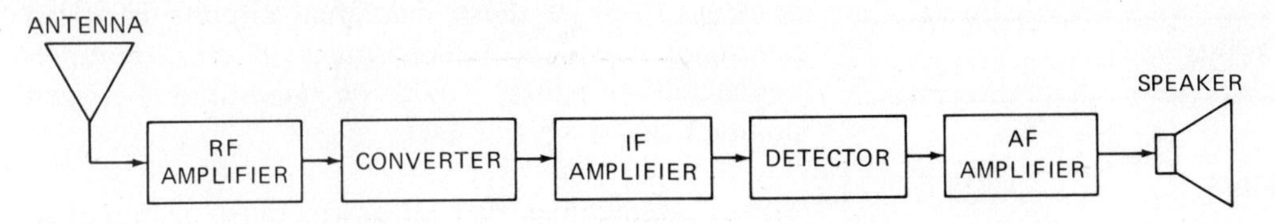

Fig. 2.3-3. Series-connected stages in a radio receiver. The radio signal moves left to right, from the antenna, through each stage in turn, and is changed to sound coming from the speaker.

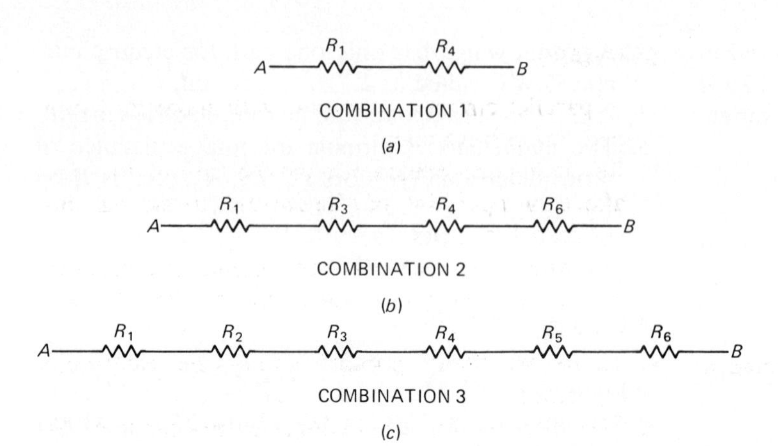

Fig. 2.3-4. Series-connected resistors. (a) Two resistors in series; (b) four resistors in series; (c) six resistors in series.

TABLE 2.3-2. Series Resistor Combinations

Resistor Combination	*Measured Resistance, Ω* R_1	R_2	R_3	R_4	R_5	R_6	*Computed Value* R_T	*Measured Value* R_T
1	330	X	X	2300	X	X	2630Ω	2629Ω
2	330	X	1170	2300	X	4690	8490Ω	8490Ω
3	330	470	1170	2300	3250	4690	12210Ω	12214Ω

QUESTIONS

1. Were the computed values and the measured values of each combination of resistors equal? If not, why not?
2. Explain, in your own words, two methods of finding the total resistance of series-connected resistors.
3. For combination 3 how does the sum of the color-coded values of the resistors compare with (*a*) the measured value of R_T in Table 2.3-2? (*b*) the computed value obtained by adding the measured values of each resistor?
4. Would there be any effect on the total resistance of combination 2 if the positions of any of the resistors were changed?
5. Do the results of your measurements in Table 2.3-2 prove that the total resistance of a series circuit is equal to the sum of the values of each of the resistors in the circuit?

Answers to Self-Test

1. 200
2. true
3. series
4. resistor
5. $R_1 + R_2 + R_3 + \cdots$

EXPERIMENT 2.4. Resistance of Parallel-Connected Resistors

OBJECTIVE

To measure the total resistance of combinations of parallel-connected resistors

INTRODUCTORY INFORMATION

Parallel-Connected Resistors

A parallel circuit is one in which there are two or more paths for electric current flow. Figure 2.4-1 shows such a circuit. Note that if any one of the resistors is removed from the circuit, paths for current flow still exist through the remaining resistors. Therefore, in a parallel circuit there must be a complete circuit or path for current flow through each individual resistor. Each of these individual circuits is called a *branch circuit*. Total resistance of these parallel-connected resistors would be measured between points X and Y (Fig. 2.4-1).

NOTE: Remember that all power must be removed from the circuit before resistance measurements are made.

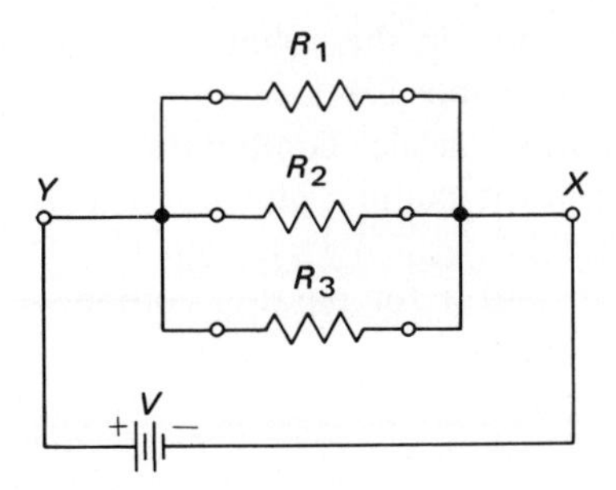

Fig. 2.4-1. Voltage applied across three resistors connected in parallel.

Total Resistance R_T of Parallel Resistors

It is reasonable to assume that more current can flow from the battery when there are several paths than when there is only one path. Now, if more current is allowed to flow from the battery with each additional branch circuit added, it is clear that the total opposition R_T to current flow is becoming smaller than if there were just one path. The total resistance to current flow from the power source does indeed decrease as more branch circuits are added. In fact, R_T is less than the resistance of the smallest branch resistor. We shall prove this with the ohmmeter later in this experiment.

Mathematically the formula for calculating parallel resistances is

$$\frac{1}{R_T} = \frac{1}{R_1} + \frac{1}{R_2} + \frac{1}{R_3} + \cdots \qquad (2.4\text{-}1)$$

The arithmetic involved is made quite simple by the use of the $1/x$ (reciprocal) button on many modern electronic calculators.

Radio and television circuits are normally connected in parallel with the power supply, as shown in Fig. 2.4-2. Defects in radio and television receivers sometimes involve parallel-connected circuits. The technician therefore needs to understand and should be able to analyze parallel circuits in order to locate the bad circuit or defective part in an electronic device, such as a radio or television receiver.

SUMMARY

1. A parallel circuit is a circuit with more than one path for current flow.
2. Removing one branch of a parallel circuit does not affect the operation of (the current in) the remaining branch circuits.
3. The total resistance of parallel-connected resistors is less than the resistance of the smallest branch resistor.
4. There are many parallel circuits in electronic equipment.
5. The formula for calculating R_T for parallel resistors is

$$\frac{1}{R_T} = \frac{1}{R_1} + \frac{1}{R_2} + \frac{1}{R_3} + \cdots$$

SELF-TEST

1. A single individual path for current within a parallel circuit is called a BRANCH circuit.
2. The R_T of parallel-connected resistors is LESS (more/less) than the value of the smallest branch resistor.
3. The formula for R_T of parallel-connected resistors is $\frac{1}{R_T} = \frac{1}{R_1} + \frac{1}{R_2} + \frac{1}{R_3} \cdots$
4. The more resistor branch circuits there are, the SMALLER (greater/smaller) is the R_T.

MATERIALS REQUIRED

- Equipment: EVM
- Resistors: ½-W 330-, 470-, and two 1200-Ω.

PROCEDURE

1. Refer to Fig. 2.4-3 and choose the resistors shown as combination (*a*).
2. Measure the resistance of each of the resistors supplied for combination (*a*). Record the meas-

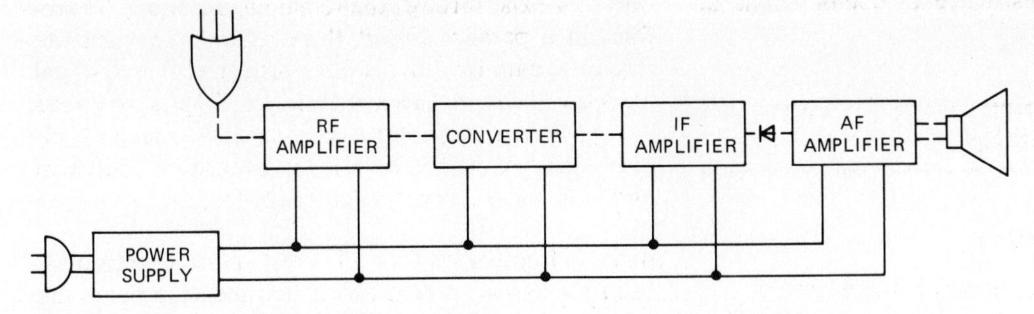

Fig. 2.4-2. Radio receiver block diagram showing all circuits in parallel with the power supply.

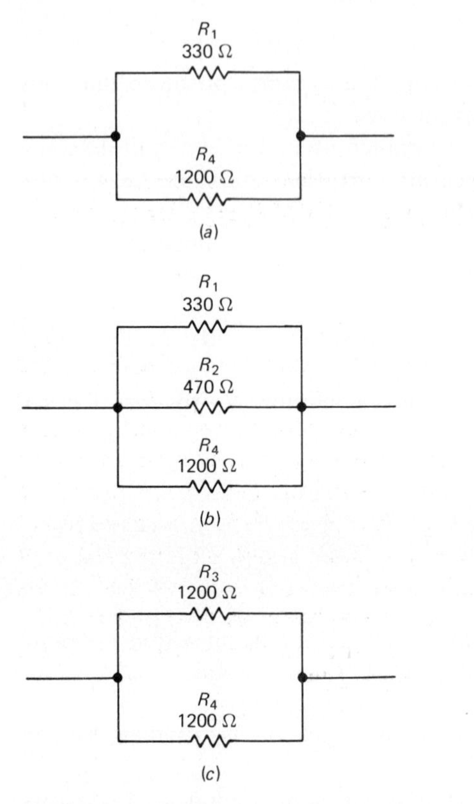

Fig. 2.4-3. Experimental parallel-resistor combinations.

ured value of each resistor in the column beneath its color-coded value in Table 2.4-1.

3. Measure the R_T of this parallel combination and record your reading in the column labeled "Measured R_T" in Table 2.4-1.
4. Repeat steps 1, 2, 3, and 4 for parallel combinations (*b*) and (*c*) in Fig. 2.4-3.

QUESTIONS

1. Was the value R_T greater or smaller than the value of the smallest branch resistor in each combination?
2. Combination (*c*) in Fig. 2.4-3 placed two resistors of equal value in parallel. From the results of measuring R_T of this combination of resistors, suggest a general rule for R_T of any two resistors of equal value connected in parallel.
3. What is the R_T of three 330-Ω resistors in parallel?
4. How can you measure the resistance of an individual resistor within a parallel circuit?

Answers to Self-Test

1. branch
2. less
3. $\dfrac{1}{R_T} = \dfrac{1}{R_1} + \dfrac{1}{R_2} + \dfrac{1}{R_3} + \cdots$
4. smaller

TABLE 2.4-1. Parallel Resistance Measurement

Parallel Combination	*Color-coded Value*	R_1 330 Ω	R_2 470 Ω	R_3 1200 Ω	R_4 1200 Ω	*Measured* R_T, Ω
(*a*)	Measured value, Ω	330Ω	X	X	1193Ω	260.89
(*b*)	Measured value, Ω	330Ω	470Ω	X	1193Ω	173.90
(*c*)	Measured value, Ω	X	X	1170Ω	1193Ω	59.10

CHAPTER 3 VOLTAGE AND VOLTAGE MEASUREMENT

EXPERIMENT 3.1. EVM/VOM Voltage Scales

OBJECTIVE

To set the zero adjust and read the voltage value at a specified point on each voltage range of the meter

INTRODUCTORY INFORMATION

Reading Linear Meter Scales

The purpose of this experiment is to familiarize the student with the reading of voltage scales of the analog meter. Though digital meters are much in use today and the prices are now comparable to those for analog meters, the analog meter continues to be used in most repair shops and in industrial applications. It is therefore necessary to be thoroughly familiar with the reading of analog voltage scales.

In general, the dc voltage scales on the EVM and VOM are linear (refer to Exp. 1.1). This means that the distance between any two equal-valued adjacent division markers on the meter scale is the same length. For example, the distance on the scale arc between 0 and 10 V is the same as the distance between 10 and 20 V or between 90 and 100 V.

Not every voltage which the meter can measure is printed on the meter scale. Therefore, it is necessary to supply numbers for unnumbered calibration marks on the scale. For example, in Fig. 3.1-1 the meter pointer is not pointing directly to a marked scale division but to an unnumbered mark. Notice that there are five equal spaces marked off between the numbered calibrations. In this case since the range switch is set to 10 V, the numbered calibrations represent a difference of 1 V, and each unnumbered division represents 0.2 V. The meter reads 4.2 V. If the range switch of the meter were set to 0 to 100 V instead of 0 to 10 V and the same scale had to be used, what would the meter read?

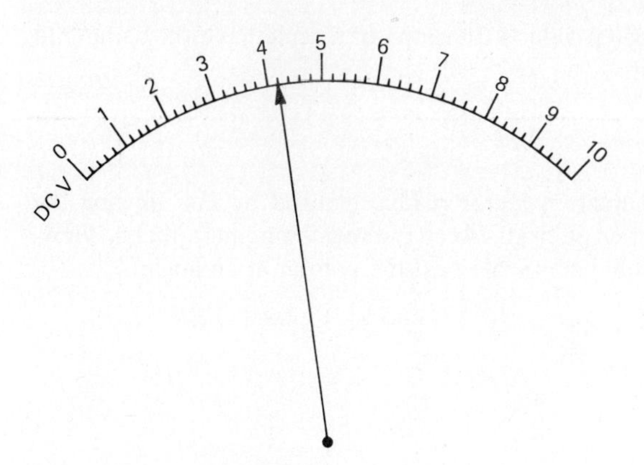

Fig. 3.1-1. A linear dc voltmeter scale.

Zero Adjust

Before making any measurement with the voltmeter, be certain that the meter pointer rests over 0 on the voltage scale. In an analog EVM, a zero-adjust control is placed on the front panel of the meter for the purpose of zeroing the meter before any measurement is made. This is the same as starting measurements with the end of a rule and not at the 1-in or 2-in mark. If you started measuring an 11-in board at the 1-in mark on the rule, the rule would show that the board was 12 in long. Of course this is *not* correct. The same is true in measuring voltages. If the meter pointer rests on (reads) 2 V before measurements begin, the measurement will be incorrect by 2 V.

Older vacuum-tube meters require a warm-up time of several minutes. If the meter is not allowed to warm up to a stable operating temperature, the voltage readings will vary and the zero control must constantly be adjusted.

Parallax Errors

When reading voltages with analog meters, care must be taken to view the scale from directly above the pointer. If this is not done, a reading error results. This error is called a *parallax error*. For example, the passenger sitting next to the driver in an automobile cannot read the speedometer accurately because of parallax error. The passenger does not view the pointer directly above the numbers of the speedometer as does the driver, and therefore may read a speed of 40 instead of 50 miles per hour.

Many manufacturers place a mirror on the meter scale to minimize parallax error. With the mirror on the scale, when you look directly past the pointer at the scale, no reflection of the pointer is seen in the mirror. However, if you are not looking directly past the pointer at the scale, a mirror image of the pointer is seen. If the pointer image is seen, an error will result in your meter reading.

SUMMARY

1. Voltage scales are linear; that is, equal arc distances represent the same voltage difference.
2. Sometimes it is necessary to assign values to unnumbered scale calibrations. A linear scale makes this process possible.
3. The voltmeter must always be zeroed before making a measurement.
4. Vacuum-tube meters require a warm-up time.
5. Parallax error is a reading error caused by not viewing the meter scale from directly above the pointer. To avoid this error, many meters use mirrored scales. If a reflection of the meter pointer is seen in the mirror, parallax error will occur.

SELF-TEST

Check your understanding by answering these questions.

1. What is meant by the term *linear?* EQUAL LENGTH DIVISIONS.
2. Refer to Fig. 3.1-1. If the range switch were set to 0 to 1000 V, what would the meter reading be? 420V
3. Refer to Fig. 3.1-1. If the range switch were set to 0 to 50 V, what would the meter reading be? 21V
4. Zeroing the voltmeter is not necessary since it will measure whatever is applied to its probes. FALSE (true/false)
5. What is parallax error? ERROR OBTAINED BY VIEWING OFF TO ONE SIDE RATHER THAN STRAIGHT ON

MATERIALS REQUIRED

- Equipment: EVM with analog scale (VTVM or TVM)

PROCEDURE

1. We are going to "simulate" the reading of voltages with the voltmeter by varying the zero adjust so that the pointer moves up-scale just as if a voltage were being applied to the meter. So secure the meter from your instructor. (If the meter is battery-powered, it may be unstable if the batteries are weak.)
2. Turn on your meter and allow it to warm up if necessary.
3. Set the function switch to +DC V. Adjust the zero adjust so the meter is set to zero.
4. Adjust the zero adjust to simulate a reading of *a.* 1 V, *b.* 11 V, *c.* 171 V, *d.* 275 V, *e.* 760 V, *f.* 0.75 V.

NOTE: For the simulation to be accurate, the range switch must also be set to the proper range.

5. Have your instructor check each reading or draw the voltmeter scale and the position of the pointer for each reading. If you draw the voltmeter scale, be sure also to write the position of the range switch for each voltage reading.

QUESTIONS

1. Why should we learn to read analog scales?
2. Why is zeroing the voltmeter important?
3. Do all EVMs have zero adjusts?
4. How many voltage ranges does your EVM have?
5. What is the maximum voltage which can be measured with your EVM?

Answers to Self-Test

1. All equal-valued divisions of a scale have the same unit lengths.
2. 420 V
3. 21 V
4. false
5. The error in meter reading caused by not viewing the meter scale from directly above the pointer, that is, viewing the meter scale past the pointer at an angle.

EXPERIMENT 3.2. Voltage Measurement

OBJECTIVE

To measure the voltage of a dry cell and a low-voltage power supply

INTRODUCTORY INFORMATION

Measuring DC Voltage

Electronics technology is inconceivable without measurement of electrical quantities. Circuit analysis and understanding of circuit function are simplified by the basic measurements of voltage, current, and resistance. Repair of electronic devices, such as TV receivers, usually requires electrical measurements.

The technician is the "know-how" person. (Of course, the technician must also "know why.") In the laboratory, the technician will be the person primarily involved in making the measurements which will test the scientists' or engineers' theories. The technician will test the operation and characteristics of an experimental circuit and in the repair shop will make measurements while fixing a device which is not working. The technician must therefore know the instruments of this technology and must understand how to use them. The technician must be aware of the effect an instrument may have on the circuit and the errors of measurement, and must also know how to interpret the results of these measurements.

Measurement of dc voltage is basic to all electronics work. In this experiment we shall be concerned with learning how to measure voltage using the dc voltage function of an EVM.

Voltage is defined as electrical pressure. It is the *difference* in electrical pressure between two points. The voltage *across* two points is measured with a voltmeter. An EVM combines the jobs of the voltmeter, the ohmmeter, and sometimes the current meter. In this experiment we shall use the EVM to measure voltage. The student should refer to the manufacturer's operating instructions for a particular EVM before using it in this experiment.

Generally, the following facts apply regardless of the make of EVM:

1. The EVM must be turned on, and if it is a VTVM, it must be allowed to warm up.
2. The meter test leads should be plugged into the proper meter jacks. The black lead is the common or ground lead, and the probe or red lead is the voltage, signal, or "hot" lead.
3. The + or − polarity switch is set to agree with the polarity of the voltage being tested. For example, for measuring voltages which are positive with respect to common, the function selector switch is set to the + DC position.
4. The dc voltage probes are shorted together and the EVM "zero" adjustment is used to set the meter pointer to zero. Zero is usually the left-hand side of the meter scale. (On zero-center meters, zero is in the middle of the scale.)
5. The common lead is connected to the return (usually negative) of the circuit under test. This test lead is usually connected first.
6. The range selector switch should be set to the highest dc voltage range in measuring an *unknown* dc voltage. This is to avoid damaging the meter. If measurement later shows that the voltage falls on a lower range, the instrument should be switched to the lower range.
7. The dc probe is then connected to the circuit under test. (The insulated probe and not the metal prod should be held when using the voltmeter.)
8. The voltage value is read from the proper voltage scale. The maximum voltage for a specific voltage range is at the right end of the scale.
9. The dc voltage scale on an EVM is separate from the *ohm* scale. The meter pointer moves in an arc from left to right.
10. The dc voltage scale is linear, with equal spacings for equal voltage changes. There may be two or three scale calibrations along the dc arc to make it possible to read voltage on all the dc ranges of the instrument. The greater the deflection toward the right, the higher the voltage that is being measured.

DC Voltage Sources

Batteries

In this experiment the student will use two types of voltage sources, the dry-cell battery and the dc power supply.

Dry batteries consist of arrangements of primary cells, called *dry cells*. The familiar flashlight "battery" is really a dry cell. Individual dry cells produce a low voltage. A battery is usually a combination of cells connected and offered in one package.

Electronic Power Supplies

Electronic, regulated, variable power supplies are used extensively in both school and industrial laboratories.

A variable voltage-regulated supply is one which can be adjusted to deliver any required voltage within its range of operation. Figure 3.2-1 shows a typical electronic, regulated, variable power supply. The voltage put out by this supply remains constant despite changes in load current, within specified limits. For example, one manufacturer states that its supply will deliver 0 to 400 V to 150 mA. This means that the output can be varied from 0 to 400 V and that the current drawn must not be greater than 150 mA. Once the supply is adjusted for a particular output voltage, the voltage will not vary appreciably so long as no more than 150 mA is drawn from the supply.

Electronic power supplies are *line*-operated devices. The usual controls on the front panel are an ON-OFF switch, which may also include a *stand-by* position; a *voltage* control which sets the output voltage level; and a meter switch which can be set to measure either the dc voltage delivered at the output terminals or the dc current delivered to the load.

A power supply may have facilities for providing two or more independent dc voltages, in which case separate controls and separate output terminals are contained on the instrument. Also, on some supplies the output current can be limited and an adjustment knob will be available on the front panel.

The polarity of the dc terminals on the supply is usually marked either −, +, and *gnd* (for ground). A red jack is conventionally used for the positive and a black for the negative terminal of the supply.

AC voltages required for electronic experiments are also available at terminals on the supply. These will be reserved for another experiment, as they are not needed at this time.

CAUTION: *The output terminals of most power supplies should never be short-circuited (shorted, or the output leads attached together), because the supply may be damaged.* To prevent the terminals from shorting, keep the leads from these terminals from making contact.

Measuring High Voltage

Sometimes it is necessary to measure voltages which are higher than the highest voltmeter range. For this purpose, special voltmeters are available which will measure up to 50,000 V. Also, many manufacturers produce special high-voltage probes which are used with their own standard voltmeters and extend the range of these meters to about 50,000 V. See Figs. 3.2-2 and 3.2-3.

These meters and probes are used by the technician in servicing television receivers. Today's color television sets may employ voltages on the picture tube close to 32,000 V. Of course the service technician must be able to measure these voltages accurately and *safely*.

High-voltage measurement requires special *safety* precautions, which the technician will learn while becoming familiar with television servicing.

SUMMARY

1. Voltage is defined as electrical pressure.
2. A battery is an arrangement of electric cells.

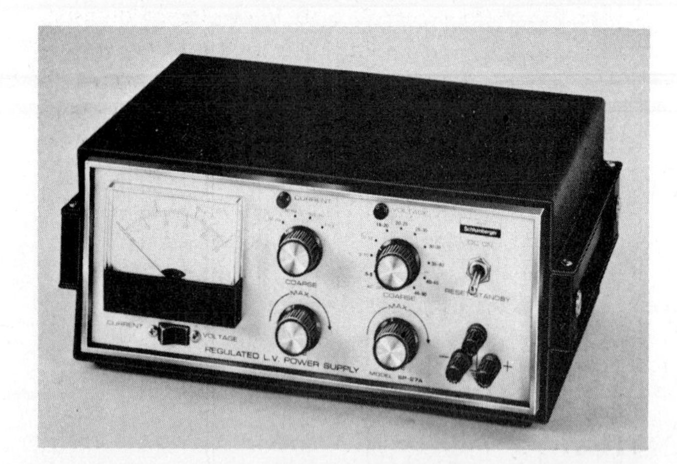

Fig. 3.2-1. A typical variable voltage-regulated power supply. (*Health Company*)

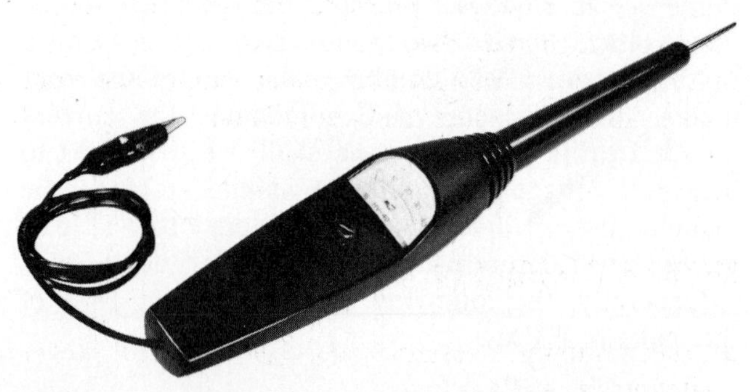

Fig. 3.2-2. A single-purpose high-voltage meter. (*B and K Dynascan*)

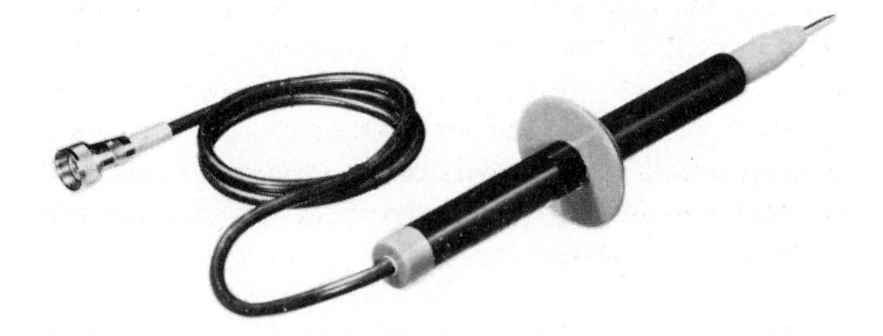

Fig. 3.2-3. A high-voltage probe for use with a standard voltmeter. (*B and K Dynascan*)

3. Two readily available sources of power are the dry battery and the variable voltage-regulated power supply.
4. Electronic power supplies are line-operated devices.
5. The two voltmeter leads used for measurement of dc voltage are normally a black lead and a red lead. The black is for the common or ground, and the red is for the voltage or signal being measured.
6. The voltmeter reads the electrical pressure across two points in a circuit.

SELF-TEST

Check your understanding by answering these questions.

1. Voltage is the DIFFERENCE in electrical pressure between two points.
2. A VTVM can be used to measure voltage. TRUE (true/false)
3. Zero is on the LEFT (right/left) hand side of the voltmeter scale.
4. The voltmeter scale is nonlinear. FALSE (true/false)
5. A variable voltage-regulated power supply will maintain the voltage at which it is set unless the current output exceeds the supply's limits. TRUE (true/false)
6. The output terminals of a power supply must not be SHORTED when the supply is on.

MATERIALS REQUIRED

- Power supply: Variable, voltage-regulated power supply; dry cells
- Equipment: EVM and proper leads for measuring dc voltages

PROCEDURE

Dry Cells

1. Measure and record in Table 3.2-1 the voltage of each of the dry cells supplied. Use the range of the voltmeter where you get the maximum pointer deflection without going off scale. Connect the negative lead of the meter to the negative terminal of the cell, positive lead to positive.

TABLE 3.2-1. Measuring Dry-Cell Voltages with an EVM: Range 1

Dry-Cell Number	*1*	2	3	4
Voltage	1.467	1.523	1.529	1.522

2. Repeat the measurements of step 1 using the next higher range of the voltmeter. Record your results in Table 3.2-2.

TABLE 3.2-2. Measuring Dry-Cell Voltages with an EVM: Range 2

Dry-Cell Number	*1*	2	3	4
Voltage	1.46	1.52	1.52	1.52

Electronic Power Supply

3. Sketch the front panel of the power supply assigned to you. Label all switches, controls, and output terminals. Read the operating instructions and proceed only after they are clearly understood.
4. Power **on.** Permit the supply to warm up. Set voltage control for maximum voltage (but no higher than 50 V) as indicated on the panel meter of the power supply.
5. Set the EVM on + DC, 50 V or closest dc range (equal to or greater than 50 V). Connect the negative lead of the meter to the negative terminal of the supply and the positive meter lead to the positive terminal. Does the EVM read the same voltage as the panel meter on the power supply? YES
6. Gradually decrease the supply voltage, measuring the output on the EVM. Reduce the EVM range when necessary to secure a better measurement. Note whether the EVM and panel meter read significantly different voltages on any setting.

QUESTIONS

1. List four precautions which must be observed in measuring voltage.

2. List the voltage ranges of your EVM.
3. How much voltage is required to give full-scale deflection of the pointer on a 300-V range?
4. What would happen to a dry cell or battery if the positive and negative terminals were shorted?
5. Why must we be careful to keep from short-circuiting the output terminals of a power supply?

Answers to Self-Test

1. difference
2. true
3. left
4. false
5. true
6. shorted

CHAPTER 4 MEASUREMENT AND CONTROL OF DIRECT CURRENT

EXPERIMENT 4.1. Direct-current Measurement

OBJECTIVE

To connect a current meter and measure current in a dc circuit

INTRODUCTORY INFORMATION

Current and the Ampere

Electric current was mentioned in previous experiments and by this time you probably have some idea what it is. In this book we shall define current as the movement of electric charges (electrons) in a circuit.

In electronics it is often necessary to measure current, that is, to determine how much current there is in a circuit. To measure current, an ammeter, a milliammeter, or a microammeter is used.

The basic unit of measure for electric current is the ampere, represented by the capital letter A. The ampere is a large quantity of current not often found in low-power electronic circuits. The most frequently used measure of current in electronics is the milliampere (mA) which is a thousandth ($^1/_{1000}$) of an ampere. In decimals 0.001 A stands for one milliampere, 0.002 A for two milliamperes, 0.010 A for 10 milliamperes, etc. Another way of writing, say, 2 milliamperes is 2 mA. The small letter m stands for milli, which means one-thousandth (0.001). The other unit employed in the measure of current is the microampere (μA). The microampere is one-millionth of an ampere, and in decimals 1 microampere is written as 0.000 001 A. In decimals 0.000 010 A equals 10 microamperes; 0.000 016 A equals 16 microamperes, etc. These may also be written as 1 μA, 10 μA, 16 μA, etc.

Current Exists only in a Complete Circuit

In earlier experiments the uses of the ohmmeter and voltmeter for measuring resistance and voltage were studied. From the nature of the experiments it was apparent that resistors have "resistance" that can be measured directly with an ohmmeter. The quantity of ohms of resistance is not dependent on the connection of that resistor in a circuit. The characteristic of resistance is associated with the component itself.

Similarly, in the measurement of voltage, it was evident that voltage is a characteristic of some voltage source and that voltage can also exist independently without the need for a complete circuit.

Electric current differs from voltage and resistance in that it cannot exist by itself. A voltage source by itself is insufficient to create current. A voltage source *and* a closed (complete) path are required for the flow of current.

Connecting the Ammeter

Current in an electric circuit can be compared with water flow in a pipe. If you wish to measure the amount of water flowing per second (rate), you have to insert a flowmeter inside the pipe. In this way all the water flowing in the pipe must pass through the flowmeter, which can then measure the rate of water flow. So it is with the measurement of electric current. Since current is the movement of electric charges, the circuit must be broken and an ammeter inserted in series in the circuit. All the electric charges then move through the ammeter, which indicates the rate of electron movement. See Fig. 4.1-1 for the placement of the ammeter in the electric circuit.

When placing the ammeter in *series* with the circuit, polarity must be observed. That is, the common (negative) meter lead must be connected to the more negative point (closer to the negative source) in the circuit. The "hot" lead (positive) is connected to the more

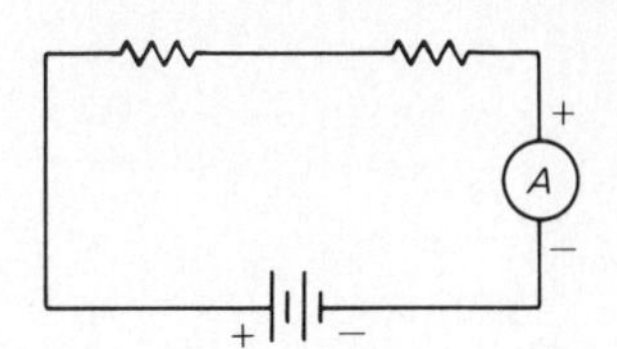

Fig. 4.1-1. The ammeter is placed in series with the circuit.

positive (further from the negative or closer to the positive) point in the circuit. When the meter is connected properly, the pointer will move from left to right. If the pointer moves in the opposite direction, the meter leads must be reversed.

Refer to Fig. 4.1-2. Notice that the circuit has been broken (opened) at points A and B. To measure the current in this circuit, the ammeter will be inserted between these two points. Point A is closer to the negative source, so the negative meter lead will be attached here. Point B is farther from the negative source (we can also say that it is closer to the positive source) and therefore will be connected to the positive meter lead.

CAUTION: The current meter must *not* be connected across (that is, in parallel with) any component. *It must always be connected in series* with the component to measure the current moving through the component. Failure to observe this rule may result in serious damage to the meter. *Never connect an ammeter directly to a voltage source.*

When the ammeter is placed in the circuit, it should always be set to its highest range and then switched to lower ranges, as necessary, for an accurate reading. This will protect the meter against damage due to overload.

Ammeter scales vary from meter to meter. Direct current is sometimes read on the same scale used for dc voltage. If so, the only difference in the reading will be that you are measuring amperes, milliamperes, or microamperes instead of volts. In some meters the scale may be a separate scale used only for current. Whatever the scale is like, you will have no trouble reading it if you have learned how to read the voltage scales in previous experiments.

Multirange ammeters have two or more current ranges which are marked on the range switch. For example, the current ranges on one multirange meter include a 5-mA, 25-mA, 100-mA, 250-mA, and 1-A range. The same meter *scale* is used for each range, but the technician must multiply by the proper range factor in reading the meter.

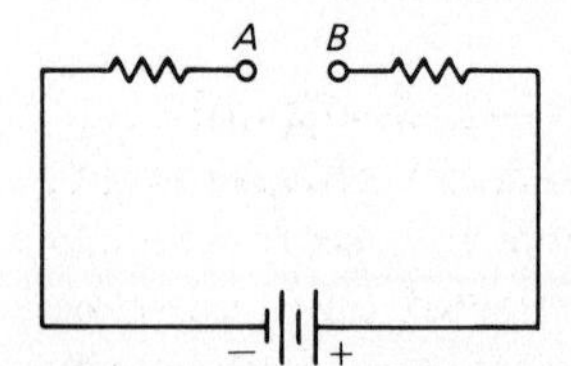

Fig. 4.1-2. The electric circuit is broken (A and B) to receive the ammeter.

NOTE: Since there is only one path for current in a series circuit, current is the same everywhere in that circuit. So it does not matter where the circuit is broken and the meter inserted.

SUMMARY

1. *Current* is the movement of electric charges in a circuit.
2. Current cannot exist by itself. Current must have a closed path (circuit) in which to move and a pressure to cause it to move (voltage).
3. The ampere is the unit of current. Its symbol is A.
4. There are smaller current units, the milliampere (mA), which equals $^1/_{1000}$ of an ampere (0.001 A), and the microampere (μA), which equals $^1/_{1,000,000}$ of an ampere (0.000 001 A).
5. Most measurements made in electronic servicing are in milliamperes.
6. The ammeter is *always* connected in series with the electric circuit.

SELF-TEST

Check your understanding by answering these questions.

1. The two components necessary for electric current flow are VOLTAGE and CIRCUIT.
2. Write the decimal fraction of an ampere for 23 mA .023, 300 μA .000300, 167 mA .167, and 5230 μA .005230
3. The ammeter is placed in SERIES (series/parallel) with the circuit.
4. Draw a simple circuit and place an ammeter in it. Mark the battery and meter with the correct polarities.
5. At the start of measurement when the ammeter is connected in the circuit, the range switch should be set to the HIGHEST (highest/lowest) range available.

MATERIALS REQUIRED

- Power supply: Variable, low-voltage dc source
- Equipment: EVM with current ranges
- Resistors: $^1/_2$-W 1000-Ω

PROCEDURE

1. Secure the components from your instructor and measure the resistance of the resistor to make sure its value is within tolerance.

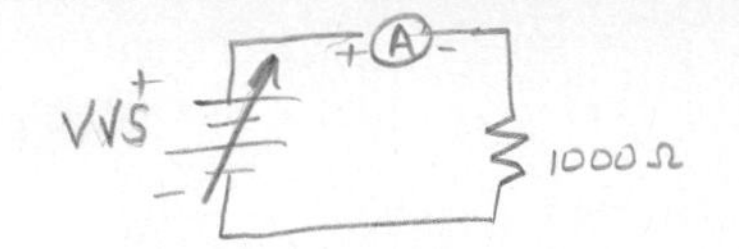

2. Draw a schematic diagram of the variable voltage source (VVS), one 1000-Ω resistor, and the ammeter connected in series. Mark all polarities.
3. Now connect the circuit drawn in step 2. Power is **off.** Voltage output control is set to zero. Set the current meter to its highest range. Make note of the units of measure printed on the range switch and remember that this is what you are measuring. Always record the unit of measure with the value.
4. **Power on.** Now adjust the voltage to 15 V. Measure the current and record its value in Table 4.1-1.

NOTE: The current range may have to be decreased for an accurate reading in this and the following measurements.

5. Adjust the voltage output of the VVS to 10 V. Measure the current and record its value in Table 4.1-1.
6. Adjust the voltage output of the VVS to 5 V. Measure the current and record its value in Table 4.1-1.

QUESTIONS

1. What is the maximum amount of current your current meter will measure?
2. What are the different ranges on your current meter?

TABLE 4.1-1. Connecting an Ammeter to Measure Current

Output of DC Supply	*15 V*	*10 V*	*5 V*
Measured circuit current	6.9mA	6.9mA	5.1mA

3. The measurements in this experiment were made in what units (A, mA, or μA)?
4. Do the ranges on your current meter "overlap"? If they do, indicate those which overlap.
5. Why must the current meter be placed in series with the circuit?
6. What danger, if any, is there in placing a current meter across a voltage supply?

Answers to Self-Test

1. a circuit; a voltage source
2. 0.023 A; 0.000 300 A; 0.167 A; 0.005 230 A
3. series
4. 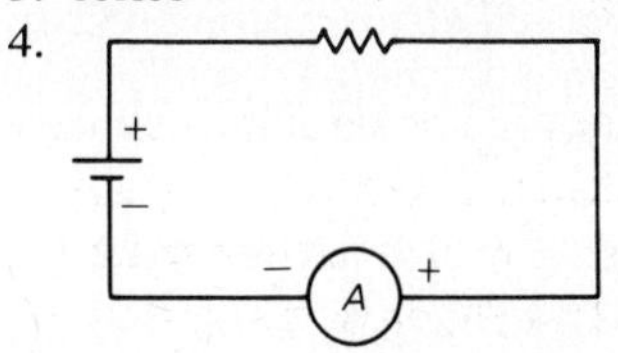

5. highest

EXPERIMENT 4.2 Control of Current by Resistance

OBJECTIVE

To measure the effect of resistance in controlling current in a dc circuit

INTRODUCTORY INFORMATION

In the preceding experiment we studied how to measure current. You will recall that *voltage* is the electrical pressure (electromotive force—emf) which causes current flow in a circuit and that *resistance* is the opposition to this current. In electronics one of the primary functions of electronic devices and components is the control of current in a circuit. The technician must therefore have a thorough knowledge of how current is controlled and must know that direct current can be controlled by the voltage *or* the resistance of a circuit, and how this control is accomplished. This experiment will be concerned with the control of current by *resistance.*

Inverse Relationships

There are many forces in nature which affect the motion of some object, just as resistance affects current in a circuit. For example, a valve in a water line can control the amount of water current flow. Observe that if the opposition to the water flow doubles, the water flow is cut in half. If it triples, water flow is cut to one-third. What is this relationship between opposition to water flow and the movement of water in a pipe? It is called an inverse relationship and deals with cause and effect which are *opposites*. In this case the cause is the increase in opposition to water flow. The effect is the decrease in water flow.

Control of Current by Resistance

It seems logical that since resistance acts to oppose electric current, that current can be controlled by resistance, just as a valve in a water pipe can control the flow of water in that pipe. And so, you can see that if the resistance in an electric circuit increases, the current decreases; if the resistance decreases, the current increases. If the resistance is cut in half, current will double, since there is an inverse relationship between current and resistance in a circuit.

The capital letter *I* (for *intensity*) is used to denote electric current, just as *V* stands for voltage and *R* stands for resistance. Can you write a formula which shows the relationship between current *I*, voltage *V*, and resistance *R* in an electric circuit? Keep in mind that *I* and *R* are inversely related, and assume that *V* is some constant voltage.

SUMMARY

1. Direct current can be controlled by resistance in a circuit.
2. Resistance in a circuit is an opposition to current flow.
3. Frequently in nature the opposition to an *action* and the effect on that action are *opposites*. This relationship between the opposition and the action is called an *inverse relationship*.
4. The relationship of current in a circuit to the resistance in the circuit is an *inverse* one.

SELF-TEST

Check your understanding by answering these questions.

1. If the resistance in a *dc* circuit increases, the current in the circuit DECREASE (increases/decreases).
2. If the resistance in a circuit has increased to FOUR times its original value, the current in the circuit will decrease to one-fourth its original value, assuming that the voltage stays the same.
3. If the resistance of a circuit changes from 2000 to 6000 Ω, and the current in the original circuit was 2 mA, the current in the 6000-Ω circuit will be .67 mA, if the voltage is the same.
4. Current can easily be controlled by changing RESISTANCE.
5. The relationship between current and resistance is a(n) INVERSE one.

MATERIALS REQUIRED

- Power supply: Low-voltage, variable dc source
- Equipment: EVM with current ranges
- Resistors: ½-W three 1000-Ω
- Miscellaneous: 2-W 5000-Ω potentiometer: SPST switch

PROCEDURE

1. Obtain the materials required from your instructor. Measure each resistor to be sure it is within tolerance.
2. Connect the circuit shown in Fig. 4.2-1. Set the voltage control of the power supply to zero. The ammeter range should be set to maximum but may have to be changed after each circuit change in this experiment. It is wise *always* to set the range to maximum after any circuit change.
3. **Power on.** Set the voltage control of the VVS to 15 V. Measure the current and record the value in Table 4.2-1. This is the original circuit current.
4. Add another 1000-Ω resistor in series in the circuit. (Power should be **off** any time a change is made in a circuit.) The total resistance is now 2000 Ω. The power supply is still set to 15 V. Measure the current. Record this value in Table 4.2-1.
5. Add another series-connected 1000-Ω resistor. The total resistance in the circuit is now 3000 Ω. The power supply is still set to 15 V. Measure the current. Record the value in Table 4.2-1.
6. Place the potentiometer connected as a rheostat in series with one 1000-Ω resistor. Figure 4.2-2 shows this connection.
7. Observe the current meter while adjusting the variable resistor. Does changing resistance cause

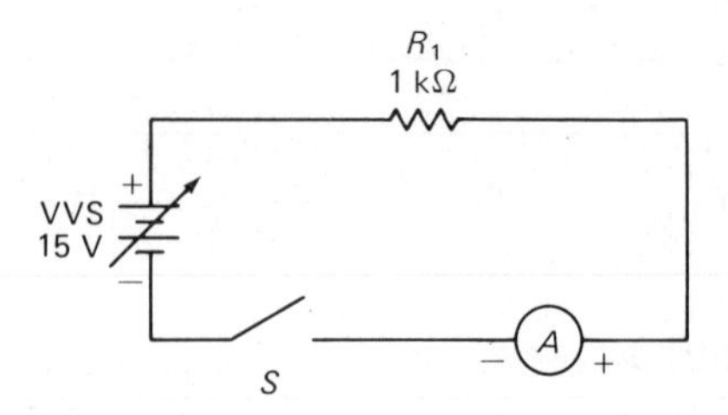

Fig. 4.2-1. Current measurement in a simple series circuit.

TABLE 4.2-1. Effect of Resistance on Controlling Current in a Circuit

Power Supply, VVS	*+15 V dc*		
Circuit resistance, Ω	1000	2000	3000
Measured circuit current *I*, mA	6.9 mA	5.6 mA	4.73 mA

the circuit current to change? YES How? VARIES 2.43 mA - 6.86 mA, DEPENDING ON POSITION OF POT

8. Remove the 1000-Ω resistor from the circuit, leaving an "open" circuit. What is the current measured? 0 mA

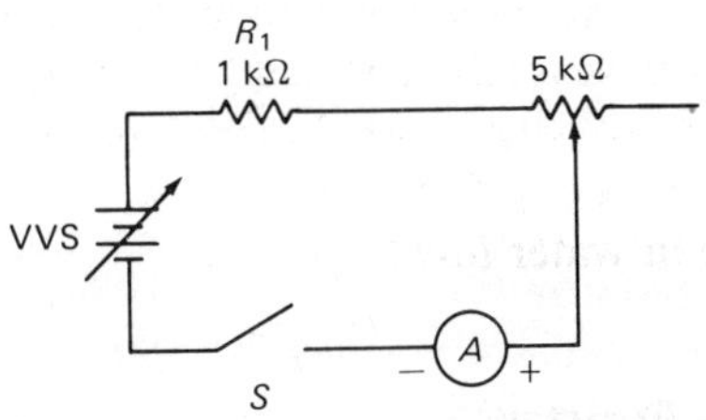

Fig. 4.2-2. Circuit with a variable resistor for current control.

QUESTIONS

1. The circuit in step 4 had its resistance doubled over the original resistance. By how much should the current have decreased over the original current? Apply the law of inverse relationships.
2. How does the calculated value in question 1 compare with the measured value in step 4? Comment on any unexpected results.
3. The resistance of the circuit in step 5 was three times greater than the resistance of the original circuit. The current in step 5 should have been ________ of the original current in step 3.
4. How does the calculated value in question 3 compare with the measured value in step 5? Comment on any unexpected results.
5. Using the measurements of this experiment, explain what is meant by the term *inverse relationship*.
6. What is meant by an open circuit?
7. Will an open circuit permit current flow?

Answers to Self-Test

1. decreases
2. four
3. 0.67
4. resistance
5. inverse

EXPERIMENT 4.3. Control of Current by Voltage

OBJECTIVE

To measure the effect of voltage in controlling current in a dc circuit

INTRODUCTORY INFORMATION

The importance of controlling current by varying the resistance in an electric circuit was discussed in the last experiment. This experiment will deal with controlling current by changing the circuit voltage.

Control of Current by Voltage

Voltage is the electrical pressure which causes current to flow. If the pressure is increased, the flow increases. If the pressure is decreased, the flow decreases. Read the last two sentences again, but this time think of them as explaining the movement of water in a pipe.

Again, we can see a relationship between electrical behavior and other natural behaviors, for example, that of water. This time the relationship is a *direct* one. If we push *more* we move *more*, or if we push *less* we move *less*, assuming the opposition or resistance remains the same. Voltage here is considered to be a push, since it is an electrical pressure. The movement, of course, is the flow of current in the circuit.

Since the relationship between voltage and current is a direct one, we can see that if the voltage doubles, the current doubles. Or if the voltage is cut in half, the current is cut in half, and so on. This relationship (law), and other circuit relationships, are often used in analyzing circuit performance when "troubleshooting" defective equipment.

SUMMARY

1. Current can be controlled by changing the circuit voltage.
2. Current flow is caused by the "push" of the electrical pressure (voltage).
3. The relationship between voltage and current is a *direct* relationship; that is, if the voltage increases, the current increases; if the voltage decreases, the current decreases.
4. The behavior of electrical voltage and current can be described by the same law as other pressures and movements in nature.

SELF-TEST

Check your understanding by answering these questions.

1. The relationship of voltage to current is a DIRECT one.
2. If the circuit voltage is 6 V and it is doubled, the effect of the voltage increase is to DOUBLE the current, if resistance remains the same.
3. If the circuit voltage is reduced from 12 to 4 V (a factor of 3), the effect of the decrease in voltage is to DECREASE current by a factor of 3, if resistance remains the same.

MATERIALS REQUIRED

- Power supply: Variable, low-voltage, regulated dc source
- Equipment: EVM with current ranges
- Resistors: ½-W 1000-Ω
- Miscellaneous: SPST switch

PROCEDURE

1. Draw the schematic of a circuit containing the VVS, a switch, a 1000-Ω resistor, and a current meter connected in series. Have your instructor check the drawing.

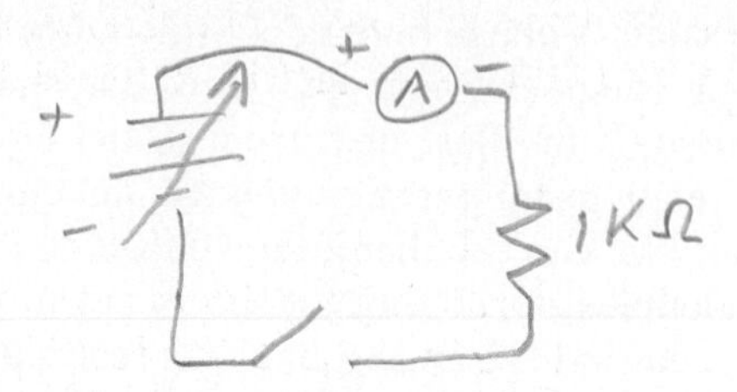

TABLE 4.3-1. Effect of Voltage in Controlling Current in a dc Circuit

Circuit Voltage	*5* V	*10* V	*15* V
Measured circuit current, *I*	5.01 mA	6.86 mA	6.86 mA

2. Obtain the materials required from the instructor. Connect the circuit drawn in step 1. Set the output of the VVS to zero and the range of the current meter to maximum.
3. **Power on.** Adjust the voltage to 5 V. Measure the circuit current and record the value in Table 4.3-1.
4. Increase the voltage to 10 V. Measure the circuit current and record the value in Table 4.3-1.
5. Increase the voltage to 15 V. Measure the circuit current and record the value in Table 4.3-1.
6. The voltage in step 4 is double the original 5 V. The current *should* be DOUBLE the original circuit current.
7. The voltage in step 5 is triple the original 5 V. The current should be TRIPLE the original circuit current.

QUESTIONS

1. Were the measurements taken in steps 4 and 5 the same as the calculations in steps 6 and 7? Explain any unexpected results.
2. What kind of relationship between voltage and current in a circuit is demonstrated by the experiment?
3. In troubleshooting a defective radio, the voltage supply is found to be much lower than it should be. What would we expect has happened to the current drawn by the radio?

Answers to Self-Test

1. direct
2. double
3. reduce; 3

CHAPTER 5 OHM'S LAW AND THE SERIES CIRCUIT

EXPERIMENT 5.1. Ohm's Law

OBJECTIVE

To determine experimentally the relationship between the current I in a resistor, the voltage V across the resistor, and the resistance in ohms R of the resistor

INTRODUCTORY INFORMATION

We have established that there is a relationship between resistance, current, and voltage. The relationship between resistance and current is an inverse one. If the resistance increases, the current in the circuit decreases. The relationship between voltage and current is a direct one. If the voltage increases, the current increases.

In a closed dc circuit, containing a resistance R and a voltage source V, the current will increase if the voltage is increased, or decrease if the voltage is decreased—as long as the resistance does not change. Also, in a closed dc circuit with R and V, I will decrease if R is increased, or I will increase if R is decreased—as long as V does not change.

Ohm's Law

These relationships are important, but up to this point they have been explained in a purely descriptive manner. Frequently technicians need to know specifically the number of ohms, the number of amperes, or the voltage in a circuit. Of course they can measure the unknown quantity. But suppose they need to compute it first. It is possible to do this because a quantitative relationship exists between I, V, and R. A quantitative relationship is one in which specific numbers are used instead of words such as *increase* and *decrease*. A quantitative statement of a relationship will not only state what changes occur but will tell us how much change will occur. First developed by George Simon Ohm, the quantitative relationship between V, R, and I is given by Ohm's law:

$$I = \frac{V}{R} \qquad (5.1\text{-}1)$$

This formula states that the current in amperes in a resistive circuit equals the applied voltage, in volts, divided by the resistance, in ohms, of the circuit. Ohm's law is simple, requiring only division or multiplication in order to calculate the unknown quantity. With the electronic calculator so readily available today, the answers can easily be computed to several decimal places when necessary.

Ohm's law can be written in any one of three ways, depending on which of the quantities, V, I, or R, is required. These are as follows:

$$I = \frac{V}{R} \qquad R = \frac{V}{I} \qquad V = I \times R \qquad (5.1\text{-}2)$$

The formula $R = V/I$ states that the resistance in ohms of a circuit equals the applied voltage in volts divided by the current in amperes flowing in the circuit.

The formula $V = I \times R$ states that the applied voltage, in volts, equals the product of the current in amperes times the resistance of the circuit in ohms.

Applying Ohm's Law

To illustrate that Ohm's law tells us not only that a change occurs but *how much change* occurs, we shall use the circuit in Fig. 5.1-1. This circuit has a 10-V source and a 10-Ω resistor. This circuit, according to Ohm's law, will have 1 A of current.

$$I = \frac{V}{R} = \frac{10\text{ V}}{10\ \Omega} = 1\text{ A}$$

We know that the current will double if we halve the resistance. Our formula will prove this. For example, change the resistance in Fig. 5.1-1 to 5 Ω.

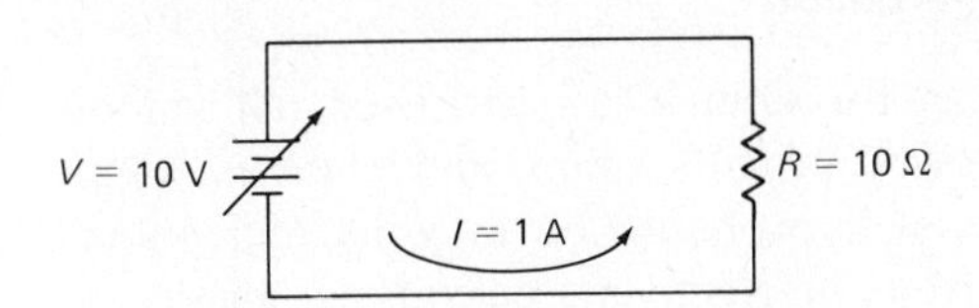

Fig. 5.1-1. A simple closed dc series circuit with all Ohm's law values given.

The current will now be 2 A.

$$I = \frac{V}{R} = \frac{10 \text{ V}}{5 \ \Omega} = 2 \text{ A}$$

In a previous experiment we found that the current doubled if the voltage doubled. Let us change the source in Fig. 5.1-1 to 20 V and leave the resistance at 10 Ω. The current will now be 2 A.

$$I = \frac{V}{R} = \frac{20 \text{ V}}{10 \ \Omega} = 2 \text{ A}$$

Ohm's law allows us to find an unknown value if any two of the elements of the law are known. We just found an unknown current above. Similarly, if the voltage is not known, but I and R are known, V can be calculated. The formula $V = I \times R$ is used for this. Refer to Fig. 5.1-1 again. Assume that the voltage is not known, but $I = 1$ A and $R = 10$ Ω. Then

$$V = 1 \text{ A} \times 10 \ \Omega = 10 \text{ V}$$

Errors of Measurement in Verifying Ohm's Law

In this experiment students will attempt to develop the equation for Ohm's law from experimental data. They can expect that the data may contain some errors of measurement and that the answers will therefore not be perfect. Errors can be brought into the calculations by incorrect meter readings, parallax errors, and meter inaccuracy.

Errors caused by incorrect meter readings can be corrected only by exercising greater care in reading the instrument. Instrument accuracy cannot be improved, but more accurate instruments can be used when necessary. Precision laboratory instruments are expensive and not usually available in most school laboratories. However, the tolerance of test equipment is given in the equipment specifications.

Another source of error results from the process of inserting an instrument into a circuit to make a measurement. If the instrument alters circuit conditions in any way, incorrect readings may be obtained. These are called "loading" errors and will be discussed in greater detail later.

Units of Measurement

When formulas for Ohm's law are used, all values must be expressed in volts, ohms, and amperes. Your measurements will probably be in volts and ohms. However, current in electronic circuits is usually in milliamperes. When this is so and the current value is to be used in the formula for Ohm's law, the value in milliamperes must be converted to amperes (see Exp. 4.1). For example, 9 mA = 0.009 A.

Verifying Value of R_T of Series-Connected Resistors by Ohm's Law

In a previous experiment the total resistance R_T of series-connected resistors was measured with an ohmmeter. It was found that

$$R_T = R_1 + R_2 + R_3 + \cdots$$

That is, the total resistance is the sum of all the individual resistances. Ohm's law can be used to verify this formula.

Assume a measured voltage V is applied to series-connected resistors R_1, R_2, etc., as in Fig. 5.1-2. By measuring circuit current I with ammeter A, and substituting the values of V and I in the equation for Ohm's law, we find R_T. Thus

$$R_T = \frac{V}{I}$$

Now, by adding the values of R_1, R_2, etc., we can show that the computed value of R_T, using Ohm's law, does indeed conform with the formula value. That is,

$$R_T = R_1 + R_2 + R_3 + \cdots$$

SUMMARY

1. The relationship between the voltage V applied to a closed circuit by some voltage source, the total resistance R, and the current I in that circuit is given by the formula $I = V/R$.
2. The relationship between the voltage drop V across a resistor R and the current I in that resistor is given by the formula $I = V/R$.
3. Measurement errors do occur and these must be considered in attempting to establish the accuracy of a formula such as Ohm's law.

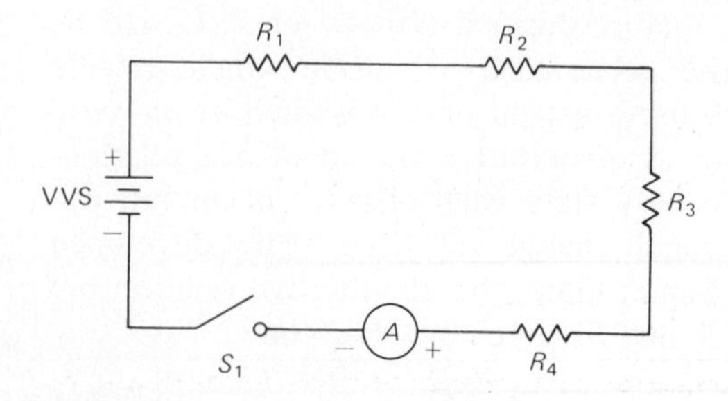

Fig. 5.1-2. Four resistors in series.

4. Among the measurement errors that may occur are: (*a*) incorrect reading of the meter scale; (*b*) incorrect reading errors due to parallax; (*c*) errors resulting from the accuracy of the equipment used; (*d*) errors caused by the insertion of the test equipment into the circuit; (*e*) errors due to component tolerance.
5. Ohm's law can be stated in forms other than $I = V/R$ for finding an unknown V or R. Thus, $V = I \times R$ and $R = V/I$.
6. When Ohm's law is applied to calculate an unknown quantity in the circuit, the units used are volts for V, ohms for R, and amperes for I.

SELF-TEST

Check your understanding by answering these questions.

1. The formula for finding an unknown V in a closed circuit when I and R are known is $V =$ I · R.
2. How much current is there in a closed circuit with a 15-V source and three 1000-Ω resistors in series? .005 A or 5 mA.
3. How much resistance is there in a closed circuit which has a 15-V source and 7.5 mA of current? 2000 Ω.
4. OHM'S LAW is a statement of the relationship between V, R, and I in a closed dc circuit.

MATERIALS REQUIRED

- Power supply: Variable, voltage-regulated power supply
- Equipment: EVM
- Resistors: ½-W 1000-Ω
- Miscellaneous: SPST switch

PROCEDURE

1. Obtain the materials required. Set the output of the VVS to 15 V, as measured by the EVM.
2. Connect the circuit of Fig. 5.1-3. Switch S_1 is open. The EVM is set on its highest direct current range. Close S_1. Measure the circuit I and record the value in Table 5.1-1.

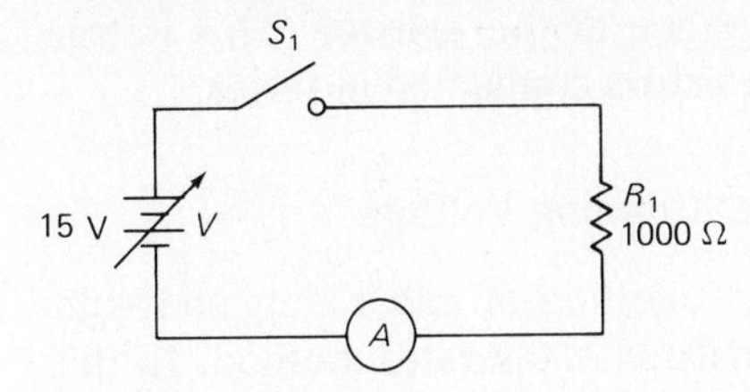

Fig. 5.1-3. A simple dc series circuit with two values (R and V) given.

TABLE 5.1-1. Verifying Ohm's Law: Steps 1 to 5

Measured Values			*Calculated Values*		
V, V	R, Ω	I, A	$I = V/R$	$V = I \times R$	$R = V/I$
15V	991Ω	.012A	.015A	15V	1000Ω

3. Remove the current meter and reconnect the circuit. Measure the voltage "drop" across R. (The voltage across R is called a *voltage drop* because the voltage or pressure used to push current through a resistance is "dropped" from use anywhere else in the circuit.) Record this value in Table 5.1-1.
4. Remove the resistor from the circuit and measure its resistance. Record this value in Table 5.1-1.
5. Complete the formulas in Table 5.1-1 by filling in the measured values for the *known* elements of the formulas and calculating the *unknown* element. For example, in the formula $I = V/R$ the unknown value is I and the known values are V and R. Take the measured values from the table and substitute them into the formula for V and R. The answer to the calculation, I, should be the same (or within tolerance) as the measured I. If they are, this would prove that if two of the elements of Ohm's law are known, the third element, in this case I, can be calculated accurately.
6. Repeat steps 1, 2, 3, 4, and 5 with the VVS output at 30 V. Record all measurements in Table 5.1-2.
7. Repeat steps 1, 2, 3, 4, and 5 for the circuit of Fig. 5.1-4. Record all measurements in Table 5.1-3. Measure V (step 3) across *both* resistors.

TABLE 5.1-2. Verifying Ohm's Law: Step 6

Measured Values			*Calculated Values*		
V, V	R, Ω	I, A	$I = V/R$	$V = I \times R$	$R = V/I$
30V	991Ω	.027A	.03A	30V	1000Ω

QUESTIONS

1. Do the measured values and the calculated values in this experiment compare within tolerances of the equipment used? If not, why not?
2. Is the voltage across a resistor directly or indirectly proportional to the current through it?

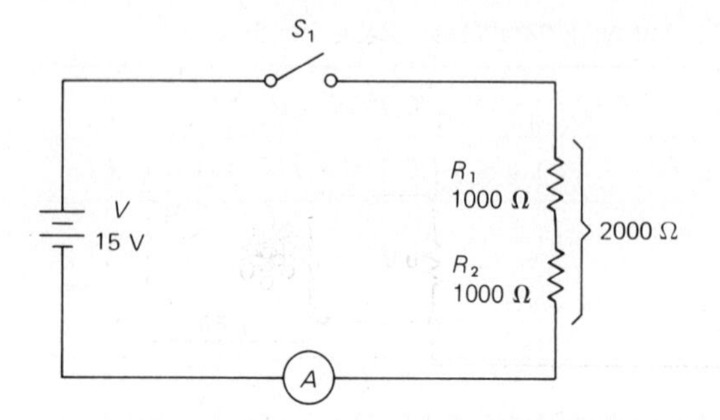

Fig. 5.1-4. A dc circuit with two resistors in series.

3. What will the voltage drop across a resistor be if its resistance is 2200 Ω and the current through it is 15 mA?
4. Explain Ohm's law in your own words.
5. In step 6 the circuit voltage was increased from 15 to 30 V and the resistance remained the same. Did the results of your measurements and calculations for this step confirm Ohm's law? Explain.

TABLE 5.1-3. Verifying Ohm's Law: Step 7

Measured Values			*Calculated Values*		
V, V	*R*, Ω	*I*, A	$I = V/R$	$V = I \times R$	$R = V/I$
15V	1970Ω	.0074A	.0075A	15V	2000

6. In step 7 the circuit resistance was increased from 1000 to 2000 Ω. The voltage was set at 15 V. Did the results of your calculations and measurements for this step confirm Ohm's law? Explain.

Answers to Self-Test

1. $I \times R$
2. 0.005 A; 5 mA
3. 2000
4. Ohm's law

EXPERIMENT 5.2. The Series Resistive Voltage Divider (Unloaded)

OBJECTIVE

To prove, by experiment, that the voltage V_1 across a resistor R_1 in a series resistive voltage divider is

$$V_1 = \frac{R_1}{R_T} \times V$$

where R_T is the total resistance of the resistors and V is the applied voltage

INTRODUCTORY INFORMATION

Voltage Drops across Series-Connected Resistors

When resistors are connected in series and a voltage source is applied across the resistors, the total voltage of the source is divided among the resistors. Remembering that voltage is an electrical pressure may help us to understand this idea. It takes some pressure to push current through each opposition (resistor). If it took all the pressure to push current through the first resistor, there would be no pressure left to push current through the remaining resistors. We must say, then, that *each resistor in the series circuit will have some voltage dropped across it if any current is flowing.*

The amount of pressure or voltage needed at each resistor to push the current along in the series circuit depends on the resistance of the resistor. The higher the resistance, the more pressure will be required to push current through it. In like manner, the smaller the resistance of a resistor, the less pressure will be required to push current through it. In fact, the voltage drop across a resistor is directly proportional to its resistance. For example, if two resistors are connected in series across a voltage source and one resistor is two times as large as the other, the voltage drop across the larger resistor will be two times as great as the voltage drop across the smaller resistor. If there are three resistors in series, each resistor will have a voltage drop across it that is in direct proportion to the resistance of the resistor. This is true for any number of resistors connected in series.

Ratio Method for Determining Voltage

Perhaps the simplest method of calculating the voltage drop across a resistor is the ratio method. In this method a ratio is set up between one resistor of the series circuit and the total resistance of the circuit.

As an example, in Fig. 5.2-1 the total resistance is 10,000 Ω. R_1 is 2000 Ω. The ratio of R_1 to R_T is 2000 to 10,000. Put as a fraction, this means that R_1 is $^{2000}/_{10,000}$ of the total resistance R_T. If this is simplified, R_1 is $^2/_{10}$ or 0.2 of the total resistance. Since resistance and voltage in a series circuit are directly proportional, this means that $^2/_{10}$ or 0.2 of the applied voltage will appear across R_1. In this case $0.2 \times 30\text{ V} = 6\text{ V}$, and this is the voltage across R_1.

R_2 is $^3/_{10}$ or 0.3 of R_T, so 0.3 of the applied voltage or 9 V will appear across R_2. How much voltage would be measured across R_3? ________

A formula is provided for the ratio method of calculating voltage drops in the series circuit.

$$V_1 = \frac{R_1}{R_T} \times V \tag{5.2-1}$$

In some instances the formula is not needed since it is often a simple matter to do the calculations mentally.

If technicians understand the ratio method and know how to calculate voltage drops in a series circuit, they can frequently speed up servicing defective equipment by analyzing the affected circuits. Knowing the voltage source and the color codes of the resistors in the circuit, they can often determine whether the circuit is behaving properly by measuring just *one* voltage drop instead of two or more.

Sum of the Voltage Drops in a Series Circuit

Another method for determining the voltage drop across a series resistor was explained in Exp. 5.1. This was $V_1 = I \times R_1$. Here, the current through the resistor in amperes is multiplied by the resistance in ohms to give the voltage drop across the resistor. If this were done for each resistor in a series circuit and the voltages were added, the sum would be the same as the applied voltage V. The sum of all the voltage drops of a series circuit *must* equal the applied voltage. It could not add to a value less than the applied voltage, for this would mean that some of the source voltage was not being used. All electrical pressure in a circuit, just like water pressure in a system, must be used. If all the pressure cannot be used in the system, it escapes in the form of a broken water line or a burned component.

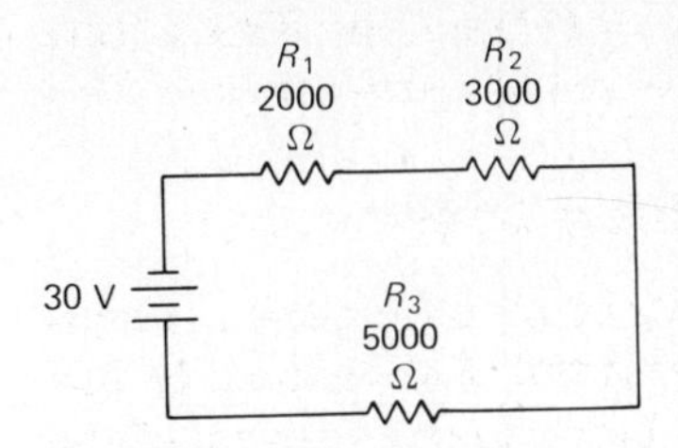

Fig. 5.2-1. A series-resistive unloaded voltage divider.

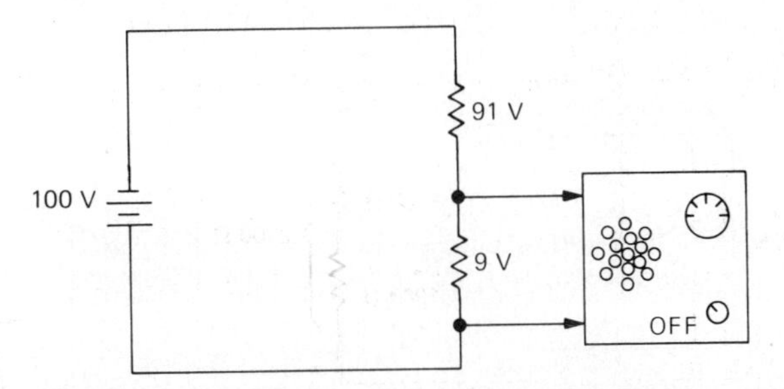

Fig. 5.2-2. A loaded voltage divider used to operate a transistor radio.

By the same reasoning the sum of the individual voltage drops cannot exceed the source voltage. If it did, it would mean that some extra voltage was acquired somewhere, and this is not possible unless another source is placed in the circuit.

Regardless of the method used in calculating voltage drops in a series circuit, if the voltage drops across each resistor are added together, the sum will be the same as the applied voltage or source. For this reason a series of resistors is often called a voltage divider. *The source voltage is divided among the resistors of the circuit.*

Sometimes the voltage divider is used to supply a particular voltage to a component or circuit; if so it is called a *loaded* voltage divider. A loaded voltage divider is shown in Fig. 5.2-2. If the voltage divider is not used to supply voltage for the operation of a component or circuit, it is referred to as an *unloaded* voltage divider. This experiment deals with the unloaded voltage divider shown in Fig. 5.2-3.

Potentiometer as a Voltage Divider

A potentiometer is often used as a voltage divider. In fact, the volume control of a radio is such a voltage divider. All the sound signal is applied to it, and only a portion is taken off and transferred to the audio or sound amplifiers. The more signal voltage taken from the potentiometer, the higher the volume. The less signal voltage taken from the potentiometer voltage divider, the lower the volume. See Fig. 5.2-4.

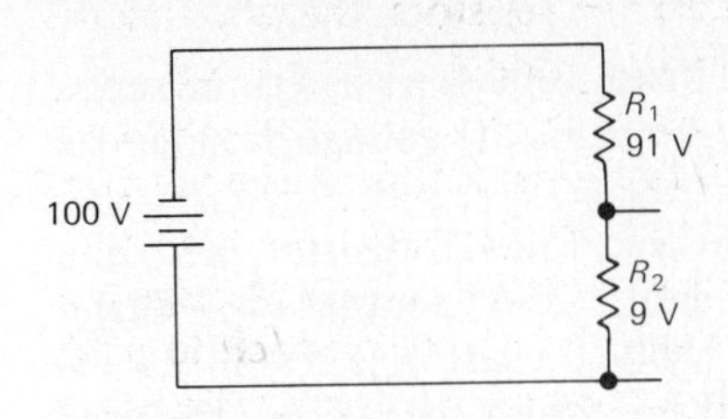

Fig. 5.2-3. An unloaded voltage divider.

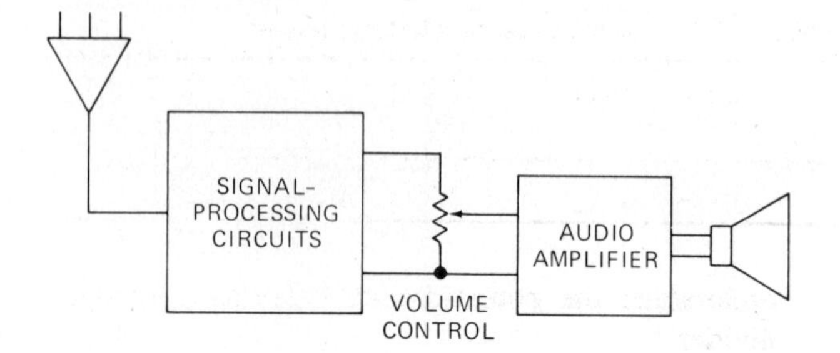

Fig. 5.2-4. The potentiometer is a voltage divider when used as a volume control in a radio.

To understand how the potentiometer, or *pot*, works as a voltage divider, assume it to be two separate resistors connected in series as in Fig. 5.2-3. The variable arm of the pot is actually the point where R_1 and R_2 are connected. Why use a potentiometer? Because by using the pot, the ratio of the resistances to each other and to the total resistance can be changed. This makes it very useful for dividing the supply voltage and for making the division of this supply voltage *variable*.

SUMMARY

1. Voltages in a series resistive circuit add to equal the source voltage.
2. There are two methods of calculating voltage drops across series resistors—the ratio method, $V_1 = R_1/R_T \times V$, and the Ohm's law method, $V_1 = I \times R_1$.
3. The voltage drop across a resistor is directly proportional to the resistance.
4. There are two basic types of voltage-divider circuits—the unloaded divider and the loaded divider.

SELF-TEST

Check your understanding by answering these questions.

1. If the resistors in Fig. 5.2-1 had their positions exchanged, would their voltage drops change? __NO__ (yes/no)
2. A voltage divider can be made into a variable voltage divider by adding a __VARIABLE__ resistor to the circuit.
3. In a circuit with three 1000-Ω resistors in series and a 30-V source, what is the voltage drop across each resistor? __10__ V
4. In a circuit with one 1500-Ω resistor, and one 1000-Ω resistor, and a 25-V source, the voltage drop across the 1500-Ω resistor is __15__ V and the voltage drop across the 1000-Ω resistor is __10__ V.

MATERIALS REQUIRED

- Power supply: Variable, low-voltage, regulated dc source
- Equipment: EVM
- Resistors: ½-W 2200-, 3300-, 4700-Ω
- Miscellaneous: SPST switch; 2-W 10-kΩ potentiometer

PROCEDURE

1. Secure the materials required from your instructor, and connect the circuit shown in Fig. 5.2-5. Set the VVS to zero.
2. **Power on.** Adjust the VVS to 10 V. Record the exact value in Table 5.2-1.
3. Measure and record in Table 5.2-1 the voltage drop across each resistor. Calculate the voltage drop across each R, using the ratio method, and record the values in Table 5.2-1.
4. Add the measured voltage drops and record in Table 5.2-1, in the column headed "Sum of Measured Voltage Drops."
5. Exchange the positions of R_1 and R_3. Measure the voltage drops across each of these resistors. Did the voltage drop across each of these resistors change with the change in position? __NO__ (yes/no)
6. Connect the circuit shown in Fig. 5.2-6. Set the VVS to zero.
7. **Power on.** Adjust the VVS to 10 V. Record the exact value in Table 5.2-2 in the column headed "Supply Voltage, V."
8. Attach one voltmeter lead to the negative *end* terminal of the potentiometer and the other lead to the *center* terminal of the pot. *Be sure to observe proper polarity when connecting the voltmeter.*
9. Adjust the pot until the voltmeter reads 3 V. This is the voltage V_1 in Fig. 5.2-6. Now move the

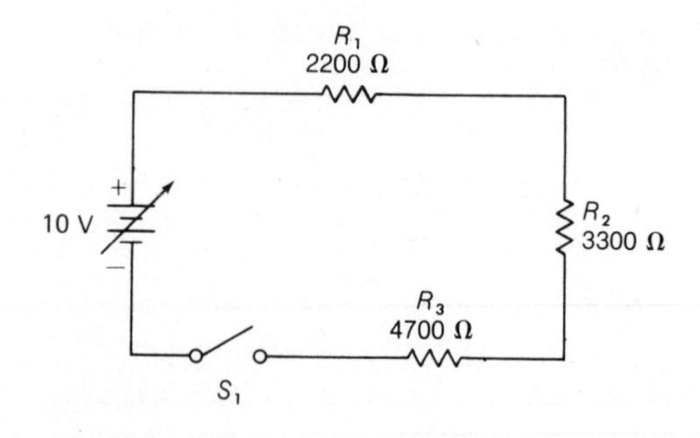

Fig. 5.2-5. A series voltage divider (unloaded).

TABLE 5.2-1. Series Voltage Divider (Unloaded)

Power Supply, V	*Resistor Number*	R_1	R_2	R_3	*Sum of Measured Voltage Drops*
10V	Voltage measured across	2.17	3.25	4.59	10.01V
10V	Voltage calculated across	2.15	3.23	4.61	9.99V

meter leads so you will read the voltage V_2 between the center terminal and the *other* end terminal. *Again be sure to observe proper polarity.* Record this value, V_2, in Table 5.2-2.

10. Add the measured voltage drops across the pot and record in column "$V_1 + V_2$." Did they add to the supply voltage? YES (yes/no)

11. Repeat steps 7 to 10 with other settings of the pot and with different source voltages until you

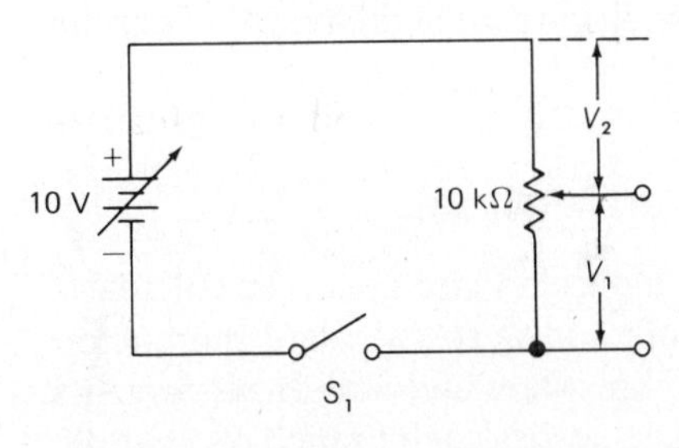

Fig. 5.2-6. A potentiometer voltage divider (unloaded).

TABLE 5.2-2. Potentiometer as a Voltage Divider

Supply Voltage, V	V_1	V_2	$V_1 + V_2$
10 V	3 V	7.02	10.02 V

understand the potentiometer's use as a voltage divider.

QUESTIONS

1. Refer to the experimental results in Tables 5.2-1 and 5.2-2. Compare the measured values with the calculated values. Were they the same? If not, why not?
2. Give a law explaining the behavior of voltage across the potentiometer regardless of the setting of the arm of the potentiometer.
3. What is meant by a loaded voltage divider?
4. Refer to Fig. 5.2-6. Explain the effect on circuit current as the potentiometer arm is varied.
5. In your own words explain why the voltage drops in a series circuit add to the source voltage.
6. Using Ohm's law, calculate the voltage drops across each resistor in Fig. 5.2-5. Show your calculations.
7. Are the voltages calculated in question 6 the same as the voltages calculated by the ratio method (Table 5.2-1)? If not, why not?

Answers to Self-Test

1. no
2. variable
3. 10
4. 15; 10

CHAPTER 6 THE PARALLEL CIRCUIT

EXPERIMENT 6.1. Current in a Parallel Circuit

OBJECTIVE

To determine experimentally that the total current I_T in a circuit containing resistors connected in parallel is:

1. Greater than the current in any branch of the circuit
2. Equal to the sum of the currents in each branch

INTRODUCTORY INFORMATION

Branch Currents

In considering a series circuit, it was established that a closed circuit is required for current; that current stops flowing when the circuit is open; that current in a series circuit is the same everywhere. What are the characteristics of a parallel circuit?

Figure 6.1-1 shows three resistors connected in parallel and a voltage V applied across them. If the line connecting the battery to the resistors is broken and an ammeter is connected in the circuit as in Fig. 6.1-2, the ammeter will measure current. This is the total or "line" current which is drawn by the three resistors from the battery.

A simple experiment will show an important characteristic of the parallel circuit. If, in Fig. 6.1-2, resistor R_1 is removed from the circuit, the line current measured by the ammeter decreases. If R_2 is then removed from the circuit, the line current decreases further. What remains is a simple series circuit of R_3, V, and the ammeter. The "line" current is now the current drawn by R_3 from the battery. This current may be computed directly by Ohm's law.

The results of this experiment prove that in Figs. 6.1-1 and 6.1-2 there are indeed three complete circuits for current flow. These are R_1, R_2, and R_3. Each of these individual paths is called a *branch* or *leg* of the parallel circuit.

One characteristic of a parallel resistive circuit, then, is that the total current I_T in the circuit is greater than the current in any branch. This also means that each branch current in the resistive parallel circuit is less than the total or line current I_T.

Total Current in a Parallel Circuit

The total current in a parallel circuit can be calculated in two ways. One method is to calculate the current in each branch by Ohm's law and add these currents. Recall that the voltage across each branch of a parallel circuit is the same. Therefore, for the current in a branch circuit the Ohm's law formula $I = V/R$ makes use of the voltage across the branch and the resistance of the branch.

The second method of calculating current in a parallel circuit is to *measure* or calculate the *total* resistance R_T and then use Ohm's law, which in this case takes the form $I_T = V/R_T$. A method for calculating R_T will be considered in the next experiment.

A simple experiment can prove that the sum of the

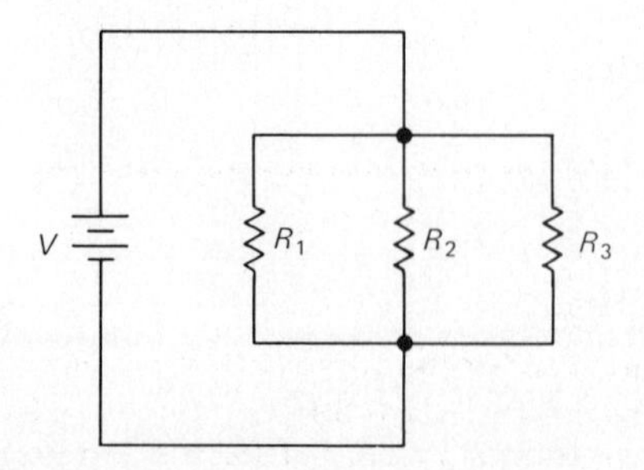

Fig. 6.1-1. Parallel resistive circuit.

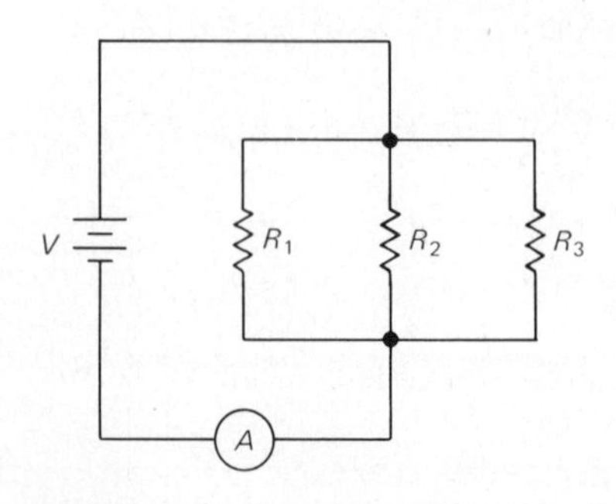

Fig. 6.1-2. Measurement of I_T.

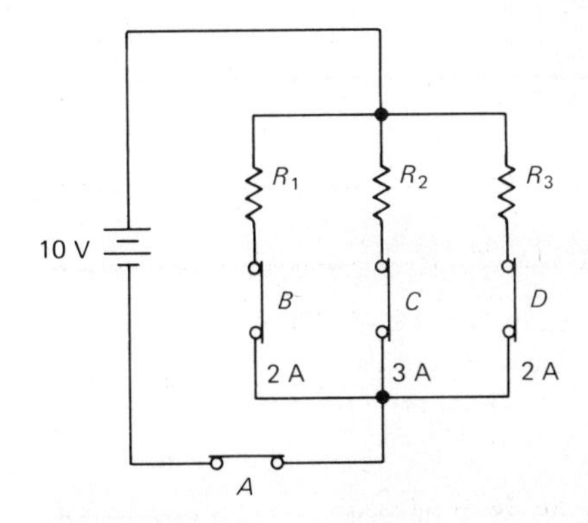

Fig. 6.1-3. Branch current measurements in the parallel resistive circuit.

branch currents *is* the total current. A circuit such as the one in Fig. 6.1-3 can be connected and the total current measured at point A. The branch currents can be measured at points B, C, and D. Add the measured values of the branch currents. The sum of these branch currents should be the same as the total current I_T, measured at point A.

SUMMARY

1. In a parallel resistive circuit each of the individual currents in each parallel branch is less than the total line current.
2. The total line current is greater than each individual branch current.
3. The total line current is the sum of all the branch currents. $I_T = I_1 + I_2 + I_3 + \cdots$
4. The voltage across each branch is the *same* voltage.

SELF-TEST

Check your understanding by answering these questions.

1. In Fig. 6.1-3 the currents are $I_1 = 2$ A, $I_2 = 3$ A, and $I_3 = 2$ A. The total current is ___7___ A.
2. In Fig. 6.1-3 the voltages across R_1, R_2, and R_3 must all be ___10___ V.
3. The voltage across each branch of a parallel circuit must be ___EQUAL___.
4. A parallel circuit represents ___MORE THAN ONE___ (one/more than one) path(s) for current.

MATERIALS REQUIRED

- Power supply: Regulated, variable dc source (VVS)
- Equipment: EVM, VOM
- Resistors: ½-W, 2200-, 3300-, 4700-Ω
- Miscellaneous: SPST switch

PROCEDURE

1. Secure the materials required from your instructor and measure the resistance of each of the resistors supplied. Record the values measured in Table 6.1-1.
2. Connect the circuit of Fig. 6.1-4. Set the VVS control to zero, close switch S_1, and set the meter to the proper range.
3. **Power on.** Adjust the VVS for 15 V. Measure and record in Table 6.1-1 the total current I_T.
4. The power supply is adjusted and should not need readjustment as the circuit is changed. The experimenter should, however, open S_1, removing power from the circuit before any changes are made in the circuit such as the removal and replacement of the current meter.
5. Now measure and record the current in each of the branch circuits.
6. Open S_1 so no power is on the circuit. Measure and record in Table 6.1-1 the total resistance R_T of the three parallel resistors.
7. Using Ohm's law, calculate the I_T by using the measured values of R_T and the source voltage. Record this value in Table 6.1-1.
8. Using Ohm's law, calculate the current in each branch circuit. Use the measured resistance of the resistors and the source voltage. Record the result in Table 6.1-1.
9. Calculate I_T by adding the individually calculated circuit currents. Record I_T in Table 6.1-1.

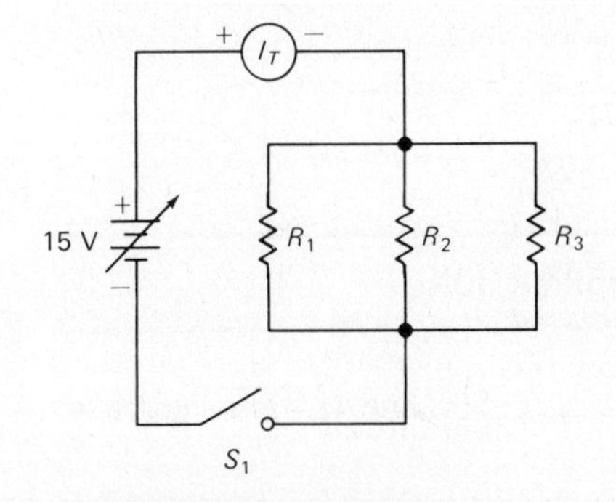

Fig. 6.1-4. Experimental circuit.

QUESTIONS

1. How do the individual measured branch currents compare with the calculated branch currents?
2. How does the total measured current compare with the total calculated current?

TABLE 6.1-1. Parallel-Circuit Measurements

Resistors	R_1 2200 Ω	R_2 3300 Ω	R_3 4700 Ω	R_T	
Measured resistances, Ω	2210	3220	4640	1034	
Measured currents, mA	1.83	1.25	.87	$I_T =$ 3.96	
Calculated currents, mA	.00678	.00454	.00319	$I_T = I_1 + I_2 + I_3$ .01451	$I_T = \frac{V}{R_T} =$ .00148

3. Explain any differences between the measured and calculated values discussed in questions 1 and 2.
4. What is the effect on total current of parallel connected resistors of: (*a*) Increasing the number of resistors in parallel? (*b*) Decreasing the number of resistors in parallel?

Answers to Self-Test

1. 7
2. 10
3. the same
4. more than one

EXPERIMENT 6.2. Total Resistance of Resistors Connected in Parallel

OBJECTIVE

To verify by experiment that the total resistance R_T of resistors connected in parallel is less than the smallest branch resistance, and that R_T may be computed using the formula

$$\frac{1}{R_T} = \frac{1}{R_1} + \frac{1}{R_2} + \frac{1}{R_3} + \cdots$$

INTRODUCTORY INFORMATION

Total Resistance in a Parallel Circuit

If a voltage source such as a battery is connected in a circuit, it "sees" the total resistance of the circuit. It is this total resistance or opposition which limits the current in the circuit. This is true whether the currents "split" into parallel branches or are held in one series path inside the circuit.

Since the total resistance R_T limits the current in the circuit, a single resistor with the value of R_T can be used to replace all the resistors in the circuit. If this replacement resistor has the same value as the R_T of the previous circuit, it will hold circuit current to the same value. We say that the single resistor is the "equivalent" of the resistive network it replaces. Figure 6.2-1 shows three resistors connected in parallel with an applied voltage V. Figure 6.2-2 shows the equivalent circuit of a single resistor connected to the applied voltage source V.

Measuring R_T of Parallel Resistors

The value of R_T for any parallel resistive network (Fig. 6.2-1) may be *measured* by placing an ohmmeter across the network (points X and Y).

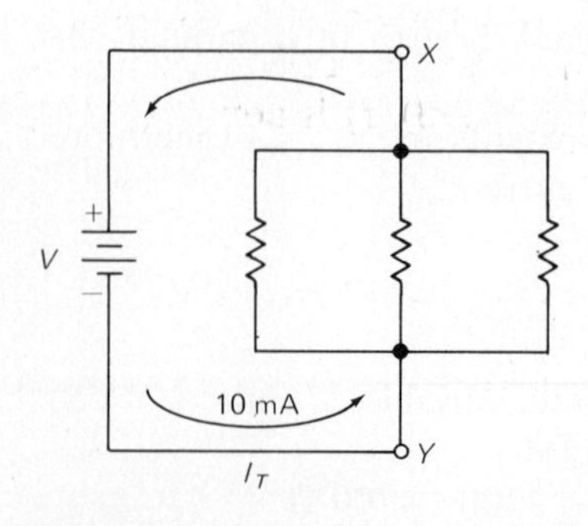

Fig. 6.2-1. Parallel resistive circuit.

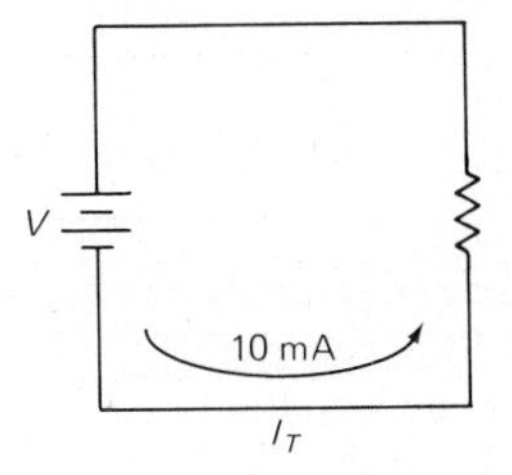

Fig. 6.2-2. Equivalent resistive circuit.

NOTE: As with any resistance measurement, either the connection to the voltage source V must be broken or the voltage source must be removed from the circuit before the ohmmeter is attached, or meter damage may result.

Calculating R_T by Ohm's Law

R_T may also be *calculated* by measuring I_T (refer to Exp. 6.1), measuring V applied to the circuit, and using Ohm's law. Here

$$R_T = \frac{V}{I_T} \tag{6.2-1}$$

R_T Smaller than the Smallest Branch Resistance

In a previous experiment we found that the total current I_T in a parallel circuit was greater than any individual branch current. We also learned that current and resistance have an inverse relationship. That is, the smaller the resistance, the greater must be the current. We can therefore conclude that since the total current in a parallel circuit is greater than any branch current, the total resistance must be smaller than the smallest branch resistance.

The fact that the R_T of a parallel circuit is smaller than the resistance of the smallest branch circuit can be easily shown by experiment. Suppose a circuit is connected which consists of a source V of 10 V and a resistor of 10 Ω. According to Ohm's law the current in the circuit is 1 A. Now suppose another resistor of 5 Ω is placed in the circuit in parallel with the 10-Ω resistor. According to Ohm's law the current through this resistor is 2 A (see Fig. 6.2-3). I_T is now 3 A because the source must supply each branch circuit with the amount of current permitted by the circuit resistance. Since more current is now coming from the source, it would seem reasonable that the total resistance has decreased. The formula for finding the total resistance, $R_T = \frac{V}{I_T}$, will prove that R_T is less than the smallest branch resistance. Substitute 10 for V, the source voltage, and 3 for I_T, and the result is 3.3 Ω, which is less than the smallest branch resistance of 5 Ω. The equivalent of this circuit would be one 3.3-Ω resistor connected to the 10-V source. According to Ohm's law the current in this equivalent circuit would be 3 A, the same as that in the parallel circuit.

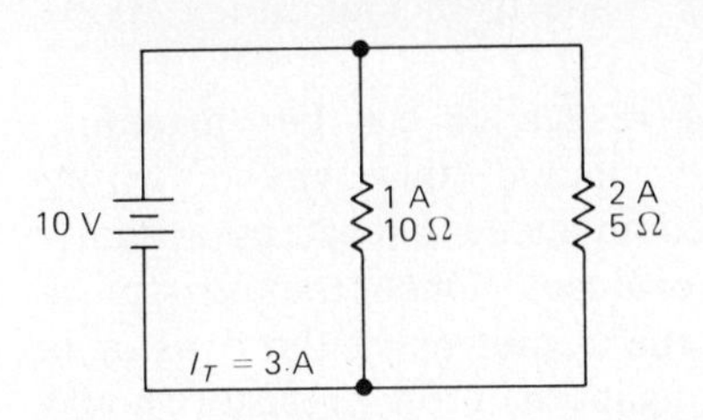

Fig. 6.2-3. Simple parallel circuit.

Calculating R_T by Parallel-Resistance Formula

A third method for finding the R_T of parallel-connected resistors requires knowing the value of each resistor in the combination. Nothing else is needed; that is, we do not need to know the voltage source or circuit current. We calculate R_T just from the values of the parallel-connected resistors, using the formula

$$\frac{1}{R_T} = \frac{1}{R_1} + \frac{1}{R_2} + \frac{1}{R_3} + \cdots \tag{6.2-2}$$

The arithmetic involved is made quite simple by use of the reciprocal ($1/x$) function found on many modern electronic calculators. An example will show how the formula is used. Suppose we wish to find the R_T of three parallel resistors, $R_1 = 100\ \Omega$, $R_2 = 200\ \Omega$, and $R_3 = 500\ \Omega$. Substitute these values in the formula and we see that

$$\frac{1}{R_T} = \frac{1}{100} + \frac{1}{200} + \frac{1}{300}$$

A calculator shows that

$$\frac{1}{100} = 0.01$$

$$\frac{1}{200} = 0.005$$

$$\frac{1}{500} = 0.002$$

Adding, we get $1/R_T = 0.017$.

Now, finding the reciprocal of each side of the last equation gives $R_T = 58.8\ \Omega$, and this value is the total resistance of the three parallel-connected resistors. Observe that the value of R_T is less than 100 Ω, the resistance of the smallest branch resistor.

A second formula

$$R_T = \frac{R_1 \times R_2}{R_1 + R_2} \tag{6.2-3}$$

can be used if there are only two parallel resistors. This formula is sometimes easier to use than the reciprocal formula. If more than two resistors are in parallel, calculate first the R_T of a pair of the resistances. Use this R_T as R_1 and the value of another resistor from the circuit as R_2. The new value of R_T will be the total resistance of the three resistors. This process can be continued until all resistance "pairs" have been calculated and the true total resistance is known.

Let us use each of the parallel-resistance formulas with the circuit of Fig. 6.2-3.

$$\frac{1}{R_T}=\frac{1}{R_1}+\frac{1}{R_2}=\frac{1}{10}+\frac{1}{5}=\frac{1}{10}+\frac{2}{10}=\frac{3}{10}$$

$$R_T=\frac{10}{3}=3.3\ \Omega$$

$$R_T=\frac{R_1\times R_2}{R_1+R_2}=\frac{10\times 5}{10+5}=\frac{50}{15}=3.3\ \Omega$$

A shortcut method of finding R_T for equal-valued parallel resistances is to divide the resistance of one resistor by the number of parallel resistors. But remember, *all the resistances must be of the same value.* Example: What is the R_T of two 1000-Ω resistors in parallel? The number of resistors is 2 and the value of any one of the resistors is 1000 Ω. So we simply divide 1000 by 2, and the answer is 500 Ω. If there were three 1000-Ω resistors, we would divide 1000 by 3, and the answer would be 333 Ω. You may wish to work each of these problems by either of the other methods explained previously. This would be proof of the method.

A method of verifying each of the formulas for parallel resistances is to select a group of resistors and connect them in parallel. Measure the resistance of the combination, then calculate the total resistance of the combination. If the measured and calculated values are equal, we have experimental evidence that the formulas are correct.

Measuring Individual Resistances in a Parallel Circuit

Suppose it is necessary to measure the resistance of R_1 in the parallel network of Fig. 6.2-1. How can this be done? Obviously, it *cannot* be done by measuring across R_1 in the network, since this would give the value of R_T. We can measure R_1 only by disconnecting it from the parallel network and measuring it outside the circuit. Or we can disconnect one lead of R_1, thus removing the effect of the network. We can then measure R_1 by placing the ohmmeter across it. See Fig. 6.2-4.

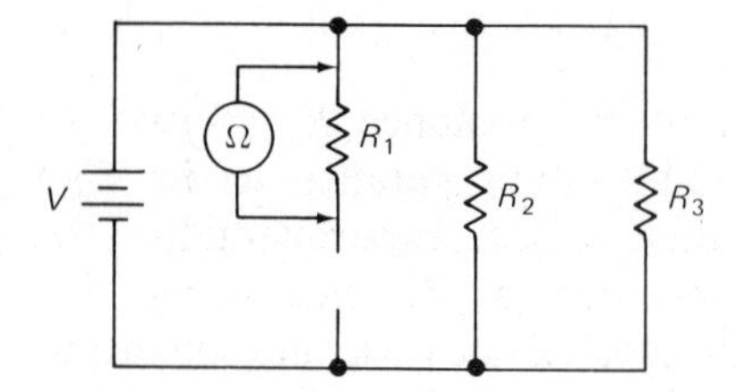

Fig. 6.2-4. Measuring the resistance of a single resistor in a parallel circuit.

Using Parallel Circuits as Single Resistances

The technician uses the theory of parallel circuits every working day. It is not uncommon for the technician to use more than one resistor to replace a single resistor. The reason for doing this is that the correct resistor is not always readily available. For example, suppose a 1000-Ω resistor is needed but is not available. However, 2000-Ω resistors are available. So two 2000-Ω resistors are connected in parallel. The total resistance of the parallel-connected resistors is, of course, 1000 Ω.

It is not always so easy to produce the needed resistance, and a special formula may be used to determine what value resistors to place in parallel, from among those available, to give the required value of resistance. That formula is:

$$R_u=\frac{R_K\times R_E}{R_K-R_E} \qquad (6.2\text{-}3)$$

R_u is the *unknown* resistance needed to be placed in parallel with the available resistance. R_K is the *known* resistance or the resistance which is available for use. R_E is the resistance to be made by the parallel circuit.

Assume 10-kΩ resistors are available and a resistor of 8 kΩ is needed. Using the above formula, we shall calculate the resistor *needed* to create an R_T of 8 kΩ when paralleled with the available 10-kΩ resistor.

$$R_u=\frac{R_K\times R_E}{R_K-R_E}=\frac{10\ \text{k}\Omega\times 8\ \text{k}\Omega}{10\ \text{k}\Omega-8\ \text{k}\Omega}=\frac{80\ \text{k}\Omega}{2}=40\ \text{k}\Omega$$

We can see that a 40-kΩ resistor must be connected in parallel with the available 10-kΩ resistor in order to produce the desired 8-kΩ resistor. Of course, the 40-kΩ resistor can be made from four *series* 10-kΩ resistors.

Though the proper resistance can be "manufactured" by the above method, this type of wiring procedure is not recommended except as a short-term solution to a problem. Once the resistor is "manufactured" and the circuit using the resistor is found to be operable again, the proper resistor should be secured.

SUMMARY

1. The total or equivalent resistance R_T of two or more resistors connected in parallel, as in Fig. 6.2-1, may be determined experimentally by measuring the total current I_T, measuring the voltage V across the parallel network, and substituting the measured values in the formula

$$R_T = \frac{V}{I_T}$$

2. Another method of determining experimentally the total resistance R_T of two or more parallel-connected resistors is to place an ohmmeter across the parallel circuit. The meter measures R_T.
3. Resistance should never be measured while power is applied to the circuit.
4. A formula which expresses the relationship between R_T and R_1, R_2, R_3, etc., of parallel-connected resistors is:

$$\frac{1}{R_T} = \frac{1}{R_1} + \frac{1}{R_2} + \frac{1}{R_3} + \cdots$$

5. Another formula for two parallel resistors is:

$$R_T = \frac{R_1 \times R_2}{R_1 + R_2}$$

6. In order to measure the resistance of one resistor in a parallel circuit, the resistance must be removed or one lead disconnected from the circuit.

SELF-TEST

Check your understanding by answering these questions.

1. In Fig. 6.2-1 with $I_T = 0.02$ A and $V = 50$ V, what is the R_T? ________.
2. For the conditions in question 1, the voltage across each resistor in the circuit is ________ V.
3. What resistance would be placed in parallel with a 1000-Ω resistor to give an R_T of 500 Ω? ________ Ω
4. What is the R_T of a parallel circuit containing three resistors of 2200, 1000, and 4700 Ω? R_T = ________ Ω

MATERIALS REQUIRED

- Power supply: Regulated, variable dc source
- Equipment: EVM, 0 to 10 milliammeter
- Resistors: ½-W 2200-, 3300-, 4700-Ω
- Miscellaneous: SPST switch and connecting wires

TABLE 6.2-1. Resistance Measurement

Rated Value, Ω	*2200*	*3300*	*4700*
Measured value, Ω	2210	3230	4640

TABLE 6.2-2. Measuring R_T of Parallel Resistors

Parallel Combination	*Rated Value*, Ω			*Measured Value* R_T, Ω	*Computed Value* R_T, Ω
	R_1	R_2	R_3		
1	2200	3300	X	1310	
2	2200	3300	4700	1030	

PROCEDURE

1. Measure the resistance of each of the resistors supplied and record the values in Table 6.2-1.
2. In Table 6.2-2 you will see that the resistors are to be arranged in two combinations. Combination 1 requires that the 2200- and 3300-Ω resistors be connected in parallel. Combination 2 requires all three of the resistors to be connected in parallel. Connect each combination, measure R_T, and record each value in Table 6.2-2.
3. Compute R_T by the parallel-resistance formula, and record the value in Table 6.2-2.
4. Connect the circuit of Fig. 6.2-5, using the resistors of combination 1 in Table 6.2-2.
5. Measure I_T and record the value in Table 6.2-3.
6. Measure the voltage across the parallel circuit and record the value in Table 6.2-3.
7. Compute R_T and record the value in Table 6.2-3.

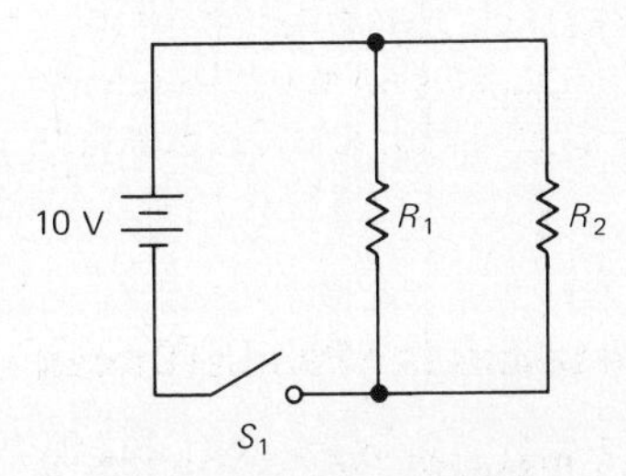

Fig. 6.2-5. Experimental circuit.

QUESTIONS

1. What is the effect on the R_T of parallel-connected resistors of: (*a*) Increasing the number of resistors in parallel? (*b*) Decreasing the number of resistors in parallel?

TABLE 6.2-3. R_T by Ohm's Law

Combination	*Measured Value*, V	*Measured Value I_T*, A	*Computed Value R_T*, Ω
1	4.93	3700	

2. Support your answers to question 1 by referring specifically to the measurements you recorded in Table 6.2-2.
3. Do the measured values of R_T in Table 6.2-2 agree with the computed values of R_T in the same table? If not, why not?
4. Do the results of the experiment prove the formulas used to calculate R_T?
5. What are the three methods you used in this experiment to determine the total resistance R_T of parallel-connected resistors?

Answers to Self-Test

1. 2500
2. 50
3. 1000
4. 600

CHAPTER 7 SERIES-PARALLEL CIRCUITS

EXPERIMENT 7.1. Characteristics of Series-Parallel Circuits I

OBJECTIVE

To determine and verify experimentally the law for total resistance R_T of a series-parallel combination of resistors

INTRODUCTORY INFORMATION

Two Methods for Finding R_T of a Series-Parallel Network

Figure 7.1-1 shows a series-parallel arrangement of resistors. In this circuit, R_1 is in series with the parallel circuit between points B and C, which in turn is in series with R_3. What is the total resistance R_T between points A and D? Obviously we can measure R_T with an ohmmeter, or R_T can be found by the voltage-current method, $R_T = V/I_T$.

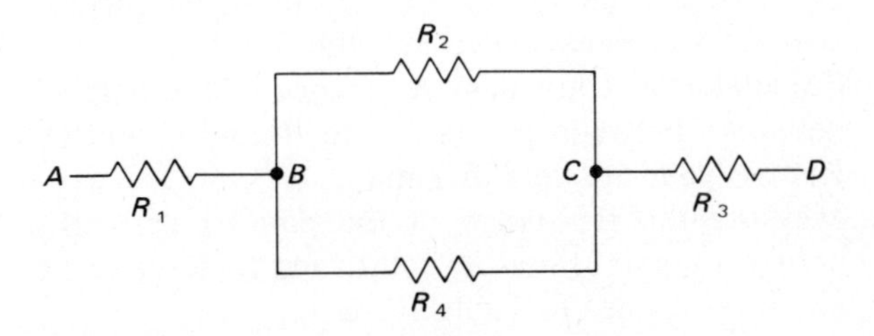

Fig. 7.1-1. Series-parallel resistive circuit.

Finding R_T by Using Series- and Parallel-Resistor Formulas

It is also possible to calculate the total resistance of a series-parallel circuit by making use of both series- and parallel-resistor formulas. As an example, see Fig. 7.1-2. The first step in solving for R_T is to combine all resistors which are in series with each other by using the series formula $R_T = R_1 + R_2 + R_3 + \cdots$ to give the equivalent resistance of the series components. For example, in Fig. 7.1-2 resistors R_6 and R_7 are in series. These are added to give the resistance $R_{6\text{-}7}$, which then replaces R_6 and R_7 as in Fig. 7.1-3. Figure 7.1-3 is the same as Fig. 7.1-2 except that R_6 and R_7 have been replaced by the single resistor $R_{6\text{-}7}$.

Step 2 is to combine all resistors which are in parallel with each other. Here, the parallel formulas are used to give the equivalent resistance of each parallel network. For example, in Fig. 7.1-3 let $R_{2\text{-}3}$ be a resistor equal to the equivalent resistance of the parallel resistors R_2 and R_3. That is, $R_{2\text{-}3} = (R_2 \times R_3)/(R_2 + R_3)$. Similarly, $R_{5\text{-}6,7}$ is the equivalent resistance of the parallel resistors R_5 and $R_{6\text{-}7}$.

Step 3 is to redraw Fig. 7.1-3 and replace each of the parallel networks with its equivalent resistance. We get the simple series circuit Fig. 7.1-4.

Step 4 is to add all the series resistances in Fig. 7.1-4 to get the total resistance. $R_T = R_1 + R_{2\text{-}3} + R_4 + R_{5\text{-}6,7} + R_8$. The total resistance of Fig. 7.1-4 is the same as that of Fig. 7.1-2, and can be considered the single resistor R_T in Fig. 7.1-5.

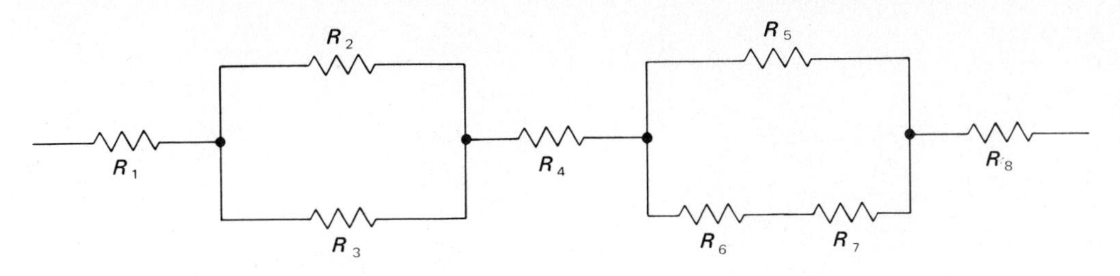

Fig. 7.1-2. Complex series-parallel resistive circuit.

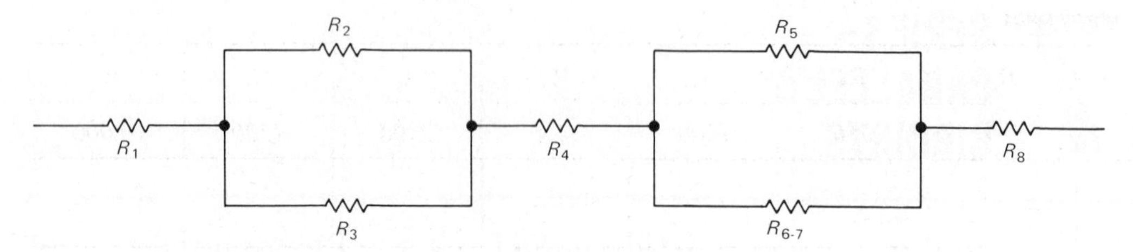

Fig. 7.1-3. Step 1 in the simplification of Fig. 7.1-2.

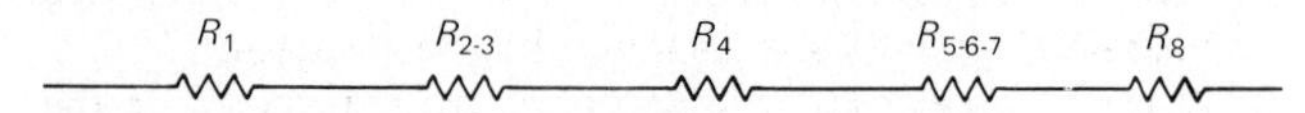

Fig. 7.1-4. Step 2 in the simplification of Fig. 7.1-2.

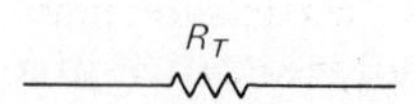

Fig. 7.1-5. Final equivalent circuit of Fig. 7.1-2.

SUMMARY

1. In a series-parallel network like that in Fig. 7.1-1, the total resistance R_T of the network can be found by measuring across the network with an ohmmeter.
2. In determining the total resistance of a series-parallel resistance network, both series and parallel formulas are used.
3. In a series-parallel network like that in Fig. 7.1-2, the total resistance measured across the end terminals of the network can be found by replacing each series combination of resistors with an equivalent resistance, then replacing each parallel combination of resistors by its equivalent resistance. A series equivalent circuit is left, and R_T can be calculated by adding the equivalent series resistors.

SELF-TEST

Check your understanding by answering these questions.

1. In Fig. 7.1-1, $R_1 = 120\ \Omega$, $R_2 = 280\ \Omega$, $R_3 = 330\ \Omega$, and $R_4 = 470\ \Omega$. The value of R_T is ________ Ω.
2. Which formula would be used to find the equivalent resistance between points B and C in Fig. 7.1-1? ________
3. The total resistance of a series-parallel network cannot be measured with an ohmmeter. ________ (true/false)
4. A(n) ________ resistance is that resistance value which can be substituted for a resistive network.

MATERIALS REQUIRED

- Equipment: EVM
- Resistors: ½-W 330-, 470-, 560-, 1200-, 2200-, 3300-, 4700-, and 10,000-Ω
- Miscellaneous: SPST switch and connecting wires

PROCEDURE

1. Measure the resistance of each of the resistors supplied and record its value in Table 7.1-1.
2. Connect the circuit of Fig. 7.1-6 using the resistance values as shown in Table 7.1-1.
3. Calculate the equivalent resistance of the parallel networks between points A and B, and C and D. Record your answers in Table 7.1-2.
4. Measure the resistance of the parallel networks between points A and B, and C and D. Record the measured values in Table 7.1-2.
5. Calculate R_T and record the value in Table 7.1-2.
6. Measure R_T and record the value in Table 7.1-2.

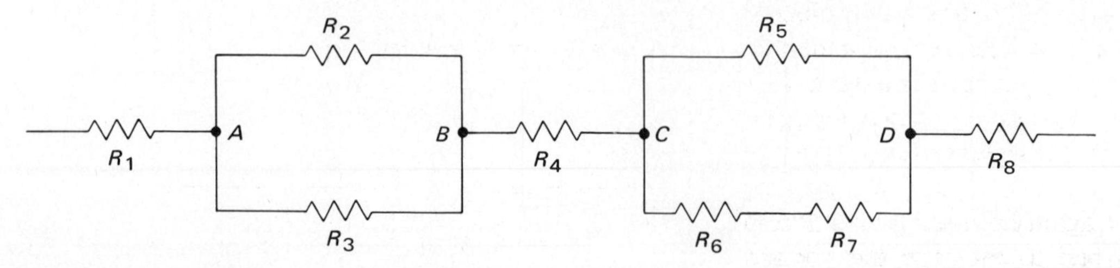

Fig. 7.1-6. Experimental series-parallel circuit.

TABLE 7.1-1. Measured Resistor Values

Rated Value, Ω	R_1	R_2	R_3	R_4	R_5	R_6	R_7	R_8
	330	*470*	*560*	*1200*	*2200*	*3300*	*4700*	*10,000*
Measured value, Ω	329	470	557	1180	2260	3260	4710	9980

TABLE 7.1-2. Experimental Measurements and Calculations

	Computed Value, Ω	*Measured Value, Ω*
A-B		255
C-D		1761
R_T		13510

QUESTIONS

1. In measuring the resistance at points *A-B* in step 4, was it necessary to break the circuit at *A* or *B*? Why?
2. Refer to Table 7.1-2. Are the measured and computed values of R_T the same? Should they be? Explain.
3. Refer to Table 7.1-2. Are the measured values of the resistances between points *A-B* and *C-D* and the computed values the same? Should they be? Explain.
4. What conclusion can you draw from your measurements in Table 7.1-2 concerning the total resistance of a series-parallel circuit? Refer to specific measurements to substantiate your conclusion.

Answers to Self-Test

1. 741+
2. parallel
3. false
4. equivalent

EXPERIMENT 7.2. Characteristics of Series-Parallel Circuits II

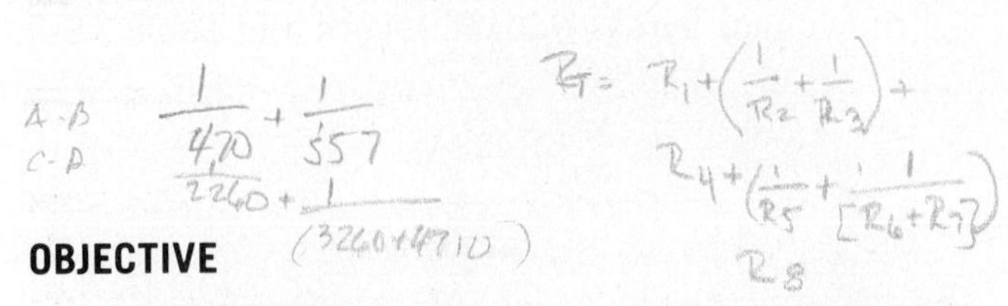

OBJECTIVE

To verify that the voltage across each leg in a parallel circuit is the same as the voltage across the parallel network

INTRODUCTORY INFORMATION

Voltage across Each Leg of a Parallel Circuit

Consider the circuit of Fig. 7.2-1. Between points *B* and *D*, R_2 is in parallel with the series-connected combination of R_4 and R_5. To find the voltage across resistor R_2, we measure across it, or we measure between points *B* and *D*. To determine the voltage across the series combination of R_4 and R_5, we also measure across them, or again between points *B* and *D*. Moreover, if we wished to measure the voltage across the parallel circuit consisting of R_2, R_4, and R_5, we would again measure between points *B* and *D*. This is so because the parallel branches or legs are joined by a common connection at point *B* and also at point *D*.

Since a voltage check across any or all legs must be made between the same two common points, the

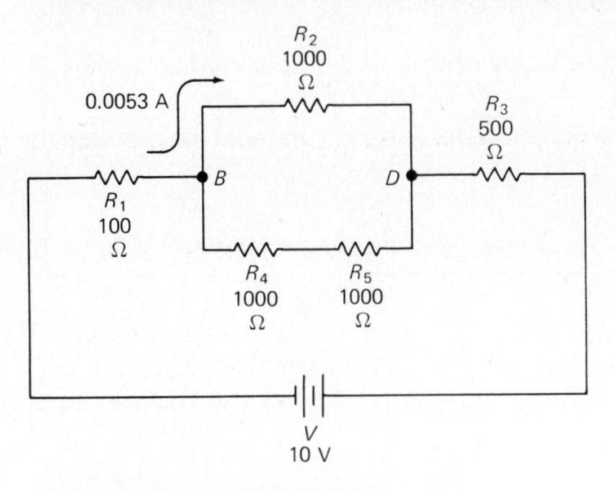

Fig. 7.2-1. Series-parallel resistive network.

voltage across any or all legs must be the same. We reached this conclusion in Exp. 6.1. We shall verify it in this experiment.

The voltage drop V across a parallel circuit may be calculated by multiplying the total current I_T in the parallel circuit by the equivalent resistance R_T of the circuit. The formula used is $V = I_T \times R_T$.

Also, since the voltage V across each branch is the same, the current through *any one branch*, I_B, and the resistance R_B of that branch can be used in the formula $V = I_B \times R_B$. By calculating the voltage drop across one branch, we have found the voltage drop across the parallel circuit. Similarly, by calculating the voltage drop across a parallel circuit, we have found the voltage across each leg of that circuit.

SUMMARY

1. The voltage across each branch of a parallel circuit is the same.
2. The voltage drop across a parallel circuit can be calculated by multiplying the total circuit current by the equivalent resistance of the circuit.
3. The voltage drop across a parallel circuit can also be calculated by multiplying the *current through a branch* by the *resistance of the branch.* Since all branches have the same voltage, the parallel-network voltage is the same as that calculated for any one branch.

SELF-TEST

Check your understanding by answering these questions.

1. Measuring the voltage across one branch of a parallel circuit is the same as measuring the voltage across the entire parallel circuit. ________ (true/false)
2. In a parallel circuit the voltage across each branch is ________.
3. In the circuit of Fig. 7.2-1 the voltage across R_2 is ________ V.
4. The voltage drop across a parallel circuit can be

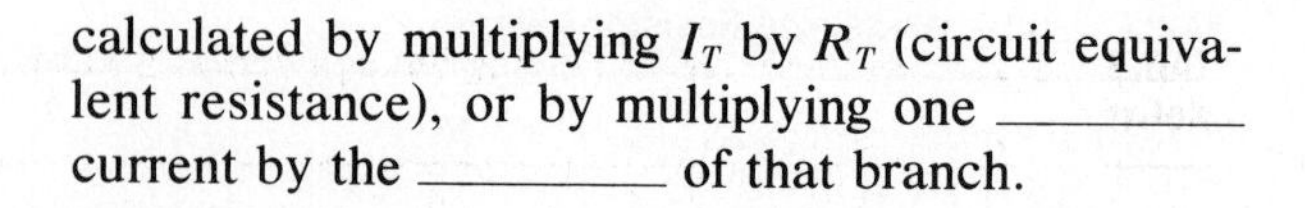

calculated by multiplying I_T by R_T (circuit equivalent resistance), or by multiplying one ________ current by the ________ of that branch.

MATERIALS REQUIRED

- Power supply: Regulated, variable dc source
- Equipment: EVM with current ranges
- Resistors: ½-W 330-, 470-, 1200-, 2200-, 3300-Ω
- Miscellaneous: SPST switch and connecting wires

PROCEDURE

1. Connect the circuit of Fig. 7.2-2. S is open. Set the output of the regulated power supply to 20 V and maintain it at this level.
2. Close S. With an EVM set on the proper range, measure the voltage across R_1, R_2, and R_3 and the series combination of R_4 and R_5. Measure also the voltage across the parallel network B-D. Record your measurements in Table 7.2-1.
3. Measure the current I_2 in R_2 and record it in Table 7.2-2. Calculate the voltage drop across R_2, using the measured value of branch current, and record it in Table 7.2-2.
4. Measure also and record the total current in the circuit. Using the value of total current, calculate the voltage across BD and record it in Table 7.2-2.

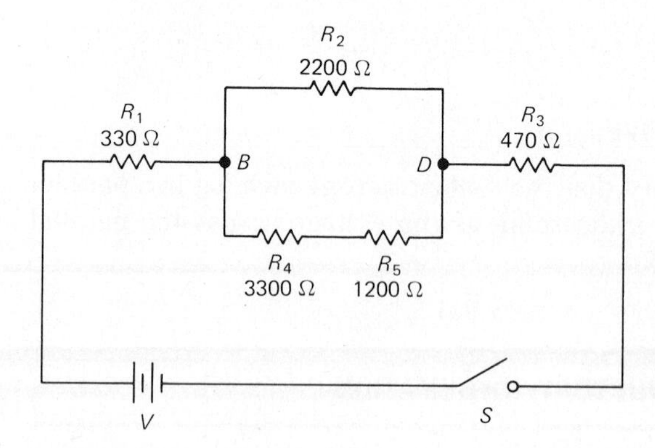

Fig. 7.2-2. Experimental circuit.

TABLE 7.2-1. Voltage Measurements of Experimental Series-Parallel Circuit

V Applied	V_1 (across R_1)	V_2 (across R_2)	$V_{4\text{-}5}$ (across R_4-R_5)	V_3 (across R_3)	V_{BD} (across points B-D)
20.0	2.8	13.0	13.0	4.1	13.0

TABLE 7.2-2. Two Methods for Finding Voltage across a Parallel Network

I_2 in R_2	V_2 across R_2 (calculated)	I_T	V_{BD} across BD (calculated)
5.61		4.13	

QUESTIONS

1. What can you say about the voltage across each leg of a parallel circuit? Refer to specific measurements to substantiate your answer.
2. What information do you need to compute the current in each leg of a parallel network?
3. What information do you need to compute the voltage drop across a parallel network?
4. Did the measurement of the voltage drop across the parallel network *D-B* equal the calculated voltage drop in steps 3 and 4? What does this tell you about voltage drops in parallel circuits?

Answers to Self-Test

1. True
2. the same
3. 5.26
4. branch, resistance

CHAPTER 8 KIRCHHOFF'S LAWS

OBJECTIVE

To verify by experiment that the sum of the voltage drops across series-connected resistors in a closed circuit is equal to the applied voltage

INTRODUCTORY INFORMATION

The solution of complex electric circuits is simplified by the application of Kirchhoff's laws. These laws were formulated and published by the physicist Gustav Robert Kirchhoff (1824–1887), and they established the basis for modern network analysis. They are applicable to circuits with one or more voltage sources. In this experiment we shall be concerned with a single voltage source.

Voltage Law

Kirchhoff's voltage law states that around any closed circuit, the sum of the applied voltage(s) must equal the sum of the voltage drops within the circuit.

This idea was explored in Exp. 5.2. There, we found that the applied voltage was completely used. None is ever lost, nor is any ever gained. If a voltage is applied to a circuit, it is used by the components in the circuit—totally. If the voltage drops across the resistors of a series circuit are added together, this sum will be the same as the applied voltage. This is Kirchhoff's voltage law.

Kirchhoff's Law in Series Circuits

Consider the series circuit in Fig. 8.1-1. The three series-connected resistors may be replaced by a single resistor R_T whose resistance is equal to the sum of R_1, R_2, and R_3. The current I_T in the circuit is such that the product of I_T and R_T equals the applied voltage V_A. That is,

$$V_A = I_T \times R_T$$

EXPERIMENT 8.1. Kirchhoff's Voltage Law

Current is the same everywhere in a series circuit. Therefore I_T is the current in R_1, in R_2, and in R_3. The voltage drop V_1 across R_1 is therefore

$$V_1 = I_T \times R_1$$

Similarly

$$V_2 = I_T \times R_2 \qquad (8.1\text{-}1)$$

and

$$V_3 = I_T \times R_3$$

Adding Eqs. (8.1-1), we get

$$V_1 + V_2 + V_3 = I_T \times R_1 + I_T \times R_2 + I_T \times R_3 \qquad (8.1\text{-}2)$$

Factoring I_T, we have

$$V_1 + V_2 + V_3 = I_T\,(R_1 + R_2 + R_3) \qquad (8.1\text{-}3)$$

Now, $R_1 + R_2 + R_3 = R_T$. Therefore we can write Eq. (8.1-3) as follows:

$$V_1 + V_2 + V_3 = I_T \times R_T \qquad (8.1\text{-}4)$$

But $I_T \times R_T = V_A$, as above. Therefore

$$V_A = V_1 + V_2 + V_3 \qquad (8.1\text{-}5)$$

Equation (8.1-5) is a mathematical statement of Kirchhoff's voltage law, namely, that the applied voltage in a series circuit is equal to the sum of the voltage drops across each of the resistors in that circuit.

Kirchhoff's voltage law is supported by common sense. Suppose, for example, that the sum of the measured voltages V_1, V_2, and V_3 in Fig. 8.1-1 were

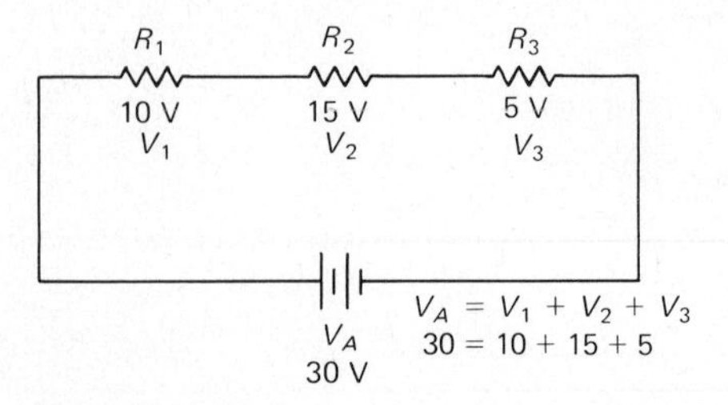

Fig. 8.1-1. Kirchhoff's law applied to the series circuit.

greater than the applied voltage V_A by, say, 5 V. Then the current in the circuit would be

$$I_T = \frac{V_A + 5}{R_T}$$

But this is contrary to fact, for the measured current $I_T = V_A/R_T$, by Ohm's law. Similarly, the sum of V_1, V_2, and V_3 cannot be less than V_A. So we can say with certainty that $V_A = V_1 + V_2 + V_3$. We express Kirchhoff's law mathematically by the formula:

$$V_{\text{applied}} = V_1 + V_2 + V_3 + V_4 + V_5 + \cdots$$

This formula says that the voltage drops across all the series-connected circuit resistances add to equal the applied voltage, that no voltage is lost, and no voltage is gained, but it is all used.

When analyzing circuits, technicians should keep this law in mind at all times. It allows them to find "misplaced" voltages and circuit faults which cause incorrect voltage distribution.

Kirchhoff's Law in Parallel Circuits

The only difference in treatment of Kirchhoff's law in the parallel circuit is that one must always remember that parallel resistances have the same voltage across them. Refer to Fig. 8.1-2. Here there are two parallel circuits connected in series with each other. Note that V_1 is the voltage appearing across both R_1 *and* R_2 because they are in parallel. Also, V_2 is the voltage appearing across both R_3 and R_4 because they, also, are in parallel.

It might help to visualize the circuit of Fig. 8.1-2 as it would look if the parallel combinations of resistors were replaced with their equivalent resistances. See Fig. 8.1-3. Now, the circuit is treated as a series circuit just like that in Fig. 8.1-1.

To use Kirchhoff's law in a series-parallel circuit, the circuit is simplified to its equivalent series circuit

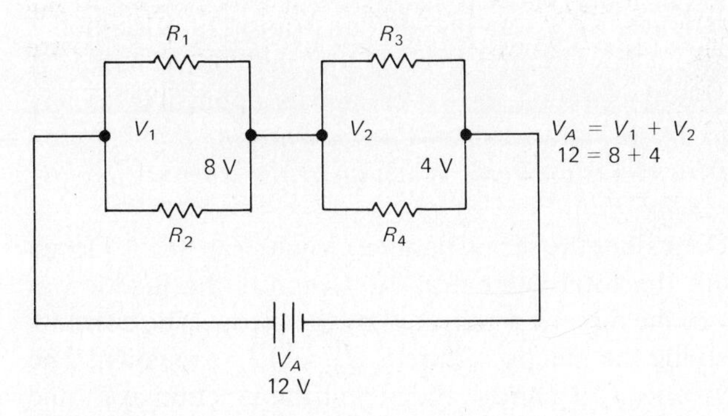

Fig. 8.1-2. Kirchhoff's law applied to the parallel circuit.

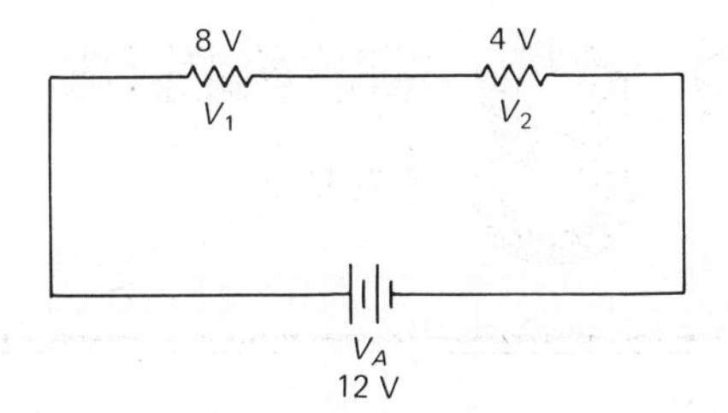

Fig. 8.1-3. Equivalent circuit of Fig. 8.1-2.

as in Exp. 7.1, and Kirchhoff's law is then applied as it would be to a simple series circuit.

SUMMARY

1. Kirchhoff's voltage law states: Around any closed circuit, the sum of the applied voltage(s) must equal the sum of the voltage drops within the circuit.
2. Kirchhoff's law is valid for series, parallel, and series-parallel circuits.
3. For parallel and series-parallel circuits the circuit is usually simplified by replacing all parallel networks by their equivalent resistances. The circuit is then a simple series circuit, and Kirchhoff's law can be easily applied.

SELF-TEST

Check your understanding by answering these questions.

1. In a circuit with four series-connected resistances, $V_1 = 12$ V, $V_2 = 18$ V, $V_3 = 15$ V, $V_4 = 3.5$ V. The applied voltage V_A must then equal ________ V.
2. In order for Kirchhoff's law to be true, a ________ circuit is necessary.

MATERIALS REQUIRED

- Power supply: Regulated, variable dc source
- Equipment: EVM
- Resistors: ½-W 330-, 470-, 1200-, 2200-, 3300-, and 4700-Ω
- Miscellaneous: SPST switch and connecting wires

PROCEDURE

1. Connect the circuit of Fig. 8.1-4 using the values shown.
2. Adjust the output of the power supply so that $V_A = 40$ V. Measure and record this voltage in

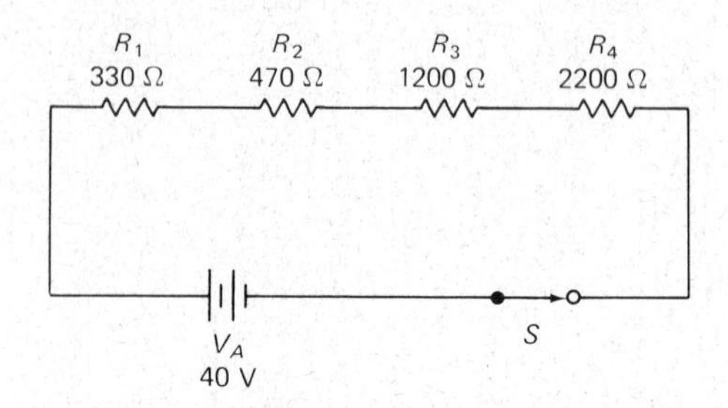

Fig. 8.1-4. Experimental circuit (step 2).

Table 8.1-1. Measure also and record the voltages V_1, V_2, V_3, and V_4 across R_1, R_2, R_3, and R_4, respectively. Add these voltage measurements and enter the sum in the proper column in Table 8.1-1.

3. Connect the circuit of Fig. 8.1-5 using the values shown.
4. Adjust the output of the power supply so that $V_A = 40$ V. Measure this voltage and record it in Table 8.1-1. Measure also and record the voltages V_1, V_2, V_3, and V_4. Add these voltages and enter the sum in the proper column in Table 8.1-1.

TABLE 8.1-1. Circuit Measurements for Kirchhoff's Voltage Law

Step	V_A	V_1	V_2	V_3	V_4	$V_1 + V_2 + V_3 + V_4$
2	40.0	3.1	4.2	11.9	20.7	39.9
4	40.0	2.8	2.8	18.8	15.6	40.0

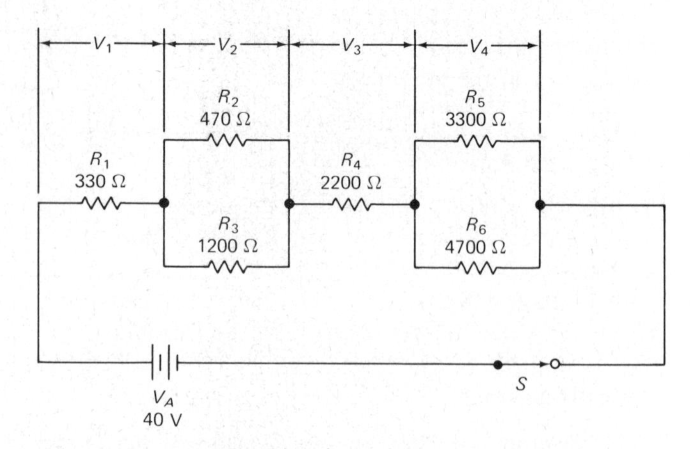

Fig. 8.1-5. Experimental circuit (step 4).

QUESTIONS

1. How does the sum of V_1, V_2, V_3, and V_4 (step 2) in Table 8.1-1 compare with V_A? Explain any difference.
2. How does the sum of V_1, V_2, V_3, and V_4 (step 4) compare with V_A? Explain any difference.
3. In your own words explain Kirchhoff's voltage law.
4. Write the equation for Kirchhoff's voltage law.

Answers to Self-Test

1. 48.5
2. closed

EXPERIMENT 8.2. Kirchhoff's Current Law

OBJECTIVE

To verify experimentally that the sum of the currents entering a junction is equal to the current leaving that junction

INTRODUCTORY INFORMATION

Current Law

In Exp. 6.1 you verified that the total current I_T in a circuit containing resistors connected in parallel is equal to the sum of the currents in each of the parallel branches. This was one demonstration of Kirchhoff's current law, limited to a parallel network. The law is perfectly general, however, and is applicable to any circuit. It states that the *current entering any junction of an electric circuit is equal to the current leaving that junction.*

Consider the series-parallel circuit, Fig. 8.2-1. Designate the total current as I_T. I_T enters the junction at A, in the direction indicated by the arrow. The currents leaving the junction A are I_1, I_2, and I_3, as shown. The currents I_1, I_2, and I_3 then enter the junction at B, and I_T leaves the junction at B. What is the relationship

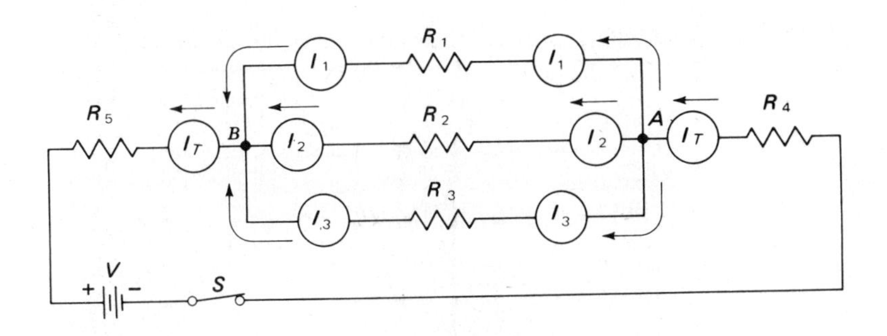

Fig. 8.2-1. Currents in a series-parallel circuit.

between I_T, I_1, I_2, and I_3? This relationship is stated mathematically by Kirchhoff's current law:

$$I_T = I_1 + I_2 + I_3 + \cdots \tag{8.2-1}$$

This means that whatever current enters a junction such as point A in Fig. 8.2-2 will leave that junction. It makes no difference whether the current splits at the junction, as in Fig. 8.2-2, or whether the split currents come together at the junction, as in Fig. 8.2-3.

Again, this is a principle which is important to electronic troubleshooting. It helps the technician evaluate "normal" circuit conditions and determine "normal" voltages in the circuit, in order to compare these with voltages measured in the suspected circuit. Abnormal voltages point to a defective circuit. Once the defective circuit is localized, the defective component can be found.

SUMMARY

1. Kirchhoff's current law states that the current entering any junction of an electric circuit is equal to the current leaving that junction.
2. The mathematical formula which states Kirchhoff's current law is $I_T = I_1 + I_2 + I_3 + \cdots$.

SELF-TEST

Check your understanding by answering these questions.

1. The current ________ any junction is ________ to the current leaving the junction.
2. In Fig. 8.2-2 the current entering junction A is

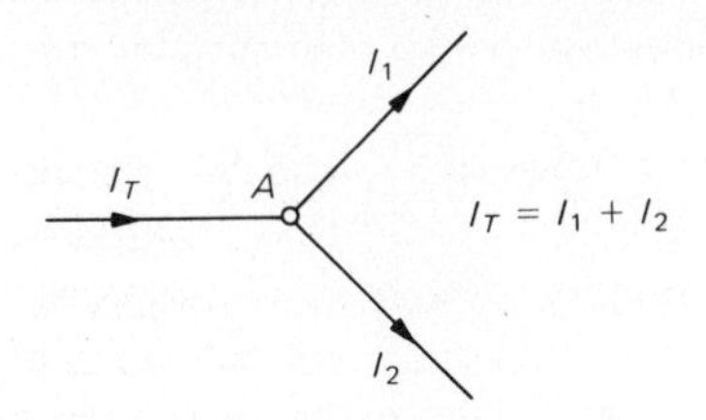

Fig. 8.2-2. Current entering and splitting at a junction.

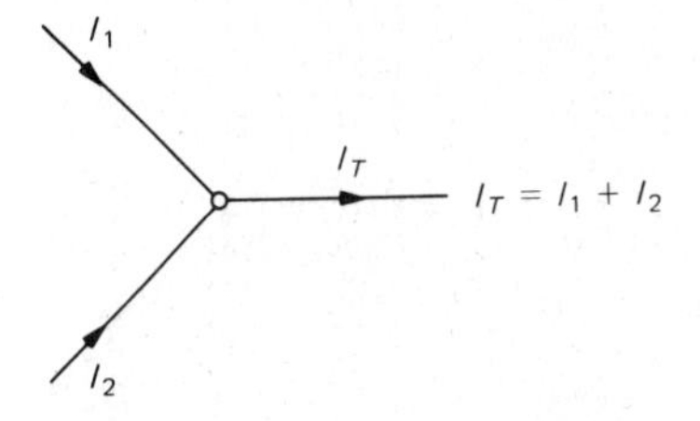

Fig. 8.2-3. Current coming together at a junction.

0.5 A and the current labeled I_1 which leaves that junction is 0.3 A. What is I_2? ________ A.

3. In Fig. 8.2-3, I_1 is 0.47 A and I_2 is 0.12 A. What is I_T? ________ A.
4. The formula which describes the relationship between the currents in Fig. 8.2-3 is ________.

MATERIALS REQUIRED

- Power supply: Regulated, variable dc source
- Equipment: EVM with current ranges
- Resistors: ½-W 330-, 470-, 2200-, 3300-, and 4700-Ω
- Miscellaneous: SPST switch and connecting wires

PROCEDURE

NOTE: Before connecting the milliammeter to measure current, disconnect the power supply by opening switch S. Follow this procedure each time the meter is moved. Observe meter polarity.

1. Connect the circuit of Fig. 8.2-4. Adjust the power source to 40 V. Use the resistor values shown in Fig. 8.2-4.
2. Measure and record in Table 8.2-1 the currents I_T at point A, I_1 through R_2, I_2 through R_3, I_T at point B, I_3 through R_4, and I_4 through R_5.
3. Add I_1 and I_2 and record the sum in the column labeled $I_1 + I_2$ in Table 8.2-1.
4. Add I_3 and I_4 and record the sum in the column labeled $I_3 + I_4$ in Table 8.2-1.

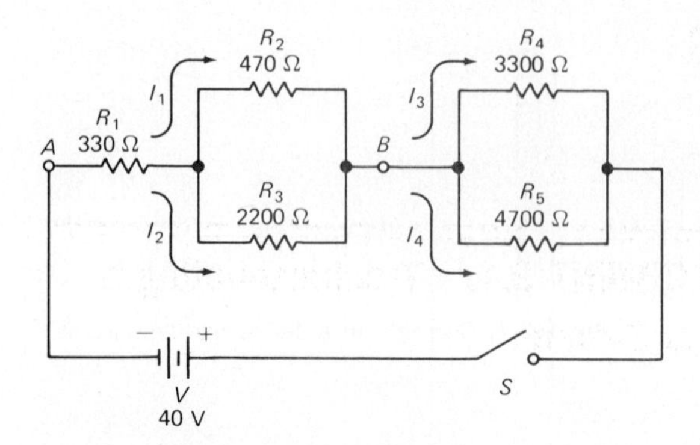

Fig. 8.2-4. Experimental circuit.

5. Calculate I_T at points A and B and record the sums in Table 8.2-1.

QUESTIONS

1. Do the sums of the branch currents (steps 3 and 4) equal I_T? Explain any difference.

TABLE 8.2-1. Measurements for Kirchhoff's Current Law

	I_T at A	I_1	I_2	I_T at B	I_3	I_4
Measured, mA	~~6.28~~ 15.05	12.82	~~2.6~~ 2.56	15.05	~~9.0~~ 8.95	6.15
Calculated, mA		$I_1 + I_2$			$I_3 + I_4$	

2. Explain how this experiment confirms Kirchhoff's current law.
3. In your own words state Kirchhoff's current law.
4. Write the equation for Kirchhoff's current law.

Answers to Self-Test

1. entering, equal
2. 0.2
3. 0.59
4. $I_T = I_1 + I_2$

CHAPTER 9 TROUBLESHOOTING RESISTIVE CIRCUITS

EXPERIMENT 9.1. Troubleshooting a Series Circuit

OBJECTIVE

To find a defect in a series circuit by voltage, current, and resistance measurement

INTRODUCTORY INFORMATION

Series Circuit Rules

In troubleshooting electronic circuits, the technician must keep in mind the rules and laws which govern the current, voltage, and resistance relationships within the circuit. These have been explored in earlier lab experiments. Let us summarize these for the *series* circuit:

1. Current is the same throughout the circuit.
2. Voltage drops add to equal the source voltage.
3. The total resistance is the sum of the individual resistances in the circuit.

NOTE: Rules 1 and 2 above are a restatement of Kirchhoff's voltage and current laws.

Troubleshooting with the Current Meter

By applying these rules, the service technician can analyze the series circuit and determine what *should* be happening, that is, what values of voltage, resistance, and current to expect at various points in the circuit. Then, if the measured current, voltage, or resistance values are appreciably different from the computed values, there must be a defect in the circuit. As an example, see Fig. 9.1-1. This circuit is basically a series circuit, since the diode (about which you will learn more in later experiments) does not conduct current as it is connected here. That is, the diode here acts like an open switch. Suppose the current is measured at point A and found to be greater than the normal value. At point B the same current is found as at A, but at point C less current is measured than at points A and B. Since there is less current at point C than at A or B, and since current should be the same everywhere according to the series-circuit current rule, the diode must be drawing current. But the diode is not supposed to conduct as it is here connected; therefore it must be defective. If the current through the diode is measured, it should be the difference between that measured at points B (I_T) and at C (I_C). The reason is that according to Kirchhoff's current law

$$I_T = I_C + I_{\text{diode}}$$

Therefore

$$I_{\text{diode}} = I_T - I_C$$

The measurement of current is often very difficult in commercial electronic equipment because the circuit must be disconnected and the ammeter inserted into the circuit. This means that wires and components within the circuit must be unsoldered so that the measurement can be made. Afterward, the wires or components must be soldered back into the circuit. This is very time-consuming for the technician who must make the repair in the shortest time possible in order to make a profit on the work. For this reason current measurements are rarely made in troubleshooting commercial electronic equipment.

Two other tests of the circuit can be performed in much less time and with less effort. Either voltage or resistance measurements can be made without removing components from the circuit, although it is sometimes necessary to disconnect one end of a component in order to make resistance measurements.

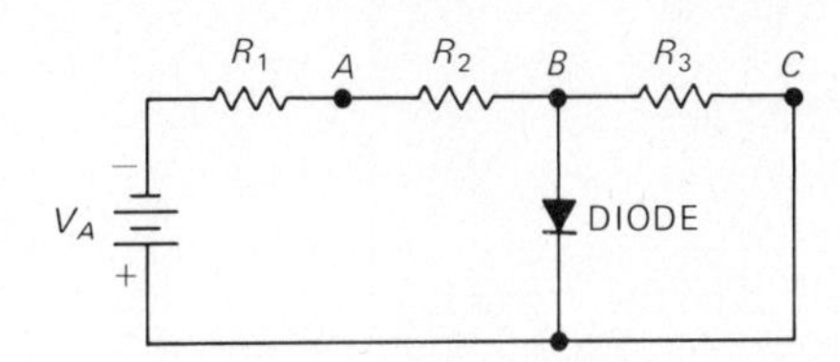

Fig. 9.1-1. Circuit for troubleshooting with the ammeter.

Troubleshooting with the Ohmmeter

Refer to Fig. 9.1-2. This is a series-resistive circuit with a defective resistor. How do we find the defective resistor? One way is to measure each with the ohmmeter. In more complex circuits, however, the ohmmeter may give an incorrect reading because other circuit components may offer parallel paths to the component which is being measured. To overcome this problem, one end of the component is disconnected from the circuit of which it is a part. Despite this problem it is still wise first to measure the resistance of each resistor while it is in the circuit and investigate a resistor that does not measure properly if the others check correctly. To make sure the suspected resistor is defective, one end of it should be removed from the circuit and another resistance check made. Of course it must be remembered that *all power must be off while resistance checks are being made.*

Troubleshooting with the Voltmeter

Again refer to Fig. 9.1-2. Now the defective resistor will be found by voltage measurement. The first measurement should be of the applied voltage V_A. It should be the same as that called for by the service literature of the equipment which is being repaired. Assume that the applied voltage for the series circuit in Fig. 9.1-2 is correct at 150 V. Assume also that current in the circuit is unknown and that breaking the circuit and measuring it would be too time-consuming. In Exp. 5.2 the ratio method of predicting voltage drops in a series circuit was discussed. (Unless the current in the circuit is known, Ohm's law cannot be used to show us how much voltage would appear across each resistor.) So, we use the ratio method to determine how much voltage *should* be measured across each resistor in the circuit.

The total resistance R_T of the circuit is 10 kΩ. R_1 is one-tenth of R_T; therefore $\frac{1}{10}$ of the applied voltage should be measured across R_1. By the same reasoning, R_2 should have $\frac{2}{10}$ of V_A across it; R_3 should have $\frac{1}{2}$ of V_A across it; and R_4 should have $\frac{2}{10}$ of V_A across it. When these are calculated, the following voltages should appear across the resistors in this circuit:

$$V_{R1} = \tfrac{1}{10} \times 150\ \text{V} = 15\ \text{V}$$
$$V_{R2} = \tfrac{2}{10} \times 150\ \text{V} = 30\ \text{V}$$
$$V_{R3} = \tfrac{1}{2} \times 150\ \text{V} = 75\ \text{V}$$
$$V_{R4} = \tfrac{2}{10} \times 150\ \text{V} = 30\ \text{V}$$
$$V_A = 150\ \text{V}$$

Note that the voltage drops add to equal the applied voltage, according to Kirchhoff's voltage law.

Now, the voltage drop across each resistor must be measured. Each should be close to the predicted voltage drop calculated above, if the circuit is normal.

Analyzing Incorrect Voltages

Assume that the measured voltage drops across the resistors in Fig. 9.1-2 are $V_{R1} = 30$ V, $V_{R2} = 60$ V, $V_{R3} = 0$ V, and $V_{R4} = 60$ V. These add to equal the applied voltage, but they are not the same as the calculated values. One resistor (R_3) has *no* voltage drop across it. How can this be? Notice also that all the other voltages are *higher* than they should be. By closer inspection, it can be seen that the 75 V which should be dropped across R_3 has been taken up by the other resistors. R_1 received 15 V of the 75 V, R_2 received 30 V, and R_4 received 30 V. If we add these extra voltages, their sum equals 75 V. This is the missing 75 V. But why did not R_3 drop this 75 V, instead of the other resistors? Because it is *shorted.* A short circuit has 0 resistance, and according to Ohm's law ($V = I \times R$), there can be no voltage drop across 0 resistance. (See Exp. 5.2.)

This method of troubleshooting a series circuit is valid for any series circuit and is not limited to resistive circuits. It should be pointed out, however, that a shorted resistor is extremely rare. Resistors are more likely to change value or open.

Again refer to Fig. 9.1-2. This time suppose the measured voltage drops are: $V_{R1} = 0$ V, $V_{R2} = 0$ V, $V_{R3} = 0$ V, and $V_{R4} = 150$ V. These are not the voltages predicted for this circuit. Only one resistor has any voltage drop, and it has the entire V_A across it. Could it be that R_1, R_2, and R_3 are all shorted? This is highly unlikely, since a short in a resistor is rare. It is even more unlikely that three resistors in the same circuit are shorted. The only other explanation would be that there is no current in the circuit, and therefore, according to Ohm's law, there is zero voltage across R_1, R_2, and R_3.

The next questions are, Why is there no current in these resistors, and why is V_A measured across R_4?

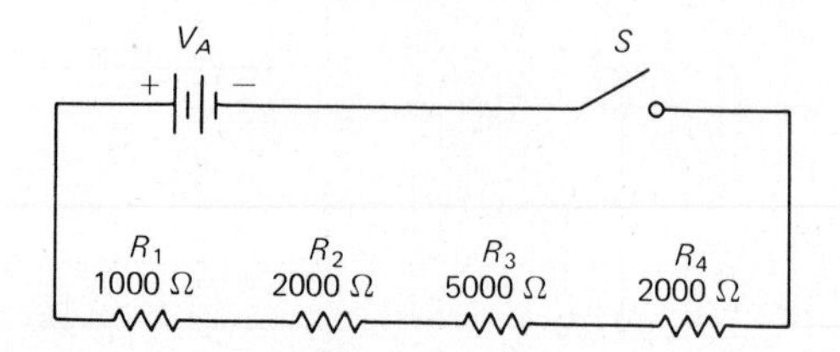

Fig. 9.1-2. Series circuit.

The answer is that R_4 is *open*. Since the series circuit is broken by the open R_4, there can be no current in the circuit. Because there is no current in the circuit, there is no voltage drop across R_1, R_2, and R_3. When the voltmeter is placed across R_4, the resistance of the voltmeter replaces R_4 and closes the circuit. Because the resistance of the voltmeter is very high, V_A is measured across it.

Another way of looking at the problem is that the resistance of R_4 is so great (it is infinite) that the other resistors are a very small part of the total resistance. By the ratio method of calculating voltage in a series circuit, since R_1, R_2, and R_3 are such a small part of the total resistance, the voltage drops across R_1, R_2, and R_3 must be a very small part of the applied voltage V_A—so small, in fact, that the voltage across them cannot be measured.

SUMMARY

1. In troubleshooting electronic equipment, the technician must keep in mind the rules and laws which govern the current, voltage, and resistance relationships within the circuit, in order to calculate, if necessary, the values to expect on measurement.
2. Troubleshooting the series circuit can be done with a voltmeter, current meter, or ohmmeter.
3. Troubleshooting the series circuit with the current meter is physically difficult, since the circuit must be broken to connect the ammeter.
4. Troubleshooting the series circuit with the voltmeter is probably simplest, since no wires or components need to be disconnected from the circuit.

SELF-TEST

Check your understanding by answering these questions.

1. The measurement of ________ is difficult in commercial electronic equipment.
2. A shorted resistor in a series circuit has 0 V dropped across it. ________ (true/false)
3. An open resistor in a series circuit will have ________ dropped across it.
4. When troubleshooting a series circuit with the ohmmeter, one end of the circuit or resistor being tested may have to be ________ in order to obtain correct readings.
5. In order to predict what the voltage drops will be in the series circuit when the current is unknown, the ________ method of calculating the voltage drops is used.

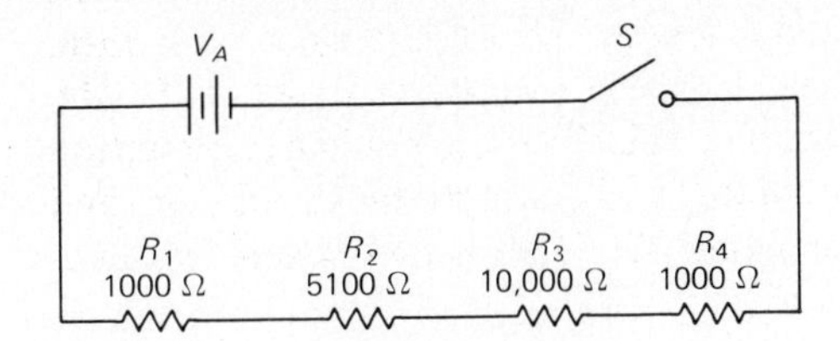

Fig. 9.1-3. Experimental circuit.

MATERIALS REQUIRED

- Power supply: Regulated, variable dc source
- Equipment: EVM with current ranges
- Resistors: ½-W two 1000-, 5100-, and 10,000-Ω
- Miscellaneous: SPST switch and connecting wires

PROCEDURE

1. Obtain the materials required and connect the circuit in Fig. 9.1-3. Adjust the power supply output for 35 V.
2. Calculate the voltage drop which *should* appear across each resistor and V_A. Use the ratio method. Record these values in Table 9.1-1.
3. Measure the voltage drop across each resistor and V_A, and record them in Table 9.1-1.
4. In order to simulate a shorted resistor, connect a short piece of bare hookup wire across R_1 as shown in Fig. 9.1-4. Now measure the voltage drops for each resistor, including R_1. Record these measurements in Table 9.1-1.
5. Add the measurements taken in step 4 and record the sum in Table 9.1-1.
6. Your instructor will give you a defective resistor which may be shorted, open, or out of tolerance. Replace R_4 with the defective resistor.

TABLE 9.1-1. Series Circuit Troubleshooting Measurements

Resistor	R_1	R_2	R_3	R_4	V_A
Calculated V					
Measured V	2.03	10.46	20.4	2.06	34.9
Measured V (R_1 shorted)	0.00	11.12	21.6	2.19	34.9
Measured V (defective R_4)	.12	60	1.18	32.9	34.9

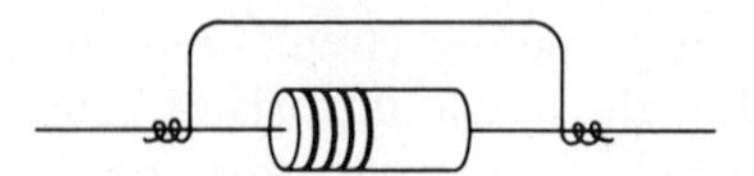

Fig. 9.1-4. Resistor shorted by hookup wire.

7. Measure the voltage drops across each resistor and record them in Table 9.1-1.
8. From the voltage drops measured in step 7, determine the defect of R_4. ________ (shorted, open, out of tolerance, etc.)
9. Troubleshoot the circuit of step 6 with the ohmmeter and ammeter.

QUESTIONS

1. Did your measurements confirm Kirchhoff's voltage law? Explain.
2. Explain how you determined the defect of R_4 from the voltage measurements in step 8.
3. Did troubleshooting the series circuit with the ohmmeter (step 9) reveal the defect? How?
4. Did troubleshooting the series circuit with the ammeter (step 9) reveal the defect? How?

Answers to Self-Test

1. current
2. true
3. V_A
4. disconnected
5. ratio

EXPERIMENT 9.2. Troubleshooting a Parallel Circuit

OBJECTIVE

To find a defect in a parallel circuit by voltage, current, and resistance measurement

INTRODUCTORY INFORMATION

Parallel Circuit Rules

In troubleshooting electronic circuits, the technician must keep in mind the rules and laws which govern the current, voltage, and resistance relationships within the circuit. These have been explored in earlier lab experiments. Let us summarize these for the *parallel* circuit:

1. Current splits into branch currents with I_T equal to the sum of the branch currents.
2. Voltage is the same across parallel branches.
3. Total resistance of a parallel circuit is less than the resistance of the branch with the least resistance.

These rules help the service technician analyze the parallel circuit and compute the voltage, current, and resistance which should appear in the circuit. If the measured values of current, voltage, or resistance do not agree with the computed values, there must be a defect in the circuit.

Troubleshooting with the Current Meter

For an example of how to troubleshoot the parallel circuit with the current meter, refer to Fig. 9.2-1. This circuit is a simple parallel circuit in which the current splits at point x. In troubleshooting this circuit it is helpful to know the total current I_T and the branch currents I_1 and I_2. The procedure, in troubleshooting with an ammeter, is first to measure I_T. If it is not correct according to manufacturer specifications, or your calculations, something in the circuit must be defective. To find which of the two branches is defective, the current in each is measured and compared with the known or calculated current. This comparison should point to the defective branch.

Suppose in Fig. 9.2-1 that I_T is lower than it should be. If the applied voltage is correct, it follows that the resistance of the circuit has increased. We suspect

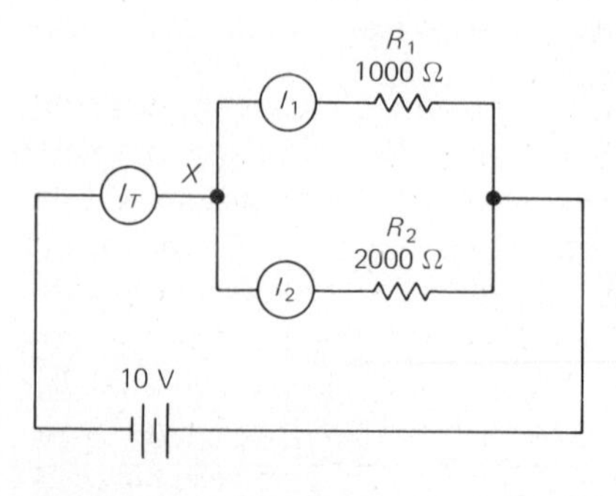

Fig. 9.2-1. Finding the defect with the current meter.

that either R_1 or R_2 has increased in value. The question is which. We can find out by measuring I_1 and I_2. Assume I_1 measures 7.5 mA and I_2 5 mA. By the use of Ohm's law the currents in each branch can be easily calculated as follows:

$$I_1 = \frac{V}{R_1} = \frac{10\text{ V}}{1000\ \Omega} = 0.010\text{ A or } 10\text{ mA}$$

$$I_2 = \frac{V}{R_2} = \frac{10\text{ V}}{2000\ \Omega} = 0.005\text{ A or } 5\text{ mA}$$

The measured and calculated currents in R_2 are the same, 5 mA. Therefore R_2 is okay. The measured and calculated currents in R_1 are not the same. The measured value is lower than it should have been, and therefore R_1 is defective. R_1 has increased in value.

Troubleshooting with the Voltmeter

In the simple parallel circuit normal defects cannot usually be found with the voltmeter. The reason is that the voltage is the same across all parallel branches, whether resistance in any one branch has or has not changed (except for a short). Voltage measurement will normally not reveal the defective part in a simple parallel circuit.

A short in a simple parallel circuit *can* be found by a voltage check, but again the measurement does not reveal in which branch the defect is. See Fig. 9.2-2. From earlier experiments we know that the total resistance of the parallel circuit is less than the smallest branch resistance. In Fig. 9.2-2, the smallest resistance is the short circuit, which has 0 resistance. Ohm's law tells us that there will be no voltage drop across a resistance of 0 Ω. Therefore, the voltage across this parallel circuit is 0 V. A reading of 0 V across a parallel circuit where we normally expect some voltage means that there is a short in the circuit but does not tell us which branch the short is in. The ohmmeter must then be used to pinpoint the defective component.

Troubleshooting with the Ohmmeter

A defective resistor in a parallel circuit can be found with an ohmmeter. If it is thought that there is a defect in the parallel circuit, the R_T of the circuit should be calculated. Once the R_T is known, it can be measured with an ohmmeter, and the two values compared. If they are appreciably different, there is trouble. An example will clarify the process. The calculated R_T of the parallel circuit in Fig. 9.2-3 is 15 kΩ. Now assume the measured R_T is 20 kΩ. The resistance of one or more resistors has increased. How do we find

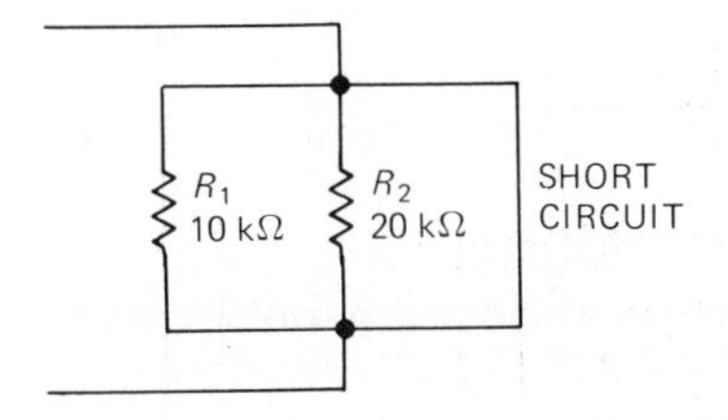

Fig. 9.2-2. Shorted parallel circuit.

the defective resistor? Each branch resistor is checked with an ohmmeter. To do this, one end of the resistor being checked must be removed from the circuit. The ohmmeter check will show how the measured value of each resistor compares with its rated value.

Whatever method is used, there will be some problem. The voltmeter cannot tell us which branch resistor is defective. For the current meter to measure branch or total current, the circuit must be broken. And the circuit must also be broken if the ohmmeter is used to measure individual branch resistances. The most commonly used method of troubleshooting the parallel circuit is the ohmmeter method because it requires no circuit power, as does the current meter method. Without power, circuit components can be safely opened and measured.

SUMMARY

1. In troubleshooting electronic equipment, the technician must keep in mind the rules and laws which govern the current, voltage, and resistance relationships within the circuit, in order to calculate the normal values if necessary.
2. Troubleshooting a parallel circuit involves voltage current or resistance measurements. Sometimes a combination of measurements is necessary.
3. A voltmeter will not indicate which resistor in a simple parallel circuit is defective.
4. An ohmmeter is popularly used in troubleshooting a parallel circuit because no circuit power is required while the tests are being made.
5. Power must be shut off before connecting an ammeter in the circuit to measure current. Power must be off all the time in measuring circuit resistance with an ohmmeter.

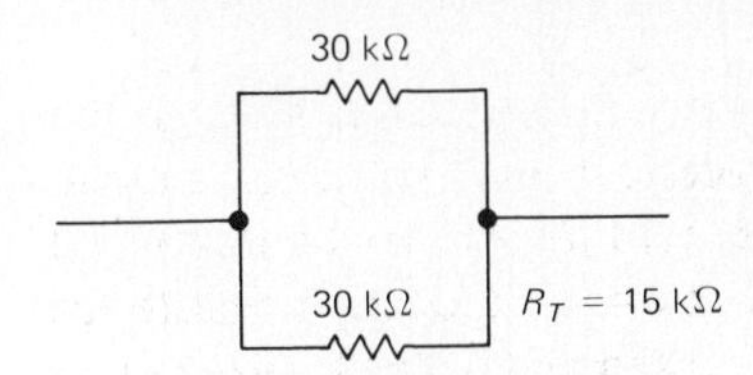

Fig. 9.2-3. Simple parallel circuit.

SELF-TEST

Check your understanding by answering these questions.

1. The R_T of a shorted parallel circuit is ________ Ω.
2. The voltage drop across a shorted parallel circuit is ________ V.
3. To troubleshoot a parallel circuit with an ohmmeter, one end of the resistor or component under test must be ________.
4. The most frequently used method of testing a parallel circuit for defects is the ________ method.

MATERIALS REQUIRED

- Power supply: Regulated, variable dc source
- Equipment: EVM with current ranges
- Resistors: ½-W three, any values between 10 and 30 kΩ, one of which is defective
- Miscellaneous: SPST switch and connecting wires

PROCEDURE

1. Obtain the materials required and connect the circuit in Fig. 9.2-4. Adjust the power supply output for 10 V.
2. Measure the voltage across the parallel circuit and record it in Table 9.2-1. How much voltage appears across each resistor? 10 V. Can the defective resistor be found in this manner? ________ (yes/no)
3. List the color-coded values of the resistors in Table 9.2-1. Using the color code on the resistors, calculate and record the R_T of the parallel circuit.
4. Disconnect the power source. With an ohmmeter measure and record R_T. Is it equal to the calculated R_T in step 3? ________ (yes/no)
5. Find the defective resistor with the ohmmeter.

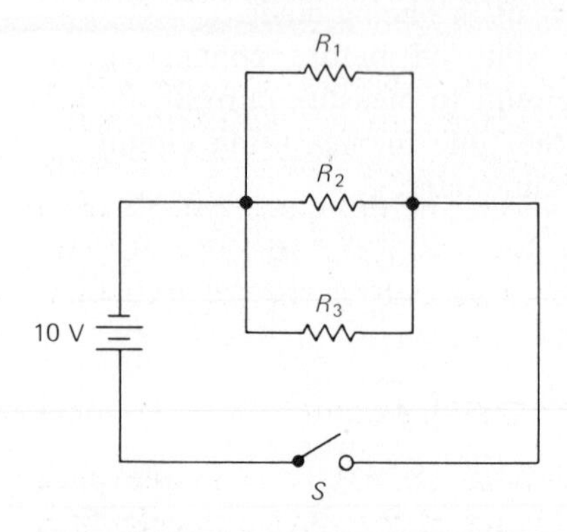

Fig. 9.2-4. Experimental circuit.

TABLE 9.2-1. Measurements in Troubleshooting a Parallel Circuit

Voltage Measured across Parallel Circuit: V	10.00			
	R_1	R_2	R_3	R_T
Rated or calculated, Ω	24K	270K	12K	
Measured, Ω	23.9K	279K	12.5K	
	I_1	I_2	I_3	I_T
Calculated, mA				
Measured, mA	.41	.03	[illegible] .79	
Defective resistor ohmmeter method				
Defective resistor ammeter method				

In Table 9.2-1 list the value of each resistor measured. List also the defective resistor (number or value).

6. Using the color-coded values of the resistors, calculate and record the current in each branch. Calculate and record I_T.
7. Measure and record I_T and the currents in each branch.
8. Compare the calculated and measured currents. Which resistor is defective (number or value)?

QUESTIONS

1. Did your measurements confirm Kirchhoff's current law? Explain.
2. Were you able to determine correctly the defective resistor by the use of the voltmeter ________, the current meter ________, and the ohmmeter ________? Answer yes or no.
3. What advantages are there in troubleshooting the parallel circuit with *each* meter?
4. What disadvantages are there in troubleshooting the parallel circuit with *each* meter?
5. Would troubleshooting the parallel circuit in a working electronic device be any different from troubleshooting it on the breadboard? Explain.
6. What is the effect of an open resistor in a simple parallel circuit? What does the voltmeter test tell you about an open circuit in a simple parallel circuit? Explain.

Answers to Self-Test

1. 0
2. 0
3. disconnected or broken
4. ohmmeter

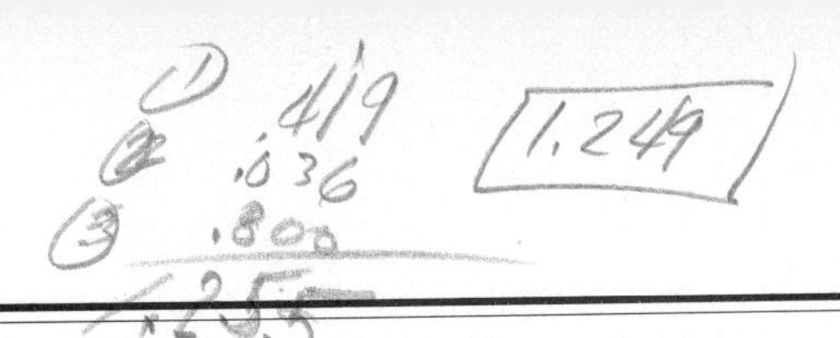

EXPERIMENT 9.3. Troubleshooting a Series-Parallel Circuit

OBJECTIVE

To find a defect in a series-parallel circuit by voltage, current, and resistance measurement

INTRODUCTORY INFORMATION

Troubleshooting the series-parallel circuit requires that the technician know the relationships between V, I, and R in both series and parallel circuits. The same techniques of troubleshooting you learned in Exps. 9.1 and 9.2 are used in troubleshooting the series-parallel circuit. The series-circuit techniques are used to troubleshoot the series portion of the series-parallel circuit, and parallel-circuit techniques are used in the parallel portion.

Troubleshooting with the Current Meter

Refer to Fig. 9.3-1. In order to troubleshoot any circuit with the current meter, the circuit current must be known. The circuit current either is given on the schematic of the device or can be calculated by Ohm's law if the supply voltage and the total resistance are known. If R_T is the total resistance and V is the supply voltage, $I_T = V/R_T$. In Fig. 9.3-1 $R_T = 30\ \text{k}\Omega$, and the supply voltage or applied voltage is 60 V. By applying Ohm's law ($I_T = V/R_T$), we find that $I_T = 0.002$ A or 2 mA. By checking at points A, B, C, or D, 2 mA should be measured, unless there is a defective resistor. This is the series portion of the circuit, and the current should be the same at all points.

If current measurements are taken at points X and Y, 1 mA should be measured for each branch. (Since the resistances are equal for each branch, the current should split in half as it divides between the two branches of the parallel portion of the circuit.) If the two currents are not equal or very close to equal, one of the resistors (R_3 or R_4) is defective.

In the parallel portion of the circuit it is possible to determine which resistor is defective by using the current meter. In the series portion of the circuit it is not possible to determine which resistor is defective by use of the current meter. Refer to Exps. 9.1 and 9.2 for a review of troubleshooting with the current meter.

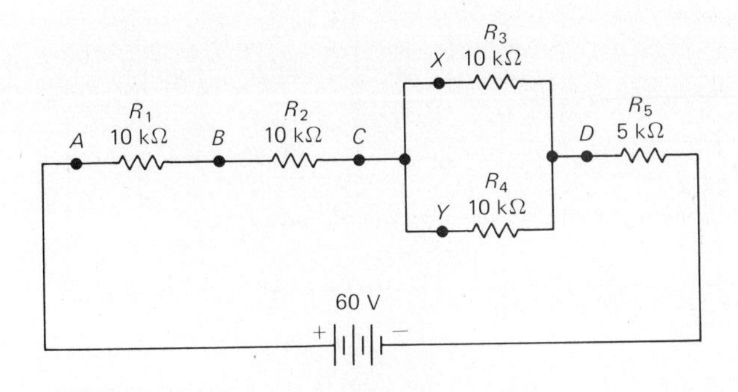

Fig. 9.3-1. Series-parallel circuit.

Troubleshooting with the Voltmeter

Again refer to Fig. 9.3-1. In order to troubleshoot any circuit with the voltmeter, the voltages to be found in the circuit should be known. The voltages either are given in the manufacturer's schematic or can be calculated by Ohm's law if the current and resistance of the circuit are known or are calculated. Since $V = I \times R$ for each individual voltage drop, $V_{R1} = 20$ V. Similarly, since R_2 has the same current through it and has the same resistance, $V_{R2} = 20$ V. V_{R5} has the same current but only half the resistance. Its voltage drop, therefore, is $\frac{1}{2}$ of 20 V, or 10 V. You may recognize the application here of the ratio method of calculating voltage drops.

The parallel portion of the circuit has a total resistance of 5 kΩ, and all the current of the circuit must go through the parallel circuit. Is this not identical to the resistance and current of R_5? Then the voltage drop across the parallel circuit is also 10 V, and the voltage is the same across each parallel resistor.

If the voltage drops, as calculated above, are added, the sum is 60 V. This is in accordance with Kirchhoff's voltage law and suggests that the Ohm's law calculations were correct. Now, the voltmeter can be used to find the defective resistor. Refer to Exps. 9.1 and 9.2 for a review of troubleshooting with the voltmeter.

Troubleshooting with the Ohmmeter

Troubleshooting with the ohmmeter is simple once it is known which circuit of the device is defective. It is troublesome, however, because it may be necessary to unsolder components before making resistance measurements.

All that is required in the series-parallel circuit is to measure directly across series components for their resistance. When measuring parallel-circuit components, either remove one end of the component being measured or calculate the resistance of the parallel combination and then measure across the entire circuit. If the R_T of the parallel circuit does not check with the calculated values, it will be necessary to remove the branch components one at a time and measure each individually.

For a complete review of troubleshooting series or parallel circuits with the ohmmeter, refer to Exps. 9.1 and 9.2.

SUMMARY

1. Troubleshooting series-parallel circuits is no different from troubleshooting series or parallel circuits. The series part of the circuit is treated as a series circuit by itself, and the parallel part of the circuit is treated as any parallel circuit would be.
2. The current meter will not indicate which component is defective in a series circuit, but it will in a parallel circuit.
3. The voltmeter will not indicate which component is defective in a parallel circuit but it will in a series circuit.
4. The ohmmeter is effective for finding defective components in either a series or a parallel circuit.

SELF-TEST

Check your understanding by answering these questions.

1. The circuit in Fig. 9.3-2 is a series-parallel circuit. In troubleshooting, resistors R_1, R_2, and R_3 may be treated as a series circuit. ________ (true/false)
2. Is the circuit in Fig. 9.3-3 a series, a parallel, or a series-parallel circuit? ________

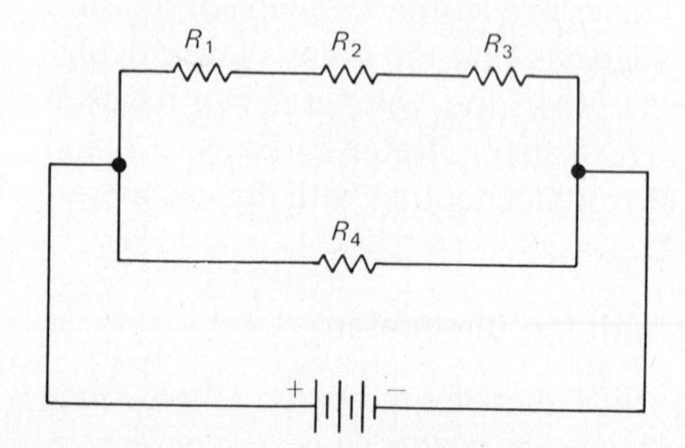

Fig. 9.3-2. Three series resistors in parallel with a single resistor.

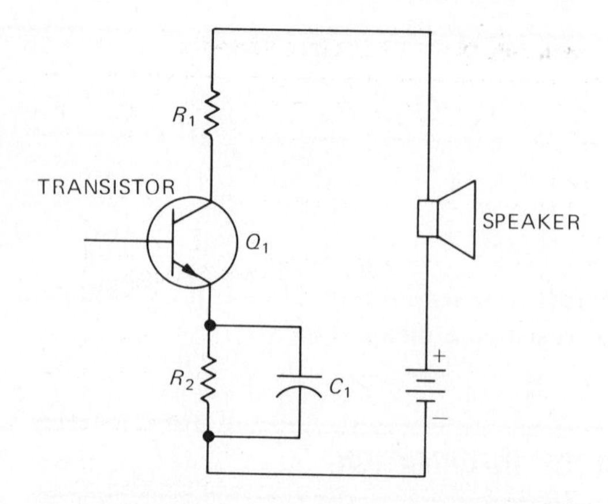

Fig. 9.3-3. Circuit with a variety of components.

3. In Fig. 9.3-3, which components are in series? ________
4. In Fig. 9.3-3, which components are in parallel? ________

MATERIALS REQUIRED

- Power supply: Regulated, variable dc source
- Equipment: EVM with current ranges
- Resistors: ½-W six, any values between 10 and 30 kΩ, with one defective
- Miscellaneous: SPST switch and connecting wires

PROCEDURE

1. Obtain the materials required and connect the circuit in Fig. 9.3-4. Adjust the power supply for 10 V output.
2. Calculate R_T. Calculate I in each resistor. Calculate V across each resistor. Record these values in Table 9.3-1.
3. Measure the current in each resistor and record

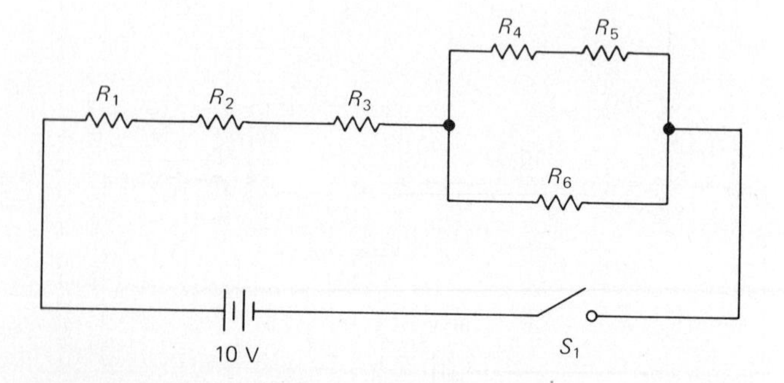

Fig. 9.3-4. Experimental circuit used in troubleshooting series-parallel circuits.

TABLE 9.3-1. Series-Parallel Circuit Measurements

		R_1	R_2	R_3	R_4	R_5	R_6	R_T
Calculated values	I							
	V	.112		.111	1.08	1.03	.008	
Measured values	I	[illegible]	[illegible]	[illegible]	[illegible]	[illegible]	[illegible]	.112
	V	3.02	2.01	2.66	1.03	1.29	2.31	
	R	28K	19K	25K	11K	13K	279K	
Color-coded values		27K	18K	24K	10K	12K	270K	

the values in Table 9.3-1. Is it now possible for you to determine which resistor is defective? Which one? ________.

4. Measure the voltage across each resistor and record the values in Table 9.3-1. Is it now possible for you to determine which resistor is defective? Which one? ________.
5. Measure the resistance of each resistor. Record these values in Table 9.3-1. Also record the color-coded values for each resistor. Compare the measured values with the color-coded values. Is it now possible to determine which resistor is defective? Which one? ________.

QUESTIONS

1. With which meter was the defective resistor most easily found? Explain.
2. Is it readily apparent from Table 9.3-1 which resistor is defective when the measured and calculated values of I are compared? When V values are compared? When R values are compared?
3. Does the troubleshooting of the series-parallel circuit differ from the troubleshooting of series or parallel circuits? Explain.

Answers to Self-Test

1. true
2. series-parallel
3. Q_1, R_1, the speaker, and the *combination* of R_2 and C_1
4. R_2 and C_1

CHAPTER 10 MAGNETISM AND ELECTRO-MAGNETISM

EXPERIMENT 10.1. Magnetic Fields about Bar and Horseshoe Magnets

OBJECTIVE

To determine experimentally the pattern of magnetic lines of force about bar and horseshoe magnets

INTRODUCTORY INFORMATION

Electricity and magnetism are closely associated. The electricity used in homes and industry is produced by rotating a coil of wire wound on an armature inside the magnetic field of a generator. Moreover, electronics makes extensive use of magnetically operated devices such as relays, meters, transformers, door openers, and alarm systems.

In the following experiments no attempt will be made to measure any magnetic quantities. This is generally outside the field of the electronics technician. Instead, our experiments will deal with magnetic effects in general.

Some Physical Properties of Magnets

Certain materials, called magnets, exhibit unique properties. A magnet attracts iron and other ferrous materials. Between two bar magnets, properly oriented (Fig. 10.1-1*a*), a force of attraction exists, and one magnet is pulled to the other. If one of the bar magnets is rotated through 180° (Fig. 10.1-1*b*), the magnets repel each other.

If a small bar magnet, such as that in Fig. 10.1-2, is freely suspended at its center of gravity, one end of the magnet turns toward the geographic north. If the magnet is spun and permitted to come to rest again, the same end points to the north. We call the north-seeking end of the magnet the north pole, designated N, and the other end the south pole, designated S. A magnetic compass, used for navigation, utilized this characteristic of a magnet as far back as the second century A.D. The earth itself is known to be a huge magnet with magnetic poles close to, but not the same as, the geographic poles. In fact, the magnetic pole *near* the north geographic pole is the *south magnetic* pole of the earth, while that *near* the south geographic pole is the *north* magnetic pole of the earth. The needle of a magnetic compass aligns itself with the magnetic poles of the earth.

If a bar magnet is cut in two (Fig. 10.1-3), each of the pieces has the properties of a bar magnet. If we then cut each piece into many more pieces, each smaller piece continues to act like a small bar magnet with a north and a south pole. We cannot isolate just a single pole to any tiny piece.

A magnet does not attract some materials. Air, wood, paper, and glass, to mention but a few, will *not* be attracted by a magnet. These materials are nonmagnetic. Other materials such as iron and its compounds exhibit marked magnetic properties, and are called *ferromagnetic*. These substances may be magnetized by placing them in contact with or in the magnetic field of a *natural magnet*. Moreover, some

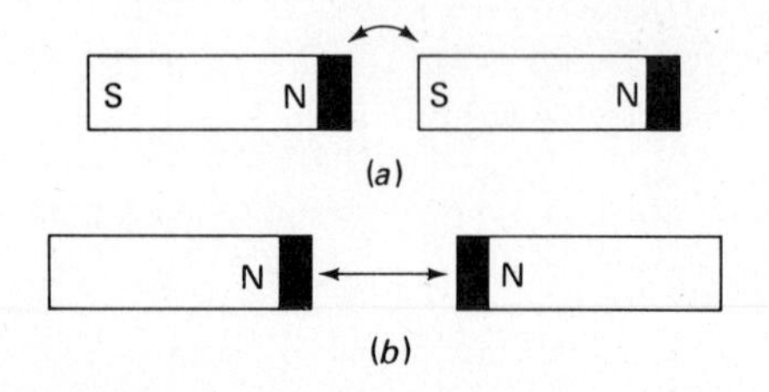

Fig. 10.1-1. (*a*) Two bar magnets properly oriented are attracted toward each other. (*b*) If one bar magnet is turned 180°, the magnets repel each other.

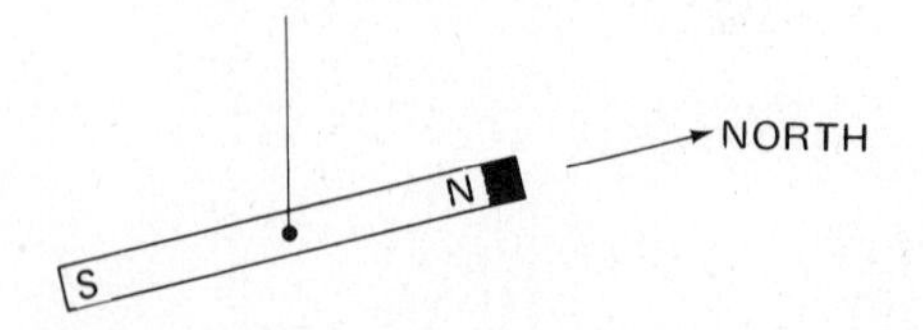

Fig. 10.1-2. A bar magnet free to turn about its center of gravity will point toward the north.

Fig. 10.1-3. If a bar magnet is cut into two pieces, each piece exhibits the properties of a bar magnet.

of these continue to act as magnets when the inducing magnet has been removed. Certain mixtures of aluminum, nickel, and cobalt are used to form very strong permanent magnets, called *alnico* magnets. These are used in radio loudspeakers and other magnetic devices. Hard steel also retains its magnetism. Soft iron is a temporary magnet, for when the inducing magnet is removed, the soft iron loses its magnetism.

Magnetic Field about a Bar Magnet

The force associated with a magnet is said to exist in the magnetic field about the magnet. This magnetic field is presumed to consist of lines of force whose concentration and configuration may be demonstrated experimentally.

Place a cardboard or a piece of glass over a bar magnet and sprinkle iron filings on the cardboard. Tap the cardboard or glass gently so that the iron filings are free to move. The iron filings arrange themselves in a unique pattern (Fig. 10.1-4). There is a noticeable concentration of the iron filings along curved lines, which we call "lines of force." The greatest concentration of these lines of force appears at the ends or poles of the magnet. As the lines of force move out from the magnet, the separation between them increases. That is, the magnetic field gets weaker the further we are from the magnet. We would therefore expect that the force of attraction of a magnet is greatest in the immediate vicinity of the magnetic poles where the lines of magnetic force are most heavily concentrated, a fact which is readily verified experimentally.

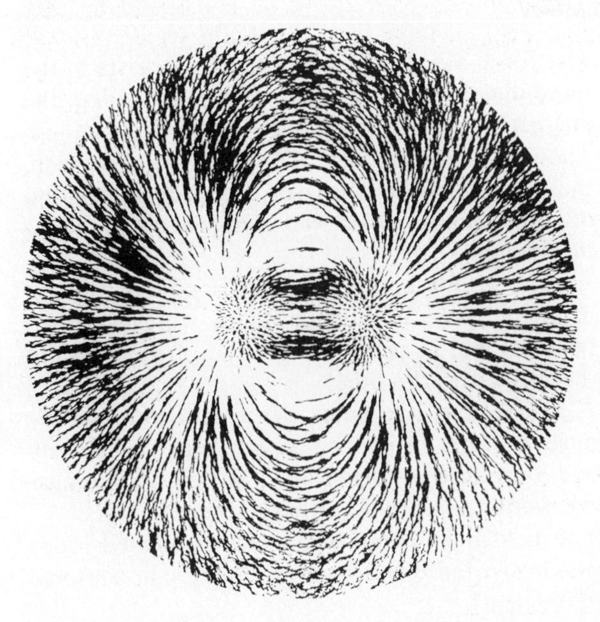

Fig. 10.1-4. Distribution of iron filings in lines of force about a bar magnet.

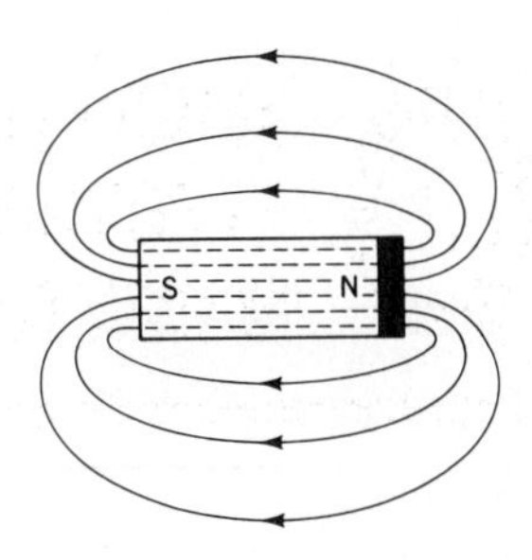

Fig. 10.1-5. Field of force indicated by magnetic field lines.

All the magnetic lines in the field about a magnet constitute the magnetic flux. The stronger the magnetic field, the more lines of force there are in the flux.

A drawing of the lines of force about a bar magnet is shown in Fig. 10.1-5. Note that each field line is a continuous closed loop, leaving the N pole, traveling around the magnet to the S pole and through the magnet back from the S pole to the N pole. The N to S direction of the field lines is purely arbitrary. Note that field lines do *not* cross each other, but push apart. Field lines are most heavily concentrated at the poles.

Figure 10.1-5 is a two-dimensional view of the magnetic field about a bar magnet. A three-dimensional view would show that the magnetic field is indeed three-dimensional.

Unlike Poles Attract

If the N pole of a magnet is brought near the S pole of a second magnet, there is a force of attraction between the two magnets; the lines of force tend to shorten and pull the two magnets together. The field pattern about the two separated magnets now appears as in Fig. 10.1-6, and resembles the pattern about a single bar magnet, except for the visibility of the lines between adjacent poles of the two magnets, which are

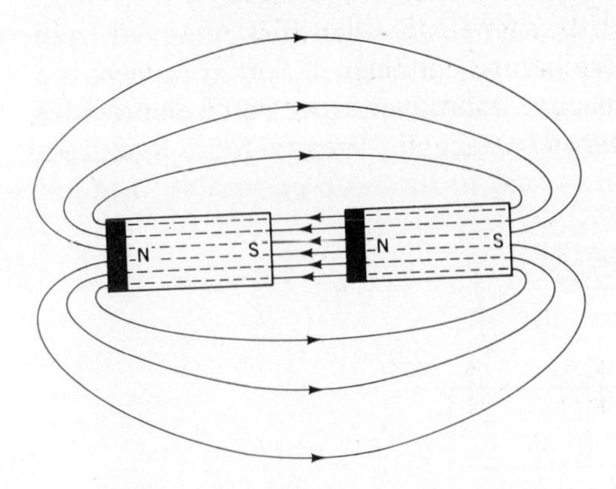

Fig. 10.1-6. Lines of force between unlike poles reinforce, and the magnets are attracted.

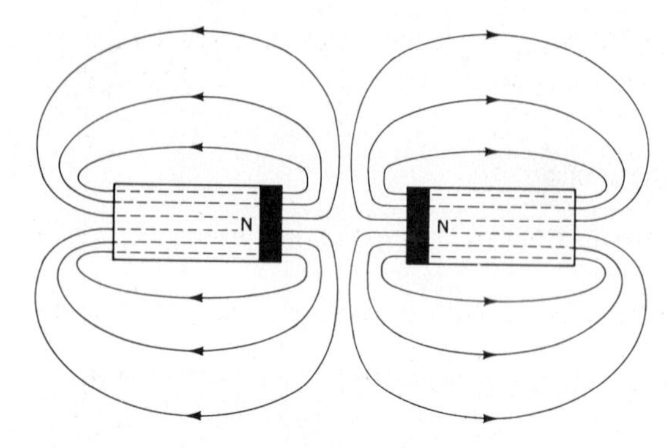

Fig. 10.1-7. Like poles repel. Field pattern about two magnets when like poles of the magnets are brought near each other.

indeed extensions of the lines which complete their loops through the magnets.

Like Poles Repel

If like poles of two magnets are brought near each other, the two magnets are repelled and push apart. The field pattern about these two magnets would appear as in Fig. 10.1-7. The lines of force close to the adjacent like poles of the two magnets are distorted and push apart.

Magnetic Induction

Magnetic materials more readily admit magnetic lines of force than nonmagnetic materials. When a magnetic material, such as soft iron, is placed in the magnetic field of a permanent magnet, this effect may be noted. The lines of force near the soft iron are attracted to the soft iron, as though it were a magnet, and pass through the soft iron rather than through the space around it (Fig. 10.1-8). In fact, the soft iron, while in the field of the permanent magnet, acts like a magnet and will attract iron filings to itself. We say that the soft iron has been magnetized by *induction*. It will lose its magnetism when it is removed from the field of the permanent magnet. Soft iron, then, is a temporary magnet. Moreover, since soft iron provides an easier path for magnetic lines of force than does air, soft iron is said to be more permeable than air.

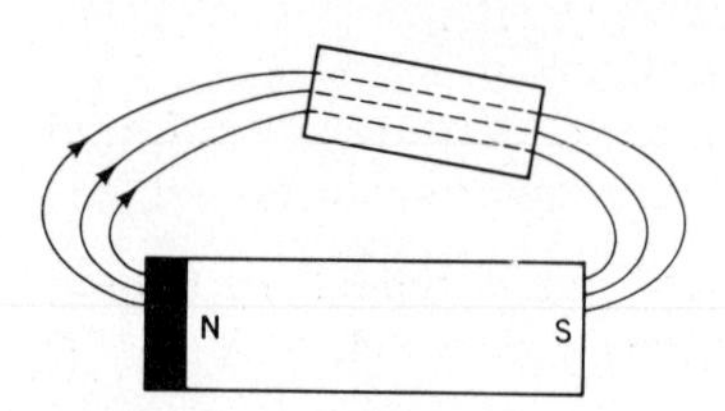

Fig. 10.1-8. Action of soft iron on the magnetic field about a permanent magnet.

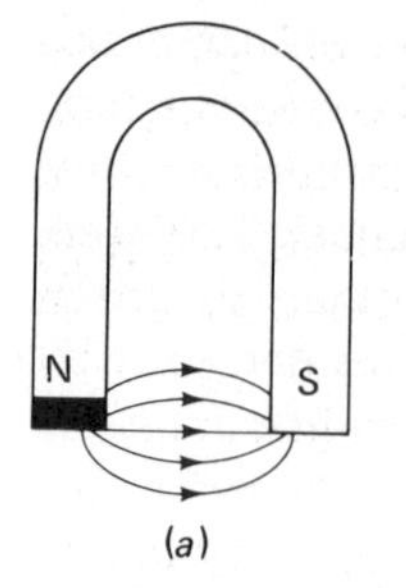

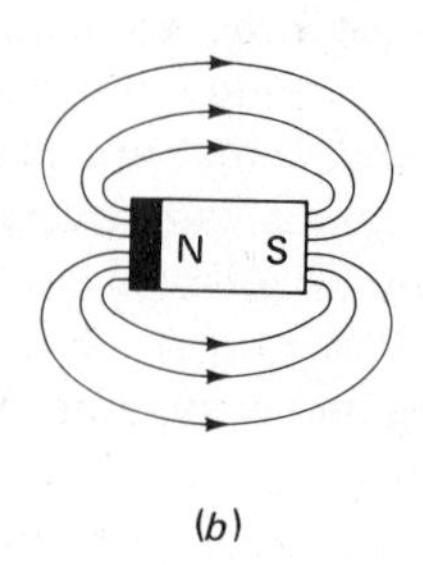

Fig. 10.1-9. (*a*) Magnetic field in the air gap of a horseshoe magnet is more concentrated than the magnetic field in (*b*) the air gap of a bar magnet whose poles are equally spaced.

Magnetic Field about a Horseshoe Magnet

A horseshoe magnet may be considered a bar magnet bent in the form of a horseshoe (Fig. 10.1-9*a*). The magnetic poles remain at the ends of the magnet, but the geometry of the magnet concentrates the lines of force in the air gap between the poles. Compare the lines of force in the air gap of a horseshoe magnet (Fig. 10.1-9*a*) with those of a bar magnet (Fig. 10.1-9*b*) whose poles are as far apart as those in the horseshoe magnet. It is apparent that the field in the air gap of the horseshoe magnet is more highly concentrated. Horseshoe magnets with semicircular soft-iron pole pieces are used in meter movements. The magnetic flux in their air gap is uniform and highly concentrated. These are two requirements for a sensitive, linear meter movement.

SUMMARY

1. The force associated with a magnet exists in the three-dimensional magnetic field surrounding the magnet.
2. Though magnetic lines of force cannot be seen, their effects can be observed by the concentration of iron filings along well-defined paths about the magnet.
3. A line of force is the path which an isolated magnetic pole (if such a thing were possible) would travel in moving from one of the poles of the magnet to the other.
4. A magnet has two poles, north and south. If the magnet were suspended at its center of gravity and were free to turn, the *north* pole of the magnet would point in a northerly direction.
5. The field lines about a magnet are closed loops moving in the direction N to S (by conventional agreement).
6. Lines of magnetic force repel each other and tend to spread apart.

7. The lines of force about a magnet do not cross.
8. Like poles of magnets repel, unlike poles attract.
9. Lines of magnetic force meet less resistance in magnetic than in nonmagnetic materials. Therefore if a piece of soft iron (magnetic) is placed in the field of a magnet, it will distort the field by channeling in itself the lines of force in the vicinity.
10. Magnetic materials may be magnetized by induction, that is, in the presence of a magnetic field. Some will retain their magnetism after the inducing magnet is removed. Others, such as soft iron, will lose their magnetism when the inducing force is removed. Soft iron is therefore in the class of temporary magnets.
11. A knowledge of magnetism is essential to the electronics technician because electronics makes extensive use of magnetically operated devices.

SELF-TEST

Check your understanding by answering these questions:

1. Magnetic lines of force are visible to the naked eye, as is evident from the pattern of iron filings about a magnet. ________ (true/false)
2. The force of a magnet exists in its ________ ________.
3. Magnetic lines will cross each other close to the poles of the magnet. ________ (true/false)
4. The north-seeking pole of a bar magnet which is freely suspended at its center of gravity and is free to turn points to the ________ (north/south) magnetic pole of the earth.
5. Some applications of magnetism in electronics are:
   (*a*) ________
   (*b*) ________
   (*c*) ________
   (*d*) ________
6. Electricity may be generated by a coil of wire ________ in a ________ ________.
7. The magnetic field in the air gap between the poles of a ________ ________ is more concentrated than the field in the air gap between the poles of a ________ ________.
8. Some nonmagnetic materials are:
   (*a*) ________; (*b*) ________; (*c*) ________; (*d*) ________.
9. Strong permanent magnets may be made from a mixture of ________, ________, and ________. These are then called ________ magnets.
10. Magnetic lines of force meet less resistance in air than in soft iron. ________ (true/false)

MATERIALS REQUIRED

- Equipment: Polaroid camera and holder, if available
- Miscellaneous: Two bar magnets, magnetic compass, horseshoe magnet, iron filings, and cardboard ($8\frac{1}{2} \times 11$ in)

PROCEDURE

1. Examine the compass assigned to you. How can you tell which is the N pole? ________ Identify the N pole of the compass.
2. The poles of the two bar magnets and the horseshoe magnet assigned to you are unmarked. How can you determine which is the N, which the S pole of each of these magnets? ________ Identify the N and S poles of each of these magnets.
3. Place a cardboard over one of the bar magnets assigned to you. Sprinkle iron filings on the cardboard, around the magnet. Gently tap the cardboard until a recognizable pattern is seen. This is the pattern of a cross section of the lines of force about the bar magnet. Photograph or draw the pattern for future use. Identify the position of the poles of the magnet and indicate the direction of the lines of force.
4. Now place the two bar magnets, in line, so that the N pole of one is adjacent to and about *one* (1) in away from the S pole of the other. Repeat step 3.
5. Turn one of the bar magnets (in step 4) 180°, so that the N pole of one is adjacent to and about *one* (1) in away from the N pole of the other. The magnets are still in line. Repeat step 3.
6. Place a cardboard over a horseshoe magnet and repeat step 3.

QUESTIONS

NOTE: Submit drawings or photographs of all field patterns you obtained in this experiment.

1. Compare the magnetic field pattern about a bar magnet, which you obtained in this experiment, with that in Fig. 10.1-4.
2. From the distribution of iron filings about the bar magnet, was it evident that lines of magnetic force are continuous

loops? If not, in which of the experimental patterns was it more evident?
3. How did the field pattern you obtained when you faced like poles of the two bar magnets compare with that in Fig. 10.1-7? Comment on the differences.
4. How did the field pattern you obtained when you faced unlike poles of two bar magnets compare with that in Fig. 10.1-6? Comment on the differences.
5. What evidence have you, if any, that magnetic lines of force repel each other?
6. How can you verify experimentally that soft iron is more permeable than air?

Answers to Self-Test

1. false
2. magnetic field
3. false
4. south
5. (*a*) relays; (*b*) transformers; (*c*) electrical generators; (*d*) speakers
6. moving; magnetic field
7. horseshoe magnet; bar magnet
8. (*a*) wood; (*b*) paper; (*c*) glass; (*d*) air
9. aluminum; nickel; cobalt; alnico
10. false

EXPERIMENT 10.2. Magnetic Field Associated with Current in a Wire

OBJECTIVES

1. To verify experimentally the existence and direction of a magnetic field about a wire carrying current
2. To determine experimentally the pattern of magnetic lines of force about a solenoid

INTRODUCTORY INFORMATION

Current Producing a Magnetic Field

The physicist Hans Christian Oersted observed that a compass needle was deflected when it was placed in the vicinity of a wire carrying an electric current. Moreover, he observed that the direction in which the compass needle pointed depended on its position relative to the wire and on the direction of the current.

These observations may be illustrated with the help of the diagram in Fig. 10.2-1. *W*, the wire carrying current, is perpendicular to the plane of the table through which the wire is drawn. A compass placed flat on the table in position 1 deflects as shown. The direction of the compass needle changes when the compass is moved to positions 2, 3, and 4.

When current in the wire is shut off, the compass needle always points to the north, regardless of its position on the table relative to the wire. Moreover, when current in the wire is turned on again but reversed, the compass needle is again deflected by the force about the wire. However, this time the needle points in the opposite direction from its original orientation in positions 1 through 4 in Fig. 10.2-1.

These findings point to the conclusions that

1. A magnetic field is developed about a wire carrying current.
2. The direction of the magnetic field depends upon the direction of current in the wire.
3. The magnetic field appears to be circular about the wire.

Subsequent researchers showed that the magnetic field lies in a plane perpendicular to the current-carrying wire, that the magnetic field developed by the current *is* circular with the current-carrying wire at the center, and that the magnetic field intensity is greatest close to the wire.

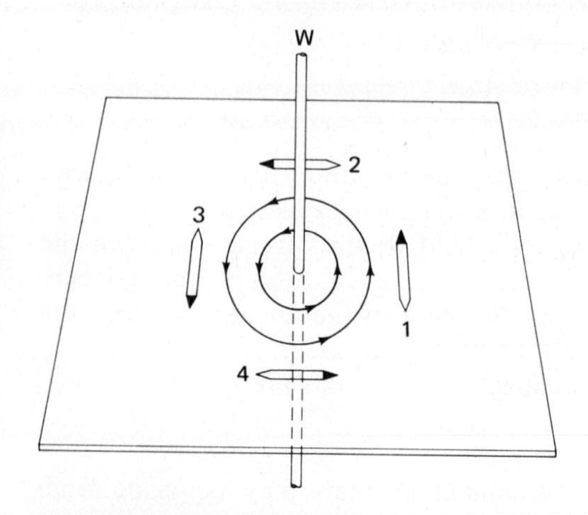

Fig. 10.2-1. Deflection of a magnetic compass in the vicinity of a wire carrying current.

One other fact may be noted. A magnetic field is developed about any moving electrical charge, whether that moving charge is a negative electron or is positive. Moreover, the motion of the electrical charge need not be restricted to a wire, for an electron beam, moving in the vacuum of a cathode ray tube, has associated with it a magnetic field which has the same characteristics as the field about a wire carrying current.

Experiments with the effects of the magnetic field about a wire lead to a rule which makes it possible to predict the direction of the lines of force. The left-hand rule states: If the fingers of the left hand encircle the wire and the thumb of the left hand points in the direction of electron flow in the wire, the fingers point in the direction of the magnetic lines of force (Fig. 10.2-2).

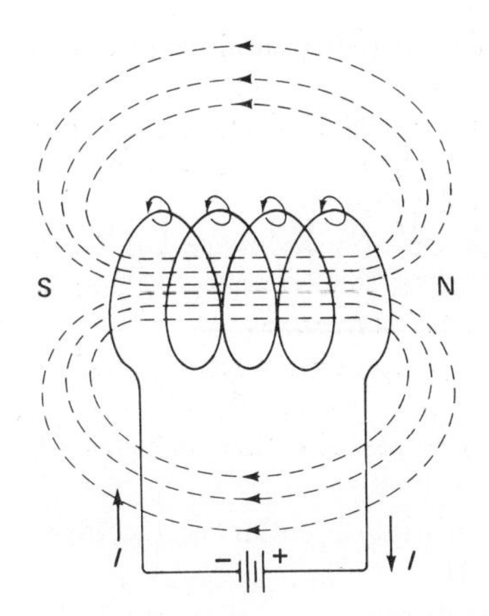

Fig. 10.2-3. Magnetic field about a solenoid.

Magnetic Field about a Solenoid

We observed in the previous experiment that when two magnetic fields were brought near each other, they interacted and the lines of force were distorted. We can extend this observation to note that the lines of force of two magnetic fields *aid each other* when their lines of force are in the *same direction,* and *oppose (cancel) each other* when their lines of force are in opposite directions. Wire shaped in the form of a solenoid coil (Fig. 10.2-3) utilizes this characteristic to form an electromagnet, when current is sent through the solenoid. The circular lines of magnetic force about the wire in the solenoid combine when they are in the same direction, and cancel when oppositely directed, to form a field pattern which is very much like that of a bar magnet. Note that the magnetic lines of force are most heavily concentrated inside the coil, just as they are in a bar magnet. Moreover the distribution of the field lines outside the coil is very similar to the field distribution about a bar magnet.

The location of the N and S poles of a solenoid may be determined by another left-hand rule. If you grasp the solenoid so that the fingers point in the direction of electron current, then the thumb points in the direction of the N pole (Fig. 10.2-4). Observe that the position of the poles may be reversed by changing the direction of current or changing the direction of the winding.

The form about which a solenoid is wound may consist of magnetic or nonmagnetic material. However, soft iron is generally used, because the iron core, having a higher permeability than nonmagnetic materials, increases the flux density inside the core. When current is turned on, the iron core is magnetized and acts like a bar magnet. The magnet loses its "strength" when current is turned off. This is another reason for using a soft-iron core: to create a temporary magnet when it is needed and eliminate it when it is not needed.

The strength of a solenoid magnet may be increased by increasing the number of turns of wire in the coil or by increasing the current in the coil.

We can understand magnetic effects better by comparing these effects with those in an electric circuit. In an electric circuit the variables are V, electromotive force; I, the resulting current; and R, the opposition to current in the circuit. In a magnetic circuit the equivalent variables are *mmf,* magnetomotive force; ϕ, the lines of magnetic force produced; and $\mathcal{R}$ (re-

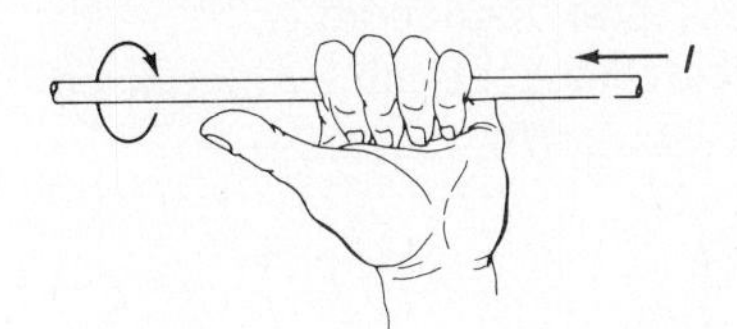

Fig. 10.2-2. Left-hand rule showing direction of magnetic lines of force about a wire carrying electron current.

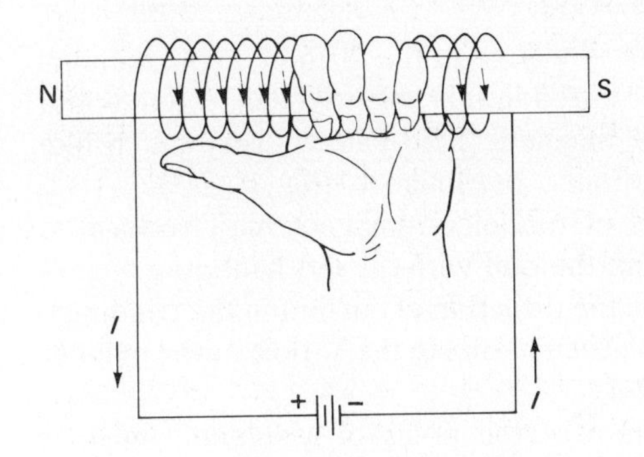

Fig. 10.2-4. Left-hand rule for locating the N pole of a solenoid magnet.

luctance), the opposition of the circuit to producing lines of force. Algebraically,

$$\text{mmf} = At \qquad (10.2\text{-}1)$$

where A equals the amperes in the circuit (coil), and t equals the number of turns in the coil. The formula which relates the three variables in a magnetic circuit is

$$\phi = \frac{\text{mmf}}{\mathcal{R}} \qquad (10.2\text{-}2)$$

This is comparable to Ohm's law for electric circuits.

Reasoning from Eqs. (10.2-1) and (10.2-2), it becomes apparent that the strength (ϕ) of a solenoid magnet is directly proportional to the number of ampere-turns (At); that if the current (A) in the coil is changing, the strength of the magnet is changing in direct proportion to A. That is, an increase or decrease in current will result in a proportionate increase or decrease in the strength of the resulting magnetic field.

Industrially, solenoids in the form of relays are used as electrical switches to turn a process on and off. Solenoids are also used as valves to permit a liquid to circulate in a system or to turn the liquid off. Bells, buzzers, and many other devices are solenoid-operated.

SUMMARY

1. A magnetic field is developed by a moving electric charge.
2. Circular magnetic lines of force appear around a wire carrying current. The magnetic field is at right angles to the wire and may be visualized as a cylindrical "sleeve" surrounding the wire.
3. The direction of the lines of force depends on the direction of current. If the wire is grasped in the left hand with the thumb pointing in the direction of electron current, the fingers point in the direction of the circular magnetic lines of force (Fig. 10.2-2).
4. If the wire is wound in the form of a coil and current is permitted in this solenoid coil, the solenoid exhibits a magnetic field pattern very much like the field about a bar magnet (Fig. 10.2-3).
5. The poles of this electromagnet may be located by grasping the coil with the left hand, the fingers pointing in the direction of current in the windings. The thumb then points to the N pole of the magnet (Fig. 10.2-4).
6. The position of the poles of a solenoid magnet may be reversed by (*a*) reversing the current or (*b*) reversing the direction of the winding.
7. Solenoids are frequently wound on soft-iron cores. When current is turned on, the iron core acts like a temporary bar magnet, but it loses its magnetism when current is turned off.
8. The strength of a solenoid magnet may be increased by increasing the current or the number of turns, or both.
9. A magnetic circuit may be compared with an electric circuit in which:
   (*a*) mmf (magnetomotive forces) is the equivalent of V (electromotive force).
   (*b*) ϕ, the resulting lines of magnetic flux, is the equivalent of I.
   (*c*) $\mathcal{R}$, the opposition to ϕ, is the equivalent of R.
10. In a magnetic circuit

$$\phi = \frac{\text{mmf}}{\mathcal{R}}$$

11. In a magnetic circuit mmf $= At$.
12. In a solenoid magnet, the strength of the magnet (ϕ) is directly proportional to A, the current in the coil; that is, ϕ increases as A increases, and ϕ decreases as A decreases.
13. Solenoid magnets are used to actuate the switching mechanism in a relay, a bell, buzzer, and other electromagnetic devices.

SELF-TEST

Check your understanding by answering these questions.

1. An electric current creates a magnetic field. ________ (true/false)
2. The magnetic field at any one point about a wire carrying current lies in a plane ________ to the wire at that point.
3. If an observer looks along a wire carrying electron current, in the direction of electron flow, the direction of the magnetic field about the wire appears ________ (clockwise/counterclockwise) to the observer.
4. Soft iron is more permeable than air. Therefore a solenoid with a soft-iron core has a more intense field in the core than does a solenoid with an air core. ________ (true/false)
5. The left-hand rule can be used to determine the identity of the ________ in a solenoid.
6. The strength of an electromagnet can be increased by increasing (*a*) ________ (*b*) ________.

MATERIALS REQUIRED

- Power supply: Variable dc source, low-voltage, high-current

- Equipment: 0- to 1-A ammeter; Polaroid camera and holder, if available
- Resistors: 25-W 15-Ω
- Miscellaneous: SPST switch, magnetic compass, iron filings, cardboard ($8\frac{1}{2} \times 11$ in), 16 ft of #18 varnish-insulated copper wire, 2-in cylindrical ($\frac{1}{2}$-in diameter) hollow coil form (cardboard, plastic, or copper tubing); 2-in round soft iron core for coil form

PROCEDURE

Solenoid without Core

1. Use #18 varnish-insulated copper wire to construct a solenoid coil. On a hollow cylindrical form, cardboard, plastic or copper tubing, 2 in long by $\frac{1}{2}$ in in diameter, wind clockwise in two or three layers a 100-turn coil. Scrape the ends of the wire free of insulation and secure one end at conductive terminal A, mounted at one end of the coil form. Similarly secure the other scraped end at conductive terminal B, mounted at the other end of the coil form. Mark ends A and B.
2. Connect the coil constructed in step 1 in the circuit of Fig. 10.2-5. End A of the coil is connected to the positive terminal of the supply; end B, through a resistor R and ammeter M to the negative terminal of the supply. Power is **off.**
3. Set a magnetic compass down on the lab table so that it is not near any magnets or magnetic materials. Orient the compass case so that the compass pointer is set on north N, as in Fig. 10.2-6. Place end A of the coil about 2 in away from the point marked E on the compass. Orient the solenoid coil so that its axis is in line with the line of the compass drawn from W to E, as in Fig. 10.2-6. The solenoid coil, with no current in it, should have no effect on the compass pointer.
4. Turn the power supply **on** and adjust its output so that there is $\frac{3}{4}$ A in the coil, as measured on M. Record in Table 10.2-1 the direction of the compass pointer. *Data for steps 4 through 14 should be recorded in Table 10.2-1.*
5. Maintaining the W-E orientation of the axis of the solenoid coil, slowly push the coil so that end A is $\frac{1}{2}$ in away from point E on the compass. The effect of the solenoid on the compass pointer should be more evident than in step 4. Record the direction of the compass pointer.
6. Reduce current in the solenoid to 100 mA. The solenoid is still $\frac{1}{2}$ in from point E on the compass,

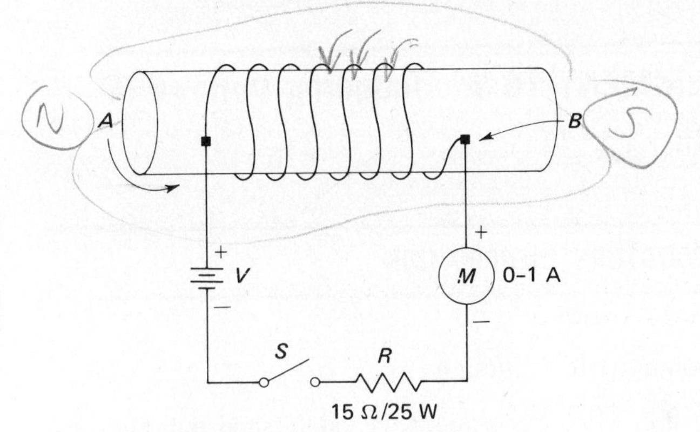

Fig. 10.2-5. Experimental circuit to determine characteristics of the magnetic field about a solenoid.

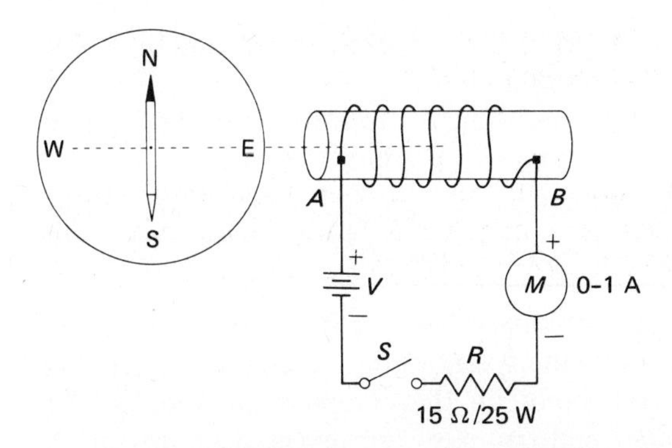

Fig. 10.2-6. Axis of the solenoid is in line with the line drawn from W to E. When there is no current in the solenoid, the compass points to the north N.

TABLE 10.2-1. Electromagnet Characteristics

Step	*Current in Solenoid,* A	*Solenoid Distance from* E, in	*Current Polarity* A	B	*Direction of Pointer*
4	$3/4$	2	+	−	80°W
5	$3/4$	$1/2$	+	−	90°W
6	$1/10$	$1/2$	+	−	59°W
7	0	$1/2$	+	−	N
9	$3/4$	2	+	−	W
10	$3/4$	$1/2$	+	−	W
11	$1/10$	$1/2$	+	−	W
12	0	$1/2$	+	−	65° [illegible]
13	$3/4$	$1/2$	−	+	70°E
14	0	$1/2$	−	+	40°E

as it was in step 5. Record the direction of the compass pointer.
7. Reduce solenoid current to zero. Record the direction of the compass pointer. Power **off.**

Solenoid with Soft-Iron Core

8. Place a 2-in-long soft-iron core inside the solenoid coil. Move the solenoid so that end *A* is again 2 in from the point on the compass marked E. The solenoid and compass are oriented as in step 3.
9. **Power on.** Adjust the power supply output until there is ¾ A current in the solenoid. Record the direction of the compass pointer.
10. Move the solenoid coil until it is ½ in away from point E on the compass, maintaining the W-E orientation of its axis. Record the direction of the compass pointer.
11. Reduce solenoid current to 100 mA. Record the direction of the compass pointer.
12. Reduce solenoid current to zero. Record the direction of the compass pointer.
13. **Power off.** Reverse polarity of current meter leads. Reverse also polarity of power supply leads. **Power on.** Adjust the supply for ¾ A current in the solenoid coil. The solenoid is still ½ in away from the point marked E on the compass, as in step 10. Record the direction of the compass pointer.
14. **Power off.** Record the direction of the compass pointer.

Magnetic Field about a Solenoid

15. The soft-iron core is still inside the solenoid coil, as in steps 8 through 14. The solenoid is connected as in step 13. **Power on.** Set the current in the solenoid to ¾ A.
16. Position a cardboard plane over the solenoid. Sprinkle iron filings on the cardboard and gently tap the cardboard until the filings settle into an identifiable pattern. Photograph or draw this pattern.

QUESTIONS

Confirm your answers to the following questions by referring specifically to your experimental data.

1. What verification have you of the fact that there is a magnetic field associated with current in a coil?
2. In step 10 which end of the electromagnet, *A* or *B*, is the N pole? Why?
3. In step 13 which end of the electromagnet is the N pole? Why?
4. Is the magnetic polarity of the electromagnet affected by the direction of current in the electromagnet? How?
5. Do the results of your experiment confirm the lefthand rule for determining the poles of an electromagnet? Explain.
6. Is the strength of the electromagnet affected by the amount of current in the solenoid? How?
7. What does the magnetic field pattern about a solenoid coil electromagnet resemble?

Answers to Self-Test

1. true
2. perpendicular
3. counterclockwise
4. true
5. poles
6. (*a*) number of turns; (*b*) amount of current

EXPERIMENT 10.3. Inducing Voltage in a Coil

OBJECTIVES

1. To verify experimentally that a voltage is induced in a coil when the lines of force of a magnet cut across its windings
2. To verify experimentally that the polarity of the induced voltage depends on the direction in which magnetic lines of force cut the coil windings

INTRODUCTORY INFORMATION

Electromagnetic Induction

In the preceding experiment we established that current, which is defined as the *movement* of electric charges, generates a magnetic field. In this experiment we shall demonstrate that a moving magnetic field which cuts across a conductor will generate a movement of electric charges in the conductor; that is, it will produce a voltage in the conductor and if the conductor is part of a complete circuit, current will flow in the conductor. This is a very important fact for the technician to understand. Antennas and radio and television tuners operate on the principle of magnetic induction. Similarly, power generation and its distribution depend on electromagnetic induction.

If a zero-center galvanometer is connected to the ends of a metal rod, and the rod is placed in the air gap between the poles of a permanent magnet, there will be no movement of the galvanometer pointer when the rod has come to rest in the magnetic field. However, if the rod is moved up, *cutting* the lines of magnetic force, the galvanometer will deflect showing current, as in Fig. 10.3-1. If the rod is then moved down, the galvanometer will again deflect, but in the opposite direction. Once the rod has come to rest, the galvanometer will register zero current. Now, if the rod is permitted to rest and the *magnet* is moved up, away from the rod, there is again current flow in the rod. Then if the magnet is brought down toward the rod, there is current flow in the opposite direction. Note that in each case the rod was perpendicular to the lines of force of the magnet, and the effect of a relative motion, that is, a motion of the rod or the magnet, was to *cut* the magnetic lines of force. The result was to induce a voltage in the conductive rod, causing current, as registered by the galvanometer. If the rod is now moved parallel to the lines of force between the poles of the magnet, there is no deflection of the galvanometer. *The conclusion we reach is that a relative motion between the conductive rod and the magnetic lines of force will induce a voltage in the rod only when the lines of force are cut by the rod.*

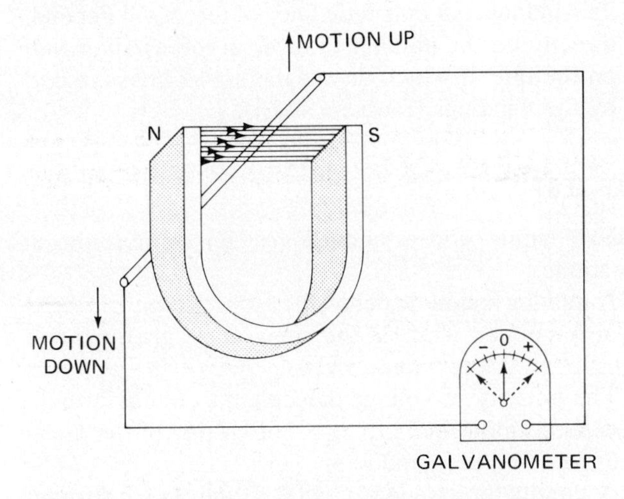

Fig. 10.3-1. The galvanometer registers current when there is motion up or down, that is, when the lines of force are cut by the rod.

Polarity of Induced Voltage

The fact that the galvanometer pointer deflected either to the left or to the right of zero, depending on the direction of cutting of the lines of force, is evidence that the polarity of the voltage induced in the conductive rod depends upon the *direction* of cutting of the lines of force. The polarity of the induced voltage can be established by Lenz' law.

Lenz' law states that the induced voltage must be of such a polarity that the direction of the resulting current in the conductor sets up a magnetic field about the conductor which will oppose the *motion* of the inducing field. An example will illustrate the meaning of Lenz' law.

Consider the solenoid wound about an air core (Fig. 10.3-2). The ends of the solenoid are connected to a galvanometer, as shown. If a bar magnet is pushed into the coil, so that the S pole of the magnet enters the left side of the coil, the current resulting in the solenoid should set up a magnetic field whose S pole will be at the left side of the solenoid. Since like magnetic poles repel, Lenz' law would be satisfied, for the induced current in the solenoid would have set up a magnetic field which opposes the inducing field. To verify that a S pole is in fact produced at the left end of the solenoid, observe the direction in which the galvanometer pointer deflects. Note that it moves to the right, indicating a voltage whose polarity is shown in Fig. 10.3-2. Applying the left-hand rule, grasp the solenoid so that the fingers point in the direction of electron current, as shown by the arrows on the solenoid core. The thumb then points to the N pole. It is apparent, then, that at the left end of the solenoid

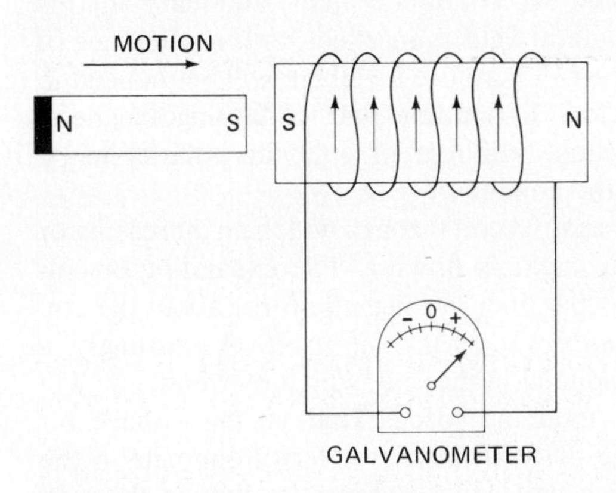

Fig. 10.3-2. Verifying Lenz' law. The polarity of the magnetic field created by the induced current in the solenoid opposes the polarity of the inducing field.

winding there is a S pole, at the right end a N pole, and we have confirmation of Lenz' law.

If the solenoid is next moved out of the core, the left end of the solenoid will become a N pole, thus attempting to restrain the bar magnet from moving out of the core. The voltage induced in the solenoid will be of the opposite polarity, and the galvanometer will deflect to the left.

Counter EMF

The discussion to this point has dealt with the observation that a voltage is induced in a wire or coil when there is relative motion between the wire or coil and the magnetic field of a permanent magnet, that is, when the wire or coil cuts the lines of force of the magnet. The facts are more general, however. The effect of induction in a coil or conductor is achieved when there is relative motion between the coil or conductor and the field of *any* magnet, including an electromagnet.

In a previous experiment it was demonstrated that a magnetic field exists about a coil carrying current. The coil carrying current is in fact an electromagnet and may be substituted for the permanent magnet in the discussion earlier in this experiment.

The term "relative motion" is not limited merely to the physical movement of a magnet near a conductor, or of a conductor in a magnetic field. Relative motion may exist without any physical movement whatsoever. Consider a coil through which an increasing or decreasing (not a steady state) current is flowing. In the case of an increasing current the magnetic field about the coil is increasing or expanding, that is, it is a *moving field.* When the current in the coil is decreasing, the magnetic field about the coil is decreasing or collapsing, again a moving magnetic field. A conductor held stationary in this moving magnetic field is in effect *cutting* the lines of magnetic force. Therefore a voltage will be induced in the conductor. Expanding and collapsing magnetic fields will cause voltages of opposite polarity to be induced in the conductor.

Consider again a coil through which an increasing or decreasing current is flowing. The expanding or collapsing (moving) lines of magnetic force about the coil *cut the windings of the coil itself.* Accordingly a voltage is induced in the coil, which by Lenz' law will oppose the inducing force. That is, the voltage induced in the coil will have a polarity opposite to the voltage which first caused current to flow in the coil. The voltage induced in the coil may therefore be termed a *counter emf.*

Magnitude of Induced Voltage

The amplitude of voltage induced in a solenoid depends directly on (*a*) the *number* of turns N of wire in the solenoid and (*b*) the *rate* at which the lines of flux are cut by the windings of the solenoid. Thus the more windings, the higher will be the induced voltage across the ends of the coil. Similarly, the faster the magnetic flux is cut by the windings, the higher will be the voltage produced across the solenoid. A mathematical formulation of this law was first given by the physicist Michael Faraday. It is suggested that the student refer to a standard text on electromagnetism for the exact formula, which is the mathematical statement of Faraday's law.

SUMMARY

1. A voltage is induced in a conductor when the conductor *cuts* the lines of force in a magnetic field. Cutting the lines of force can be achieved by moving either the conductor or the magnetic field.
2. The polarity of the voltage induced in the conductor is determined by the direction of cutting of the lines of force. Thus if a positive voltage is induced, say, by a conductor cutting the lines of force in a *downward* direction, a negative voltage will be induced by an *upward* cutting of these same lines of force.
3. The polarity of the induced voltage can be predicted by Lenz' law, which states: The polarity of the voltage induced in a conductor must be such that the *magnetic field set up by the resulting current* in the conductor will oppose the motion of the inducing magnetic field.
4. The amount of voltage induced in a solenoid when its windings cut magnetic lines of force will depend directly on the number of turns in the winding and on the rate at which the magnetic flux lines are cut by the windings (Faraday's law).

SELF-TEST

Check your understanding by answering these questions.

1. A moving magnetic field will ________ a ________ in a copper wire, if the lines of magnetic force ________ the wire.
2. The polarity of voltage induced in a conductor will depend upon the ________ of cutting of the lines of force by the conductor.
3. A conductor moving parallel to the lines of force in a magnetic field ________ (will/will not) have a voltage induced in it.

4. The more turns there are in a solenoid whose windings cut the lines of force in a magnetic field, the greater will be the voltage induced in it, all other things being equal. ________ (true/false)
5. By means of ________ law, it is possible to predict the polarity of voltage induced in a coil.

MATERIALS REQUIRED

- Equipment: Galvanometer or very sensitive electronic microammeter which can be zero centered
- Miscellaneous: Solenoid – 100 turns of # 18 varnish-insulated copper wire wound in three layers on a 3 × 1 in (diam.) hollow cardboard or plastic cylindrical form

PROCEDURE

1. Connect a galvanometer or sensitive zero-centered dc microammeter, set on its lowest range, to the terminals of the air core solenoid as in Fig. 10.3-2.
2. Position a bar magnet near the solenoid, with the S pole closest to the left end of the solenoid, as in Fig. 10.3-2.
3. With the bar magnet stationary, does the meter show any induced voltage? Record your observation in Table 10.3-1.
4. Now push the bar magnet into the solenoid so that the south pole of the magnet enters the air core first. Does the meter show any induced voltage? What is the polarity of the voltage? What is the current amplitude? Record these observations in Table 10.3-1.
5. Now pull the bar magnet out of the solenoid. Does the meter show any induced voltage? What is the polarity of the voltage? Current amplitude? Record these observations in Table 10.3-1.
6. Now reverse the bar magnet so that the N pole is closest to the left end of the solenoid. Insert the bar magnet into the air core. Does the meter show any induced voltage? Polarity? Current amplitude? Record your observations in Table 10.3-1.
7. Repeat step 5.
8. Again plunge the magnet into the air core, but more rapidly than in step 6. Observe and record in Table 10.3-1 the amplitude of current in the circuit.
9. Remove the magnet and again plunge it into the air core, but more rapidly than in step 8. Observe and record in Table 10.3-1 the amplitude of current in the circuit.
10. Now position the bar magnet vertically on the table, with the N pole on the table and the S pole in the air above the table. The magnet will remain stationary.
11. With the meter still connected to the solenoid, plunge the solenoid down over the magnet. Ob-

TABLE 10.3-1. Magnetic Induction

Step	Condition	*Voltage*	*Current*
		Polarity	*Amplitude, μA*
3	Magnet stationary.	0	0
4	S pole of magnet near left end of solenoid. Magnet is pushed into solenoid.	—	.2 mA
5	Magnet pulled out of solenoid.	+	.2 mA
6	N pole of magnet near left end of solenoid. Magnet is pushed into solenoid.	+	.2 mA
7	Magnet pulled out of solenoid.	—	.2 mA
8	N pole of magnet near left end of solenoid. Magnet is pushed into solenoid more rapidly than in step 6.	X	.8mA
9	Same as step 8 but more rapidly.	X	>1 mA
11	Solenoid plunged down over magnet.	—	X
12	Solenoid pulled up, away from magnet.	+	X

serve and record in Table 10.3-1 polarity of voltage induced.

12. Pull the solenoid out of the magnet. Observe and record in Table 10.3-1 polarity of voltage induced.

QUESTIONS

1. What are the conditions for electromagnetic induction of a voltage in a solenoid?
2. What verification have you that a voltage is in fact induced in a solenoid when a bar magnet is plunged into it or pulled out of it? Refer specifically to your data to support your answer.
3. What verification have you that it is the *relative* motion involved in cutting of lines of force which induces a voltage in the solenoid? Refer specifically to your data to support your answer.
4. What is Lenz' law?
5. What verification have you that Lenz' law is true? Refer specifically to your data to support your answer.
6. What is Faraday's law?
7. Do you have any verification of all or part of Faraday's law? Explain.

Answers to Self-Test

1. induce; voltage; cut
2. direction
3. will not
4. true
5. Lenz'

CHAPTER 11 VOLTMETER CIRCUIT AND LOADING EFFECTS

EXPERIMENT 11.1. Sensitivity of a DC Meter Movement

OBJECTIVE

To determine experimentally the sensitivity of a dc meter movement

INTRODUCTORY INFORMATION

The Moving-Coil Meter Movement

DC voltage and current in electric and electronic circuits are usually measured by meters employing a moving-coil meter movement. This type of movement is based on the principle of the galvanometer developed by the French physicist Arsène d'Arsonval. D'Arsonval's galvanometer was a basic meter which measured very small currents. Though it was substantially modified later by Edward Weston and others, the basic moving-coil meter movement in use today is still known as the d'Arsonval movement.

Construction

The moving-coil meter consists of a horseshoe magnet with semicircular pole pieces made of soft iron added at the ends. A circular soft-iron core is positioned in a uniform magnetic field provided by the magnet and the pole pieces (Fig. 11.1-1*a*). Very fine phosphor bronze nonmagnetic wire wound on a rectangular aluminum frame (Fig. 11.1-1*b*) makes up the moving coil. The coil assembly is centered in the air gap between the soft-iron core and the magnet (Fig. 11.1-1*c*). Each end of the coil is connected to a spring. The two springs are symmetrically mounted on the coil, one at the top, the other at the bottom. At the center of the springs are pivots (Fig. 11.1-1*d*) which fit into jeweled bearings to reduce friction, thus permitting the coil to move freely in the air gap. The free ends of the two springs are brought to the two meter terminals. These serve as the terminals of the coil. A pointer is attached to the coil assembly and moves with it. The pointer can be positioned so that its rest or zero position is at the center or at the left side. A zero-adjust screw is used to compensate for minor changes in coil position. Stops are provided at the left and right sides of the magnet to limit the degree of movement of the pointer.

Operation

When a potential difference is placed across the meter terminals from an external circuit, current flows through the meter coil, setting up a magnetic field around it. This magnetic field interacts with the field of the permanent magnet, and the torque developed causes the coil to turn on its pivots. The number of degrees of rotation of the moving coil is determined by the strength of the magnetic field about the coil, which in turn is proportional to the amount of current in the coil. The greater the current, the greater the rotation, and therefore the greater is the deflection of the pointer attached to the moving-coil assembly. A scale calibrated in suitable units is used to read the measurement made in the circuit.

The direction of current through the coil determines the polarity of its magnetic field and the direction of rotation of the coil assembly. If current direction is reversed in a meter whose zero position is at the left, the pointer attempts to move downward to the left, off the scale. It is restrained by the *stop* on the left. Since the system requires a clockwise movement of the pointer, improper direction of current makes measurement impossible. It is apparent, therefore, that the moving-coil meter is a polarized device, and *polarity must be observed* in measuring voltage or current with it.

The springs play an important role in the operation of the meter. First they act as a restraining force on the rotation of the coil. Hence, they must be precision-wound to permit accurate measurements. The springs also act as conductors to carry current to and from the coil. Their action brings the pointer back to zero rest after the measurement is completed and there is no longer current in the coil. Finally, they are wound in opposite directions to compensate for temperature changes, because a temperature change affects both

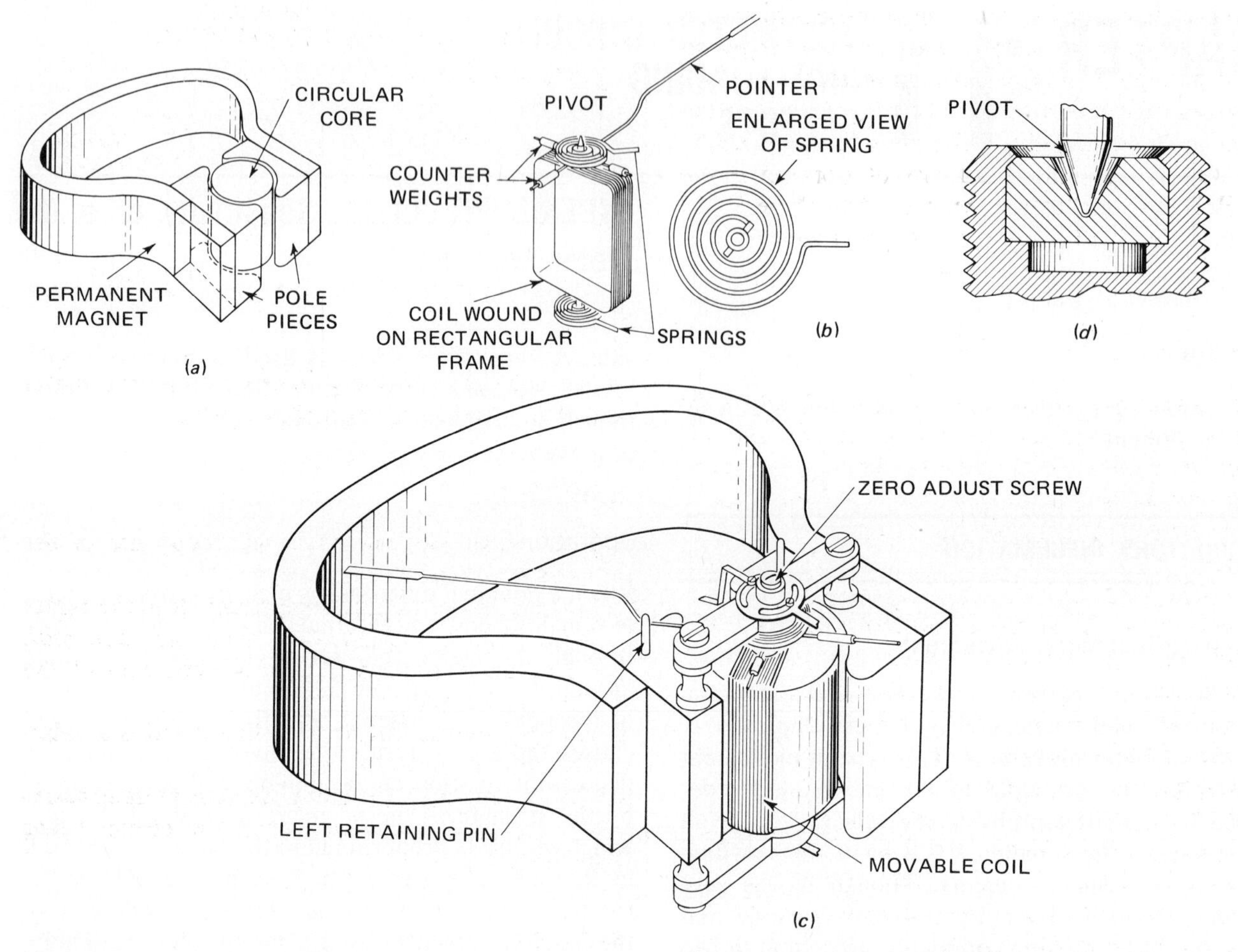

Fig. 11.1-1. Construction and arrangement of parts in a moving-coil meter movement. (*Reprinted from Army Technical Manual TM11–664*)

springs in the same manner. Since they are wound in opposite directions, the effect of an expansion or contraction of one is counteracted by an expansion or contraction of the other acting in the opposite direction. Hence, the net change is zero.

Sensitivity of the Meter Movement

The amount of current I_m required to produce full-scale deflection of the pointer is called the *sensitivity* of the meter movement. Meters have sensitivities ranging from a few microamperes to many milliamperes. The smaller the current required to produce full-scale deflection, the greater is the sensitivity of the movement.

The sensitivity of the movement depends on the strength of the permanent magnet and the number of turns of the coil. The more turns on the coil, the smaller is the coil current required for full-scale deflection. Of course, the more turns, the greater is the internal or dc resistance of the coil.

The characteristics which are therefore associated with a meter movement are its sensitivity and internal resistance. Another characteristic derived from these two is the ohms-per-volt rating. The significance of this characteristic will be more evident when the loading effects of voltmeters are considered in a later experiment.

Internal Resistance R_m of a Meter Movement

The internal resistance R_m of a dc meter movement is the resistance which the meter coil offers to the flow of direct current in the movement. R_m is one of two important characteristics. The other is I_m. If these two values are known, the meter movement can be used to design a multifunction meter for the measurement of voltage, current, and resistance.

Meter Linearity

In the design of the meter movement great pains are taken to provide a uniform magnetic field in the air gap where the coil is located. The purpose is to ensure linear changes in deflection with linear changes in current through the coil. In a linear meter the number of degrees of rotation is directly proportional to the current in the coil. Thus, full-scale deflection results when 100 percent of the rated current is present in the coil. Half-scale deflection occurs when there is 50 percent of the rated current, etc. A linear meter movement makes possible a linear scale.

Meter Accuracy

Meter movements can be built with a high degree of precision and hence can be made accurate within a fraction of a percent. These are intended for very precise laboratory measurements. General-purpose meters, however, are usually rated at 2 percent accuracy.

The percentage of accuracy is based on the full-scale reading of the meter. Hence, for readings lower than full scale, the percentage of error is greater than the rated value. The significance of this fact is more apparent when the movement is converted into a voltmeter. Thus, a meter rated at 2 percent on the 100-V range is accurate within 2 V for any value measured on this range. The percent of error in a 10-V reading on this range could conceivably be as high as 20 percent. It is evident, therefore, that meter readings close to full-scale deflection are the most accurate, if circuit loading, or the effects of the meter on the circuit, is ignored.

Determining the Characteristics of a Meter Movement

The technician must know the characteristics of the meter being used in order to know what effect it will have on the circuit being measured. Sometimes it is also necessary to "make" a meter or to add another range to an existing meter. Thus, in this experiment the meter sensitivity will be explored.

Sensitivity

One method of measuring the sensitivity I_m of a meter movement is shown in the circuit in Fig. 11.1-2. The meter movement under test is placed in series with a rheostat R, a limiting resistor R_1, a "standard" current meter M, and a battery V. Rheostat R is adjusted until the pointer of the meter under test reaches full-scale deflection. The current required to cause full-scale deflection is read from the "standard" current meter. This is the required *sensitivity* I_m of the meter movement under test.

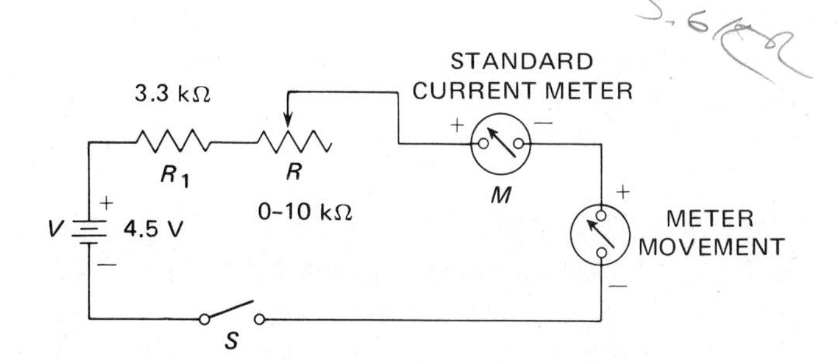

Fig. 11.1-2. Circuit for measuring sensitivity I_m of a meter movement.

The accuracy of the "standard" current meter limits the accuracy of the measurement.

SUMMARY

1. Most meter movements in use today are of the d'Arsonval type.
2. In the d'Arsonval movement, current in the meter coil sets up a magnetic field which interacts with the permanent magnet, causing deflection of the meter pointer.
3. The d'Arsonval movement is linear and is a polarized device.
4. The sensitivity of the meter movement determines the effects of the meter upon the circuit being measured.

SELF-TEST

Check your understanding by answering these questions.

1. What determines the accuracy of your measurement of I_m? __________
2. The amount of current I_m required to produce full-scale deflection of the pointer is called the __________ of the meter.
3. The sensitivity of the movement depends on the strength of the __________ __________ and the number of __________ of the coil.
4. The percentage of accuracy is based on the __________ __________ reading of the meter.

MATERIALS REQUIRED

- Power supply: 4.5-V battery or a regulated, variable dc source
- Equipment: EVM with current ranges; meter movement with $I_m = 1$ mA; rheostat (resistance decade box) 0 to 10 kΩ, adjustable in 1-Ω steps
- Resistors: ½-W 3.3-kΩ, 15-kΩ
- Miscellaneous: SPST switch; connecting wires; 10-kΩ potentiometer, if decade box is not available

PROCEDURE

Measuring Sensitivity of a Meter Movement

1. Connect the meter movement whose sensitivity you wish to measure in the circuit Fig. 11.1-2. S is open. V is a 4½-V battery or power supply set to 4.5 V. Rheostat R is set to 10,000 Ω, at the start.

CAUTION: Check to see that the "standard" current meter (an accurate commercial current meter may be used as the standard) and the meter movement under test are connected with the proper polarity.

2. Close S and slowly reduce the resistance of R until full-scale deflection is indicated on the meter movement. Read the value I_m from the standard meter and record it in Table 11.1-1.

Linearity Check

3. Increase the resistance R until the pointer of the meter movement (under test) is deflected three-fourths of the way on its scale. Measure the current (using the standard meter) at this level, and record it in Table 11.1-1 in the column labeled I (¾).

Measure also the currents required to cause half-scale deflection and quarter-scale deflection of the meter-movement pointer and record them in Table 11.1-1, under columns "I (½)" and "I (¼)," respectively. For quarter-scale deflection it will be necessary to replace R_1 with a 15,000-Ω resistor. If the pointer of your meter came to rest at exactly ¾ scale with $\frac{3}{4}I_m$ and exactly ½ scale with $\frac{1}{2}I_m$, etc., the meter movement was a linear one. Was your movement linear? ________

TABLE 11.1-1. Sensitivity and Linearity

Measured Values			
I_m	I (¾)	I (½)	I (¼)
.372mA	.260mA	.180mA	.081

QUESTIONS

1. Is the scale on the meter movement under test linear or nonlinear? What do your measurements indicate concerning meter-movement linearity?
2. What error is introduced in measuring I_m of a meter movement using the circuit of Fig. 11.1-2?
3. Why was the error, discussed in number 2 above, disregarded in this experiment?

Answers to Self-Test

1. the "standard" meter
2. sensitivity
3. permanent magnet, turns
4. full-scale

EXPERIMENT 11.2. Voltmeter Multipliers

OBJECTIVE

To determine by experiment the value of the resistance required to convert a dc meter movement into a voltmeter of a specified range

INTRODUCTORY INFORMATION

Nondigital meters used in electronics consist of basic dc meter movements connected in various circuit arrangements. The basic movement has a moving element to which a pointer is attached. Current in the meter movement results in a force which acts on the moving element. The pointer is deflected and moves in an arc along a scale calibrated in the units being measured. These units may be volts, ohms, amperes, etc., as in a VOM which can perform all these measurements.

In a previous experiment you learned that a meter movement has two characteristics associated with it:

1. Sensitivity I_m—the current required to cause full-scale deflection of the meter pointer.
2. Resistance of the movement R_m.

Voltmeter Multipliers

The circuit components of a voltmeter depend on these meter-movement characteristics. For example, suppose it is required to construct a 0- to 100-V meter from a basic 0- to 1-mA movement which has a resistance of 200 Ω.

Figure 11.2-1 shows the basic circuit arrangement for this voltmeter. A and B are the terminals of the meter where the test leads are inserted. A and B are coded (−) and (+), respectively, from the polarity of the meter movement. When measuring voltage, A is placed on the negative side of the voltage source and B on the positive terminal. Otherwise, current through M is reversed and the movement may be damaged; M is the 200-Ω meter movement. It is shown as a resistance R_m, with a pointer, connected in series with a resistor R_{mult} called the *multiplier*. The purpose of the multiplier resistor R_{mult} is to limit the current in this circuit to 1 mA when 100 V is applied across the meter leads (A and B). This will cause full-scale deflection of the pointer, and the scale can be marked or calibrated 100 V at this point.

If a linear meter movement is used, deflection of the pointer is directly proportional to the current in the movement. The current depends on the voltage across terminals A and B of the meter (Fig. 11.2-1). The meter scale can therefore be calibrated in volts, although the movement itself responds to current. This is another example of Ohm's law where $V = I \times R$.

Ohm's law is used to find the value of the multiplier resistor R_{mult}. Consider the experimental voltmeter of Fig. 11.2-1. It consists of the series combination of R_{mult} and R_m. Therefore the total resistance R_T of the meter is $R_T = R_{mult} + R_m$. We wish to construct a 100-V meter, that is, a meter whose scale may be calibrated 100 V at full scale. Assume that the meter movement has an internal resistance R_m of 200 Ω and a sensitivity of 1 mA. If Ohm's law is applied, it is found that R_T must be 100,000 Ω to allow 100 V at A and B to push 1 mA of current through the circuit for full-scale deflection. For

$$R_T = \frac{V}{I_m} \quad \text{or} \quad R_T = \frac{100}{0.001} = 100{,}000\ \Omega$$

Since $R_T = R_{mult} + R_m$, it follows that $R_{mult} = R_T - R_m$. If R_T must be 100,000 Ω to limit the current in the circuit to 1 mA when 100 V is applied, the multiplier resistor must be 100,000 − 200, or 99,800 Ω. Our formula for finding the multiplier resistance is

$$R_{mult} = R_T - R_m \tag{11.2-1}$$

Some examples will demonstrate how multipliers for different ranges may be calculated.

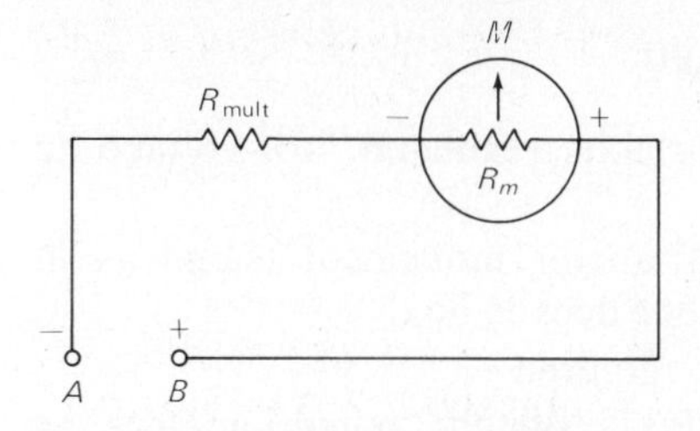

Fig. 11.2-1. Single-range voltmeter.

Example 1

Suppose we use the same meter movement as in Fig. 11.2-1, but this time we shall construct a 10-V meter. By Ohm's law, find the R_T necessary to limit the current to the movement's rating (here, 1 mA). Thus

$$R_T = \frac{V}{I_m} = \frac{10}{0.001} = 10{,}000\ \Omega$$

The total resistance required in the voltmeter circuit to limit the current to the meter rating of 1 mA when the 10 V is applied is 10,000 Ω. The meter movement has an R_m of 200 Ω. Therefore

$$R_{mult} = R_T - R_m = 10{,}000 - 200 = 9800\ \Omega$$

and the multiplier must have a resistance of 9800 Ω.

Example 2

Convert a meter movement of 150-Ω R_m and an I_m of 1 mA to a 500-V meter.

$$R_T = \frac{V}{I_m} = \frac{500}{0.001} = 500{,}000\ \Omega$$

The total resistance required in the voltmeter circuit to limit the current to the meter rating of 1 mA when the 500 V is applied is 500,000 Ω. The meter has an R_m of 150 Ω. Therefore

$$R_{mult} = R_T - R_m = 500{,}000 - 150 = 499{,}850\ \Omega$$

and the multiplier must have a resistance of 499,850 Ω.

Meter Accuracy

Resistors are not commonly available with values such as those found for the multipliers in the examples above. But resistors close to these values are commonly available, for example, 10 and 500 kΩ. In making an experimental voltmeter, where accuracy is not the most important factor to consider, we can use the 10- and 500-kΩ resistors. If these resistors are used, what will be the percentage of error of the meter reading? In example 1, the needed value of R_{mult} was 9800 Ω and the total resistance 10,000 Ω. Instead, we shall be using a circuit whose total resistance is 10,200 Ω. The error, then, is of 200 parts in 10,200 parts; written as a fraction, it is an error of $^{200}/_{10,200}$. With the aid of the pocket calculator, divide 200 by 10,200, and the percent of error is shown to be approximately 2 percent. This is an acceptable error. However, the *overall* accuracy of the voltmeter will depend on *both* the accuracy of the movement *and* the tolerance of the multiplier.

In example 2, the difference between the required resistance for R_{mult} and the readily available resistance of 500 kΩ is 150 Ω. The error, then, is an error of 150 parts in 500,150 parts; written as a fraction, it is an error of $^{150}/_{500,150}$. To find the percentage of error, divide 150 by 500,150, and the percentage of error is shown to be about 0.03 percent. This would be an error so small that it would not be detectable in the meter reading. Again, however, remember that the error of the voltmeter consists of the error of the movement *and* the multiplier.

Generally the most precise resistors available are used as multipliers in general-purpose meters. Normally 1 percent resistors can be found at most electronics suppliers.

Calibrating the Voltmeter

Suppose Fig. 11.2-1 is the experimental 100-V meter. The scale of this voltmeter must be calibrated. If 0 V is applied ($V_{AB} = 0$) across the voltmeter, there is no current in the meter movement and hence no deflection of the pointer. This is the 0-V position of the pointer on the scale. If 50 V is applied, there will be ½ mA of current in the movement, and therefore there will be half-scale deflection of the pointer. This will be the 50-V position on the scale. If 25 V is applied, there will be ¼ mA of current, and therefore there will be quarter-scale deflection of the pointer. This will be the 25-V position on the scale, and so on.

The entire 100-V scale is calibrated in this manner and appears as in Fig. 11.2-2. It is a linear scale because the meter movement used is linear.

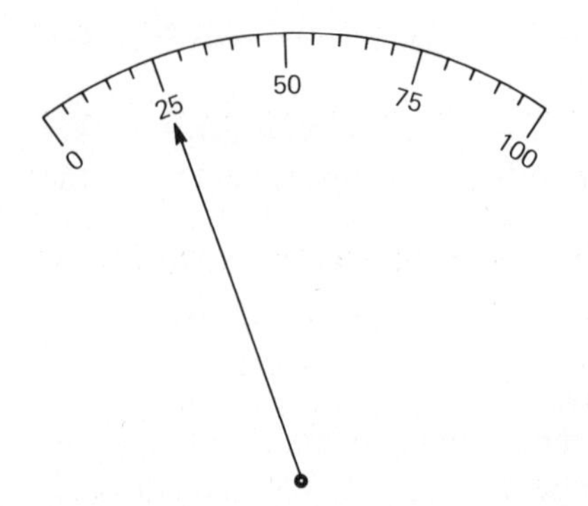

Fig. 11.2-2. Voltmeter scale.

SUMMARY

1. A basic dc meter movement may be converted into a voltmeter by adding a multiplier resistor R_{mult} in series with the meter movement, as in Fig. 11.2-1.
2. The value of the multiplier R_{mult} for a specified range must be such that it will permit full-scale deflection of the pointer when the voltmeter leads are placed across the specified range voltage.
3. For full-scale deflection of the meter, the total resistance of the meter, $R_{\text{mult}} + R_m$, must limit the current through the movement to I_m, the sensitivity of the movement.
4. Though the scale of a voltmeter is calibrated in volts, the meter itself responds to current in the movement, that is, it is a current-actuated device.
5. If a linear meter movement is used for a voltmeter, the scale of the voltmeter will also be linear.

SELF-TEST

Check your understanding by answering these questions.

1. A dc voltmeter may be made by connecting a precision resistor called a ________ in ________ with a dc meter movement.
2. If the sensitivity of the meter movement used to build a dc voltmeter is 50 μA, then the current in a 50-V meter at full-scale deflection is ________.
3. The total resistance of a dc voltmeter consists of the ________ of the multiplier resistance and the resistance of the ________ ________.
4. A 100-V meter is made from a meter movement whose sensitivity is 50 μA and whose internal resistance is 2000 Ω.
   (*a*) The total resistance of the meter is ________ Ω.
   (*b*) The resistance of the multiplier is ________ Ω.
5. In the voltmeter of question 4, half-scale deflection of the pointer indicates ________ V; quarter-scale deflection indicates ________ V.

MATERIALS REQUIRED

- Power supply: Variable, regulated, low-voltage dc source
- Equipment: EVM; meter movement (same as in Exp. 11.1); resistance decade box
- Resistors: ½-W as required
- Miscellaneous: SPST switch; potentiometers as required

PROCEDURE

(Show all your computations.)

Constructing a Voltmeter of Specified Range (30 V)

1. The meter movement you will use will be the same as that whose sensitivity I_m was measured in a previous experiment. Determine R_m. Record I_m and R_m in Table 11.2-1.
2. Compute and record in Table 11.2-1 the multiplier resistance R_{mult} required to convert your meter movement into a 30-V meter. Choose a resistor (or combination of resistors) whose total resistance R_{mult} is equal or close to the computed value. With an ohmmeter measure and record in Table 11.2-1 the resistance of the multiplier R_{mult}.

   NOTE: A resistance decade box may be used to select the proper value of R_{mult}. When this is used, read the value of R_{mult} from the calibrated dials.

   Using the selected value of R_{mult}, construct a voltmeter as in Fig. 11.2-1.
3. Connect an EVM (V_1) and the experimental voltmeter (V_2) across the low-voltage supply as shown in Fig. 11.2-3 and adjust the output voltage until the experimental voltmeter reads full scale. If your calculations and circuit are correct, the meter pointer should deflect to full scale at exactly 30-V output from the supply. Also, if your meter is to be calibrated, "30" would be written at the point where the pointer comes to rest at full scale (with 30 V applied).
4. If exact full-scale deflection is not reached on the *experimental voltmeter,* adjust the voltage supply to obtain full scale and record that voltage under "Measured Range" in Table 11.2-1.

Experimental Voltmeter (10 V)

5. Repeat steps 2 through 4, constructing a 10-V range voltmeter. Record your data in Table 11.2-1.

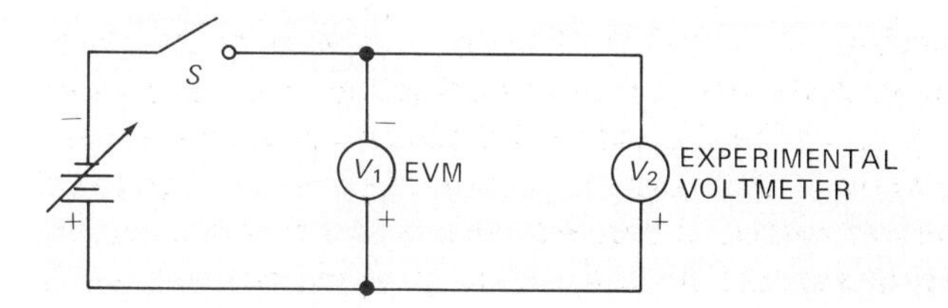

Fig. 11.2-3. Experimental voltmeter circuit.

Internal Resistance R_{in} of the Voltmeter

6. Compute the input resistance R_{in} of the experimental voltmeters and record the values in Table 11.2-1.
7. With an ohmmeter measure the input resistance R_{in} of the experimental voltmeters. Does the voltmeter pointer move? If so, does it move upscale or downscale? If it moves downscale, reverse the ohmmeter leads. Record this measurement in Table 11.2-1.

QUESTIONS

1. In Table 11.2-1 how do the required voltage ranges compare with the measured ranges for the experimental voltmeter on (*a*) 30-V range, (*b*) 10-V range? Explain any differences between the required voltage range and the actual full-scale value of voltage for each range.
2. How do the computed and measured values of input resistance in Table 11.2-1 compare? If they are not the same, explain why.
3. Without using specially designed precision resistors, explain how you obtained the required value R_{mult} for use as a multiplier in your single-range experimental voltmeter.
4. Considering the meter movement 100 percent accurate, what would have been the accuracy of the experimental voltmeter had you used a resistor R_{mult} with the same value as R_{in}?

Answers to Self-Test

1. multiplier; series
2. 50 μA
3. sum; meter movement
4. (*a*) 2M; (*b*) 1,998,000
5. 50; 25

TABLE 11.2-1. Voltmeter Multipliers

Meter-Movement Measured Values			R_{mult}, kΩ			*Resistance R_{in} of Experimental Voltmeter*	
R_m	I_m	*Required Range V*, V	*Computed*	*Measured*	*Measured Range V_1*, V	*Computed*	*Measured*
2.1 kΩ	.372 mA	30	78.1	77.0	30	80.2 kΩ	79.1 kΩ
2.1 kΩ	.372 mA	10	24.6	24.6	10	26.7 kΩ	26.7 kΩ

EXPERIMENT 11.3. DC Voltmeter Input Resistance and Ohms-per-Volt Rating

OBJECTIVES

1. To determine the relationship between the input resistance of a dc voltmeter and the sensitivity I_m of the meter movement from which it is constructed
2. To verify, by experiment, the relationship between the ohms-per-volt rating of a dc voltmeter and the sensitivity I_m of the meter movement

INTRODUCTORY INFORMATION

Input Resistance versus I_m

In Exp. 11.2 it was shown that the total resistance R_T of a dc voltmeter, on any voltage range, is equal to $R_{\text{mult}} + R_m$, where R_{mult} is the resistance of the multiplier for the specified range and R_m is the resistance of the meter movement. This total resistance R_T is called the *input resistance* R_{in} of the voltmeter; that is, it is the resistance which the voltmeter presents to the circuit where voltage is being measured.

You will recall that in calculating the *multiplier* resistance for any range, we first found R_T by using the formula

$$R_T = \frac{V}{I_m} \tag{11.3-1}$$

where V is the voltage range, and I_m is the sensitivity of the meter movement. It is clear therefore that the input resistance of a dc voltmeter on any one range is the ratio of the voltage V of that range and the sensitivity I_m of the meter movement. Since I_m remains the same (that is, since the sensitivity of the meter movement does not change), *the higher the voltage range, the higher must be the input resistance of the voltmeter*. Let us see if this is so.

If the meter movement from which the dc voltmeter is constructed has a sensitivity of 0.001 A (1 mA), then the input resistance of a voltmeter on the 10-V range is

$$R_{\text{in}} = \frac{10}{0.001} = 10{,}000\ \Omega$$

On the 50-V range it is

$$R_{\text{in}} = \frac{50}{0.001} = 50{,}000\ \Omega$$

Similarly on the 100-V and the 300-V ranges, R_{in} is, respectively, 100,000 and 300,000 Ω. These values are listed in Table 11.3-1. So it *is true* that the higher the voltage range, the higher the input resistance of the voltmeter.

What would be the effect on the input resistance of a voltmeter of using a more sensitive meter movement? For example, suppose a 50-μA meter movement is used, that is, $I_m = 50\ \mu\text{A}\ (50 \times 10^{-6}\ \text{A})$. This movement is much more sensitive than a 1-mA movement, in fact, 20 times more sensitive because 1 mA = 20 × 50 μA. The input resistance of a voltmeter on the 10-V range, constructed from a 50-μA movement, is

$$R_{\text{in}} = \frac{10}{50 \times 10^{-6}} = 200{,}000\ \Omega$$

You will recall that the input resistance of a 10-V meter made from a 1-mA movement is 10,000 Ω. It is clear that R_{in} of the 10-V meter using a 50-μA movement is 20 times higher than that of the 10-V meter using a 1-mA movement.

Several facts appear from our discussion. These are:

1. The input resistance of a dc voltmeter is directly proportional to its voltage range; the higher the voltage range, the higher the input resistance.
2. The input resistance of a dc voltmeter depends on the sensitivity I_m of the meter movement from which it is constructed; the greater the sensitivity, the higher the input resistance.

Ohms-per-Volt Characteristic

Observe in Table 11.3-2 that if the input resistance of the voltmeter (using a 1-mA movement) on any one

TABLE 11.3-1. Input Resistance of Voltmeter Using a 0- to 1-mA Movement

Voltage Range, V	*Input Resistance,* Ω
10	10,000
50	50,000
100	100,000
300	300,000

TABLE 11.3-2. Ohms-per-Volt Characteristic of Voltmeter Using a 0- to 1-mA Movement

Voltage Range	*Input Resistance*	*Ohms-per-Volt:* $\frac{\textit{Input Resistance}}{\textit{Voltage}}$
10	10,000	1000
50	50,000	1000
100	100,000	1000
300	300,000	1000

range is divided by the voltage of that range, the result is 1000, regardless of the range. This constant ratio is called the ohms-per-volt (Ω/V) characteristic of the meter, and in the case of a voltmeter using a 1-mA movement it is 1000 Ω/V. This characteristic is also called the *voltmeter* sensitivity.

If the range of the voltmeter (using a 1-mA movement) is 1 V, then the ohms-per-volt rating is

$$\frac{1}{I_m} = \frac{1}{0.001} = 1000\ \Omega/\text{V}$$

We can now see that the Ω/V rating of a voltmeter is the *inverse* of the sensitivity of the meter movement, that is,

$$\frac{\Omega}{\text{V}} = \frac{1}{I_m} \qquad (11.3\text{-}2)$$

where I_m is in amperes. The formula gives us a fast way to calculate the Ω/V rating of a voltmeter. For example, what is the Ω/V characteristic of a dc voltmeter using a 50-μA movement? It is

$$\frac{1}{I_m} = \frac{1}{50 \times 10^{-6}} = 20{,}000\ \Omega/\text{V}$$

Therefore, the voltmeter constructed from a more sensitive meter movement has a *higher* Ω/V characteristic than one constructed from a less sensitive movement.

Input Resistance versus Ohms-per-Volt Rating

Looking again at Table 11.3-2, we see clearly that if the Ω/V characteristic of a voltmeter is multiplied by the voltage range, the result is the input resistance of the voltmeter on that range. Try it.

Multiply 1000 by 10 and the result is 10,000. The table shows that 10,000 Ω is the input resistance of a voltmeter (using a 1-mA movement) on the 10-V range. The same is true for each range.

This is therefore another way of finding the input resistance of a *dc* voltmeter on any range: Multiply the Ω/V rating of the meter by the voltage of the range.

The ohms-per-volt characteristic of the voltmeter in a VOM is usually printed on the face of the VOM. The input resistance of the voltmeter can then be calculated for each range.

This discussion has dealt with voltmeters constructed from dc meter movements. In the case of electronic (EVM) or digital voltmeters, the input resistance of the meter is specified by the manufacturer and is usually the same on each range.

SUMMARY

1. The input resistance of a dc voltmeter on any one range is the resistance measured across the leads of the meter on that range. It is the sum of the multiplier resistance and meter-movement resistance.
2. The higher the range, the higher is the input resistance.
3. The input resistance of a dc voltmeter depends on the sensitivity, I_m, of the meter movement from which it is constructed; the more sensitive the movement, the higher the input resistance.
4. The ohms-per-volt characteristic of a voltmeter is the ratio of the input resistance on any range and the voltage of that range. The Ω/V rating remains the same for a dc voltmeter, regardless of the range.
5. The Ω/V rating of a dc voltmeter may be calculated from the formula

$$\frac{\Omega}{\text{V}} = \frac{1}{I_m}$$

where I_m is the sensitivity in amperes of the meter movement.

6. The input resistance of a dc voltmeter on any one range may also be calculated if the Ω/V rating of the meter is known.

$$\text{Input resistance} = \frac{\Omega}{\text{V}} \times \text{voltage range}$$

SELF-TEST

Check your understanding by answering these questions.

1. A dc voltmeter constructed from a 10-μA movement would have a ________ (higher/lower) input resistance than one made from a 100-μA movement.
2. The input resistance of a dc voltmeter on the 25-V

range, constructed from a 10-μA movement, is _______ Ω.

3. The input resistance of the dc voltmeter in question 2, on the 5-V range, is _______ Ω.
4. The Ω/V characteristic of the meter in question 2 is _______.
5. A 10-μA movement is _______ (more/less) sensitive than a 100-μA movement.

MATERIALS REQUIRED

- Power supply: Variable, regulated, low-voltage dc source
- Equipment: 1-mA movement used in previous experiments; 50-μA movement, if available
- Resistors: As calculated in experiment

PROCEDURE

1. Draw the schematic diagram for a 10-V meter using the 1-mA movement.
2. Calculate R_{mult} necessary for the meter. Have your instructor check your work.
3. Connect the circuit for the voltmeter using the correct multiplier resistor.
4. Apply 10 V to the meter. Full-scale deflection should be attained.
5. Disconnect the voltage source and, using a standard ohmmeter connected as shown in Fig. 11.3-1, measure R_T of the experimental meter. 57 K Ω.

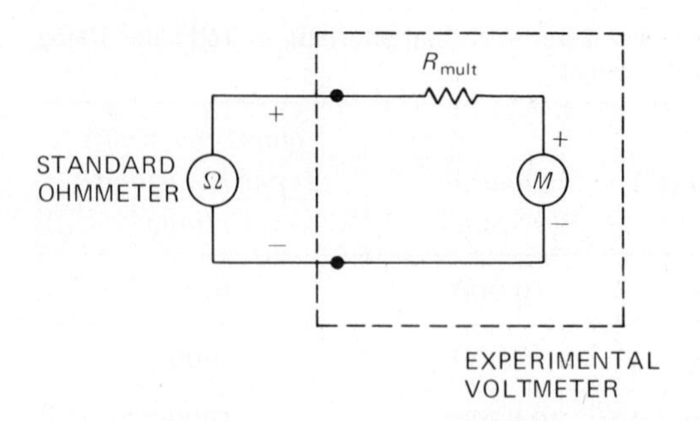

Fig. 11.3-1. Measuring input resistance of the experimental voltmeter.

6. Calculate the Ω/V rating of the meter. _______ Ω/V.
7. Repeat steps 1 to 6 for the 50-μA movement. R_T = _______ Ω, Ω/V = _______.

QUESTIONS

1. How did your calculated values of R_T compare with the measured values?
2. Which movement required the largest R_{mult}?
3. Which movement had the higher Ω/V rating?
4. Explain in your own words how I_m influences the Ω/V rating.

Answers to Self-Test

1. higher
2. 2,500,000
3. 500,000
4. 100,000
5. more

EXPERIMENT 11.4. Loading Effects of a DC Voltmeter

OBJECTIVES

To determine experimentally the effect on circuit voltage measurements caused by voltmeter (resistance) "loading"

INTRODUCTORY INFORMATION

DC Voltmeter Circuit Loading

The ohms-per-volt rating and the input resistance of a voltmeter are significant to the technician because they determine the extent to which the voltmeter will "load" the circuit when used in voltage measurement. Consider the circuit of Fig. 11.4-1. Suppose it is desired to measure the voltage from A to B, using the 100-V range of a 1000-Ω/V meter. Here the meter places 100,000 Ω resistance in parallel with the 4 megohms (MΩ) of the divider, and the total resistance R_{T2} of the parallel circuit from A to B now becomes slightly less than 100,000 Ω. Suppose, however, that the figure 100,000 Ω is used. The equivalent circuit (Fig. 11.4-2) shows what happens when the meter is placed across the circuit.

The voltage across R_{T2}, and hence across the volt-

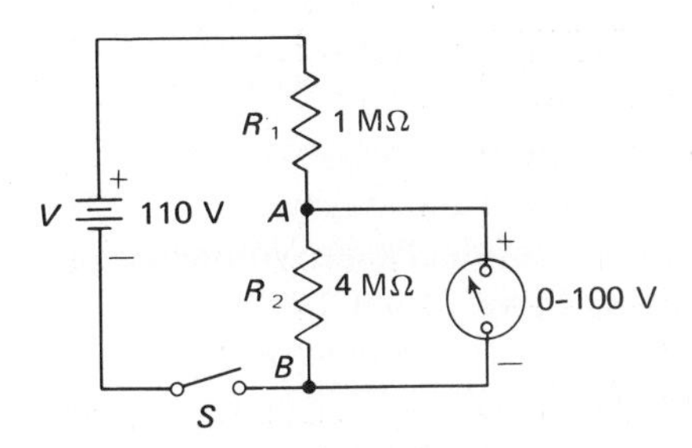

Fig. 11.4-1. Voltmeter loading.

meter, is $\frac{1}{11} \times 110 = 10$ V, and the meter would read 10 V. But this would not be the true voltage from A to B in the absence of the meter, since a computation shows that the voltage from A to B before the meter was connected would be $\frac{4}{5} \times 110 = 88$ V. The voltmeter has "loaded" the circuit. Obviously this is undesirable because it can lead to serious errors. To minimize this loading, use a voltmeter with a much higher ohms-per-volt rating which is set on the highest range which will still give sufficient pointer deflection to make a voltage reading possible. In electronics 20,000 Ω/V meters are frequently used.

It is possible with the nonelectronic voltmeter (such as the VOM) to eliminate the loading effects of the voltmeter by making two measurements on different ranges and treating the results mathematically. The interested student will find this procedure outlined in Paul B. Zbar, "Basic Electricity," 4th ed., McGraw-Hill Book Company, New York, 1974.

Normally, electronic voltmeters (those which contain transistor or tube amplifiers) have input resistances which are high enough so that loading is not a problem. Such voltmeters will have an input resistance of 10 MΩ or higher.

SUMMARY

1. When a voltmeter is placed across a resistance in a circuit, it loads the circuit by changing the effective resistance across which the voltage is being measured. The "new" resistance is the parallel combination of the resistor and the input resistance of the meter.
2. As a result of meter loading, the voltage measured may be very much *lower* than the voltage which would appear at the measurement point without the meter.
3. Loading may be minimized by using voltmeters made from high-sensitivity meter movements, that is, meter movements whose I_m is so low that the ohms-per-volt rating of the voltmeter is very high. For most work, 20,000-Ω/V meters are used.
4. Loading may also be minimized by using the highest possible range where the reading can be accurately made.

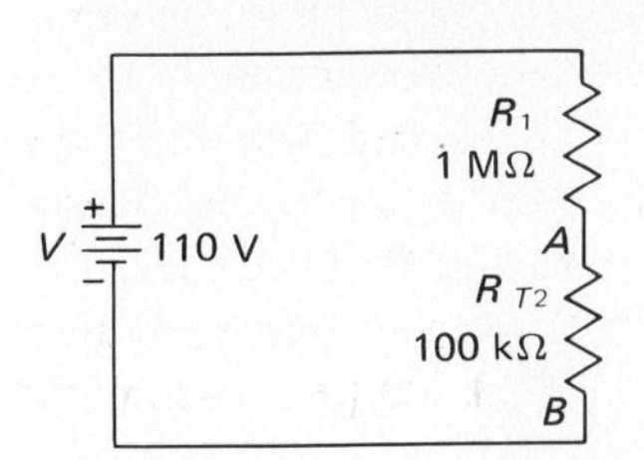

Fig. 11.4-2. Effect of a voltmeter in the circuit.

SELF-TEST

Check your understanding by answering these questions.

1. The input resistance of a 20,000-Ω/V meter on the 50-V range is ________; on the 300-V range its input resistance is ________ Ω.
2. A meter movement whose sensitivity $I_m = 50\ \mu$A is used to construct a nonelectronic voltmeter. The ohms-per-volt characteristic of that meter is ________ Ω/V.
3. In the circuit of Fig. 11.4-1, $R_1 = 1$ MΩ, $R_2 = 1$ MΩ, and $V = 100$ V. If a 1000-Ω/V meter on the 100-V range is placed across R_2, the effective resistance R_{AB} between points A and B becomes ________ Ω.
4. The voltmeter in question 3 would read ________ V, instead of the unloaded value of ________ V.

MATERIALS REQUIRED

- Power supply: Variable, regulated, low-voltage dc supply
- Equipment: EVM, VOM
- Resistors: ½-W 1.5-MΩ, 2.2-MΩ
- Miscellaneous: One SPST switch

PROCEDURE

1. Inspect the VOM given you. Its dc ohms-per-volt characteristic is ________ Ω/V.
2. Connect the circuit shown in Fig. 11.4-3. Close switch S_1 and adjust the power supply for an output of 20 V. Record this value in Table 11.4-1.
3. Calculate the input resistance of the VOM on the 10-V range. ________ Ω
4. With the VOM set on the 10-V range, measure the

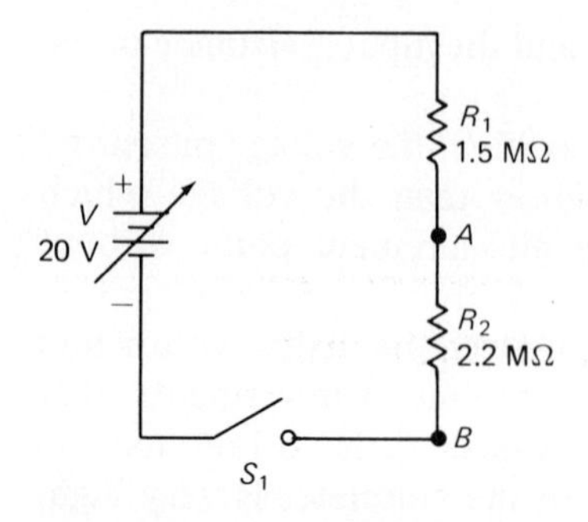

Fig. 11.4-3. Experimental circuit.

TABLE 11.4-1. Voltmeter Loading

Measurement Made by:	V_{R1}	V_{R2}	V
VOM (10-V range)			
VOM (high range)			
EVM	7.39	10.83	19.90
Calculated			

voltage across each resistor and record the measurement values in Table 11.4-1.

5. Repeat step 3 using the next higher voltage range and record the values in Table 11.4-1.
6. Repeat step 3 using an electronic voltmeter and record the values in Table 11.4-1.
7. Calculate the voltage which should appear across each resistor (without the meter connected) and record the values in Table 11.4-1.

QUESTIONS

1. Which meter loaded the circuit more, the VOM or the electronic voltmeter? How do you know? Refer to the data in your experimental tables to support your answer.
2. Which range of the VOM loaded the circuit more? Why?
3. Do the sum of the voltages across R_1 and R_2 in step 4 equal 20 V? In step 5? If not, why not?

Answers to Self-Test

1. 1 MΩ; 6 MΩ
2. 20,000
3. 90,900 (approximately)
4. 8.3 (approximately); 50

CHAPTER 12 POWER AND ITS MEASUREMENT

EXPERIMENT 12.1. Power and Its Measurement

OBJECTIVE

To verify by experiment that $W = V \times I$

INTRODUCTORY INFORMATION

Horsepower and Steam Power

Power has been defined as the rate of doing work. James Watt (1736–1819) found it difficult to sell his newly invented steam engine unless he could compare its power, that is, its rate of doing work, with the power of a horse, since horses did most of the heavy work at that time. As a result, he devised a method of comparison.

Watt's measure was called the *horsepower* (hp), and it was found by determining the average distance a horse could pull a 165-lb coal bucket up a mineshaft in 1 min. The average horse in Watt's experiment pulled the coal bucket 200 ft in 1 min. He called this rate of doing work one *horsepower*. He now had the means of stating the power developed by the steam engine in horsepower. The power of electric motors is still given in terms of horsepower. If an electric motor were substituted for the horse and pulled the 165-lb coal bucket 200 ft up the coal shaft in 1 min, it would be a 1-hp motor. This method of determining power leads to the relationship 1 hp = 550 ft-lb/s.

Electric Power—the WATT

Electrically, the unit of power is the watt (W), a term coined in honor of James Watt. Watt's law is a simple mathematical law, much like Ohm's law, and the power in a dc circuit can be calculated if you know any two of the major circuit forces, I, R, and V. Here are the three formulas for determining the power dissipated in a dc circuit.

$$W = V \times I$$
$$W = \frac{V^2}{R} \qquad (12.1\text{-}1)$$
$$W = I^2 \times R$$

Horsepower and watts are related by the formula

$$1 \text{ hp} = 746 \text{ W} \qquad (12.1\text{-}2)$$

The service technician should understand power and its measurement since all electronic devices consume energy in order to perform some type of work. The work performed may be the production of heat, light, sound, a picture, etc. Whenever work is done, power is involved. In electric devices this power is measured in watts (W).

Measuring Power by Wattmeter

The methods of measuring power depend on the equipment available. The wattmeter offers the simplest method. It uses fixed coils to indicate the amount of current in the circuit and movable coils to indicate voltage (Fig. 12.1-1). The amount of deflection of the pointer is then proportional to the power which is being measured in the circuit.

The wattmeter is connected as both an ammeter and a voltmeter. The terminals brought out from the movable coil are connected in parallel with the device whose power dissipation you wish to measure. The terminals brought out from the stationary (current) coils are connected in series with the device whose power dissipation you wish to measure. In this way both the I and V are actuating the wattmeter at the

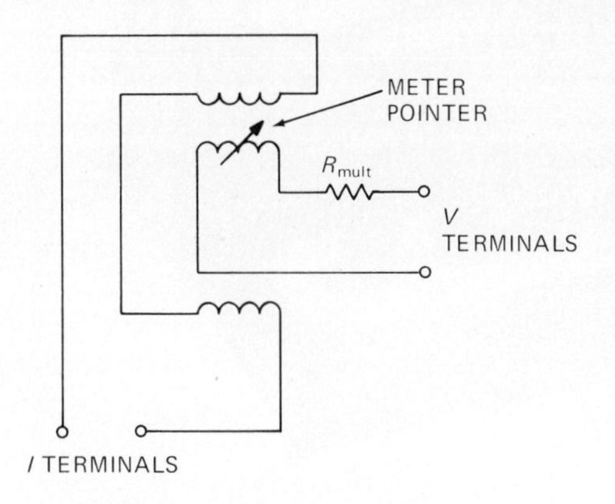

Fig. 12.1-1. Wattmeter schematic.

same time, and in effect the meter is multiplying V by I. Figure 12.1-2 shows a typical wattmeter.

Measuring Power by Voltmeter-Ammeter

A second method of measuring power requires the use of the ammeter and voltmeter. With this method, the current in the device is measured. The voltage across the device is also measured, and these current and voltage values are "plugged" into the formula $W = V \times I$, which is used to compute power.

If the resistance of the device, dissipating the power to be measured, is known, the formula $W = I^2 \times R$ can be used. Or the formula $W = V^2/R$ may be used if the voltage across the device can be measured, and if the resistance R of the device is known.

SUMMARY

1. If the voltage is V (volts) across a resistor and current is I (amperes), the wattage dissipated by the resistor is given by the formula $W = V \times I$.
2. The power W (watts) dissipated by a resistor R (ohms), in which there is direct current I (amperes), is given by the formula $W = I^2 \times R$.
3. The power W (watts) dissipated by a resistor R (ohms), across which there is a dc voltage V (volts), is given by the formula $W = V^2/R$.
4. Power is the rate of doing work, and electric power is measured in watts.

Fig. 12.1-2. Wattmeter. (*Hickok Teaching Systems, Inc.*)

SELF-TEST

Check your understanding by answering these questions.

1. The current in a 120-Ω resistor is 0.1 A. The power in watts dissipated by the resistor is ________ W.
2. The voltage across a resistor is 12 V, and the current in the resistor is 0.05 A. The power dissipated by the resistor is ________ W.
3. Power is the rate of doing ________.
4. The ________ is a meter specially designed to measure electric power.

MATERIALS REQUIRED

- Power supply: Variable, voltage-regulated dc supply
- Equipment: VOM (2) or a current meter and a voltmeter; low-power wattmeter
- Resistors: 2-W 100-Ω; 5-W 50-Ω
- Miscellaneous: SPST switch and connecting wires

PROCEDURE

1. Obtain the materials listed above from the instructor and wire the circuit shown in Fig. 12.1-3. Use the 2-W, 100-Ω resistor for R.
2. Adjust the meters for their highest ranges. Set power supply to 0. **Power on.** Switch S_1 on. Adjust the power supply for 10-V output.
3. Set the meters to the proper scales for accurate readings. Measure current in the circuit and voltage across R. Record these values in Table 12.1-1.
4. Using the values of current and voltage from step 3, calculate the power W dissipated by R. Record.
5. Repeat steps 2 through 4 using a 5-W 50-Ω resistor for R.
6. Connect the circuit shown in Fig. 12.1-4. Use the 2-W, 100-Ω resistor. Measure the power dissipated by R using the wattmeter. Record this value in Table 12.1-1.

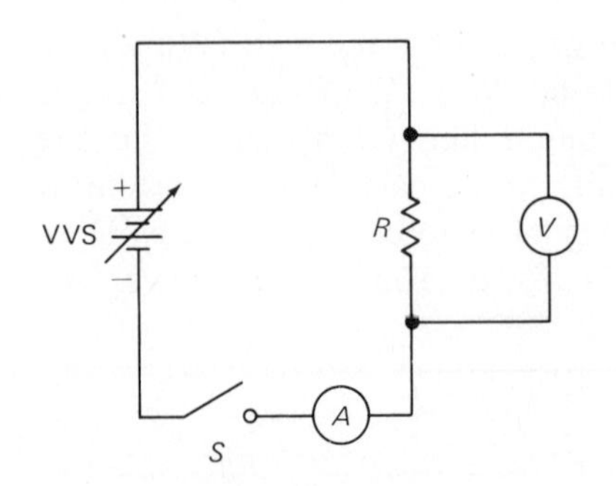

Fig. 12.1-3. Experimental circuit.

TABLE 12.1-1. DC Power Measurement

R, Ω	I Measured, A	V Measured, V	Power Calculated, W	Power Measured, W
100	5.11	.543	2.6	X
50	5.19	.284		

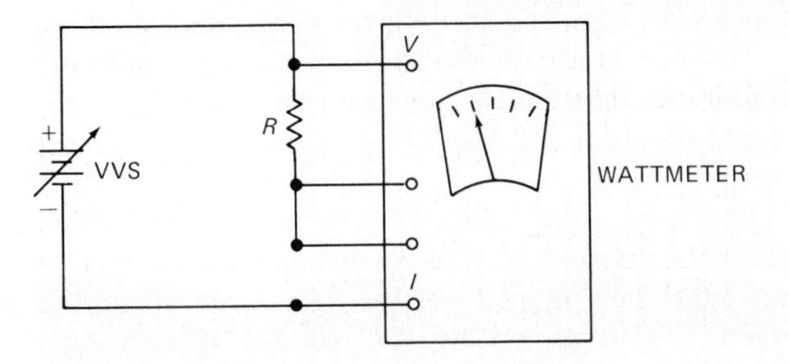

Fig. 12.1-4. Experimental circuit.

7. Repeat step 6 using the 5-W, 50-Ω resistor.
8. Touch the resistor. Is it warm?

QUESTIONS

1. What are the advantages and disadvantages of each of the two methods used to find *W* in this experiment?
2. Compare the calculated and measured values of power in this experiment.
3. If the calculated values were appreciably different from the measured values, explain why.
4. How can you tell that power was dissipated by the resistors in this experiment without measurement? (HINT: Try one of the five senses.)

Answers to Self-Test

1. 1.2
2. 0.6
3. work
4. wattmeter

CHAPTER 13 OSCILLOSCOPE OPERATION AND USE: VOLTAGE AND FREQUENCY MEASUREMENT

EXPERIMENT 13.1. Oscilloscope Operation—Triggered Scope

OBJECTIVES

1. To identify the operating controls of a triggered scope and adjust them to observe a trace
2. To view an ac waveform on the screen

INTRODUCTORY INFORMATION

The cathode-ray oscilloscope (CRO) or "scope," as it is familiarly known, is the most versatile instrument in electronics. The technician must therefore be able to operate this instrument and understand how and where it is used.

For purposes of this book, oscilloscopes will be typed as triggered or nontriggered. Triggered oscilloscopes are the more sophisticated of the two, can do more, and generally are used in industrial laboratories and plants, in engineering and technical school laboratories, and in any application requiring the study of low- and high-frequency waveforms, precise measurement of time, and timing relationships. At this point this very brief statement of the applications of triggered oscilloscopes must suffice. As the student progresses in the study of electronics, additional uses of this laboratory instrument will become evident.

The nontriggered scope is generally used in servicing work where some waveform error can be tolerated and bandwidth requirements are limited to a few megahertz. Increasingly the triggered scope is finding use in larger color TV service shops because of its greater versatility.

NOTE: In this experiment we shall be concerned with triggered oscilloscopes. If the oscilloscope you have in the laboratory is nontriggered, it is suggested that you study the Introductory Information for background, and then go on to Experiment 13.3, which deals with nontriggered scopes.

What an Oscilloscope Does

An oscilloscope automatically graphs a time-varying voltage; that is, it displays the instantaneous amplitude of an ac voltage waveform versus time. Most triggered scopes also measure dc voltages. The indicator in an oscilloscope is a cathode-ray tube (CRT). Inside the cathode-ray tube are an electron gun assembly, vertical and horizontal deflection plates, and a fluorescent screen.

The electron gun emits a beam of electrons, which strikes the fluorescent screen and causes the screen to emit light. The intensity of the light given off by the screen is determined by the voltage relationships between the elements in the electron gun assembly. The manual control of brightness is effected by a control located on the oscilloscope panel.

The motion of the beam over the CRT screen is controlled by a deflection system which includes deflection voltages generated in electronic circuits outside of the CRT, and the deflection plates inside the CRT to which the deflection voltages are applied.

Figure 13.1-1 is an elementary block diagram of an oscilloscope. The CRT serves as the indicator on which electrical waveforms are viewed. These "signal" waveforms are applied to the vertical input on the oscilloscope and are processed by "vertical" amplifiers in circuitry outside the CRT. Since the oscilloscope must handle a wide range of signal voltage amplitudes, a vertical attenuator, a variable voltage divider, acts to set up the proper signal level for viewing. The signal voltage applied to the *vertical* deflection plates causes the electron beam of the CRT to be deflected vertically. The resulting up-and-down trace is significant in that *the extent of vertical deflection is directly proportional to the amplitude of signal voltage applied to the V input.*

To make it possible for the oscilloscope to graph a time-varying voltage, a linearly changing (time-base) deflection voltage is applied to the horizontal deflection plates. This voltage is developed, in electronic circuits external to the CRT, by a time-base or sweep generator. It is this sweep generator which is either triggered or nontriggered. We have classified an oscilloscope with a triggered time-base generator as a "triggered" scope, and one whose time-base generator is free-running (nontriggered) as a "nontriggered" scope.

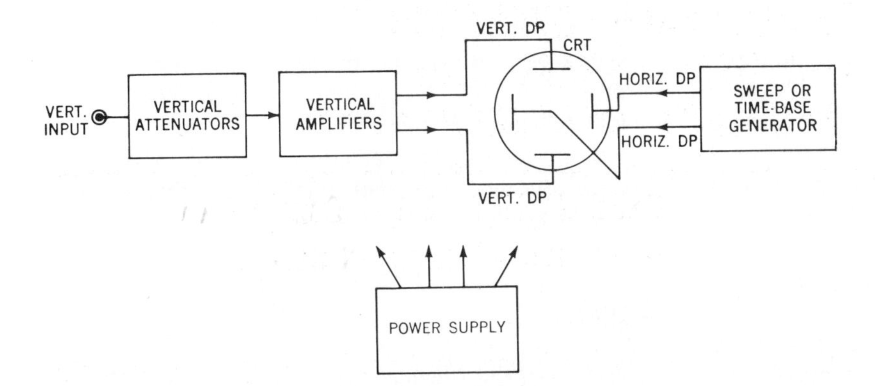

Fig. 13.1-1. Elementary block diagram of an oscilloscope.

Dual-trace Oscilloscopes

Triggered oscilloscopes with *two* traces are in common use. By means of an electronic switching arrangement two traces are developed on the screen of the scope. Dual-trace oscilloscopes make it possible to observe two time-related waveforms at different points in an electronic circuit. Familiarity with the operation of dual-trace oscilloscopes will be helpful to the student in the study of electricity and electronics.

Triggered Oscilloscope—Elementary Considerations

Some triggered scopes utilize a single-frame construction; that is, the electronic circuits external to the CRT are mounted on a chassis. This chassis, together with the CRT mount and the front panel, make up a single assembly. The sweep generator and the vertical amplifier and all the other electronic circuitry which go to make up the oscilloscope are self-contained in this single unit. There are other triggered oscilloscopes which utilize a multiframe construction. These have separate assemblies for the sweep generator and the vertical amplifier which plug into the main frame. The main frame holds the CRT and its associated circuits.

Manual Operating Controls

Intensity. This control sets the level of brightness or intensity of the light trace on the CRT. Rotation in a clockwise (CW) direction increases the brightness. Intensity should not be set too high to prevent damage to the CRT screen.

Focus. This control is adjusted in conjunction with the "intensity" control for the sharpest trace on the screen. There is interaction between these two controls, so adjustment of one may require readjustment of the other.

Astigmatism. This is another beam-focusing control found on some oscilloscopes which operates in conjunction with the focus control for the sharpest trace. The astigmatism control is sometimes a screwdriver rather than a manual control.

Horizontal and Vertical Positioning or Centering. These are trace-positioning controls. They are adjusted so that the trace is positioned or centered both vertically and horizontally on the screen. In front of the CRT screen is an etched faceplate called the *graticule*. The etchings appear in the form of horizontal and vertical graph lines. Calibration markings are usually placed on the center vertical and horizontal lines on this faceplate.

Volts/Div. (also called Volts/cm). There are two concentric controls which act as attenuators of the vertical input signal waveform (which is to be viewed on the screen). The center control marked *Variable* is continuously variable for setting the height (vertical amplitude) of the signal on the screen. Its completely clockwise position is *calibrated* for making peak-to-peak voltage measurements of the vertical input signal. Volts/Div. is the outer of the two concentric vertical attenuators. It is a switched control. A dot on the control can be thrown to the calibrated voltage markings on the panel around the control. Thus when the variable control is set to its calibrated position, the setting of the Volts/Div. control determines the voltage which is equivalent to every division of vertical signal deflection on the screen. The manner in which voltage measurements are made will be discussed in detail in a later experiment.

Time/Div. (also called Time/cm). There are two concentric controls which affect the timing of the sweep or time-base generator. The inner control is marked *Variable* and is continuously adjustable over each range of the Time/Div. control. The complete clockwise (CW) position of the variable control is calibrated for making time measurements of waveforms displayed on the screen. Time/Div. is the outer of the two concentric time-base generator controls. It is a switched control. A dot on the control can be thrown to the calibrated time markings on the panel around the control. When the variable control is in its calibrated position, the settings of the Time/Div. control determine the time it takes the trace to move horizontally across one division of the graticule.

Triggering Controls. A simple calibrated time base usually has four triggering controls associated with it. Thus one oscilloscope has a:

1. *Level control.* There is a switch position, associated with this control, labeled *Auto*(matic). In this position the trigger circuit is free-running and on each cycle triggers the sweep generator. Hence a trace always appears on the screen. The oscilloscope is frequently used in this mode of operation. When not in the automatic mode, triggering depends on some external or internal signal, and the setting of the level control determines the stability or synchronization of the sweep. In the nonautomatic mode there will be *no trace* on the screen in the absence of a triggering signal.
2. *Slope.* This switch is marked + and −, and its setting determines whether triggering of the sweep is caused by the positive or negative portion of the triggering signal.
3. *Coupling.* This selects the manner in which trigger coupling is achieved. The particular oscilloscope we are describing has three coupling modes: AC slow, AC fast, and DC.
4. *Source.* The trigger signal may be *Ext*(ernal), *Int*(ernal), or *Line.* In this experiment this switched control will be set on Int.

NOTES:

1. The controls described may have other names, depending on the manufacturer and oscilloscope model. However, once the operation of a triggered oscilloscope is understood, it will be relatively easy to operate other triggered oscilloscopes.
2. A triggered oscilloscope usually has facilities for switching off the internal sweep generator. A horizontal input jack will then receive some external sweep voltage, apply it to the horizontal processing circuits, and thus cause a horizontal trace. Of course the calibrated Time/Div. controls of the oscilloscope do not operate for this external sweep, and the time base is uncalibrated.
3. There is a vertical signal input jack on the panel which receives the input signal, via a shielded coaxial cable terminated in a probe. There are various types of oscilloscope probes, for example, direct probes and low capacitance probes. We shall be using a *direct* probe in this experiment.
4. Additional features and controls are found on many oscilloscopes. However, knowledge of their operation is not needed at this point.

SUMMARY

1. A triggered oscilloscope can be used for the measurement of dc, as well as low- and high-frequency ac waveforms and time.
2. A nontriggered scope is normally used to observe low-frequency waveforms but cannot measure time directly.
3. An oscilloscope displays a graph of the amplitude of an ac waveform versus time.
4. A cathode-ray tube is the indicator or screen of an oscilloscope.
5. The purpose of the electron gun in a CRT is to emit an electron beam which strikes the screen and causes it to give off light.
6. The signal voltages applied to the vertical deflection plates of a CRT cause the beam to be deflected up and down.
7. The horizontal deflection plates receive the linearly changing deflection voltage which generates the time base.
8. The intensity control is used for setting the brightness of the trace.
9. The focus control is used to narrow the beam into the sharpest trace. There may be an auxiliary astigmatism control for focusing.
10. The horizontal and vertical centering or positioning controls are used to position the trace on the CRT screen.
11. The etched faceplate in front of the CRT face which appears as vertical and horizontal graph lines is called the *graticule.* Linear calibration markers (height and width) are frequently etched on the graticule.
12. The Volts/Div. or Volts/cm control is calibrated for the measurement of the amplitude of signal waveforms along the vertical axis.
13. The Time/Div. or Time/cm control is calibrated for the measurement of time along the horizontal axis.
14. The triggering controls determine the manner in which a trigger pulse is initiated to start the sweep generator.
15. The trigger can be run automatically, the mode which is frequently used.
16. The trigger circuit can be actuated on Int. by the

signal waveform from within the oscilloscope circuits.

17. The trigger circuit can also be actuated on Ext. by an external signal voltage applied to an input jack labelled Ext. trigger.
18. Also, the trigger circuit can be actuated on Line position of the trigger switch by a line-derived voltage from within the oscilloscope circuits.

SELF-TEST

Check your understanding by answering these questions:

1. Both triggered and nontriggered scopes are used for the observation of ac waveforms. ________ (true/false)
2. Precise time measurements can be made with both triggered and nontriggered scopes. ________ (true/false)
3. The waveform seen on the screen of a CRT is a ________ of amplitude versus ________.
4. The ________ deflection plates of a CRT are the signal plates; the ________ deflection plates are for the time-base voltage.
5. Of the controls on an oscilloscope, those that affect the height of the signal are called ________.
6. Those controls that affect the sharpness of the trace are called ________ and ________.
7. The etched faceplate in front of the face of the CRT is called the ________.
8. A frequently used triggering mode of a trigger oscilloscope is ________.
9. The controls that affect the up-and-down movement of the trace are called ________.
10. The height of the waveform displayed on the oscilloscope screen is directly proportional to the ________ of the waveform.

MATERIALS REQUIRED

- Power supply: Source of 120-V alternating current
- Equipment: Triggered-type oscilloscope with calibrated time base, with calibrated vertical amplifier, with internal voltage calibrator, and with direct probe

PROCEDURE

NOTE: Before attempting the experiment, the student should read and become thoroughly familiar with the operating instructions of the oscilloscope.

CAUTION: Do not operate the scope with trace intensity too high.

Operating the Controls Which Affect the Trace

1. List each manual control and switch on your oscilloscope and state its function in Table 13.1-1. Include also the input jacks.
2. Turn the oscilloscope **on.** Some scopes contain a protective time-delay relay. Wait until you hear the relay click in. This will occur after the oscilloscope is warmed up and ready for operation (about 1 to 2 min). Set the Time/Div. control to 1 ms.
3. If a trace does not appear on the screen, check to see that the triggering switch is on Auto (matic). If it is not, set in on Auto.
4. If there is still no trace, turn the Intensity control completely clockwise.
5. If there is still no trace, try the Positioning (centering) controls until a trace does appear.
6. Adjust the focus, astigmatism, and brightness controls for a clear, sharp trace. Center the beam both vertically and horizontally. Set the triggering *Slope* to +, triggering *Coupling* to AC or AC Fast, triggering *Source* to Int. The oscilloscope is now ready for viewing ac waveforms.
7. Throw off the controls you adjusted in steps 2

TABLE 13.1-1. Manual Controls and Switches and Their Functions

Control or Switch	*Function*
BEAM FINDER / INTENSITY / POWER (ON-OFF)	
FOCUS	
*HORIZONTAL POSITION	
+VERTICAL POSITION	
*SWEEP TIME µsec/cm mill sec/cm	
*Vernier w/calibr pos.	
TRIGGER INT +, -, LINE, EXT	
SWEEP MAGN. X1, X5	
sensitivity +V/cm 1, 10	
mV/cm 100, 10 off, cal	
vernier w/calib	

through 6 and repeat the entire procedure for setting up a trace. When you are satisfied that you can operate the controls properly, notify your instructor.

Viewing a Waveform

8. Connect the vertical input leads of the oscilloscope to the output of the voltage calibrator on the scope. Set the calibrator to 2 V output (approx.). The calibration waveform will be seen when the vertical attenuators (Volts/Div.) and Time/Div. controls are properly set.
9. Set the variable vertical attenuator on *Calibrated* and vary the Volts/Div. control for about three divisions of signal height.
10. Now set the Variable Time/Div. control to Calibrated and vary the Time/Div. switch until three cycles (approx.) of the calibration waveform appear on the screen.
11. Now, leaving the Height/Div. controls as set, change the setting of the calibrator output to ½ V (approx.). What happens to the height of the waveform on the screen?
12. Reset the calibrator output to 5 V. What happens to the height of the waveform?
13. Reset calibrator output to 2 V, as in step 8. The waveform should now have a height of three divisions as in step 9.
14. *Increase* the sweep speed (Time/Div.) one setting on the Time/Div. switch (for example, if the switch was set to 0.5 ms/div., set it to the next lower time calibration, say, 0.2 ms/div.). What happens to the number of cycles on the screen?
15. *Decrease* the sweep speed one setting past its position in step 10. What happens to the number of cycles on the screen?
16. When you are satisfied that you understand the operation of all the manual controls on the scope, have your instructor throw them off. Readjust the operating controls until you can see the calibrator waveform, properly centered, on the screen, as in step 10.
17. Repeat step 16 until you are completely satisfied that you can operate the oscilloscope controls properly.

QUESTIONS

1. After the Volts/Div. controls were set to view the calibrator waveform, what happened to the height of the waveform when the calibrator voltage output was increased? Why?
2. After the Time/Div. controls were set to view three cycles of the calibration waveform, what happened to the number of cycles seen when the sweep speed was increased? decreased? Why?
3. Refer to steps 14 and 15. Was the frequency of the calibration waveform affected when the sweep speed was increased or decreased? What was changed?
4. What is the relationship, if any, between the number of cycles of waveform and the setting of the Time/Div. control?
5. List the controls on your scope which affect the (*a*) height of the waveform; (*b*) brightness of the trace; (*c*) sharpness of the trace; (*d*) position of the trace; (*e*) triggering of the sweep generator.

Answers to Self-Test

1. true
2. false
3. graph; time
4. vertical; horizontal
5. Volts/Div. or Volts/cm
6. focus, astigmatism
7. graticule
8. automatic
9. positioning or centering
10. amplitude (voltage)

EXPERIMENT 13.2. Voltage and Frequency Measurements with a Triggered Scope

OBJECTIVES

1. To observe the output waveforms of an audio frequency (AF) signal generator
2. To measure the peak-to-peak voltage and frequency of an AF sine-wave signal, using a triggered scope

INTRODUCTORY INFORMATION

In servicing electronic equipment such as radio and television receivers, tape recorders, etc., the service technician frequently must "follow" an ac signal through the various stages of the electronic device being checked. To do this, the technician uses the oscilloscope to *observe* the ac waveform and may also find it necessary, in determining if a stage is operating properly, to measure the amplitude of the input to, and output signal from, the stage. The technician may also need to measure the frequency of the signal(s) in the stage under test. Again, a triggered oscilloscope with a calibrated vertical input, and a calibrated time base, is used to make these measurements. How this is done will be discussed later in this experiment, and you will have an opportunity to make these measurements.

Audio Oscillator (AF Signal Generator)

In checking audio equipment such as amplifiers, an audio frequency signal source is frequently required. The AF sine-wave signal generator supplies the audio signal needed to test audio frequency circuits. But there are signal generators which provide ac signals over a much wider range of frequencies than the audio frequencies. These generators develop signals whose frequency is variable over specified ranges.

AC frequencies cover a very wide spectrum, from a fraction of a cycle per second to thousands of millions of cycles per second. No one instrument has been designed to cover this extensive range. Many generators are available commercially to provide the various frequency needs in electronics.

One of the characteristics by which signal generators are identified is the frequency range which the generator covers. The AF generator supplies frequencies from several hertz up to 20,000 Hz approximately. It is called "audio" (which means sound) because this is the range of frequencies (approximately) to which the ear responds. This is not to say that the ear can hear the electrical signal delivered by the generator. But the electrical signal can be converted to a sound signal (a vibration in the air) by a suitable device such as a loudspeaker.

An AF generator often covers a range much wider than the audio frequencies. Thus, one "audio" generator supplies a signal whose frequency can be varied from several hertz up to 600 kHz. However, this coverage is achieved in several ranges.

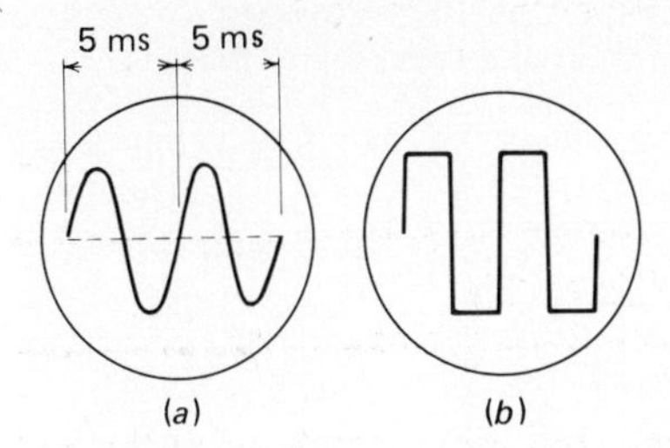

Fig. 13.2-1. AF generator develops (*a*) sine wave; (*b*) square wave.

Another characteristic which identifies a signal generator is the shape of the waveform it develops. Thus, there is a sine-wave generator whose output is sinusoidal (Fig. 13.2-1*a*), a square-wave generator whose output is the square waveform (Fig. 13.2-1*b*), and other waveform generators with which the student need not be concerned at the moment.

The usual controls and switches found on the front panel of a signal generator are:

1. ON-OFF switch for applying power to the generator.
2. *Range.* A coarse control to select a specified frequency coverage (range); for example, 10 to 1000 Hz or 1000 to 100,000 Hz, etc.
3. *Frequency.* This is a continuously variable control to select a specified frequency within any range. Associated with this control is a calibrated frequency scale.
4. *Level (output) control.* This is used to set the voltage of the output signal. There may be two types of level controls on the panel, a coarse control and a continuously variable output control. Both types are voltage-divider networks.

The output of a general-purpose sine-wave generator is not metered. Its level must be measured by an external ac voltmeter. Some laboratory-standard signal generators include a metering circuit which measures the signal delivered at the output terminals. General-purpose AF generators can deliver a signal from a fraction of a volt to many volts. Thus, the approximate output range of one instrument is from a millivolt to 20 V.

A shielded cable is used to deliver the generator signal to the external circuit. This cable is detachable from the generator.

Characteristics of an AC Signal Voltage

Alternating-current voltages are identified by certain characteristics. These are:

Waveform or Shape of the Voltage

This characteristic pertains to the manner in which the voltage varies between maximum and minimum. Thus, if the voltage is varying in a sinusoidal fashion, the scope will display the voltage as a sine wave (Fig. 13.2-2).

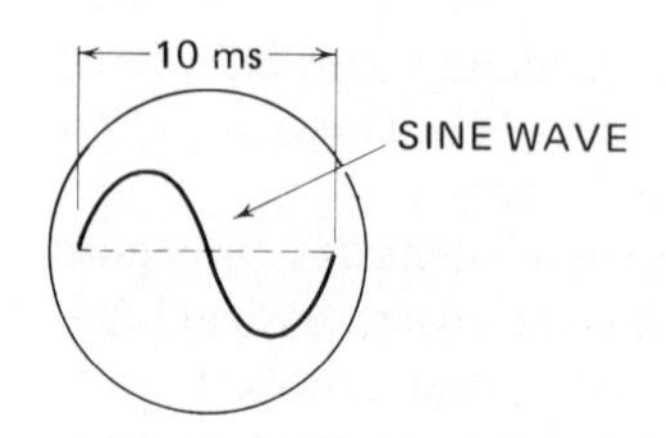

Fig. 13.2-2. Sine wave seen on scope screen.

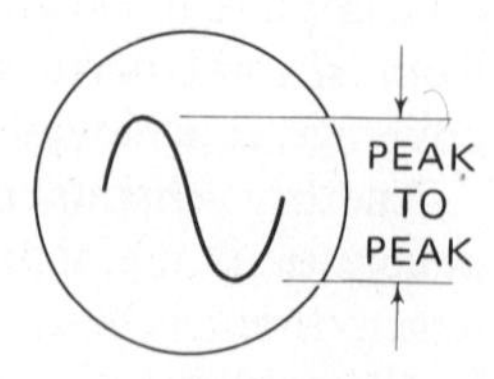

Fig. 13.2-3. Measuring peak-to-peak voltage.

Amplitude of the Voltage

This characteristic describes the difference between the positive and negative peaks of the voltage and is expressed in peak-to-peak volts (Fig. 13.2-3).

Frequency of the Voltage

An ac periodic wave completes a number of cycles every second. The time interval for each of these cycles is called the period t. Thus the period or time in seconds of one complete cycle of an ac voltage is found from the relationship

$$t = \frac{1}{f}$$

where f is the frequency of the voltage in hertz. Thus, if the voltage of Fig. 13.2-3 has a frequency of 60 Hz, the period is $\frac{1}{60}$ s. Hence, the time required for the scope to "graph" one cycle of the waveform is $\frac{1}{60}$ s.

Measuring Voltages with the Scope

The height of the voltage waveform displayed on the oscilloscope screen is directly proportional to the peak-to-peak amplitude of the voltage. Thus, for the same settings of Volts/div., or vertical-gain controls, a 100-V signal will have twice the height of a 50-V signal. A 25-V signal will have one-fourth the height of a 100-V signal. Moreover, if we suppose that for this same fixed setting of the vertical controls, every division of vertical deflection corresponds to 25 V of input, then a 25-V signal will have 1 division of height, a 50-V signal will have 2 divisions of height, a 75-V signal will have 3 divisions of height, and so on. Two and one half ($2\frac{1}{2}$) divisions of vertical deflection (height) will correspond to $62\frac{1}{2}$ V, etc. This characteristic of the signal circuits makes it possible to use the oscilloscope for measuring ac voltages.

To measure dc voltage, the method described above is also used. The ac-dc switch is placed in the dc position, and the probe is connected to the point in the circuit where the dc voltage is. The ground lead of the scope is connected to the ground of the circuit. The number of divisions the trace rises above or falls below the zero base setting is a measure of the + or − dc voltage.

NOTE: In this experiment we shall refer to the number of centimeters (cm) or the number of divisions of height of the voltage waveform, because these two markings are found on the graticules of oscilloscopes. The triggered oscilloscope frequently uses a graticule whose division markings are in centimeters. The triggered-scope graticule is usually rectangular and appears as in Fig. 13.2-4. It is shown here as 8 × 6 cm, though not all graticules have these dimensions.

In measuring voltages with the oscilloscope, the procedure is to apply the signal voltage to the vertical input of the oscilloscope and to measure on the etched faceplate or graticule the number of divisions of height of the waveform. This number is then multiplied by the calibration factor of the oscilloscope, giving the peak-to-peak voltage of the waveform. For example, if the scope is calibrated for 20 V/div., 3 div. of height measures a 60-V signal.

It should be noted that there are no voltage scales on the graticule. There are simply linear markings whose value is determined by the calibration factor to which the volts/div. vertical-gain controls of the oscilloscope are calibrated. Thus the volts/div. or vertical-gain controls correspond to the range controls of a voltmeter; the graticule markings correspond to the scale of the voltmeter.

Oscilloscope Calibration

Triggered Oscilloscope. The triggered oscilloscope has a calibrated vertical amplifier. There are two concentric input controls which affect, by attenuation, the height of the waveform displayed on the screen. The inner control is continuously variable. The switch-type volts/div. control changes the signal at-

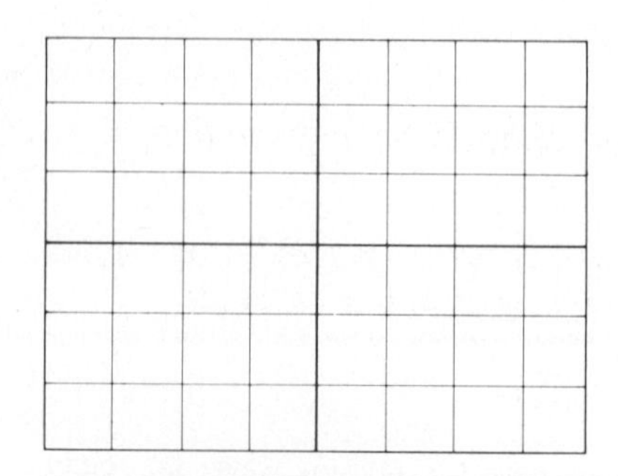

Fig. 13.2-4. Rectangular graticule (8 × 6 cm) on triggered scope.

tenuation by a precalibrated factor. The inner control must be turned to its "calibrated" position, which is completely clockwise. Then, the marking to which the switch-type volts/div. control is set is the calibration factor of the signal-measuring vertical amplifier of the oscilloscope. Thus with the inner control set to its calibrated position, and the volts/div. control set to 0.5 V, every 1 div. of height corresponds to 0.5 V.

NOTE: These rules apply to a scope with a direct probe. If a low-capacitance probe is used, the reading must be multiplied by the attenuation factor of the probe.

The volts/div. control is frequently a decade-type attenuator; that is, it switches the voltage-range calibration by a factor of 10. For this type of attenuator the calibration marker settings might read: 0.01 V, 0.1 V, 1.0 V, 10 V, and so on.

Many triggered oscilloscopes contain a warning light which goes on when the inner control is not set to its calibrated position, that is, when the scope is uncalibrated. Measurements cannot be accurately made when this light is on. Before measurement, the inner control must be turned completely clockwise to its calibrated position. The warning light will then go out and the oscilloscope is again calibrated.

Most makes of triggered oscilloscopes have a self-contained voltage calibrator for checking the calibration of the volts/cm selector. The switch-type calibrator delivers a measured voltage at its output terminals. The measured outputs usually correspond to the positions of the volts/cm selector. Thus for the volts/cm selector described above, the calibrator would deliver 0.01 V, 0.1 V, 1.0 V, and so on, depending on the switch setting of the calibrator.

To check the calibration accuracy of the volts/cm selector, the technician would apply the oscilloscope probe to the calibrator output jack and note the height of the waveform displayed. Thus, if the calibrator were set for 1.0 V output, the calibrator signal should be deflected 1 cm in height when the volts/cm selector is set to 1 V/cm.

Measuring the Period of an AC Voltage with an Oscilloscope

Since the period t and frequency f of an ac signal waveform are inversely related; that is, since

$$t = \frac{1}{f}$$

and

$$f = \frac{1}{t}$$

it is possible to calculate the frequency of a waveform if its period is known. If its frequency is known, its period may be computed.

The triggered oscilloscope has facilities for direct measurement of the period of an ac signal voltage. The characteristic which makes this possible is the calibrated sweep or time base of this type of oscilloscope.

In Exp. 13.1 you noted that to display a signal waveform on the screen, it was necessary to set the Time/cm control properly. It is this control which determines the rate at which the cathode-ray tube beam is moving across the screen to generate the trace. An example will show how the screen can be used to measure the period of the waveform displayed on the CRT.

Suppose the Time/cm control is set to 1 ms and suppose the calibrated face of the oscilloscope, called the graticule, is 10 cm wide. Then it takes 10 ms for the electron beam to move across the 10-cm screen. Now, if the width of a waveform is exactly 10 cm, as in Fig. 13.2-2, then the period of the waveform is 10 ms. That is, $t = 10 \times 10^{-3}$ s. The frequency of the displayed waveform can now be calculated, for

$$f = \frac{1}{t} = \frac{1}{10 \times 10^{-3}} = 100 \text{ Hz}$$

If two waveforms had been displayed on the 10 ms trace (Fig. 13.2-1), then the period of each cycle is 5 ms and the frequency is

$$f = \frac{1}{5 \times 10^{-3}} = 200 \text{ Hz}$$

NOTE: The trace width of a *triggered sweep* is fixed, as is the scanning rate.

SUMMARY

1. A signal generator supplies ac voltages for ac circuits.
2. An AF signal generator provides ac signals in the audio range of frequencies, that is, 10 to 20,000 Hz (approximately).
3. There are other generators which develop signal voltages over other frequency ranges.
4. An AF generator may provide a sinusoidal voltage waveform, a square wave, or some other waveshape, depending on the type and purpose of the generator.
5. An ac voltage waveform may be identified by
   (*a*) Its waveform—is it a sine wave, square wave, triangular wave, etc.?
   (*b*) The amplitude of the wave—usually measured in peak-to-peak volts.

(c) Its frequency (f) in hertz.
(d) The period or time (t) in seconds of one cycle of the waveform.

6. Triggered oscilloscopes contain calibrated signal amplifiers. These are used directly in the measurement of ac voltage waveforms and dc voltages.
7. Before a triggered oscilloscope is used to measure voltage, its calibration factors, volts/div. setting, should be checked by applying the measured calibration voltages available on the oscilloscope to the input terminals of the scope. If they check, it is ready to make measurements.
8. Frequency and period are related by the formulas

$$f = \frac{1}{t}$$

$$t = \frac{1}{f}$$

9. A triggered oscilloscope has a calibrated time base which may be used for measuring the period t of a cycle of a periodic waveform. Having measured t, f can be calculated.

SELF-TEST

Check your understanding by answering these questions.

1. The instrument which supplies a test ac signal is called a ________ ________.
2. The audio frequencies are in the range ________ Hz to ________ Hz (approximately).
3. The general-purpose AF signal generator has a metered output. ________ (true/false)
4. The characteristics which identify a periodic ac signal are:
   (a) ________
   (b) ________
   (c) ________
   (d) ________
5. If the period of an ac waveform is known, the frequency of that waveform may be calculated from the formula: ________ = ________.
6. A triggered oscilloscope has a ________ time base, which can be used to measure the ________ of an ac waveform.
7. Triggered-type oscilloscopes ________ (do/do not) have calibrated vertical amplifiers.
8. The vertical amplifiers of a triggered scope are calibrated at 10 V/cm. At this setting the height of a waveform is 2.6 cm. The voltage of the waveform is ________ V.

MATERIALS REQUIRED

- Power supply: ac voltage source
- Equipment: AF signal generator; oscilloscope with direct probe; EVM

PROCEDURE

Familiarization with AF Signal Generator

1. Familiarize yourself with the signal generator issued to you. Study the instruction manual and learn the function and use of the operating controls on the front panel.
2. **Power on.** Connect the output leads of the signal generator to the vertical input of the oscilloscope. Set the output controls of the generator in the middle of their range. Adjust the frequency control to 100 Hz.
3. Set the oscilloscope volts/div. selector until the waveform is deflected 3 divisions vertically (approximately). Set scope on automatic triggering and sync on Int. +. Adjust the oscilloscope Time/div. control for a display of two sine waves (approximately).
4. Reduce the output level of the generator signal and observe the effect on the height of the signal displayed on the oscilloscope. Is the stability of the presentation affected as the signal is reduced below a certain level? NO
5. Increase the level of the generator signal to maximum. Readjust, if necessary, the volts/div. control until the entire waveform is displayed on the screen. Is the volts/div. factor now higher or lower than in step 3? HIGHER
6. Reset the frequency of the signal generator to 200 Hz. There are now __4__ cycles on the screen.
7. Reset the frequency of the signal generator to 300 Hz. There are now __6__ cycles on the screen.

Measuring Frequency

8. Reset the oscilloscope Time/div. control until one or two cycles are displayed on the screen. Measure the width of one sine wave. ________ div.
9. Multiply the Time/div. setting for this display by the width (number of div.) of the waveform. This is the period of the sine wave. ________ s

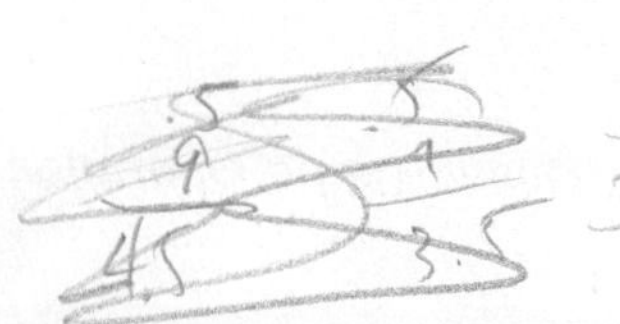

10. Using the formula $f = 1/t$, compute the frequency of the sine wave. (NOTE: f is in Hz, when t is in seconds.) $f =$ .3125 Hz

Checking Calibration Accuracy of Oscilloscope Vertical Amplifiers

11. Locate the calibration voltage jack and the calibrator-voltage selector switch (if any) on your oscilloscope. Set the vertical amplifier gain control on "calibrated."
12. Connect the direct probe of the oscilloscope to the calibration voltage jack, and check the accuracy of the volts/cm or volts/div. selector on each of its ranges (if possible). Record the results of your check in Table 13.2-1. If your oscilloscope is not properly calibrated, notify your instructor. If calibration is proper, proceed to step 13.

Measuring AC Voltages with Oscilloscope

13. Set sine-wave generator frequency to 1000 Hz. Adjust output of the generator to 1 V p-p, as measured by an EVM set to measure ac volts, peak to peak. *Do not change generator output.*
14. Measure and record generator output with your oscilloscope.
15. Repeat steps 13 and 14 for the generator voltage levels listed in Table 13.2-2.

TABLE 13.2-1. Oscilloscope Vertical Amplitude Calibration

Volts/cm setting	*Calibration Voltage* V p-p	*Height of Calibration Voltage on Screen*
CALIB	—	6 cm

TABLE 13.2-2. AC Voltage Measurements

AF Generator Output V p-p *Measured by*	
EVM	*Oscilloscope*
1 V	1 V (1 cm)
~~2~~ 10 V	10 V (10 cm)
3	

QUESTIONS

1. What is the range of frequencies which your AF generator can deliver? 10 – 100,000 Hz
2. Does there seem to be any relationship between the number of cycles displayed on the screen and the frequency setting of the AF generator, assuming there is no change in the Time/div. setting of the scope? If so, what is the relationship?
3. How did the signal frequency, computed in step 10, compare with the frequency setting of the generator? If they were not the same, explain why.
4. What are the markings (settings or calibrations) of the vertical gain selector on your scope? 1 V 10 V cm 100 mV 10 mV cm
5. Which of the vertical gain settings (see question 4) is the most sensitive; that is, which requires least voltage for 1 division of vertical deflection?
6. How did the ac voltage readings of your oscilloscope compare with those of the EVM?
7. List the advantages, if any, in using the scope as an ac voltmeter?
8. What are the disadvantages of using the scope as an ac voltmeter?

Answers to Self-Test

1. signal generator
2. 10; 20,000
3. false
4. (*a*) waveform; (*b*) amplitude; (*c*) frequency; (*d*) period
5. $f = 1/t$
6. calibrated; period
7. do
8. 26

EXPERIMENT 13.3. Oscilloscope Operation – Nontriggered Scope

OBJECTIVES

1. To identify the operating controls of a nontriggered oscilloscope
2. To set up the oscilloscope and adjust the controls properly to observe an ac voltage waveform

INTRODUCTORY INFORMATION

The triggered oscilloscope is used in industrial and research laboratories, in school laboratories, and for general industrial applications as well as for radio and TV servicing. In some small radio and TV shops, and in some technical school laboratories, the nontriggered oscilloscope may still be found. This experiment is intended for those schools which use nontriggered oscilloscopes.

The nontriggered service-type oscilloscope is less expensive than the laboratory standard triggered oscilloscope, does not have a calibrated time base (sweep), and does not have calibrated vertical amplifiers. It is generally less sophisticated than the triggered oscilloscope. The controls have different names, though they are similar in function and are relatively simple to adjust and operate. The nontriggered oscilloscope has a free-running time-base generator.

Operating Controls

At this point the student is not expected to understand the circuitry or theory of operation of the nontriggered cathode-ray oscilloscope. However, the student is expected to learn how to use it properly, and for this reason must understand the function and operation of each of the controls. These are:

Intensity. This control sets the level of brightness or intensity of the light trace on the cathode-ray tube. Rotation in a clockwise (CW) direction increases the brightness. Usually associated with this control is the ON-OFF switch, which applies power to the scope.

Focus. This control is adjusted in conjunction with the "intensity" control for sharpest trace on the cathode-ray tube. There is interaction between these two controls. Hence adjustment of one may require readjustment of the other.

Horizontal and Vertical Centering. These are positioning controls for positioning or centering the trace. They affect the horizontal and vertical position of the trace.

Sweep Selector. This control in conjunction with the fine-frequency control makes it possible to set the scope for viewing waveforms of different frequencies. They are adjusted to the position which "stops" the waveforms on the screen.

Sweep Vernier. It operates in conjunction with the coarse frequency as described.

Sync. The sweep selector and vernier controls cannot usually "stop" a waveform completely. At their closest setting the waveforms may still drift (move). The sync control is adjusted to stop this drift. It is preferable to set the sync control at approximately 10 to 20 percent of its range. This control synchronizes the scope sweep frequency with the ac signal-voltage frequency.

Sync Selector. This switch is usually set to Int. for most applications.

Vertical-gain Selector. This control operates in conjunction with the vertical-gain vernier to adjust the height of the waveform by changing the overall vertical scope gain. The "V" selector is a coarse adjust. It is usually a calibrated attenuator.

Vertical-gain Vernier. This continuous-gain control sets the height of the waveform over each range of the vertical-gain selector. Thus the "V" selector is set for the approximate height of the waveform required. The "V" vernier is then used to set the waveform to the exact height required.

Horizontal Selector. This switch is used for selecting the horizontal sweep or for setting the width of the trace in conjunction with the horizontal-gain vernier. This switch may not be found on all scopes.

Horizontal-gain Vernier. This control provides a continuous-width adjustment. It is a fine adjustment.

Control Names	*Alternate Names*
Intensity – **off**	Intensity; intensity line **off;** intensity ac **off;** beam **off**
Focus	
Horizontal centering	H centering; horizontal position
Vertical centering	V centering; vertical position
Sweep selector	Sweep; sweep range; sweep frequency; coarse frequency; saw-tooth sweep; frequency range; range
Sweep vernier	Range frequency; vernier; frequency vernier; frequency
Sync	Sync adj.; sync lock; sync signal
Sync selector	Sync; function; synchronizing input control
Vertical-gain selector	V range; vert. attenuator; V sensitivity; vert. input control
Vertical-gain vernier	V vernier; vertical gain
Horizontal selector	H sel; horiz. sens; horizontal input control; H gain selector
Horizontal-gain vernier	H gain; horizontal gain

The horizontal trace is usually adjusted to occupy $\frac{4}{5}$ the width of the screen.

Additional features and controls are found on many oscilloscopes. However, knowledge of their operation is not needed at this point. It should be noted here that similar controls on different oscilloscopes may have other names than those given. For this reason a chart is included here which lists some of the names by which each of the controls is known.

The oscilloscope has certain input jacks with which technicians should be acquainted. At this time they will be concerned only with the vertical input and ground jacks. A shielded cable is usually attached to these two jacks. The waveform to be viewed is brought into the oscilloscope by this shielded cable.

SUMMARY

1. Nontriggered oscilloscopes are sometimes used by the radio and TV service technician for the repair and adjustment of radio and TV receivers.
2. The controls which affect the trace on the cathode-ray tube are: *intensity, focus, horizontal centering,* and *vertical centering.* The name of the control is descriptive of its function.
3. The controls which affect the gain of the vertical or *signal* amplifiers are: *vertical-gain selector* and *vertical-gain vernier.* These are used to adjust the height of the signal waveform on the oscilloscope screen. The *selector* is a *coarse* attenuator; the *vernier* is a continuous or fine attenuator.
4. The controls which make it possible to view waveforms of different frequencies are: *sweep selector* and *sweep vernier.* The selector is a *coarse* control; the vernier is a continuously variable *fine* control. They operate in conjunction and are adjusted to the position which sets up the required number of cycles of the waveform to be viewed.
5. The *sync control* is adjusted to *stop* the waveform from drifting across the screen. The synchronizing source is normally the signal voltage which is being viewed on the oscilloscope when the *sync selector* is switched to Int.
6. The *horizontal selector* (not found on all scopes) and *horizontal-gain vernier* are used to adjust the width of the trace. The selector, when used, is a coarse control; the *vernier* a fine control.

SELF-TEST

Check your understanding by answering these questions:

1. An oscilloscope with a free-running time base is called a(n) ________ oscilloscope.
2. This type of oscilloscope ________ (has/does not have) a calibrated time base.
3. The time base generates the ________ on the screen.
4. The control which affects the brightness of the trace is the ________ control.
5. The ________ control is adjusted for the sharpest (clearest) trace.

6. The vertical and horizontal ________ controls are used to position the trace.
7. The controls which make the viewing of signals of a wide range of frequencies possible are: ________ selector and ________ vernier.
8. To prevent a signal waveform from drifting across the screen, the ________ control is used.
9. To set the amplitude or height of the waveform on the screen, we use the ________ ________ selector and ________ ________ vernier.
10. To set the width of the trace we use the ________ selector and ________ ________ vernier.

MATERIALS REQUIRED

- Power supply: Source of 120 V ac
- Equipment: Nontriggered oscilloscope with direct probe; AF sine-wave generator

PROCEDURE

NOTE: The student should read and become thoroughly familiar with the operating instructions of the oscilloscope being used before starting this experiment.

CAUTION: *To prevent screen burnout, do* not *leave a stationary spot on the screen.*

Getting Acquainted with Operating Controls

1. Set the sweep selector and sweep vernier in the middle of their range.
2. Set the vertical and horizontal centering controls in the middle of their range.
3. Set sync selector to INT and sync completely off.
4. Turn scope **on** and advance intensity control about three-fourths of the way up. Set vertical-gain vernier completely counterclockwise.
5. Set horizontal gain about three-fourths of the way up (clockwise).
6. Let scope warm up for a few minutes and then adjust vertical and horizontal centering controls and note effect on trace. Center the trace properly.
7. Adjust intensity and focus controls for sharpest trace and satisfactory brightness.
8. Vary the horizontal-gain vernier and note effect on trace width. Adjust this control for satisfactory trace width. Thus on a 5-in screen, adjust for a trace 4 in wide. Center the trace or beam.
9. Throw all the controls off and go through the entire procedure until a satisfactory trace has been obtained on the scope. When you are satisfied that you can operate the controls, notify your instructor.
10. The instructor will throw all the controls off and the student will readjust them properly. The student may not proceed until the instructor has approved this work.

Viewing a Waveform

11. Connect the vertical input cable of the oscilloscope to an ac signal source. There may be a test-signal or a calibrating-signal source on the oscilloscope. This ac signal is satisfactory.
12. Adjust the vertical-gain selector and vertical-gain vernier until the signal fills 80 percent of the screen in height. Recenter the beam if necessary.
13. Adjust the sweep selector and sweep vernier until there are just two cycles of signal voltage on the screen.
14. Turn the sync control up until the signal is stationary on the screen. This should be no higher than 20 percent of its range. Readjust the sweep vernier if necessary to stop the signal. There is interaction between the sweep vernier and sync control.
15. After the signal has been "stopped," turn the sync control all the way up and note the effect. Return sync to its previous setting (step 14).
16. Adjust vertical and horizontal centering and note the effect. Recenter the waveform properly.
17. Adjust focus and brightness controls and note the effect on the waveform. Reset these controls properly.
18. Vary the vertical-gain vernier and note the effect on height and synchronization (stability). Reset the height as in step 12.
19. When you are reasonably certain that you know how to operate these controls, throw them all off and go through the entire procedure again. Use a 1000-Hz sine wave from an AF generator. You should end up with two cycles on the screen.
20. Readjust the sweep vernier (and sweep selector if necessary) until you have stationary (*a*) one cycle, (*b*) three cycles, (*c*) four cycles.
21. Obtain your instructor's signature for your test setup.

QUESTIONS

1. After the waveform has been synchronized, what is the effect on synchronization, if any, of varying the vertical-gain vernier on either side of its setting?

2. Is the frequency of the test waveform changed as the sweep vernier is varied?
3. List the controls on your scope which affect the: (*a*) height of the waveform; (*b*) width of the trace; (*c*) brightness of the trace; (*d*) sharpness of the trace; (*e*) position of the trace; (*f*) steadiness of the waveform.
4. Why should you never leave just a bright spot on the screen?
5. List the sets of controls which interact with one another and indicate the effects of the interaction.

Answers to Self-Test

1. nontriggered
2. does not have
3. trace
4. intensity
5. focus
6. centering
7. sweep; sweep
8. sync
9. vertical-gain; vertical-gain
10. horizontal; horizontal-gain

EXPERIMENT 13.4. Nontriggered Oscilloscope Voltage Measurements

OBJECTIVE

To calibrate the vertical amplifiers of a nontriggered scope and measure ac and dc voltages with it

INTRODUCTORY INFORMATION

The most versatile instrument the electronics technician has is the oscilloscope. With the triggered or nontriggered scope the technician can look at a signal waveform in a circuit and can follow the progress of the signal from the input to the output, a technique called *signal tracing*. Signal tracing is commonly used in servicing electronic equipment to locate a defective stage or section.

With the triggered scope the technician can measure the peak-to-peak voltage of a waveform and can determine the period and therefore the frequency of a periodic waveform.

The nontriggered scope was the forerunner of the triggered type, and because it was the earliest type of oscilloscope, it can not do as much as a triggered model. For example, the vertical amplifiers of most nontriggered scopes are not calibrated. Moreover the time base (sweep) of nontriggered scopes is not calibrated. Calibrating the time base is generally not possible for a nontriggered scope, and so it cannot be used to make time and frequency measurements. However, calibrating the vertical amplifiers *is* possible, and when so calibrated, the nontriggered scope can be used to make peak-to-peak *ac* voltage measurements. If the oscilloscope uses direct-coupled (dc) vertical amplifiers, it is also possible to make dc voltage measurements with it. Such a scope is called an ac-dc type, and usually has an ac-dc switch with which to choose either one or the other mode of operation.

Oscilloscope Calibration

Experiment 13.2 demonstrated the use of the triggered scope for voltage measurements. The same method is used with a nontriggered scope. The student is referred to Exp. 13.2 for a review of the discussion on how an oscilloscope is used to measure voltages. But before the oscilloscope can be used for measuring voltage, its vertical amplifiers must be calibrated. A method for calibration is described below, a method you will follow in calibrating your oscilloscope in this experiment.

A nontriggered scope can be easily calibrated by proper setting of the two *vertical controls,* namely, the *vertical-gain selector* (or simply vertical selector) and the *vertical-gain vernier* (frequently labeled vertical gain). It is possible to adjust these controls for any desired volts-per-division setting within the capabilities of the oscilloscope.

The vertical-gain vernier is a continuously variable control, which in conjunction with the switch-type vertical selector can be set to a desired vertical sensitivity. For example, suppose a 1-V calibration signal is delivered to the input leads of an oscilloscope, and the vertical switch selector is set to the position labeled ×1. The vertical vernier can now be varied until the height of the signal on the screen is, say, 1 in. For these *specific settings* of the two input attenuator controls, the deflection sensitivity of the oscilloscope is now 1 V/in.

It may also be possible to find another position of the two controls for which the deflection sensitivity of the oscilloscope is, say, 3 V/in.

The vertical-gain selector, where found, is usually a calibrated divider or attenuator. The range "factor" is frequently 10. Thus, if the vertical-gain selector and vertical-gain vernier have been set for a vertical sensitivity of 10 V/in, switching to the next "range" lower (i.e., the range where less signal voltage is required to give a similar deflection) will result in a deflection sensitivity of 1 V/in. Switching back to the original range and then going one range higher will result in a deflection sensitivity of 100 V/in.

This characteristic of the vertical-gain selector makes it fairly simple to calibrate the scope for ac voltage measurements. As a matter of fact, it is possible to make almost all ac circuit voltage measurements by calibrating the scope at just two points, a deflection setting of 1 V/in and a deflection setting of 2 V/in.

To eliminate the need for using a tape measure to determine the height of the waveform, scope manufacturers use a graticule, an etched faceplate placed over the screen of the cathode-ray tube. The etching is in the form of graph markings (Fig. 13.4-1). The heavy horizontal and vertical lines are ½ or 1 in apart. The lighter graph lines divide the space between heavy lines into equidistant markings. Some etched faceplates use centimeter markings. In that case, calibrations are read in terms of volts per centimeter rather than volts per inch.

This graphed faceplate is useful for ac measurements. If the vertical-gain controls have been set for a deflection range of 10 V/in, and if there are five spacings in the inch, each spacing counts 2 V.

The procedure for using the nontriggered scope for ac measurements then requires that it be calibrated at several points, using a known voltage

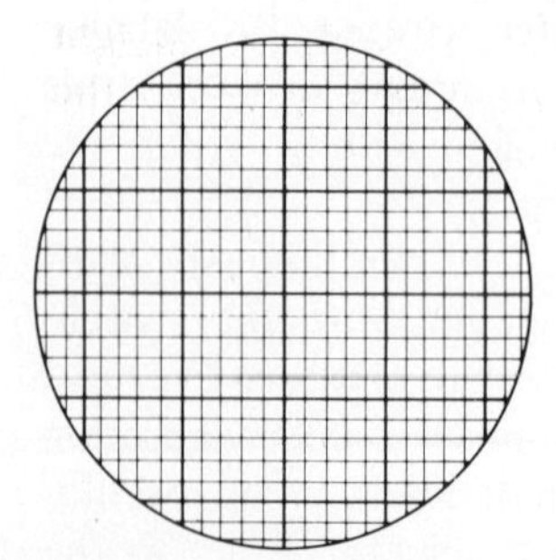

Fig. 13.4-1. Graticule on nontriggered-type oscilloscope measures 5 in in diameter.

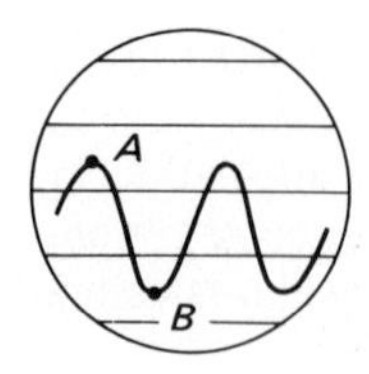

Fig. 13.4-2. Sine wave on scope screen.

source. The setting of the vertical-gain vernier and vertical-gain selector is noted for each of these calibrated points. The vertical-gain vernier is left at the determined setting and is not changed. However, the vertical selector may be switched from range to range. When it is switched, its divider factor must be considered in determining the volts-per-inch sensitivity. The unknown ac voltage to be measured is applied to the vertical input of the oscilloscope, and the height of the waveform deflection is measured. This measurement in inches or fraction of an inch is then converted into volts by multiplying inches by the V/in factor for which the scope has been calibrated.

For ease in measuring the height of the waveform, the horizontal trace may be narrowed or collapsed completely by setting the horizontal-gain control to zero. The horizontal and vertical centering controls may also be varied to orient the waveform. This facilitates vertical height measurements.

Figure 13.4-2 shows a sine wave whose peak-to-peak amplitude must be measured. Only the heavy horizontal lines, spaced 1 in apart, are shown. The scope has been calibrated for 10 V/in. The horizontal trace is collapsed so that only a vertical line is seen. The ends of this line correspond to the peaks of the waveform *A* and *B*. The line is then centered as in Fig. 13.4-3. It is evident now that it is 2 in high. Therefore, the voltage being measured is 20 V peak-to-peak.

NOTE: Scope calibration and measurement is usually peak-to-peak. The following is a sample calibration procedure. Assume a known voltage source of 25 V p-p for calibration and a range factor of 10 for the vertical-gain selector. The calibration voltage

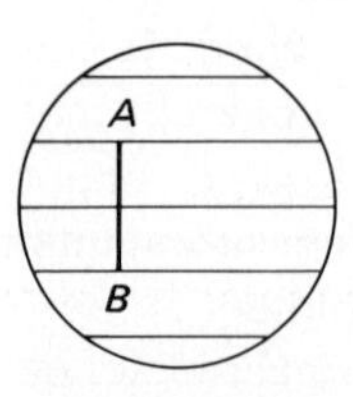

Fig. 13.4-3. Collapsing scope trace on nontriggered scope, to measure vertical waveform.

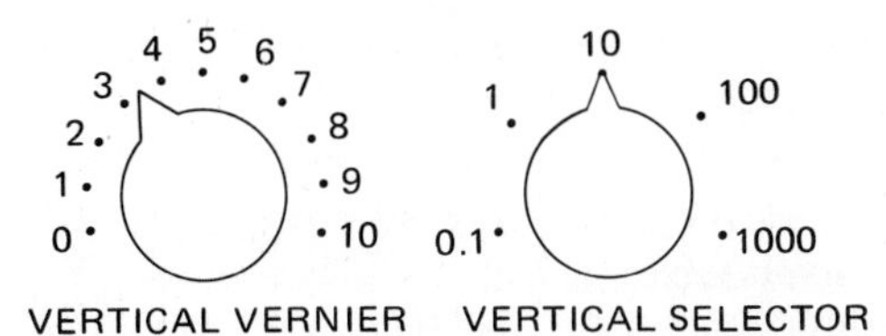

Fig. 13.4-4. Vertical attenuator (gain) controls on nontriggered scope.

is fed into the scope binding posts labeled "vertical input." Since it is desired to obtain a convenient V/in relationship, the vertical vernier and vertical-gain selector are set up for 2.5 in of vertical waveform deflection. The vertical-vernier (continuously variable) control is set at least 20 percent up from its minimum position when possible. The positions of the two controls are noted (see Fig. 13.4-4) and recorded for future use. See line 1, Table 13.4-1.

It is now a simple matter to complete the table for various settings of the vertical-gain selector. The vertical-gain selector is shifted one position and the effect upon the height of the waveform is noted. If it becomes larger, it indicates greater overall scope sensitivity. Thus, less signal voltage is required for the same 1 in of deflection. If this is the case, the new selector position and new sensitivity position are noted as in line 2 of the table. If the opposite effect is noted upon the height of the waveform, the table is filled in as in line 3. On all ranges, however, the vernier position remains unchanged, for this particular calibration.

Voltage Calibration Sources for Nontriggered Scopes

Internal. Most oscilloscopes provide a self-contained measured ac voltage source for calibration purposes. This is usually a sampling of the line voltage and may be either fixed or adjustable. Some oscilloscopes have facilities for automatically feeding a measured voltage to the vertical input terminals by means of a function switch set to a "calibrate" position. On other oscilloscopes, the voltage is available at a test jack on the front panel. If a variable voltage is provided, either it is metered or the value may be read from the calibrated dials associated with the "calibration" level control. Whatever the calibration voltage, its peak-to-peak value will be indicated on the oscilloscope panel, or in the oscilloscope instruction manual.

External. For nontriggered oscilloscopes which do not have self-contained facilities, external voltage calibrators are available. These are line-derived adjustable voltage sources. The peak-to-peak output voltage level may be read on a self-contained meter or from calibrated dials. The output of the voltage calibrator is fed to the vertical input of the oscilloscope for calibration.

If an external voltage calibrator is not available, a sine-wave generator and an accurately calibrated ac voltmeter can be used. The sine-wave generator frequency control is set at 1000 Hz, and the output control is set for a specified rms voltage as measured by the ac voltmeter used here as a standard. The rms value is then converted to peak-to-peak by multiplying by 2.82. This measured signal is fed to the input leads of the nontriggered scope. Scope calibration is then accomplished as previously described.

SUMMARY

1. Either a triggered or a nontriggered oscilloscope can be used for ac voltage measurements.
2. Triggered oscilloscopes contain calibrated signal amplifiers. These are used directly in the measurement of ac voltage waveforms and dc voltages.
3. Nontriggered oscilloscopes are not calibrated and must be calibrated if they are to be used as voltmeters.
4. An accurate, known voltage source, whether provided by the oscilloscope calibrator or by an external calibrator, is used as an input to the nontriggered scope.
5. The vertical-gain selector of the nontriggered scope is switched to a desired setting, and the gain vernier is then adjusted to the desired height for the required V/in factor. The position of the

TABLE 13.4-1. Vertical-Gain Controls and Scope Calibration

Known Calibrating Voltage	*Deflection, in*	*Vertical Setting*		*Volts/in, A/B*
		Vernier	*Selector*	
25	2.5	3.4	10	10
		3.4	1	1
		3.4	100	100

vernier is marked on the oscilloscope. It must not be moved from this calibrated setting. However, the vertical-gain selector can be switched to its various positions, and it acts as a range multiplier. The scope is now calibrated for signal-voltage measurements.
6. The signal voltage is then applied to the vertical input of the calibrated triggered or nontriggered oscilloscope. The vertical height of the signal is measured and multiplied by the volts/cm or V/in factor to which the oscilloscope is set. The result is the required voltage of the signal waveform being measured.
7. Oscilloscope signal amplifiers are normally calibrated to make peak-to-peak measurements.
8. There are instruments called voltage calibrators which may be used as external calibration sources. They deliver a measured signal voltage for calibration purposes.
9. For the nontriggered scope, if neither an internal nor an external calibrator is available, an ordinary sine-wave generator may be used, in conjunction with an accurate ac voltmeter. A measured 1000-Hz signal can be used as the calibration source.

SELF-TEST

Check your understanding by answering these questions.

1. Nontriggered-type oscilloscopes ________ (do/do not) have calibrated vertical amplifiers.
2. A nontriggered scope has been calibrated for 10 V/in on the ×10 range of the vertical-gain selector. At this setting the height of a waveform is 3.6 in. The voltage of the waveform is ________ V.
3. For the oscilloscope in question 2, the position of the vernier control is kept at the same point as it was when calibrated, but the vertical-gain selector is switched to the ×100 position. The height of a waveform on the screen is 1.5 in. The waveform voltage is ________ V.
4. Calibration voltages are available on most oscilloscopes, whether they are triggered or nontriggered. ________ (true/false)
5. If an oscilloscope does not have self-contained facilities for voltage calibration, an ________ ________ may be used.

MATERIALS REQUIRED

- Power supply: Known ac voltage source of 18 V p-p; fused line cord
- Equipment: Oscilloscope with direct probe, AF signal generator, electronic voltmeter or VOM
- Resistors: ½-W 5100-, 10,000-, and 15,000-Ω

PROCEDURE

Calibrating Oscilloscope with Noncalibrated Vertical Amplifiers

Calibration Using Internal Calibrator

1. Adjust oscilloscope controls for viewing the calibration voltage.
2. Apply the calibration voltage to the input terminals and adjust the controls for a display of two cycles. Set the calibration level controls (if any) and the vertical-gain selector and vertical-gain vernier so that the deflection sensitivity of the oscilloscope is 1 V/in on the ×1 vertical range. In Table 13.4-2 record the level of the calibration voltage and the settings of the vertical-gain selector and vernier for 1 V/in. If the vernier does not have a numbered scale, mark its position with a pencil.

Next, adjust the calibration level controls and vertical-gain selector and vernier for a deflection

TABLE 13.4-2. Oscilloscope Calibration

Calibration Source	*Calibration Voltage Level*	V/in	*Settings of Vertical*	
			Gain Selector	*Gain Vernier*
Internal		1		
		2		
External		1		
		2		

sensitivity of 2 V/in. In Table 13.4-2 record the level of the calibration voltage and the settings of the vertical-gain selector and vernier for 2 V/in. If the vernier does not have a numbered scale, mark its position.

Calibration Using External Calibrator, or AF Sine-Wave Generator and AC Voltmeter

NOTE: If an external voltage calibrator is used, set its output controls first for 1 V, then for 2 V as required in Table 13.4-2. If an AF sine-wave generator is used, set its output controls so that the 1000-Hz sine wave as measured on an ac voltmeter is first 0.354 V rms (corresponding to 1 V peak-to-peak), then 0.708 V rms (corresponding to 2 V peak-to-peak).

3. Repeat steps 1 and 2, recording your results in Table 13.4-2. Then proceed to step 4.

Measuring AC Voltages with Oscilloscope

4. Remove the vertical input cables from the calibrating voltage source and apply 18 V peak-to-peak (as measured with an ac voltmeter) across the circuit of Fig. 13.4-5. Using the oscilloscope calibrated at 2 V/div., measure, and record in Table 13.4-3, the peak-to-peak voltages from *A* to *G*, *B* to *G*, and *C* to *G*. Compute, and record in Table 13.4-3, the peak-to-peak voltages appearing at the measured points. Show your computations.

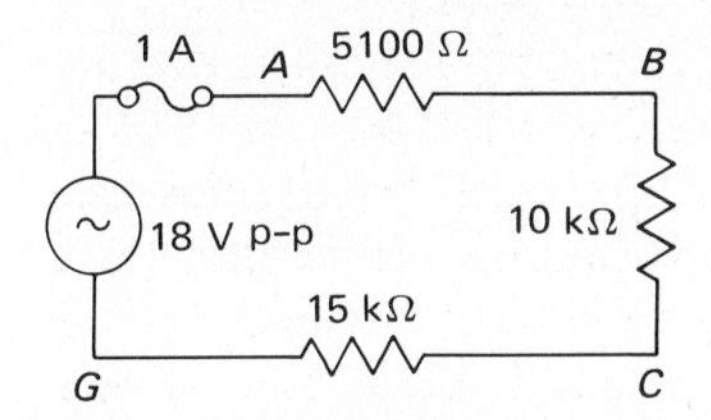

Fig. 13.4-5. AC voltage across divider network.

5. Reset the vertical controls for a deflection sensitivity of 1 V/div. Again measure and record in Table 13.4-3 the voltages at the indicated test points.
6. With the oscilloscope calibrated at 1 V/div., measure, and record in Table 13.4-3, the maximum signal voltage delivered by the audio-signal generator set at 1000 Hz. If possible, also measure and record the minimum output of the AF generator at 1000 Hz.

Measuring DC Voltages with Oscilloscope

7. If you are using an ac-dc oscilloscope, switch to dc. Remove the 18-V ac source from Fig. 13.4-5, and connect a dc voltage source across the points *AG* of Fig. 13.4-5. Set the output of the dc supply at 6 V, as measured with an EVM. You now have a dc voltage divider. Set the calibrated vertical controls on your scope at 1 V/div.
8. Center the trace vertically on the screen. This is the zero reference line. With the oscilloscope measure and record in Table 13.4-4 the dc voltage across each set of test points listed. Now measure the same voltages with an EVM, and record your measurements in the appropriate column of Table 13.4-4.

QUESTIONS

1. (*a*) Does your oscilloscope have calibrated or noncalibrated vertical amplifiers? (*b*) Is it a triggered or nontriggered scope? (*c*) Is it a wideband scope?

TABLE 13.4-3. Voltage Measurements

Test Points	Computed Volts Peak-to-Peak	Measured Volts Peak-to-Peak: Scope Calibrated 1 V/div.	Measured Volts Peak-to-Peak: Scope Calibrated 2 V/div.
A to *G*			
B to *G*			
C to *G*			

Audio Oscillator Output	Maximum Volts Peak-to-Peak	Minimum Volts Peak-to-Peak

TABLE 13.4-4. DC Voltage Measurements

	Voltage Measured, V	
Test Points	*Oscilloscope*	*Voltmeter*
A to *G*		
B to *G*		
C to *G*		

2. What are the markings (settings or calibrations) of the vertical-gain selector found on your scope? Which of these is the most sensitive, that is, which requires least voltage for 1 division (cm or in) of vertical deflection?
3. How do the computed and measured values in Table 13.4-3 compare for:
   (*a*) Scope calibration of 1 V/div.?
   (*b*) Scope calibration of 2 V/div.?
4. What are the disadvantages of using the scope as an ac voltmeter? Advantages?
5. If you were unable to get a measurement of the minimum-level signal voltage delivered by the audio-signal generator, explain why. Can you suggest a method for measuring the minimum-level voltage?
6. How would you calibrate the oscilloscope for rms measurements?
7. Which provided a more accurate dc voltage measurement, the scope or the voltmeter? Justify your answer by referring to your readings in Table 13.4-4 and to the computed voltages.

Answers to Self-Test

1. do not
2. 36
3. 150
4. true
5. external calibrator

CHAPTER 14 RMS AND PEAK-TO-PEAK AC VALUES

EXPERIMENT 14.1. RMS and Peak-to-Peak AC Values

OBJECTIVE

To verify experimentally the relationship between rms and peak-to-peak values of a sinusoidal (sine wave) voltage

INTRODUCTORY INFORMATION

Alternating Voltages and Currents

Direct current and voltage (dc) sources do not change their polarity, and they maintain a constant output level for a particular load. The direction of the current does not change. The output of alternating voltage and current (ac) power sources is continually changing. Moreover, the polarity of the voltage, and the direction of ac load current, changes periodically.

Since there is such a difference between alternating and direct current, the question might naturally be asked, why bother with alternating current? Why not just use direct current? The answer is that alternating current is easy to generate and is less expensive to transmit to the user, and can be used just as effectively as direct current.

How is alternating current generated? It is beyond the scope of this book to discuss in detail the methods of generating or the mathematics of alternating current. The reader is referred to a text on electricity for a more detailed discussion. It will be sufficient to say that the simplest ac generator consists of a one-loop armature rotating at a constant velocity in a uniform magnetic field. The terminals of the single loop of wire make contact with slip rings, as in Fig. 14.1-1. As the armature rotates through a uniform magnetic field, an alternating voltage is induced in the wire. Carbon brushes bring the ac voltage to the load. Alternating current flows in the load.

Of course, a practical *ac* generator consists of thousands of windings of wire around a slotted armature. Also there are many magnetic poles through which the armature moves. However, the single-loop armature is sufficient for purposes of illustration.

One fact should be clarified in connection with this discussion: Energy is used to rotate the armature. This energy can come from moving water flowing over a dam, turning the armature, or it can be the pressure of steam. Or a gasoline motor may be used to turn the armature. Whatever the source of energy used, it is converted by the generator to electrical energy in the form of alternating current.

AC Waveform

Figure 14.1-2 is a graph of the ac voltage induced in the one-loop armature. The graph shows how the

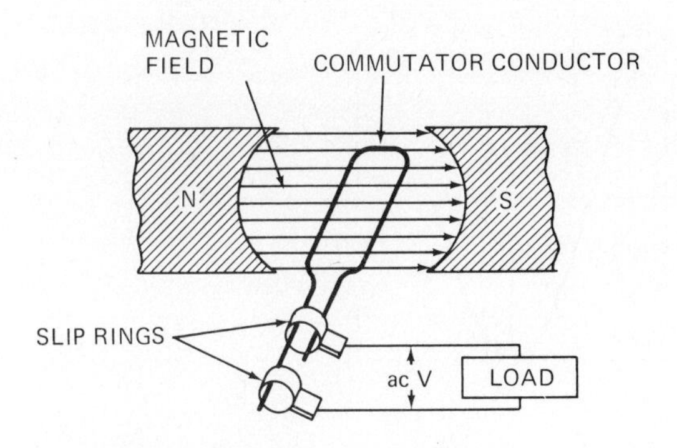

Fig. 14.1-1. Elementary ac generator.

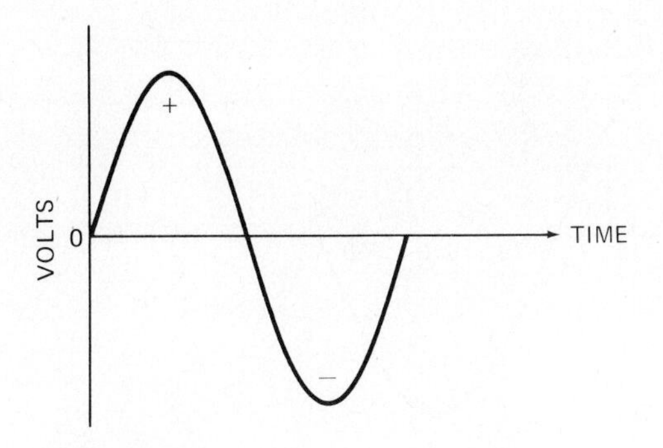

Fig. 14.1-2. One cycle of ac voltage induced in armature of one-loop generator.

output voltage varies with time. The graph is a plot of instantaneous values of voltage induced in the loop with respect to the time of rotation of the armature as it turns through the magnetic field. The waveform of voltage is called a sine wave, after the mathematical (trigonometric) function, the sine of an angle. Scientists have worked out ac mathematics, using trigonometry and algebra. But we shall not be concerned with ac mathematics here.

Observe in Fig. 14.1-2 that the ac voltage waveform rises from 0 V to a peak positive (+) value, then falls back to zero. This is the positive *alternation* of the waveform. The voltage then rises to a peak negative (−) value, and again falls back to zero. This is the negative alternation of the waveform. These two alternations make up *one cycle* of the wave. The time it takes to generate one cycle of the waveform is called the period (t) of the wave, and t is the time it takes the one-loop armature to turn completely through 360°. For example, if it takes the armature 1 s to complete a revolution, then $t = 1$ s. However, if the armature makes 60 revolutions per second, then $t = 1/60$ s.

Figure 14.1-3 is another way in which the graph of an alternating voltage is represented. The vertical is the *voltage* axis, the horizontal is the degree axis; that is, the graph is a plot of instantaneous values of voltage versus degrees of rotation of the armature. So at 0° the voltage induced in the armature loop is 0 V. It rises to maximum positive level at 90° and falls back again to zero at 180°. Between 180° and 360° the first alternation is repeated, but the output voltage is negative. So, between 180 and 270° (a 90° difference) the voltage rises to maximum negative value, and between 270 and 360° it falls back to 0 V. This completes one cycle of this periodic waveform. It is evident that degrees of rotation can be easily converted to the *time* that it takes the armature to move through those *degrees*. So Figs. 14.1-2 and 14.1-3 really represent the same ac output but in a slightly different way.

An oscilloscope can be used to measure the *peak* value of each alternation. Note that both positive and negative peak voltages are the same in value but opposite in polarity. Or the oscilloscope can measure the peak-to-peak voltage of the waveform, as in Fig. 14.1-4. At any moment in time it is also possible to measure the instantaneous voltage value of the ac waveform with an oscilloscope.

Effective (rms) versus Peak AC Values

The graph of an alternating voltage or current simply shows the way in which the voltage is varying instantaneously with respect to time (or degrees of rotation). It is not useful for measuring purposes because an ac voltmeter cannot follow these rapid changes. Some equivalence must be determined between the *effect* of alternating and direct current in a circuit. For example, what level of alternating current flowing in some fixed value resistor R, for a specified time t, will cause the same heating effect in the resistor as direct current I, flowing through the same resistor, in the same period of time? If this equivalence can be established, then ac volt and current meters can be calibrated to read effective or equivalent dc values. This has been accomplished.

It was found that alternating current is equivalent in heating effect to direct current which is 0.707 times the peak of the ac waveform, as shown in Fig. 14.1-5. Putting it in a formula, we have

$$\text{Effective (rms) value of ac} = 0.707 \times \text{peak value} \tag{14.1-1}$$

So, if the *peak* value of an ac voltage is 100 V, its *effective* (or equivalent dc) value is 70.7 V. Now, if we know or have measured with an ac voltmeter the *rms* value of an ac voltage, we can easily convert it to peak

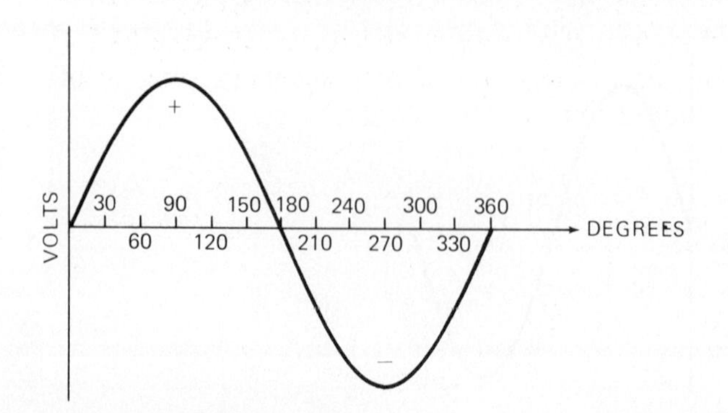

Fig. 14.1-3. One cycle of alternating voltage. Here voltage is graphed versus degrees of rotation of the armature.

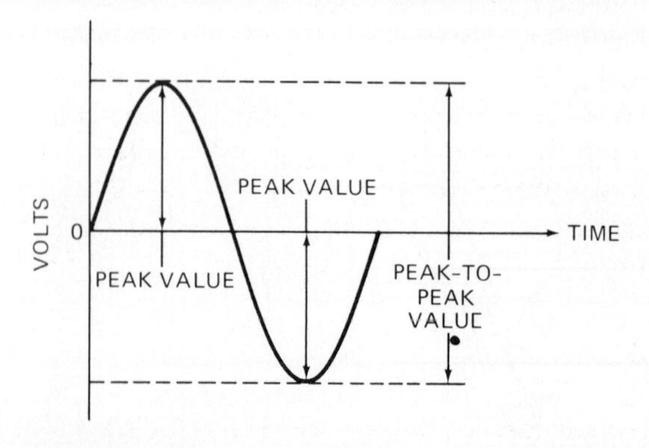

Fig. 14.1-4. Peak-to-peak voltage of an ac waveform measured by oscilloscope.

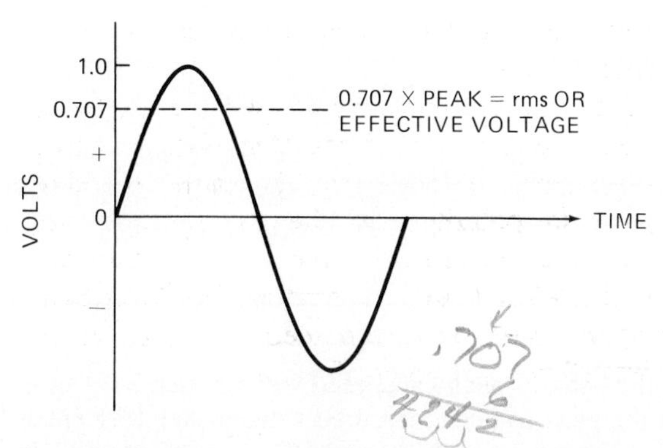

Fig. 14.1-5. RMS or effective value of alternating voltage or current is 0.707 peak value.

value. For the example just given, to change an *rms* value of 70.7 V to its peak value, we find we must multiply 70.7 by 1.414 to get back to the peak value of 100 V.

The two formulas can be given as follows:

$$\begin{aligned} \text{Peak} &= 1.414 \times \text{rms} \\ \text{rms} &= 0.707 \times \text{peak} \end{aligned} \qquad (14.1\text{-}2)$$

Oscilloscopes are usually calibrated to measure peak-to-peak voltages. In the case of the ac sine wave, the peak-to-peak value is *twice* the peak value. So to change peak-to-peak values to peak values, we must divide by 2. Also to change peak values to peak-to-peak values, we multiply by 2. This gives us two other simple formulas:

$$\begin{aligned} \text{Peak} &= \frac{\text{peak-to-peak}}{2} \\ \text{Peak-to-peak} &= \text{peak} \times 2 \end{aligned} \qquad (14.1\text{-}3)$$

Now, if it is necessary to change oscilloscope peak-to-peak values to rms values, divide the peak-to-peak by 2 [Eq. (14.1-3)], then multiply the peak value by 0.707 [Eq. (14.1-2)].

The scales of an ac voltmeter (or current meter) are calibrated to read the *rms* value of ac sine-wave voltage or current, respectively. There may also be a separate scale on the meter calibrated in peak-to-peak values. If there is not, peak-to-peak values can be found from the rms values by multiplying the rms value by 1.414 to give the peak value [Eq. (14.1-2)], then multiplying the peak value by 2 to give the peak-to-peak value [Eq. (14.1-3)].

SUMMARY

1. Oscilloscopes are calibrated to measure peak-to-peak ac voltages.
2. AC voltmeters and current meters are calibrated to measure rms (effective) values of sine-wave voltage or current.
3. Peak-to-peak values can be converted to peak values by dividing by 2.
4. Peak voltage or current can be converted to rms values by multiplying by 0.707.
5. RMS values can be converted to peak values by multiplying by 1.414.

SELF-TEST

Check your understanding by answering these questions.

1. Which will produce more heat, 120 V direct current or 120 V (rms) alternating current? ________
2. How much peak alternating current will be required to produce in a bulb as much light as 24 V direct current? ________ V.
3. The oscilloscope is calibrated to measure ______ ac voltages.
4. The ac voltmeter is calibrated to measure ______ ac voltages.
5. A sine wave of voltage is induced in a one-loop armature when the armature of the generator rotates through ________ degrees, moving at a constant velocity in a uniform magnetic field.
6. 24 V p-p is equal to ________ V rms.

MATERIALS REQUIRED

- Equipment: Oscilloscope; EVM; AF sine-wave generator
- Miscellaneous: 6-V filament transformer; fused line cord; SPST switch

PROCEDURE

Measuring Line-Derived (60 Hz) Voltages

1. Connect the circuit shown in Fig. 14.1-6.
2. With an EVM measure and record in Table 14.1-1 the rms voltage across the secondary winding of the transformer.

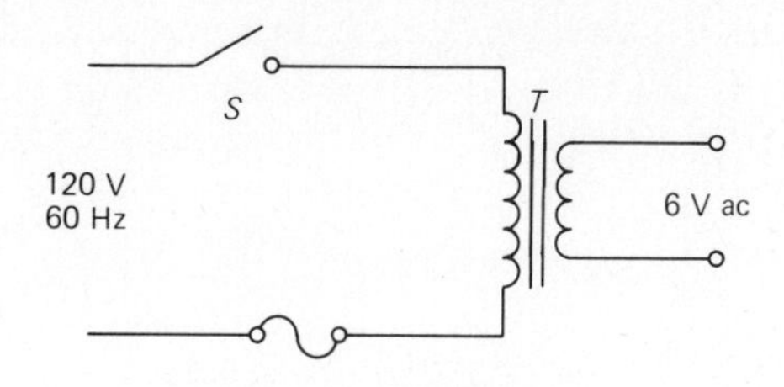

Fig. 14.1-6. Experimental circuit.

TABLE 14.1-1. RMS and Peak-to-Peak AC Voltages

	Measured Voltage		*Calculated Voltage*	
Steps	*EVM (rms)*	*Scope Peak-to-Peak*	*Peak*	*Peak-to-Peak*
2–4				
6–8				

3. Calculate and record in Table 14.1-1 the (*a*) peak, and (*b*) peak-to-peak voltage across the secondary of the transformer. Show your calculations.
4. With an oscilloscope observe the waveform and measure and record in Table 14.1-1 the peak-to-peak voltage across the secondary winding.

Measuring AF Generator Output

5. Set the AF sine-wave generator frequency at 1000 Hz, maximum voltage output.
6. With an EVM measure, and record in Table 14.1-1, the rms voltage at the output terminals of the generator.
7. Calculate, and record in Table 14.1-1, the (*a*) peak, and (*b*) peak-to-peak output voltage of the generator.
8. With an oscilloscope observe the 1000-Hz waveform and measure, and record in Table 14.1-1, its peak-to-peak output voltage.

QUESTIONS

1. Did the peak-to-peak calculated value in step 3 (*b*) agree with the peak-to-peak measured value in step 4? Explain the reasons for any differences.
2. Did the peak-to-peak calculated value in step 7 (*b*) agree with the peak-to-peak measured value in step 8? Explain the reason for any difference.
3. (*a*) What is the result of dividing the peak-to-peak value measured in step 4 by the rms value measured in step 2? What does this show, if anything?
   (*b*) What is the result of dividing the peak-to-peak value measured in step 8 by the rms value measured in step 6?
4. What is the range of frequencies of sinusoidal voltages which your EVM can measure accurately on the ac voltage range?
5. How is the secondary winding of the transformer you used rated, in rms or peak-to-peak values?
6. A manufacturer's schematic diagram for a particular device indicates that there should be 3.5 V ac across a specific resistor, R_{14}. When measured with a scope, the voltage is almost 10 V. Assume that the device is operating normally. Is there a difference between the measured and rated values? Explain.

Answers to Self-Test

1. Neither; the heat produced is the same.
2. 34 (approximately)
3. peak-to-peak
4. rms or effective
5. 360°
6. 8.5 (approximately)

CHAPTER 15 CAPACITOR TESTING

EXPERIMENT 15.1. Capacitor Testing by Capacitor Checker

OBJECTIVE

To measure the capacitance, leakage, and power factor of a capacitor

INTRODUCTORY INFORMATION

Capacitance Testers

Different types of capacitance testers are available to the technician. There is the tester which checks only the capacitance of a capacitor. This may be used to check the rated value of a capacitor or to determine the capacitance of a capacitor whose value is unknown. This type of tester will also detect open and shorted capacitors.

Other types, in addition to acting as capacitance bridges, also determine all or some of the following capacitor characteristics.

1. *Insulation Resistance for Paper, Mica, and Ceramic Capacitors.* Insulation resistance R_I is effectively a resistance in parallel with the capacitor. The value of insulation resistance considered normal varies with the type of capacitor. This characteristic is considered only for electrostatic and not electrolytic capacitors. Approximate values of insulation resistance are shown below.
   a. *Molded Paper Capacitors.* Insulation resistance R_I is inversely proportional to capacitance. The product of $R_I \times \mu F$ should equal about 1000 $M\Omega \cdot \mu F$ when new, depending on the ambient temperature.
   b. *Mica Capacitors.* When new, these will have a value R_I greater than 3000 $M\Omega$.
   c. *Ceramic Capacitors.* When new, these will have a value R_I greater than 7500 $M\Omega$.
2. *Leakage Current of Electrolytic Capacitors.* This test for electrolytic capacitors takes the place of the R_I test for electrostatic capacitors. Leakage current I is directly proportional to capacitance. Maximum leakage current permissible for electrolytics is usually included in the instruction manual of the capacitor checker. Values of leakage current are read directly on the calibrated meter scale of the tester.
3. *Power Factor of Electrolytic Capacitors.* The power factor (PF) of an electrolytic capacitor is a measure of the power losses in that capacitor. The losses occurring in a capacitor may be indicated as a resistance R in series with the capacitor. Acceptable PF values for new electrolytic capacitors are usually given in the instruction manual accompanying the capacitance tester.

NOTE: A capacitor should be replaced if any of its measured characteristics do not come within the range of acceptable norms.

For proper testing, the entire capacitor, or at least one lead of the capacitor, should be disconnected from the circuit. An exception to the rule is for a tester on the market which checks shorts and opens of capacitors wired in the circuit.

The residual capacitance of a checker should be determined before it is used to measure the values of very small capacitances. Moreover, for accurate readings the pigtails of the capacitor should be placed directly across the terminals of the checker. Long clip leads should be avoided, since their capacitance will add to that of the capacitor under test.

The student or technician should be fully familiar with the operating instructions of the capacitance checker in use in order to derive full benefit from it and, moreover, should perform all the checks which the instrument is capable of making. Figure 15.1-1 shows a typical service-type capacitor checker.

SUMMARY

1. Many types of capacitance checkers are available.
2. At least one lead of a capacitor under test must be removed from the circuit.
3. The residual capacitance of the capacitance

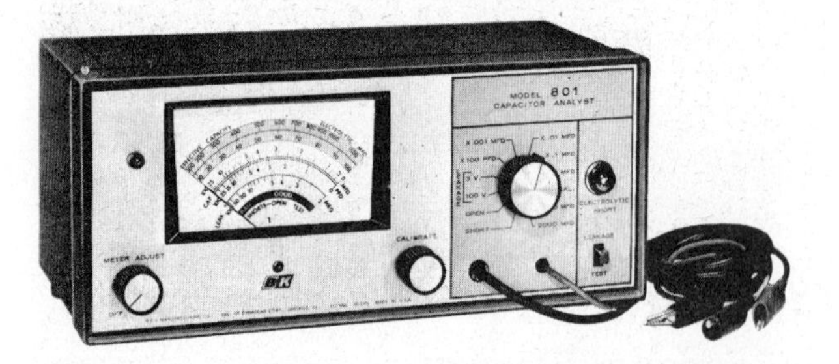

Fig. 15.1-1. Typical service-type capacitance checker. (*B and K Dynascan Corporation*)

checker must be known if accurate measurements are to be made of small-value capacitors.

4. Long test leads on capacitor testers should be avoided, because the capacitance of the leads may add significantly to the capacitance measurement of smaller-value capacitors.

SELF-TEST

Check your understanding by answering these questions.

1. The more inexpensive capacitance checker may check the rated value of the capacitor and may be used to detect shorts and ________.
2. More expensive capacitance checkers may allow testing of insulation resistance, leakage current of electrolytics, and ________.
3. It is generally not necessary to remove a capacitor from the circuit for testing, if one lead can be removed. ________ (true/false)
4. ________ of the capacitance checker must be known if accurate measurement of small-value capacitors is to be made.

MATERIALS REQUIRED

- Equipment: Capacitance checker
- Capacitors: Assortment of paper, mica, ceramic, and electrolytic

PROCEDURE

1. Determine the value of each capacitor supplied, from its color code (see Appendix). Fill in the information required in Table 15.1-1.
2. Refer to the instruction manual of the capacitance checker for procedure to measure capacitance values.
3. Determine and record the residual capacitance of the checker.

TABLE 15.1-1. Testing Capacitors with Capacitance Checker

	Capacitor					
Type	1	2	3	4	5	6
1st color						
2d color						
3d color						
4th color						
Coded value, μF						
Tolerance, %						
Measured value						
Insulation resistance						
Leakage current						
Power factor						

4. Measure each capacitor and record the results in Table 15.1-1 in the row labeled "Measured value." If the capacitor is very small, compensate for the residual capacitance of the checker by subtracting residual capacitance from measured value.
5. If the checker has facilities, measure the insulation resistance of the molded mica and ceramic capacitors supplied, and record your results in Table 15.1-1.
6. If the checker has facilities, measure the leakage current and power factor of electrolytic capacitors supplied, and record the results in Table 15.1-1.

QUESTIONS

1. What is meant by the residual capacitance of a capacitance checker?
2. How did the measured values of the capacitors in Table 15.1-1 compare with the rated values? Explain any discrepancies.
3. What indication is there on your capacitance checker of an open capacitor? A shorted capacitor?
4. When is it necessary to consider the residual capacitance of the capacitor checker?

Answers to Self-Test

1. opens
2. power factor of electrolytics
3. true
4. Residual capacitance

CHAPTER 16 *RC* TIME CONSTANTS

EXPERIMENT 16.1. *RC* Time Constants — Charging a Capacitor

OBJECTIVE

To determine experimentally the time it takes a capacitor to charge through a resistance

INTRODUCTORY INFORMATION

Charging a Capacitor

In electronic circuits, capacitors are used for many purposes. They are employed to store energy, to pass alternating current, and to block direct current. They act as filter elements and as components in "tuned" or resonant circuits, they are used in timing circuits, and they have many other functions.

Capacitors carry out their function by charging and discharging. A capacitor can store and hold a charge of electrons, a process known as charging. To charge a capacitor a voltage source is required, V in Fig. 16.1-1*a*. When switch S is closed, and the circuit is completed (Fig. 16.1-1*b*), electrons leave the negative terminal of the battery and enter the lower plate of capacitor C. At the same time, electrons leave the upper plate of C and return to the positive terminal of the battery. This process continues until the capacitor is fully charged, which occurs when the voltage across the capacitor equals the applied voltage V. When charged, the lower plate contains an excess of electrons and the upper plate a deficiency of electrons. The quantity of excess electrons on the lower plate is exactly equal to the quantity of electrons which left the upper plate. The voltage built up on the capacitor results from this difference in electron charge between the upper and lower plates. The polarity of charge on the capacitor plates is as shown in Fig. 16.1-1*b*. If, after charging, C is removed from the voltage source, as in Fig. 16.1-1*c*, it will continue to hold its charge as long as there is no complete circuit through which it can discharge.

The relationship between the charge Q on a capacitor, the size C of capacitance, and the voltage V across C is expressed by

$$Q = C \times V \tag{16.1-1}$$

In this formula Q is in coulombs, C in farads, and V in volts. It is apparent from Eq. (16.1-1) that the larger C is, the more coulombs it will store when fully charged from a given voltage source V.

Time Required to Charge a Capacitor

Consider the circuit of Fig. 16.1-2. We see a series circuit consisting of a capacitor C, a resistor R, a dc voltage source V, and an ON-OFF switch S_1. A second switch S_2 is connected across the RC combination. Both S_1 and S_2 are open.

Assume that capacitor C has no charge on it. Now close S_1, leaving S_2 open, as in Fig. 16.1-3*a*. Closing S_1 completes the charging circuit for C. If a voltmeter were connected across C, the meter would read zero volts at the instant S_1 is closed. As C charges, the meter reads an increasing voltage. This experiment simply demonstrates that a capacitor does not charge

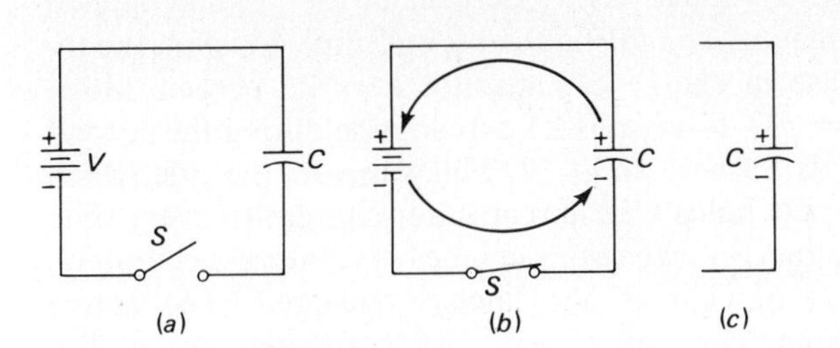

Fig. 16.1-1. Charging a capacitor.

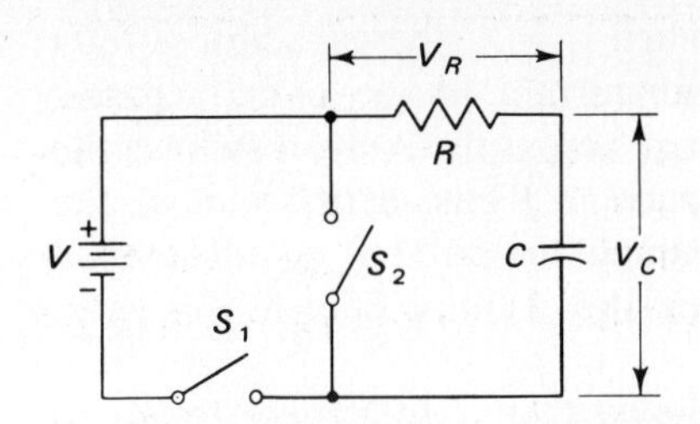

Fig. 16.1-2. *RC* circuit for charging a capacitor.

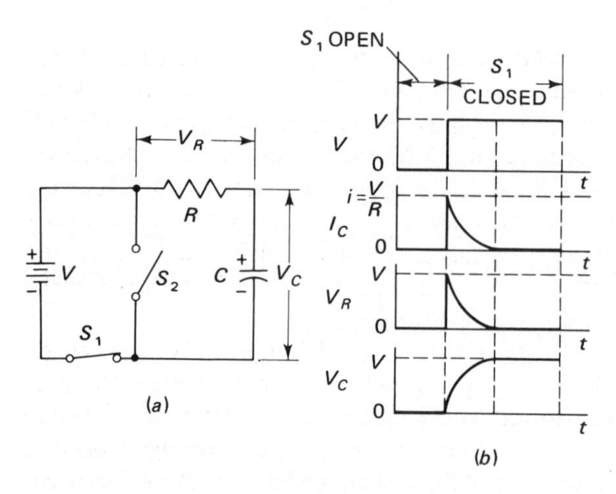

Fig. 16.1-3. Capacitor C charging in an RC circuit.

instantaneously. It takes *time* to charge it. Our concern here is with the question, "How much time does it take to charge a capacitor?"

The manner in which C charges is shown in the accompanying graph (Fig. 16.1-3*b*). While S_1 was open, C did not charge because the battery voltage was not applied across the RC circuit. Thus, with S_1 open, the applied voltage $V = 0$, the charging current $i_c = 0$, the voltage across R, $v_R = 0$, and the voltage across C, $v_c = 0$. At the *instant* switch S_1 is closed, the battery "sees" only the resistance R in the circuit. That is, C acts like a short circuit, because it offers no opposition to current. Hence, all the applied voltage V appears across R (at the instant S_1 is closed). There is maximum charging current limited only by the size of R, and there is zero voltage across C. But as S_1 remains closed, charging current causes a voltage to build up across C, with the polarity shown in Fig. 16.1-3*a*. This is a "bucking" voltage which opposes the battery voltage and acts to reduce the charging current.

At any instant of time after the start of the charging interval, the voltage which causes current to flow in the circuit is the voltage source V minus the voltage v_c to which the capacitor has already charged. If we call this the active voltage V_A, then

$$V_A = V - v_c \qquad (16.1\text{-}2)$$

It is apparent that when $v_c = V$, the capacitor is fully charged, $V_A = 0$, and current in the circuit ceases.

The decay in voltage across the resistor follows the decay of charging current. Thus, at the end of the charging interval, though the battery is still applied across the RC circuit, the charging current i_c is zero, $v_R = 0$, and $v_c = V$.

Figure 16.1-3*b* shows that it took time for C to charge. From a mathematical analysis of this circuit, which is beyond the scope of this book, we find that it takes one time constant to charge C to 63.2 (approximately) percent of the applied voltage. The time required to charge a capacitor to 63.2 percent of the applied voltage is known as the *time constant* of the circuit. The value of the time constant t in seconds is equal to the product of the resistance R in ohms and the capacitance of C in farads. Thus

$$t = R \times C \qquad (16.1\text{-}3)$$

A "time constant" is not a fixed unit of time such as a second, but is a *relative* unit of time whose actual value in seconds is determined by the values of R and C in the charging circuit. Thus, in Fig. 16.1-3, if $R = 1\ \text{M}\Omega$ and $C = 1\ \mu\text{F}$, $t = 1$ s. But if $R = 100{,}000\ \Omega$ and $C = 0.1\ \mu\text{F}$, then $t = 0.01$ s.

The student may find the following relationships a useful reference in calculating time constants:

R (in ohms) $\times$ C (in farads) $= t$ (in seconds)
R (in megohms) $\times$ C (in microfarads) $= t$ (in seconds)
R (in ohms) $\times$ C (in microfarads) $= t$ (in microseconds)
R (in megohms) $\times$ C (in picofarads) $= t$ (in microseconds)

Charge Rate of a Capacitor

An accurate graph of the rise of voltage across a capacitor which is charging is given in the "charge" graph of Fig. 16.1-4. This shows how the voltage rises across a capacitor C charging through a resistance R toward a voltage source V. The horizontal axis is the time axis set off in time constants (RC). The vertical axis is a percentage axis, with 100 percent representing the total voltage to which a capacitor can charge.

Examination of this graph shows that capacitor C charges approximately 63 percent in one time constant, 86 percent in two time constants, 95 percent in three time constants, 98 percent in four time constants, 99 percent (practically 100 percent) in five time constants. Let us consider these figures further. We note that at the end of the first time constant the "active voltage" is 37 percent of the original applied voltage. Also, during the second time constant the increase in charge of capacitor C is 23 percent ($86 - 63 = 23$). Now, this 23 percent rise during the second RC is 63 percent of 37. This suggests the rule which actually holds true for capacitor charge: In every time constant, a capacitor charges 63 (more accurately, 63.2) percent of the "active voltage." The active voltage can, of course, be computed using Eq. (16.1-2).

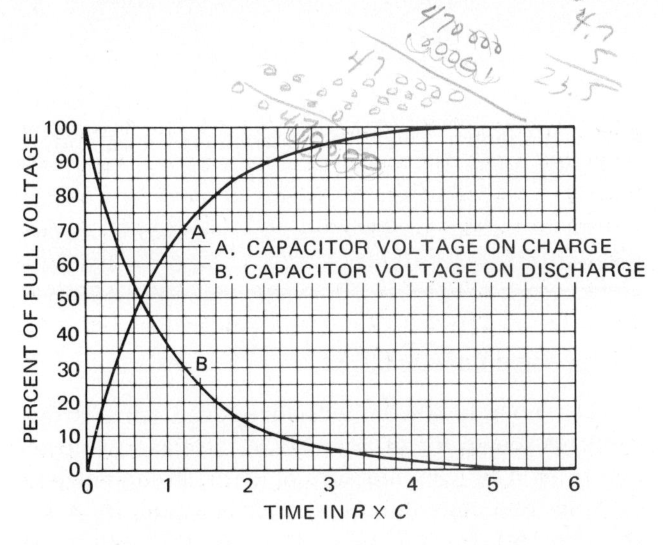

Fig. 16.1-4. Universal chart for charge and discharge of a capacitor.

Though a capacitor can theoretically "never" charge fully to the applied voltage, since there is always a remainder of 37 percent of the active voltage, to all intents and purposes we assume that C is fully charged at the end of five time constants.

Experimental Techniques in Observing Capacitor Charge

We can set up an experiment using simple equipment to demonstrate how long it takes a capacitor to charge. However, our results will be approximate because we shall not be using laboratory precision timing or measuring equipment.

Refer to Fig. 16.1-5*a*. By selecting the values of R and C, we can find the length of time it takes C to charge and discharge, using an ordinary watch with a seconds hand. In determining the charge time for C we use a large nonelectrolytic capacitor, say 1 μF, a 1-MΩ resistor, an EVM, a variable dc voltage source, and a watch. When we close switches S_1 and S_2, C will start charging toward the dc supply voltage. The EVM will read v_c, the voltage across C. We shall find that in approximately 5 RCs (5 s in this case) the capacitor will be fully charged.

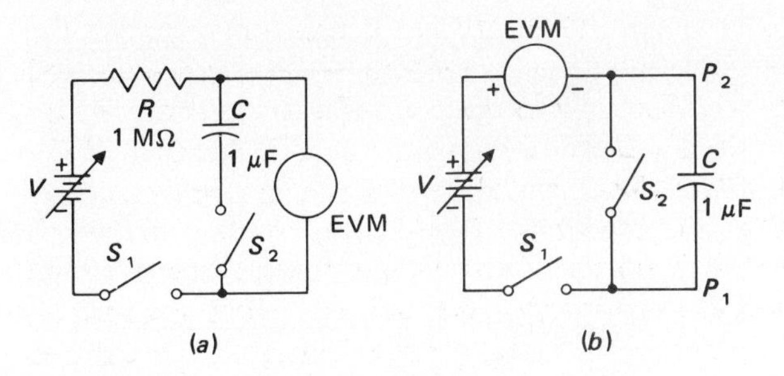

Fig. 16.1-5. Experimental circuit for the charge and discharge of a capacitor.

We should note here that C will not actually charge up to the total applied voltage, but to a value determined by the voltage divider, consisting of the internal resistance of the EVM and the value of R. Thus, if we use an EVM with an input resistance of 11 MΩ and if R is a 1-MΩ resistor, C will charge up to $^{11}/_{12}$ of the applied dc voltage. Thus if we wish to have C charge to 100 V, we would set the output of our dc supply to 109 V (approximately).

If, in the circuit of Fig. 16.1-5*a*, R is replaced by the EVM, as in Fig. 16.1-5*b*, another means is provided for determining the charge time of C. When switch S_1 is closed, C will charge through the input resistance R_{in} of the EVM. At the moment S_1 is closed, the full-battery voltage V appears across R_{in}, and the EVM measures the full voltage V. There is, of course, *zero* voltage across C at that instant. As C charges, the voltage v_c developed across C increases, while the voltage v_R developed across the input resistance of the EVM decreases. At the end of 1 RC, only 37 percent (100 − 63) of the applied voltage V will be measured on the meter. At the end of 2 RCs, 14 percent (100 − 86) will be measured by the EVM. At the end of the third, fourth, and fifth RCs, the EVM will show respectively, 5, 2, and 1 percent (approximately) of the applied voltage V. Of course the charging time constant in this circuit is $R_{in}C$.

Using this method to determine the length of time it takes C to charge, we would connect the circuit of Fig. 16.1-5*b* and set V to some value, say 100 V. The start of the timed interval would begin when S_1 is closed, and would end when the meter voltage dropped to 1 V (approximately). If the meter reading stabilizes at a measurable value of voltage, say 3 V, it does so because of capacitor leakage. In that event, readings should be discontinued just when the voltage reaches this stable level.

NOTE: As the voltage measured on the EVM falls, we can switch the meter to a lower range without disturbing the charging time constant.

By this method we can measure the charging time of C somewhat more accurately than by the preceding means. Hence it will be used in this experiment.

SUMMARY

1. Capacitors have the ability to hold a charge (store electrons).
2. The amount of time necessary to charge a capacitor varies according to the size of the capacitor and the resistance in the circuit with the capacitor.

3. Five *time constants* are necessary to charge a capacitor to 99 percent of the applied voltage (full charge).
4. A time constant is the amount of time required for a capacitor to charge to 63.2 percent of the applied voltage. The formula for a time constant is $t = R \times C$.
5. A time constant is not a fixed unit of time such as a second but is a relative unit of time whose actual value is determined by the values of R and C in the charging circuit.

SELF-TEST

Check your understanding by answering these questions.

1. The formula $t = R \times C$ is the formula for one ________.
2. For a capacitor to be considered fully charged, it must charge for ________ time constants.
3. Theoretically, a capacitor is never fully charged. ________ (true/false)
4. What factor in a circuit besides the size of the capacitor determines the length of time it takes the capacitor to charge? ________
5. The time constant for a series circuit containing a 1-MΩ resistor and a 1-μF capacitor is ________ s.
6. For the circuit of question 5, how long will it take for the capacitor to be considered fully charged? ________

9.25 MΩ

MATERIALS REQUIRED

- Power supply: Regulated, variable dc supply
- Equipment: EVM
- Resistor: ½-W 1-MΩ
- Capacitor: 1-μF
- Miscellaneous: Two SPST switches; watch with a seconds hand

PROCEDURE

Charge Time Constant

NOTE: Your instructor will advise you what is the input resistance R_{in} of your EVM, on the dc voltage range, or you can measure it with an ohmmeter. Compute and enter the charge time constant, $R_{in} \times C$, in Table 16.1-1.

1. Connect the circuit of Fig. 16.1-5*b*. Set the EVM on the 50-V or next higher range. Close S_1 and S_2 and set V at 30 V, as measured on the voltmeter.
2. Open S_2. Observe and record v_R, the voltage across the EVM, at every charge time-constant interval shown in Table 16.1-1.

   Record readings in the column labeled "v_R, first trial." Compute the voltage v_c across the capacitor at every time-constant interval and record in the column "v_c, first trial." Use the equation

$$v_c = V - v_R = 30 - v_R \tag{16.1-4}$$

3. Close S_2, thus discharging capacitor C. V should be 30 V. Open S_2 and repeat step 2. Record the measured voltage in the column "v_R, second trial," and the computed values in the column labeled "v_c, second trial."
4. Repeat step 3. Record the measured voltage in the column "v_R, third trial," and the computed values in the column "v_c, third trial."
5. Average the three values of v_c at every time-constant interval and record the results in column

TABLE 16.1-1. Charge Time Constant of a Capacitor

Charge Time Constant, $R_{in} \times C$ 9.25 s	v_R, V			v_c, V				
	1st Trial	*2d Trial*	*3d Trial*	*1st Trial*	*2d Trial*	*3d Trial*	*Average*	*Computed*
1	16.3	16.3	16.3	13.7	13.7	13.7		
2	9.3	9.2	8.9	20.7	20.8	21.1		
3	5.1	4.8	4.5	24.9	25.2	25.0		
4	3.1	2.7	2.7	26.9	27.3	27.3		
5	2.3	1.8	1.7	27.7	28.2	28.3		
10	1.8	1.3	1.3	28.2	28.7	28.7		

"v_c, average." Draw a graph of v_c average versus RC.

6. Compute, and record in Table 16.1-1, the voltage v_c which should appear across C at the end of 1 RC, 2 RCs, etc. For comparison with the graph in step 5, draw a graph of the computed values of v_c versus RC (in ink).

QUESTIONS

1. What is meant by the term *time constant* for RC circuits?
2. How can we use the charging curve of a capacitor to measure time? Relate your answer to the universal time-constant chart.
3. What factors limit the accuracy of your measurements in this experiment?
4. What is the purpose of averaging the three trial values of v_c?
5. How do the average values of v_c compare with the computed values for the charge of a capacitor? Explain any discrepancies between computed and average values.

Answers to Self-Test

1. time constant
2. five
3. true
4. the resistance
5. 1
6. 5 s

EXPERIMENT 16.2. *RC* Time Constants —Discharging a Capacitor

OBJECTIVE

To determine experimentally the time it takes a capacitor to discharge through a resistance

INTRODUCTORY INFORMATION

Discharging a Capacitor

If a charged capacitor is placed across a resistor R (Fig. 16.2-1*a*), a discharge path is provided for C. Electrons now leave the negative plate of C, move through R, and enter the positive plate of C. This process continues until the excess electrons on the lower plate have replaced the deficiency of electrons in the upper. When this exchange of electrons has taken place, both plates return to their neutral or uncharged state, and the capacitor is said to be discharged (Fig. 16.2-1*b*). A voltmeter placed across C would now show *zero* voltage across the capacitor. The movement of electrons (charges) during the charge and discharge interval, in effect, constitutes electron current flow.

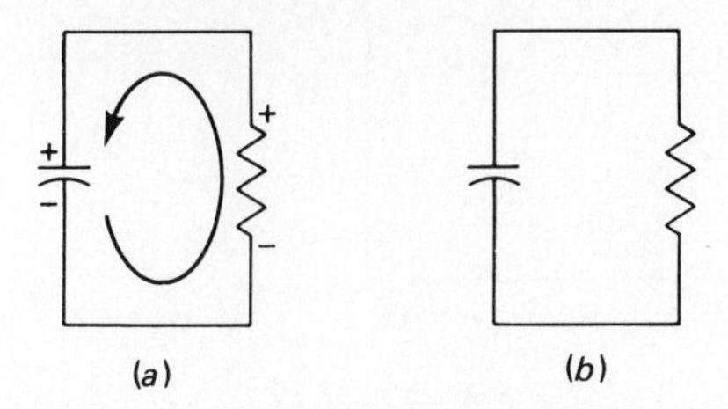

Fig. 16.2-1. Discharging a capacitor.

Capacitor Discharge—Time Required

Now how does a capacitor discharge? Assume that C in Fig. 16.2-2 has charged to the full applied voltage, with S_1 closed and S_2 open. We now open S_1 and close S_2, as in Fig. 16.2-3*a*, permitting C to discharge through R. The variation of discharge current i_d, the change in voltage v_R across R, and the decay in voltage v_c across C are shown in the graphs of Fig. 16.2-3*b*. We note that the discharge current i_d is opposite in direction to the charge current i_c. Therefore, the voltage v_R across R has the opposite polarity to that in Fig. 16.1-3*b*.

Figure 16.2-4 shows the discharge curve for C. We

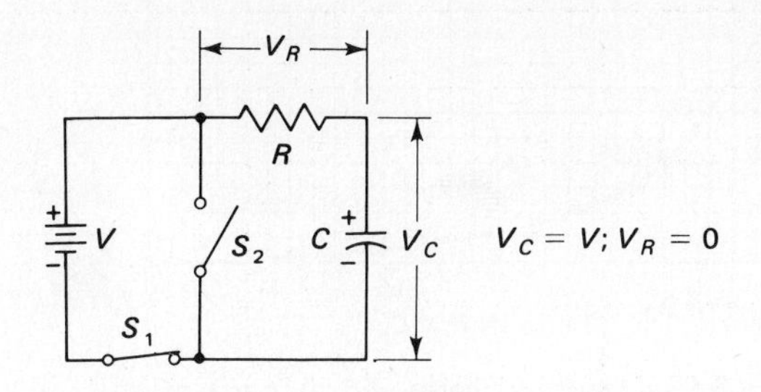

Fig. 16.2-2. Charged capacitor.

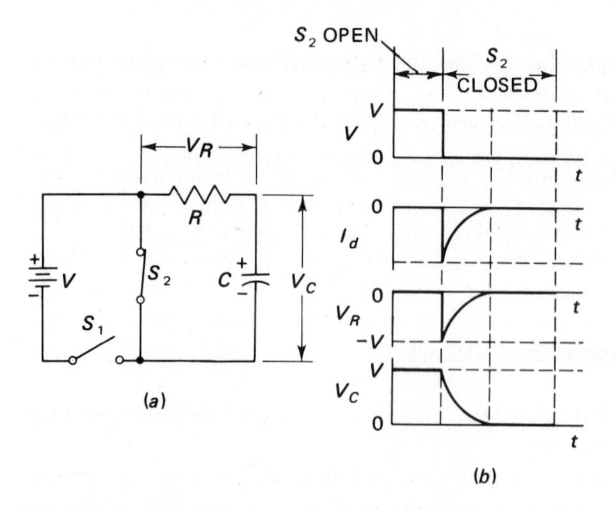

Fig. 16.2-3. Capacitor discharging in an RC circuit.

note the following significant similarities between the discharge of C and the charge of C.

In 1 RC, C discharges 63.2 percent of its total voltage. In 2 RCs, C discharges 86 percent of its total voltage, etc. The rule for discharge is therefore similar to the rule for charge: A capacitor discharges 63.2 percent of its remaining voltage in one time constant.

If we refer to the graphs in Figs. 16.1-3b and 16.2-3b we note that Kirchhoff's law applies to both charge and discharge circuits. In each case, at any instant of time, the applied voltage V is equal to the sum of the voltage across the capacitor and that across the resistor. That is,

$$V = v_c + v_R \qquad (16.2\text{-}1)$$

Experimental Techniques in Observing Capacitor Discharge

Just as we did in Exp. 16.1, we can set up an experiment using simple equipment to demonstrate how long it takes a capacitor to discharge. However, again our results will be approximate because we shall not be using laboratory precision timing or measuring equipment.

Refer to Fig. 16.2-5. By selecting the values of R (in this case the input resistance R_{in} of the EVM) and C, we can find the length of time it takes C to discharge, using an ordinary watch with a seconds hand. In determining the discharge time for C, we use a large nonelectrolytic capacitor, say 1 μF, an EVM whose input resistance R_{in} we know, a variable dc voltage source, and a watch. When we close switch S_2 and open S_1, C will start to discharge through the input resistance R_{in} of the meter. We shall assume here that the input resistance of the meter is 11 MΩ. The discharge time constant is C times the resistance of the meter, and in this case it would be 11 s. We can verify the discharge voltage time constant at (RC) intervals, 11-s intervals in this case. Our results should roughly conform to the capacitor discharge curve (Fig. 16.2-4).

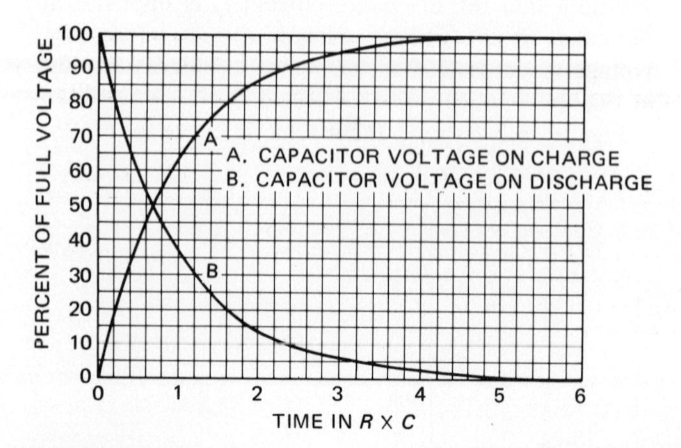

Fig. 16.2-4. Universal chart for charge and discharge of a capacitor.

SUMMARY

1. A capacitor is discharged when the charge distribution on both its plates has been neutralized, that is, when the negatively charged plate places its excess electrons on the positive plate.
2. A path for current flow is necessary for capacitor discharge.
3. A capacitor in a circuit will discharge 99 percent of its charge in five time constants.
4. A capacitor in a circuit will discharge 63.2 percent of its charge in one time constant.
5. The discharge time constant of a capacitor C discharging through a resistor R is $R \times C$.

SELF-TEST

Check your understanding by answering these questions.

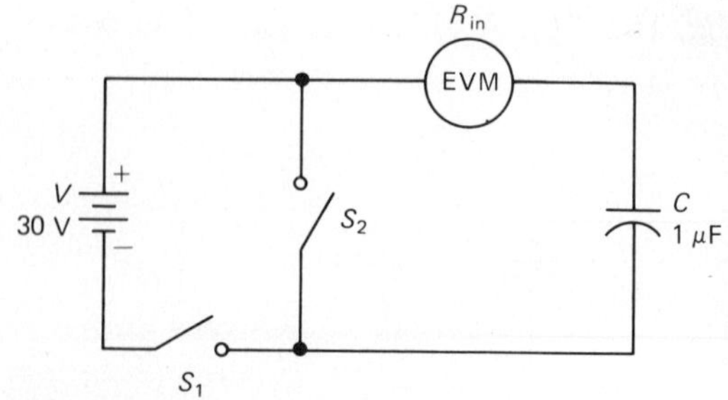

Fig. 16.2-5. Experimental circuit for verifying capacitor discharge time.

1. A capacitor will discharge ________ percent of its charge in one time constant.
2. A discharge time constant is calculated the way a charge time constant is calculated. ________ (true/false)
3. ________ time constants are necessary to discharge a capacitor fully for all practical purposes.
4. The formula for a time constant is $t =$ ________.

MATERIALS REQUIRED

- Power supply: Regulated, variable dc source
- Equipment: EVM
- Capacitor: 1-μF
- Miscellaneous: Two SPST switches; watch with a seconds hand

PROCEDURE

Discharge Time Constant

1. Determine the input R of your EVM. Use the manufacturer's specifications or measure with an ohmmeter.
2. Connect the circuit of Fig. 16.2-5. Set V at 30 V. S_1 is closed, S_2 is open.
3. Compute the discharge time constant $R_{in}C$ and record in Table 16.2-1. Use the rated value of R_{in}.
4. Open S_1 and close S_2. Observe and record v_c, the voltage across C, at every discharge time-constant interval shown in Table 16.2-1. For example, if $R_{in}C = 11$ s, readings should be made at 11-s intervals and recorded in column labeled "v_c, first trial" until C is completely discharged.

NOTE: When the voltage read on the meter stabilizes at some level, you are reading a voltage produced by the dielectric strain within the capacitor. This may occur at about the tenth time constant.

5. Open S_2. Close S_1 and permit C to charge fully (30 V).
6. Repeat step 4. Record your readings in the column labeled "v_c, second trial."
7. Repeat steps 5 and 4, recording your readings in the column labeled "v_c, third trial."
8. Average the three values of v_c for 1 RC and record in column "v_c, average." Average also and record the value of v_c for 2 RCs, 3 RCs, etc. Draw a graph (in red pencil) of v_c (average) versus RC. Let RC be the horizontal axis.
9. Compute, and record in Table 16.2-1, the voltage v_c which should appear across C after 1 RC, 2 RCs, etc. For comparison with step 8, draw a graph (in green) of the computed values of v_c versus RC.

QUESTIONS

1. What factors limit the accuracy of your measurements in this experiment?

TABLE 16.2-1. Discharge Time Constant of a Capacitor

V, V	R_{in}, Ω		Discharge Time Constant, $R_{in}C$	Voltage across *C*
	Computed	*Rated*	9.25 s	
				30

Discharge time, *RC*	v_c, V				
	1st Trial	*2d Trial*	*3d Trial*	*Average*	*Computed*
1	17.1 14.7	16.3	16.4	16.1	16.2 15.6
2	9.7 8.2	8.2	8.2	8.3	8.0
3	5.1 4.4	4.3	4.3	4.2	4.7 4.2
4	2.3	2.2	2.3	2.3	2.4 2.3
5	1.4	1.2	1.3	1.3	1. 1.4
10	.8	.7	.7	.7	.7

2. What is the purpose of averaging the three trial values of v_C?
3. How do the average values of v_c compare with the computed values for the discharge of a capacitor?
4. Why is it important to know how a capacitor discharges?

Answers to Self-Test

1. 63.2
2. true
3. five
4. $R \times C$

CHAPTER 17 CAPACITIVE REACTANCE AND PHASE MEASUREMENT

EXPERIMENT 17.1. Capacitive Reactance (X_C)

OBJECTIVE

To verify by experiment that the capacitive reactance of a capacitor may be given by the formula

$$X_c = \frac{1}{2\pi f_c}$$

INTRODUCTORY INFORMATION

Reactance of a Capacitor

A capacitor opposes alternating current (we have previously found that direct current will not flow through a capacitor). The opposition shown by the capacitor to alternating current is not a resistance, since current does not flow *through* the capacitor. This opposition to ac flow in the capacitive circuit is a reactance, so-called because it is a reaction to the current flowing into and out of the capacitor plates. Since this reaction is caused by the nature of the capacitor, it is called *capacitive reactance*. The symbol for capacitive reactance is X_c.

The following is a simplified explanation of how a capacitor works in a circuit. As a capacitor charges (see Fig. 17.1-1), one of its plates gains an excess of electrons. Since like charges repel, the more electrons are forced onto the plate, the more this negative force will oppose current flow onto that plate. This is not a resistance to electron flow but a reaction caused by the electrons which have already been forced onto the plate.

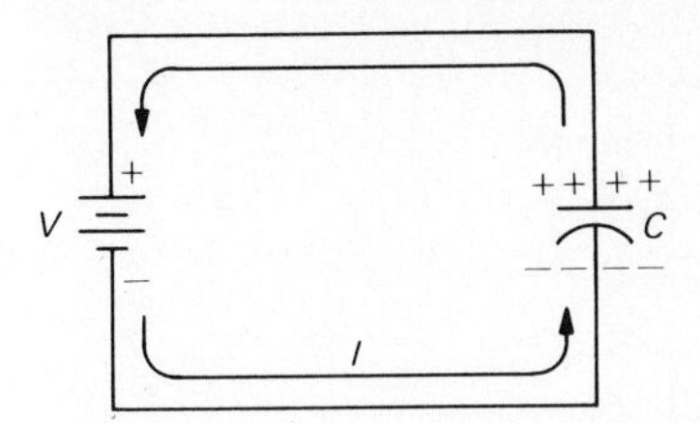

Fig. 17.1-1. As a capacitor charges, one plate takes on an excess of electrons (−), the other a deficiency of electrons (+).

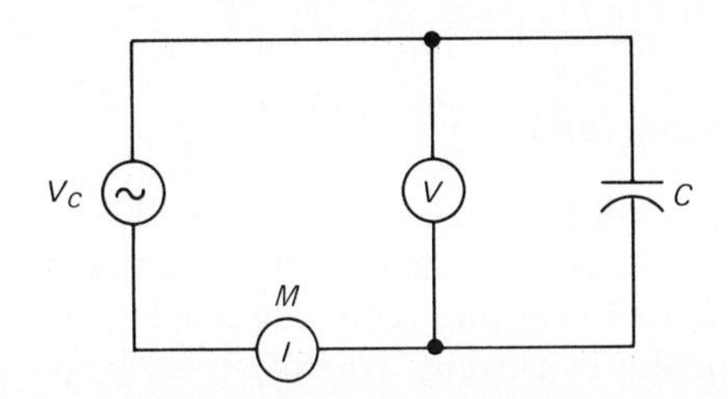

Fig. 17.1-2. A sinusoidal voltage V_c causes current I to flow in the circuit.

Alternating current reverses its direction of flow periodically. Therefore the capacitor plates are periodically being charged and discharged. As this happens, there is a buildup and decay of electrons on the alternate plates of the capacitor.

A capacitor can be charged with direct current, but once charged, the voltage across the capacitor equals the applied voltage. Therefore there is no further current flow. The *reactance* of the capacitor, after it is fully charged, is infinite. The capacitor therefore does not allow direct current to flow in its circuit. On the other hand, in the ac circuit, as there is a periodic charge and discharge of the capacitor, there *is* flow of current in the circuit attached to the capacitor (Fig. 17.1-2).

This explanation suggests that the size of the capacitor (determining the amount of charge it can hold) and the frequency of the charging current (determining how fast the capacitor has to charge) are both factors in opposing the charge and discharge of the capacitor. Therefore both C and f determine the capacitive reactance of a capacitor.

It was previously said that the capacitive reactance X_c of a capacitor is the amount of opposition it offers in an ac circuit. The unit of capacitive reactance is the ohm. However, the X_c of a capacitor cannot be measured with an ohmmeter. Rather, capacitive reactance must be measured indirectly from its effect on current in an ac circuit.

X_c is dependent on frequency and capacitance and is given by the formula

$$X_c = \frac{1}{2\pi f C} \qquad (17.1\text{-}1)$$

Here X_c is in ohms, C in farads, and f is frequency in Hz (cycles per second). The value of C in microfarads (μF) may be substituted directly in the formula

$$X_c = \frac{10^6}{2\pi f C} \qquad (17.1\text{-}2)$$

For example, suppose it is required to find the reactance of a 0.1-μF capacitor at a frequency of 1000 Hz. Substituting in Eq. (17.1-2),

$$X_c = \frac{10^6}{(6.28)(1000)(0.1)} = 1592\ \Omega$$

It is apparent from Eqs. (17.1-1) and (17.1-2) that the higher the frequency, the smaller the reactance of a capacitor, and the lower the frequency, the higher the reactance. For direct current, where $f = 0$, X_c is infinite. Therefore, direct current will not flow through a capacitor.

The reactance of a capacitor may be determined by measurement. In the circuit of Fig. 17.1-2 a sinusoidal voltage V_c causes a current I to flow in the circuit. Ohm's law, applied to ac circuits, states that

$$I = \frac{V}{Z} \qquad (17.1\text{-}3)$$

where I is measured in amperes, V in volts, and Z in ohms. The symbol Z stands for impedance, which is the opposition to alternating current. In Fig. 17.1-2, the reactance of C is the impedance of the circuit since there is *no other opposition* to current in this circuit.

$$X_c = Z \qquad (17.1\text{-}4)$$

Hence, the capacitive current I is given by the equation

$$I = \frac{V_c}{X_c} \qquad (17.1\text{-}5)$$

We see, therefore, that the amount of alternating current in the capacitor, across which an ac voltage is applied, is directly proportional to the amount of voltage across the capacitor (V_c). Also, the current in the capacitor is inversely proportional to the amount of opposition (here, X_c).

We may also write Eq. (17.1-5) as follows:

$$X_c = \frac{V_c}{I} \qquad (17.1\text{-}6)$$

We are saying that the opposition of a capacitor to alternating current, that is, its reactance, can be determined by knowing the voltage across the capacitor V_c and the current I in it. Recall Ohm's law for dc circuits where $R = V/I$. These ac formulas are basically the same as dc formulas for Ohm's law and follow the same forms. The difference is that the opposition is no longer just a resistance but is a reactance or impedance.

Equation (17.1-6) may be used to determine the reactance of a capacitor at a given frequency f. This is done by measuring the voltage V_c across the capacitor with an ac voltmeter and by measuring the current I in the capacitive circuit with an ac ammeter. The values of V_c and I are then substituted in Eq. (17.1-6).

Another method of measuring alternating current does not require an ac ammeter.

Consider the circuit of Fig. 17.1-3. This is a series RC circuit. Thus the currents flowing through the resistor and the capacitor are the same. The current in the circuit can be determined by measuring the voltage V_R across R and substituting V_R in the formula:

$$I = \frac{V_R}{R} \qquad (17.1\text{-}7)$$

where R is the resistance in ohms of the resistor.

Knowing I, next measure the voltage V_C across C. Now, substitute the values of I and V_C in Eq. (17.1-4), that is, in

$$X_C = \frac{V_c}{I}$$

and solve for X_c. The units of measurement in this formula are: X_c is in ohms, when V_c is in volts and I is in amperes.

SUMMARY

1. Capacitive reactance is the opposition of a capacitor to the charging and discharging in an ac circuit, that is, to alternating current.
2. Capacitive reactance is influenced by the frequency of the charging current and the capacitance of the capacitor.
3. The formula for determining X_c is $X_c = 1/(2\pi f C)$.

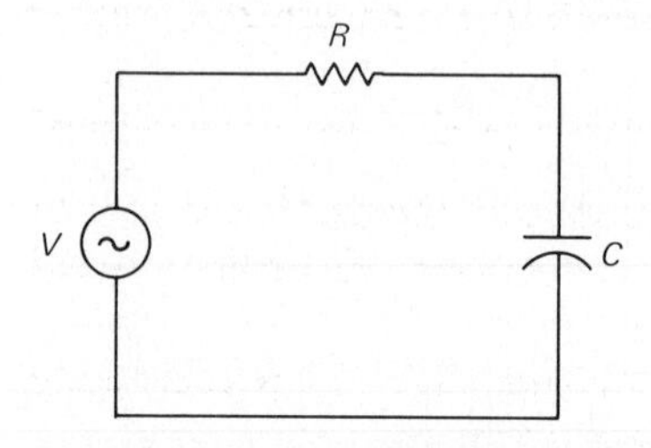

Fig. 17.1-3. Series RC circuit with an ac source V.

TABLE 17.1-1. Capacitive Reactance

$C, \mu F$		R, Ω		*V*, *ac Source rms*			I, A	X_c, Ω	
Rated	*Measured*	*Rated*	*Measured*	*Measured*	V_c, V *rms*	V_R, V *rms*	$I=\frac{V_R}{R}$	$X_c=\frac{V_c}{I}$	$X_c=\frac{10^6}{2\pi fC}$
0.5									
0.1									

4. Impedance Z is the opposition of a circuit to alternating current.
5. Ohm's law for ac circuits is $I = V/Z$.

SELF-TEST

Check your understanding by answering these questions.

1. The opposition of a circuit to alternating current is ________.
2. The opposition of a capacitor to alternating current is ________.
3. What is the formula for calculating X_c? ________
4. X_c can be measured with an ohmmeter. ________ (true/false)
5. The frequency of the applied voltage has nothing to do with X_c. ________ (true/false)
6. The reactance of a capacitor whose value is 0.1 μF in a circuit where a 100-Hz signal is applied is ________ Ω.

MATERIALS REQUIRED

- Power supply: Source of 120 V/60 Hz
- Equipment: Capacitance tester; EVM; AF generator capable of delivering 6.3 V/60 Hz rms (if a filament transformer is not available)
- Resistors: ½-W 5600-Ω
- Capacitors: 0.5-μF, 0.1-μF
- Miscellaneous: Filament transformer: 120-V/60-Hz primary, 6.3-V or 12.6-V rms secondary; fused line cord; SPST switch

PROCEDURE

1. With a capacitance checker, measure and record in Table 17.1-1 the capacitance of each C listed in Materials Required.
2. Measure and record in Table 17.1-1 the resistance of R listed in Materials Required.
3. Connect the circuit of Fig. 17.1-4.
4. Measure and record in Table 17.1-1 the rms voltage of your ac source.
5. Measure also and record in Table 17.1-1 the voltage across the capacitor V_c and the voltage across the resistor V_R.
6. Compute circuit current I ($I = V_R/R$) and record it in Table 17.1-1.
7. Compute X_c by substituting the measured value of V_c and the computed value of I in the formula $X_c = V_c/I$. Record it.
8. Compute X_c by substituting the measured value of C and the known frequency f in the formula $X_c = 10^6/(2\pi fC)$. Record it in Table 17.1-1.
9. Repeat steps 5 through 8 for $C = 0.1$ μF.

QUESTIONS

1. In your own words state Ohm's law for ac circuits.
2. What information must you know in order to determine the current in a capacitor?
3. On what factors does X_c depend?
4. How does the measured value of $X_c = (V_c/I)$ compare with the formula value $X_c = 1/(2\pi fC)$. Comment on any discrepancy.
5. List the factors which limit the accuracy of the measurements in the experiment.

Answers to Self-Test

1. Impedance, Z
2. Capacitive reactance, X_c
3. $X_c = \dfrac{1}{2\pi fC}$
4. false
5. false
6. 15,920

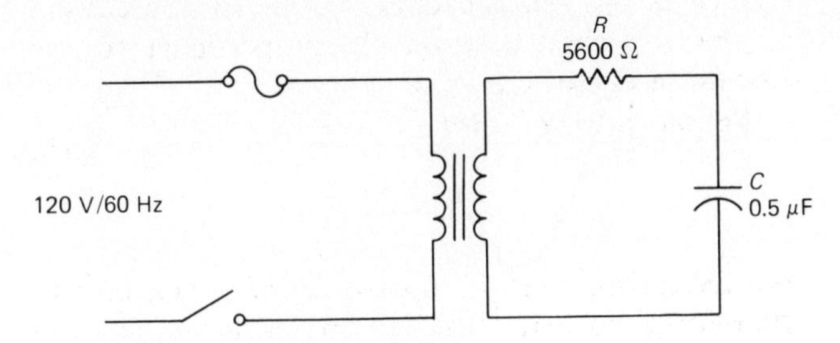

Fig. 17.1-4. Experimental circuit.

EXPERIMENT 17.2. Phase Relations in an *RC* Circuit

OBJECTIVE

To measure with an oscilloscope the phase angle between the current and voltage in a capacitive circuit

INTRODUCTORY INFORMATION

Phase Relations between *v* and *i* in a Capacitor

The maximum voltage *does not* appear across a capacitor at the same time there is maximum current flow into the capacitor. When the current and voltage do not increase or decrease at the same time, they are said to be "out of phase."

In a resistor, when the voltage across the resistor increases, the current increases. When the voltage decreases, the current decreases. Therefore in a resistor voltage and current are *in phase*. In a capacitor, current and voltage are 90° out of phase. In Exp. 14 it was shown that the graph of an ac voltage is a sine wave in which the horizontal axis units are in time or degrees. A full sine wave is completed in 360°, and the difference between the maximum sine-wave voltage and the minimum (zero) voltage is 90° (see Fig. 17.2-1). If the voltage and current in a capacitor are 90° out of phase, it means that one is maximum while the other is minimum. The following discussion will describe why the voltage across and current in a capacitor are 90° out of phase.

At the moment an ac voltage source is first applied to a capacitive circuit, the capacitor is uncharged; that is, there is no difference in the charge distribution across the plates. *Thus there is no voltage across them.* However, as the source is applied, current can immediately flow into the capacitor, for there is nothing to oppose the flow. At the instant when the voltage across the capacitor is zero, the current is maximum. As the current flows into the capacitor, that is, as the capacitor starts to charge, a voltage builds up across the capacitor plates which "bucks" the applied voltage, and so the current flow in the capacitor decreases. Thus, as the capacitor becomes charged, the charge distribution across the plates can be measured as a voltage across them. When the capacitor is fully charged, that is, when there is maximum voltage across it, there is zero current flow because the "bucking" voltage is now equal to the applied voltage, and the "effective" charging voltage is zero (see Exp. 16.1).

The current was the first to appear as the source was connected to a capacitive circuit. Voltage across the capacitor appeared later. Since the current flow into the capacitor came before voltage appeared across it, current is said to *lead* the voltage. And, it leads the voltage by 90°. This is shown in graph form in Fig. 17.2-2.

Phase Relations in an *RC* Circuit

If the circuit in which the capacitor is connected has a resistor, the circuit is called an *RC*, or resistor-capacitor, circuit. This circuit will no longer behave like a purely capacitive circuit. What happens is that both the capacitor and the resistor try to control current in the circuit. If the capacitor offers a larger opposition X_c to the alternating current than the resistor, the circuit will behave more like a capacitive circuit. If the resistor is larger than X_c, the circuit will behave more as a resistive circuit. Keep in mind that in a resistive circuit *v* and *i* are in phase, and in a

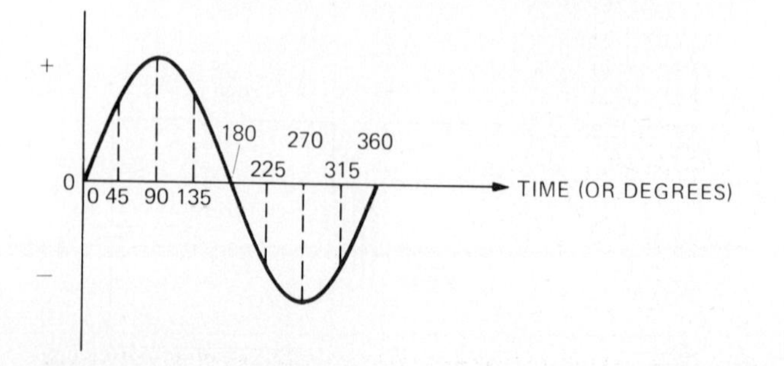

Fig. 17.2-1. Sine wave with degrees plotted on the time axis.

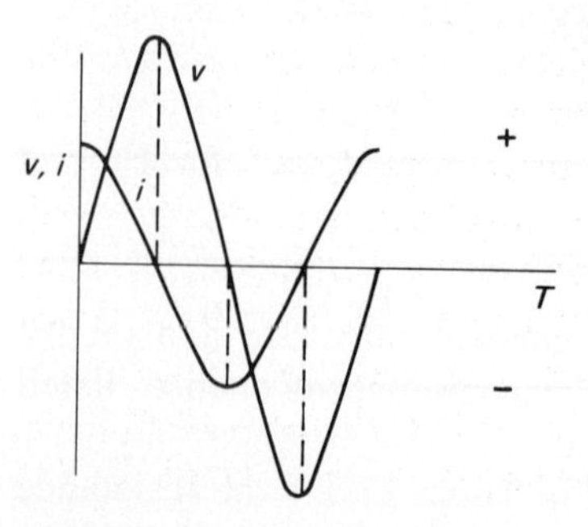

Fig. 17.2-2. Graph showing *v* and *i* out of phase by 90°, with *i* leading *v*.

capacitive circuit they are 90° out of phase. What if the resistor and capacitor offer equal opposition to current flow in the circuit? Then voltage and current are out of phase by 45°, halfway between 0° (in phase) and 90°. Of course, by varying the size of the resistor and capacitor, the phase difference between the voltage and current can be any value from 0 to 90°. This phase difference is referred to as the phase "angle."

The study of phase relationships is extremely important to the service technician. Many products, both industrial and consumer, rely on phase "shifting" for their operation. The motor speed control and color television are well-known examples. One type of motor speed control operates because the phase of a small control voltage can be made to vary in relation to the phase of the line voltage, to turn on a semiconductor device just at the right time, for the desired speed. In a color television receiver the color signals are recovered from the television signal by phase-sensing networks. It is important to understand what these circuits do, if you are going to diagnose and repair them properly.

Direct-Phase Measurements Using External Synchronization or Triggering

Phase angles between two waveforms having the same frequency, but appearing at different points in a circuit, may be measured directly with an oscilloscope. Either a triggered or a nontriggered oscilloscope may be used. With the triggered type, *triggering* must be external; that is, the trigger selector switch must be set to Ext. With the nontriggered oscilloscope, *synchronization* of the sweep must be external, that is, the sync selector switch must be set to Ext. Since it is anticipated that this text-laboratory manual will be used by students with both triggered and nontriggered oscilloscopes, in the discussion which follows the term "synchronize/trigger" will be employed to cover both types.

The technique requires that the sine wave v applied to the series RC circuit (Fig. 17.2-3) be fed to the external sync/triggering jack of the oscilloscope and the sync/trigger selector switch set to Ext. The oscilloscope-sweep circuit is then synchronized/triggered at the same time, regardless of the point in the circuit where the waveform is being observed. The applied voltage v fed to the vertical input of the oscilloscope serves as the reference for phase measurements. The reference sine wave is conveniently centered on the graticule, so that the start of the sine wave coincides with a major vertical graph line A (Fig. 17.2-4).

If the oscilloscope sweep is linear, equal distances on the time base correspond to equal angular displacements. Thus in Fig. 17.2-4, one division is divided into five equal graduations. Since one alternation (180°) covers two graticule divisions, each graduation corresponds to 18°.

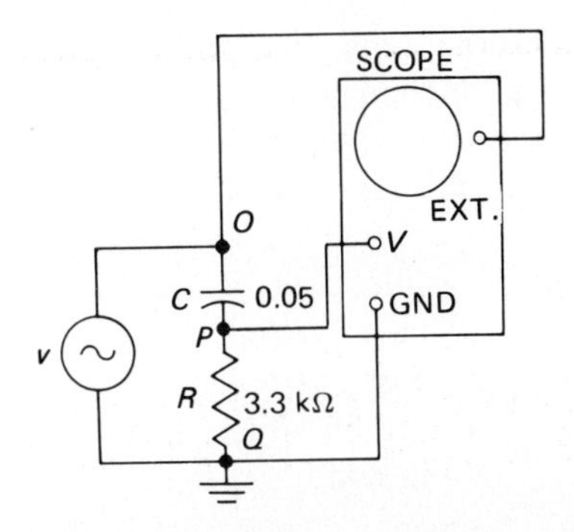

Fig. 17.2-3. Oscilloscope connections for demonstrating directly phase relations between the applied voltage v and the current i in a capacitive circuit. The sync/trigger selector of the oscilloscope is set to Ext. The applied voltage v is first displayed on the oscilloscope and acts as the reference waveform.

Once the reference waveform has been properly centered and its width in divisions noted, the number of degrees per graduation can be calculated. Then the vertical input (hot) lead of the oscilloscope is connected to point P (Fig. 17.2-3). The voltage waveform across R will now appear on the oscilloscope. The displacement (number of graduations) of the observed waveform, from the reference voltage, is measured on the time base. Multiplying the calculated *degrees per graduation* by the *number* of graduations of displacement, we get the phase difference between the applied voltage and current.

In a series dc and ac circuit, current is everywhere the same. Therefore, in Fig. 17.2-3, current in the RC circuit is the same in R as in C. Moreover, the voltage across R is in phase with the current through R. So the phase of current in the RC circuit is the same as the phase of voltage across R.

Therefore, in an RC circuit (Fig. 17.2-3) the position of the waveform across R relative to the reference

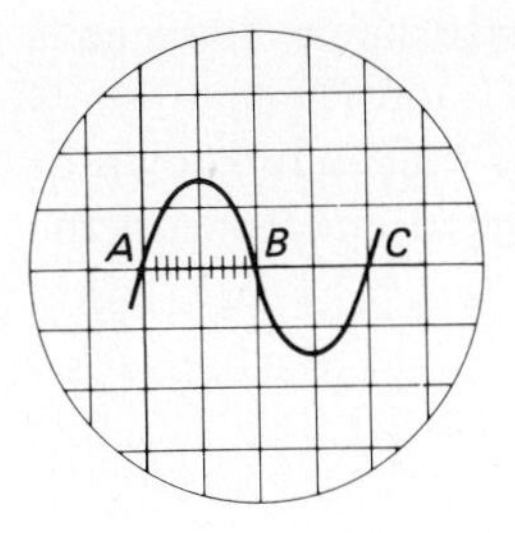

Fig. 17.2-4. Horizontal sweep used directly for phase measurements.

(applied voltage) waveform tells us whether the current i leads or lags the applied voltage v. Thus in Fig. 17.2-2 the current waveform i leads the reference waveform v by 90°. Waveform v lags i by 90°.

NOTE: If a nontriggered oscilloscope is used, the width of the time base may be adjusted as desired for convenient measurements. For example, the width of the sweep in Fig. 17.2-4 may be adjusted so that the waveform ABC is 2, 3, or 4 in wide.

Measuring Phase Angle between *v* and *i* in a Capacitor

If we wish to measure the phase angle between voltage and current in a capacitor, we can do so directly with an oscilloscope. However, an external resistor R must be connected in series with C, as in the circuit of Fig. 17.2-3. Care must be taken to select a resistor of such a value that the ratio of X_c/R is greater than 10. This circuit will act almost like a pure capacitance. Current I in the series circuit of Fig. 17.2-3 is the same throughout the circuit. Moreover, the voltage V_R across R is in phase with the current I. Hence, by measuring the phase angle between voltage V_R across R, and the applied voltage V, we are in fact measuring the phase angle between v and i in the capacitor.

NOTE: If the ratio of X_c/R is equal to 10, the error in phase angle introduced by R is approximately 6°. As X_c/R becomes greater than 10, the error decreases. In the experimental circuit for this experiment, the error introduced by R is 3.6°, approximately.

Calculating the Phase Angle of an *RC* Circuit (optional)

The phase difference, or phase angle theta (θ), between circuit current I and the applied voltage in an RC circuit can also be calculated. Some trigonometry must be used, but with trig tables or a pocket calculator available, the amount of actual math involved is minimal.

In order to calculate the phase angle between v and i, it is necessary to know X_c and R, or to know V_c and V_R. If either set of these values is known or can be measured, they are "plugged into" one of the following formulas:

$$\tan\theta = \frac{X_c}{R} \qquad \tan\theta = \frac{V_c}{V_R} \qquad (17.2\text{-}1)$$

Once the division of X_c by R or V_c by V_R is completed, refer to the trig tables and find that *number* in the column labeled *tangent*. Beside that number will be the phase angle θ in degrees.

If a calculator with trig functions is available, it is only necessary to divide X_c by R or V_c by V_R, and once the answer is displayed, to press the arc button and then the tangent button or the arctan ($\tan^{-1}$) button; the answer in degrees will be displayed.

SUMMARY

1. The voltage V across a capacitor and the current I in a capacitor are not in phase, but are 90° out of phase.
2. I leads V by 90° in a capacitor.
3. In an RC circuit, the phase difference between the applied circuit voltage V and circuit current I is determined by the ratio X_c/R.
4. If reactance of C (X_c) in an RC circuit is greater than the resistance R, it will behave as a capacitive circuit, and the phase angle will be greater than 45° but less than 90°.
5. If R is greater than X_c in an RC circuit, it will behave more like a resistive than a capacitive circuit, and the phase angle will be less than 45° but greater than 0°.
6. The phase angle can be calculated or displayed on an oscilloscope.
7. The phase angle θ is found mathematically by simple trigonometry using either of the following two equations:

$$\tan\theta = \frac{X_c}{R} \qquad \tan\theta = \frac{V_c}{V_R}$$

SELF-TEST

Check your understanding by answering these questions.

1. V and I are out of phase by ________ in a capacitor.
2. V and I are out of phase by ________ in a resistor.
3. In an RC circuit with X_c equal to R, the phase angle is ________ degrees.
4. The only way to determine the phase angle of an RC circuit is by mathematics. ________ (true/false)
5. In measuring the phase angle in an RC circuit, it is necessary to synchronize/trigger the oscilloscope ________.

MATERIALS REQUIRED

- Equipment: Oscilloscope; audio frequency sine-wave generator

- Capacitor: 0.05-μF
- Resistors: ½-W 3300-Ω; 56-kΩ

PROCEDURE

Direct-Phase Measurement Using External Sync/Triggering

1. Connect the circuit of Fig. 17.2-3. Set the sync/trigger selector switch to Ext. Feed the AF generator output signal to the Ext. sync/trigger jack on the oscilloscope. Connect scope vertical input lead to point O. Adjust the generator output 60-Hz signal to 20 V p-p. Set oscilloscope vertical-gain controls to two or three divisions of vertical deflection. Set oscilloscope controls for a two- or three-cycle display.

2. Center the waveform vertically with respect to the horizontal graticule axis. Center the waveform horizontally so that the start of the positive alternation coincides with one of the major vertical graph lines on the graticule, as in Fig. 17.2-4. This is the reference voltage waveform. Determine the degree per (horizontal) graduation factor of the time base. ________

3. Now connect the vertical input lead to point *P*, the junction of *R* and *C* in Fig. 17.2-3. Readjust the vertical-gain controls for two or three divisions of vertical reflection on the screen. Observe the current waveform. Does it lead or lag the reference waveform? ________

4. Measure the number of graduations of horizontal displacement of the waveform across *R* from the reference and convert into degrees of phase difference. ________ degrees

5. (Optional) Using Eq. (17.2-1), calculate the phase angle. You may use the values for *R* and X_c (calculated), or you may measure V_c and V_R. The calculated phase angle is ________°.

6. Use the same circuit as in step 1 but replace *R* with a 56-kΩ resistor and repeat steps 3 and 4. The measured phase angle is ________°. The calculated phase angle is ________°.

QUESTIONS

1. What is the X_c at 60 Hz of the 0.05-μF capacitor used in this experiment?
2. In step 1 of this experiment, what does the ratio X_c/R equal?
3. Why can the experimental circuit in step 1 be considered a pure capacitance (approximately)?
4. What does the measured phase angle in steps 1 to 4 indicate about the experimental circuit? Why?
5. What was the effect on phase angle of substituting a 56-kΩ resistor for the 3.3-kΩ resistor? Be specific.

Optional

6. Did the phase angle measured on the scope compare favorably with the calculated phase angle for the *RC* circuit with *R* = 3300 Ω? Discuss any discrepancies.
7. Did the phase angle measured on the scope compare favorably with the calculated phase angle for the *RC* circuit with *R* = 56 kΩ? Discuss any discrepancies.

Answers to Self-Test

1. 90°
2. 0°
3. 45
4. false
5. externally

CHAPTER 18 INDUCTANCE AND INDUCTIVE REACTANCE

EXPERIMENT 18.1. Characteristics of an Inductance

OBJECTIVE

To observe the effect of an inductance on current in a dc and ac circuit

INTRODUCTORY INFORMATION

Inductance and Reactance of a Coil

In dc circuits, resistors limit the amount (amplitude) of current. Resistors also oppose current in ac circuits. In addition to resistors, reactive components, namely, inductors and capacitors, impede currents in ac circuits.

The unit of inductance is the henry (abbreviated H), named in honor of the scientist Joseph Henry. The number of henrys in an inductor can be measured by means of an inductance bridge, an instrument with which you are not yet familiar. Inductance L is that characteristic of a coil which *opposes* a change in current. The *amount* of opposition offered by an inductor is called inductive reactance X_L.

The inductive reactance of a coil is not constant but is a variable quantity. X_L depends on the inductance L and on the frequency f of an alternating current, just as X_c depends on f and C. X_L may be computed from the formula

$$X_L = 2\pi fL \qquad (18.1\text{-}1)$$

where π is the constant 3.14, f is the frequency in hertz of the alternating current in the inductance, and L is the inductance in henrys.

Some important facts may be deduced from Eq. (18.1-1). The first is that X_L varies directly with frequency, that is, it is directly proportional to f. This variation may be shown graphically by a specific example.

Problem. Draw a graph of X_L versus f for an inductance $L = 1.59$ H.

Solution. Find X_L at each of a series of frequencies. Thus, when $f = 0$,

$$X_L = 2\pi fL = 2(3.14)(0)(1.59) = 0$$

When $f = 100$ Hz,

$$X_L = 6.28(100)(1.59) = 1000\ \Omega$$

Similarly, it can be shown that when

$$f = 200\text{ Hz}, X_L = 2000\ \Omega$$
$$f = 300\text{ Hz}, X_L = 3000\ \Omega$$
$$f = 400\text{ Hz}, X_L = 4000\ \Omega$$
$$f = 500\text{ Hz}, X_L = 5000\ \Omega$$

Figure 18.1-1 shows the variation of X_L versus f for $L = 1.59$ H. It is a straight-line graph. Hence we say that the relationship between X_L and f is a *linear one.*

A second fact which may be deduced from Eq. (18.1-1) is that for direct current, that is, for $f = 0$, $X_L = 0$. This fact is consistent with the definition of inductance as that characteristic of a coil which opposes a *change* in current. A steady-state direct current involves no change. Hence an inductance has no effect on a direct current which is constant in amplitude; that is, $X_L = 0$ for such a direct current.

Equation (18.1-1) also gives the relationship between X_L and L. If f is kept constant, X_L increases or decreases as L is increased or decreased, respectively. Again, if X_L were plotted as a function of L, a linear graph would result.

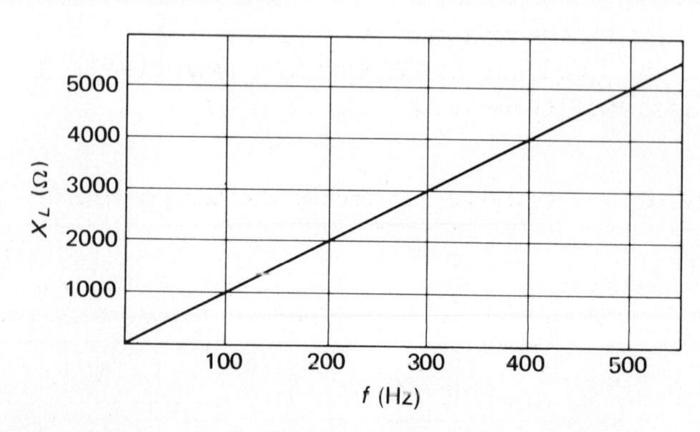

Fig. 18.1-1. Variation of X_L with f for a 1.59-H inductor.

Inductive reactance is a reaction just as capacitive reactance is a reaction. In the inductor the reaction is against increasing or decreasing current flow. As current *begins* to flow into a coil, magnetic lines of force build out around the coil. These moving lines of force cut the turns of wire of the coil and induce a voltage in the coil. The polarity of the voltage which is created is in opposition to the direction of current flow, so the induced voltage opposes that current flow (Lenz' law). When current flow is *cut off*, the magnetic field around the coil begins to collapse. This collapsing magnetic field induces a voltage in the coil, but now of the opposite polarity. The effect of the induced voltage is to keep the current flowing in the same direction it had prior to cutoff, thus *opposing* the stopping of the current flow through the coil. This reaction against changes in current is an opposition and is measured in ohms.

X_L, though measured in ohms, cannot be directly measured by the ohmmeter, since it is a reaction and not a physical resistance.

SUMMARY

1. Inductance L is the ability of a coil to oppose changes in current through itself.
2. Inductance is measured in henrys.
3. Inductive reactance X_L is the amount of opposition an inductor offers to changing current.
4. X_L is a variable quantity just as X_c is. Its value depends on f and L.
5. X_L is greater for a specific coil as the f increases.
6. X_L is greater as the L increases.
7. X_L is given in ohms but cannot be measured directly by an ohmmeter.

SELF-TEST

Check your understanding by answering these questions.

1. Inductance is measured in ________.
2. If the frequency of the applied voltage doubles, what happens to the X_L? ________________.
3. An inductance offers no opposition to pure direct current in a coil. ________ (true/false)
4. The opposition to current flow in a coil is greater if the applied voltage is alternating instead of direct current. ________ (true/false)

MATERIALS REQUIRED

- Power supply: Line-isolation transformer and variable autotransformer operating from the 120-V/60-Hz line
- Equipment: Variable-regulated dc source; 0- to 25-mA ac meter or a multirange ac milliammeter covering this range; EVM; 0- to 25-mA dc meter
- Resistors: As required
- Inductors: 8 H at 50 mA with internal resistance $R = 250\ \Omega$

PROCEDURE

Effect of an Inductance on Direct and Alternating Current

1. Measure, and record in Table 18.1-1, the resistance R of the 8-H choke.
2. Connect the circuit of Fig. 18.1-2*a*. V is a variable-regulated dc source. L is the 8-H choke; R is a resistor equal in value to the resistance of the 8-H choke; S_2 is a single-pole double-throw switch (which selects either L or R to complete the circuit); M is a 0- to 25-mA dc meter; V_1 is an EVM set on dc voltage; S_1 is a switch to apply voltage to the circuit. S_1 is open. V is set for *zero* V.
3. Throw S_2 to position 1. R is in the circuit. Close S_1. **Power on.** Adjust V until the current meter measures 20 mA. Measure and record the voltage across the resistor. *Do not vary the supply voltage.*
4. Throw S_2 to position 2. *L is in the circuit.* The voltage across L should be the same as across R. Measure the current in L and record it in Table 18.1-1. **Power off.** Disconnect and remove the dc source, the milliammeter M, and the EVM.

TABLE 18.1-1. Effect of an Inductance on Alternating Current

Resistance, Ω	*S_2 Position*	*Component in Circuit*	*DC*		*AC*	
			V, V	*I*, A	*V*, V	*I*, A
$L =$	1	*R*		0.02		0.02
$R =$	2	*L*				

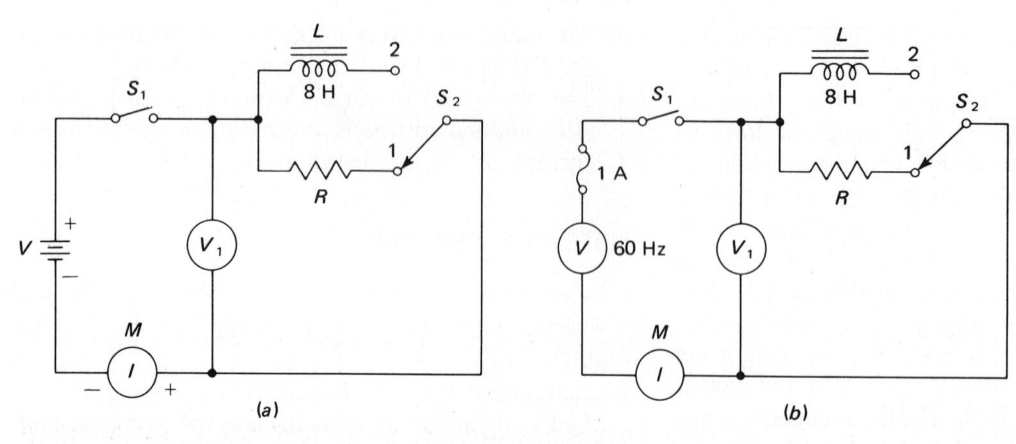

Fig. 18.1-2. Experimental circuit to demonstrate the effects of an inductance on (*a*) direct, and (*b*) alternating current.

5. Connect the circuit of Fig. 18.1-2*b*. V is an ac source consisting of a variable autotransformer plugged into an isolation transformer. M is a 0- to 25-mA ac meter, V_1 is an EVM set on ac voltage. The other components are the same as in the preceding circuit. S_1 is **off.** V is set at 0 V.
6. Throw S_2 to position 1, placing R in the circuit. **Power on.** Adjust V until the current meter measures 20 mA. Measure and record the voltage across the resistor. *Do not vary the supply voltage.*
7. Throw S_2 to position 2, *placing L in the circuit.* Readjust V if necessary to place the same voltage across L as there was across R in step 6. Measure and record the current in L.

QUESTIONS

1. From the data in Table 18.1-1, is there any significant difference between direct current in L and R, with the same applied voltage?
2. Identify any differences between the circuit current in L and R when alternating current was applied to the experimental circuit.
3. What does the experiment prove concerning the behavior of L when alternating current is applied?

Answers to Self-Test

1. henrys
2. It doubles
3. true
4. true

EXPERIMENT 18.2. Inductive Reactance

OBJECTIVE

To verify by experiment that the X_L of a coil is given by the formula $X_L = 2\pi fL$

INTRODUCTORY INFORMATION

As verified in Exp. 18.1, inductance is the ability of a coil to oppose a change in current; inductive reactance X_L is the amount of opposition of the coil to that change. It is a variable quantity measured in ohms. The value of X_L offered by a coil depends on the inductance L of the coil and the frequency of the changing current. X_L increases as L increases, decreases as L decreases. X_L increases as f increases, decreases as f decreases. X_L equals the product of $2\ \pi$, f, and L. The amount of inductance the coil has is determined by the number of turns of wire in the coil, the length and diameter of the coil, and the material upon which the wire is wound.

Measurement of X_L

The reactance of a coil X_L may be determined by measurement. In the circuit of Fig. 18.2-1, a sinusoidal voltage V causes a current I to flow. Ohm's law extended to ac circuits states that the voltage V_L across the coil is $V_L = I \times Z$. Where the resistance of the coil is small compared with X_L, $Z = X_L$ and

$$V_L = I \times X_L \tag{18.2-1}$$

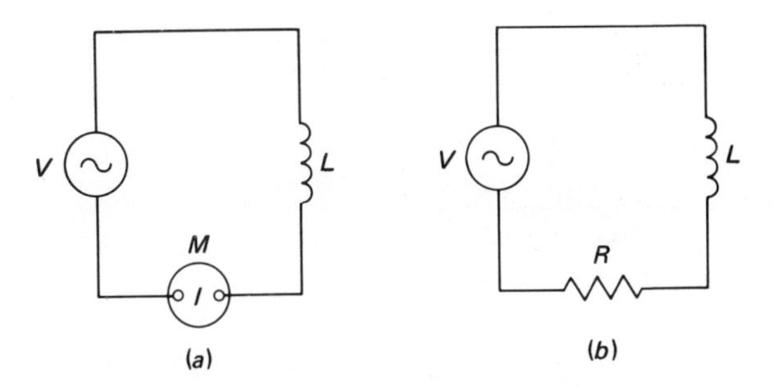

Fig. 18.2-1. Circuits used to determine X_L.

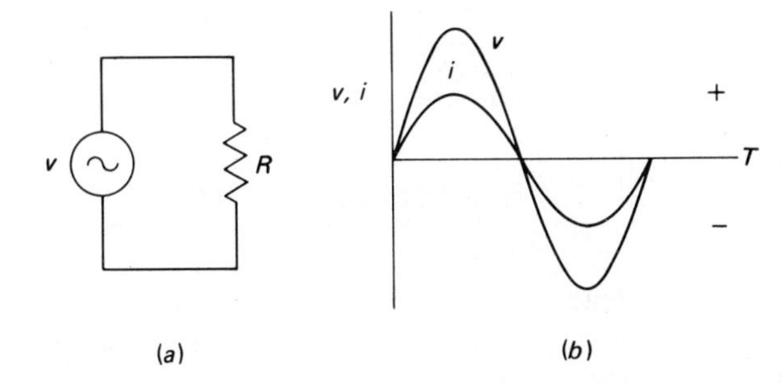

Fig. 18.2-2. In a resistive circuit, voltage and current are in phase.

This equation may also be written as

$$X_L = \frac{V_L}{I} \qquad (18.2\text{-}2)$$

Equation (18.2-2) can be used to determine X_L of a coil at a given frequency f. This is accomplished by using an ac voltmeter to measure the voltage V_L across L, and by measuring the current I with an ac ammeter. The measured values of V_L and I are then substituted in Eq. (18.2-2), and X_L is computed.

Note that a slight error is introduced by this procedure, for L has resistance in addition to inductance. However, if the value of X_L is appreciably larger than the resistance R of the coil, R may be ignored.

An alternative method for measuring the current I does not require the use of an ammeter. Instead of the meter M, a resistor R of known value is placed in series with L, as in Fig. 18.2-1*b*. In a series circuit, current is the same everywhere. Hence the current in R is the same as the current in L. The voltage V_R is then measured across R. Current is next determined by substituting the measured value of V_R and the known value of R in the equation

$$I = \frac{V_R}{R} \qquad (18.2\text{-}3)$$

Next, V_L is measured, and X_L is computed by substituting the values of V_L and I in Eq. (18.2-2).

In Eqs. (18.2-1) to (18.2-3), either the rms values V_L, V_R, and I or the instantaneous values v_L, v_R, and i may be used.

NOTE: X_L can be calculated by the formula $X_L = 2\pi fL$. No measurements are necessary to find X_L by this method if f and L are known.

Phase Relations between Voltage and Current in a Coil

In a dc circuit, as the voltage across a resistor increases, the current in R increases. This is also true in a resistive ac circuit, for if a sinusoidal voltage v is applied across R, the instantaneous variations of current i in R follow exactly the instantaneous changes in voltage v. Thus, at the instant v is going through zero, i is going through zero. When v is at maximum, i is at maximum. Voltage and current in R are said to be in phase. This relationship is shown graphically in Fig. 18.2-2.

In a pure inductive circuit, voltage and current are *not* in phase, but current lags voltage by 90°. A graph of voltage and current (Fig. 18.2-3) shows this relationship. A phase meter or an oscilloscope acting as a phase indicator may be used to verify the phase relationship between voltage and current in an inductor.

SUMMARY

1. Inductive reactance can be calculated by the formula $X_L = 2\pi fL$ or by Ohm's law for ac circuits.
2. In Ohm's law for ac circuits, Z (opposition to the alternating current) is used in place of R. Sometimes either X_L or X_C will be the only opposition in the circuit and as such is substituted for Z.
3. Current and voltage are 90° out of phase in an inductor.
4. In an inductor the voltage leads the current.

SELF-TEST

Check your understanding by answering these questions.

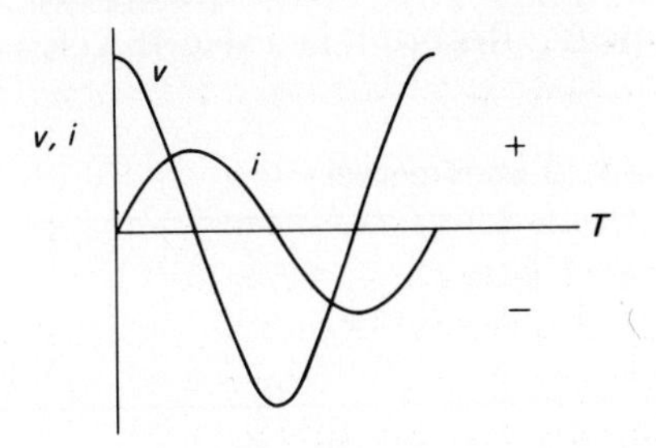

Fig. 18.2-3. In a pure inductive circuit, voltage and current are out of phase, and current lags voltage by 90°.

1. ________ is the ability of a coil to oppose changes in current.
2. ________ is the amount of opposition to current change in an ac circuit.
3. X_L is measured in ________.
4. X_L can be easily measured with an ohmmeter. ________ (true/false)
5. ________ leads ________ in an inductor.
6. Voltage and current are ________° out of phase in an inductor.

MATERIALS REQUIRED

- Power supply: Line-isolation transformer and variable autotransformer operating from the 120-V ac power line
- Equipment: EVM; AF sine-wave generator; oscilloscope
- Resistors: ½-W, 1000-Ω, and as required in step 2
- Inductor: 8 H at 50 mA with internal resistance R = 250 Ω; 30-mH choke
- Miscellaneous: SPST switch; fused line cord

PROCEDURE

Measurement of X_L at Line Frequency

1. Calculate, by the formula $X_L = 2\pi fL$, the X_L of the 8-H coil at 60 Hz. Record it in Table 18.2-1.
2. Connect the circuit of Fig. 18.2-4. R has the same value in ohms as X_L. The autotransformer is plugged into the line-isolation transformer. S_1 is open, the power source is adjusted for 0 V.
3. Close S_1, **power on.** Adjust the source output for 10 V rms.
4. Measure the voltage across R, and record it in Table 18.2-1.
5. Measure and record the voltage across L.
6. Use Eq. (18.2-3) to find the current in the circuit. Record it in Table 18.2-1.
7. Use the measurement in step 5 and the circuit current to calculate X_L. Record it in Table 18.2-1.

TABLE 18.2-1. Measuring X_L at Line Frequency

$X_L = 2\pi fL$ Ω	V_R, V (rms)	V_L, V (rms)	$I = \frac{V_R}{R}$, A	$X_L = \frac{V_L}{I}$, Ω

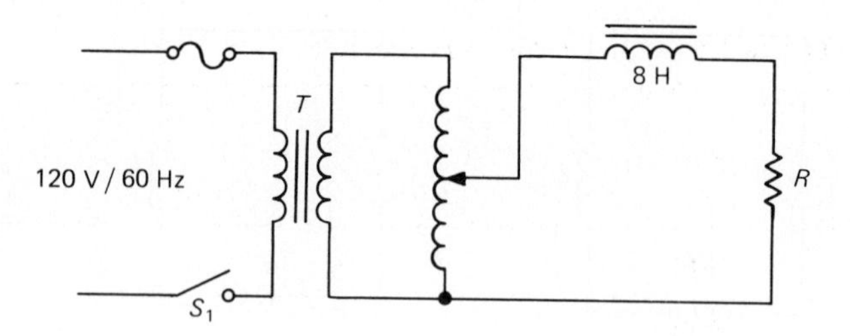

Fig. 18.2-4. Experimental circuit using 60-Hz source.

X_L versus Frequency f

8. Connect the circuit of Fig. 18.2-5. V_1 is an oscilloscope calibrated to measure peak-to-peak voltage.
9. Set the signal generator in turn to each of the frequencies shown in Table 18.2-2. Maintain the output v of the generator at 15 V p-p.
10. With the oscilloscope, measure and record the voltage v_L and v_R for each frequency.
11. Compute and record the value of i (v_R/R) for each frequency, and the value of X_L (using the formula $X_L = v_L/i$).
12. Compute also and record the formula value ($2\pi fL$) of X_L.

QUESTIONS

1. Did the value of X_L found in step 7 compare favorably with the value of X_L found in step 1? Discuss any differences.
2. Did the value of X_L found in step 11 compare favorably

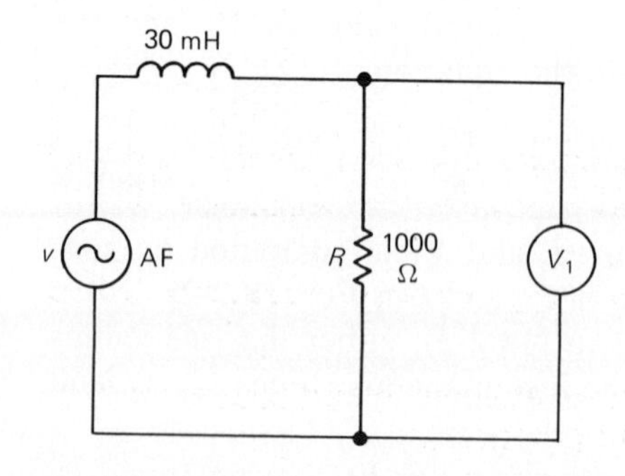

Fig. 18.2-5. Circuit to determine variation of X_L with frequency.

TABLE 18.2-2. Effect of Frequency on X_L

f, Hz	v, V p-p	v_L, V p-p	v_R, V p-p	I, A $\left(\frac{v_R}{R}\right)$	X_L, Ω	
					v_L/i	$2\pi fL$
2000						
3000						
5000						
7000						
8000						
9000						
10,000						

with the value of X_L found in step 12? Discuss any differences.

3. What factors would cause the two methods of finding X_L to produce different answers?
4. In steps 1 through 7, since the calculated X_L and the resistor were supposed to have the same opposition, what relationship would V_R have to V_L? Did it? How does this prove $X_L = 2\pi fL$?
5. How does X_L vary with frequency? Refer to your data in Table 18.2-2.

Answers to Self-Test

1. Inductance L
2. Inductive reactance X_L
3. ohms
4. false
5. Voltage; current
6. 90

EXPERIMENT 18.3. Testing an Inductor and Transformer

OBJECTIVES

1. To make a continuity check of an inductor and transformer and measure the resistance of the winding(s)
2. To voltage-check the windings of a transformer

INTRODUCTORY INFORMATION

Resistance of a Coil

An inductor consists of a number of turns of insulated wire wound around a core. There may be only several turns, or there may be thousands of turns. The more turns there are on a particular core, the greater is the inductance of the coil. The diameter of the wire used in winding a coil depends on the maximum current which that coil will be required to pass. The larger the diameter of the wire, the greater the current capabilities of the coil. A smaller-diameter wire will sustain less current. If more current is permitted to flow in a coil than that for which it is rated, the coil will overheat and may burn away the insulation around the windings, thus shorting windings together. Or an overheated coil may burn out; that is, the wire may open. In either event the coil will become defective.

In electronics we generally deal with relatively small currents. Hence, very small diameter wire is normally used in the construction of coils. Since the resistance of a wire is directly proportional to its length, the resistance of inductors consisting of many turns of fine wire will be relatively high. Thus, the characteristics of inductance *and resistance* are associated with an inductor. Though the resistance of a coil cannot be separated from the inductance, a schematic representation of a coil with inductance L and resistance R (Fig. 18.3-1) shows these two characteristics as though they were separate units acting

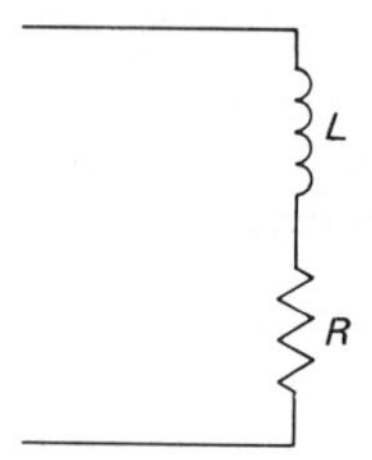

Fig. 18.3-1. Series equivalent form of an inductance.

in series. The resistance can be measured with an ohmmeter. A coil may therefore be represented by its series equivalent form (Fig. 18.3-1).

The resistance associated with the length of wire in the winding suggests that one method of testing an inductor is to place an ohmmeter across the two terminals and compare the measured with the rated resistance. If they are the same, the winding may be assumed to be continuous, that is, neither open nor shorted. If the resistance is infinite, the inductor is open. If the resistance is very much smaller than the rated value, say 20 instead of 500 Ω, it may be assumed that a large part of the coil winding is shorted. In practice, resistance checks of an inductor are made to determine if it is continuous or open. Special instruments are required to identify coils with shorted turns.

The Power Transformer

The transformer consists of two or more coupled windings, used to perform several functions. The transformer windings are separate coils wound on the same core. There are many types of transformers. One is a *power transformer,* which may step up voltage or step down voltage to produce a required voltage derived from the ac line. As a power transformer it may have windings to supply the voltage required to heat vacuum-tube filaments and to supply other voltages for various circuits. Power transformers are also used in the power supplies for transistor and other solid-state circuits.

For resistance testing, each winding of a transformer behaves as a single winding coil. The reason for this can be seen from Fig. 18.3-2, the schematic of a power transformer. The primary of the transformer is connected to the ac line. The main secondary winding may be a center-tapped winding. The voltage induced in the secondary depends on the turns ratio of the secondary winding to the primary windings. And if the secondary is center-tapped, the voltage rating will be given from the center tap to each of the outside leads. For example, a transformer winding with the rating of 260–0–260 would produce 260 V rms from the center tap to each of the end windings or 520 V from end to end.

As a rule the higher the voltage produced by a transformer winding, the lower the current which will be available from that winding. Therefore, high *V* windings can be made of wire of smaller diameter. The resistance of each of the windings depends on the number of turns and upon the diameter of the wire. Since the high-voltage windings of a transformer will have the most wire (and usually the thinnest), the resistance will be the highest of all the windings of a transformer, as high as several hundred ohms.

The filament windings of a transformer used in vacuum-tube circuits or other low-voltage windings of transformers used in transistor power supply circuits usually must produce high currents. Larger-diameter wire will be used in these high-current windings, whose resistance will therefore be low, typically less than 1 Ω for filament windings.

The primary winding must carry high current but also has a large number of turns of wire. Its resistance is usually about 5 Ω in vacuum-tube radio receivers and somewhat less in television receivers.

These "ball park" figures help the service technician identify the windings on a power transformer when the color code is nonstandard or faded so as to be unidentifiable. The EIA color code for power transformers used in vacuum-tube circuits is given in Fig. 18.3-3.

One cannot determine from the color code or from the resistance checks what the output of a transformer will be. It may be several volts or several hundred volts. It is not safe to plug an unknown transformer into the ac line without some knowledge of the transformer. Resistance checks will normally determine if there are any open circuits in the transformer and may help spot shorts. Resistance checks will usually give the technician a fair idea which leads are the primary and which are the secondary. Yet, because of possible high voltages, it is not safe simply to plug the unknown transformer into the ac power line.

A simple method of determining the approximate

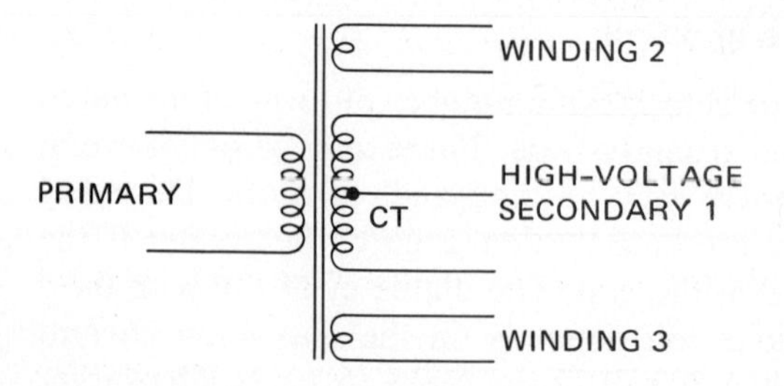

Fig. 18.3-2. Power transformer.

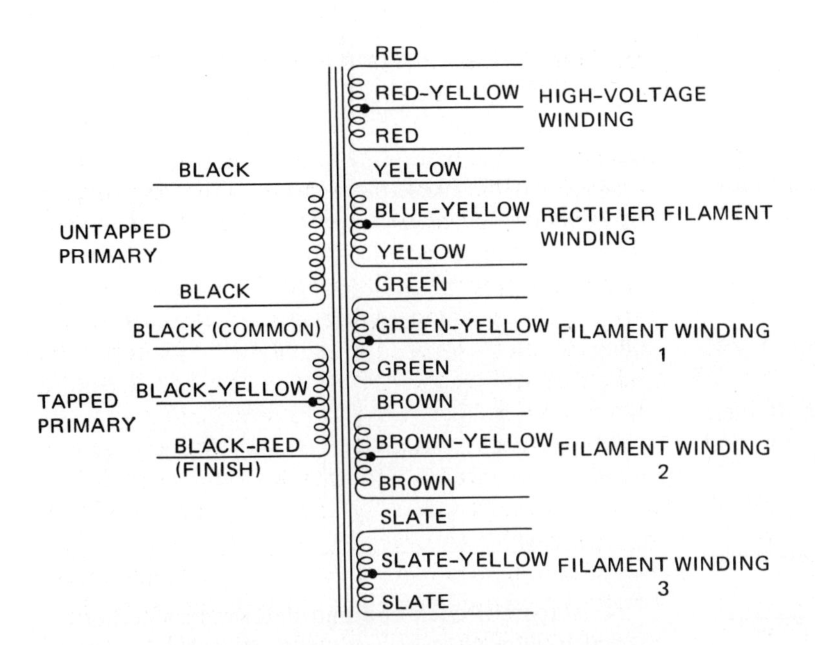

Fig. 18.3-3. EIA color code for power transformer used in vacuum-tube circuits.

output of each winding is as follows. Connect what is believed to be the primary to a sine-wave signal generator set for a frequency of 60 Hz. Adjust the generator output for a low ac rms output, say 10 V. Now, with a voltmeter, check the output of each secondary winding. If one of the windings measures 20 V, we know that this winding is a stepup winding, and the output will be twice the input or 240 V rms when 120 V rms is applied. Another winding may measure 1 V rms. This is one-tenth the input, so this winding is a stepdown winding which reduces the input voltage of the transformer by a factor of 10. When the transformer is plugged into 120 V rms, the output of this winding will produce approximately 12 V.

Not all power transformers have a high-voltage stepup secondary winding. There are power transformers with a primary and one or more *stepdown* secondary windings, such as those used in transistor and other solid-state power supplies. In these, the main secondary winding will draw *more current* than the primary, and its resistance will be *lower* than that of the primary. The main secondary winding may be center-tapped, or may have more than one tap.

With this information the technician knows what to expect when the transformer is plugged into the ac power line for further tests.

SUMMARY

1. An inductor is a coil of insulated wire wound around a core.
2. The more turns of wire in the coil, the greater the inductance.
3. The diameter of wire used determines the amount of current the coil can pass.
4. A coil of small-diameter wire can sustain less current than a coil wound of larger-diameter wire.
5. Most coils used in electronics have many turns of thin wire and therefore have high resistance.
6. The ohmmeter is used to determine whether a coil is open. In some cases it may also be helpful in determining whether it is shorted.
7. A transformer is made up of two or more coupled coils wound on the same core.
8. Resistance checks of a power transformer are made in the same manner as the resistance checks of a simple coil.
9. There are approximate figures which can be used when measuring resistance of power transformers which help identify the primary and secondary windings.

SELF-TEST

Check your understanding by answering these questions.

1. An ohmmeter will give a reading of ________ if the coil being tested is open.
2. To increase the inductance of a coil, ________ (more/fewer) turns are added to the coil.
3. Since the currents used in electronics are generally low, the diameter of the wire from which the coil is wound can be ________ (large/small).
4. Resistance of a wire is directly proportional to its length. ________ (true/false)
5. A coil with shorted turns will measure a ________ (higher/lower) resistance than its rated resistance.

6. The continuity of transformer windings cannot be tested by an ohmmeter because transformers have more than one winding. ________ (true/false)
7. Which winding of a vacuum-tube power transformer normally has the greatest resistance? ________

MATERIALS REQUIRED

- Equipment: AF sine-wave signal generator; EVM or VOM
- Inductors: An assortment of inductors with their rated dc resistance, including an *open* inductor, and a power transformer

PROCEDURE

Coil Resistance

1. Obtain the inductor assortment from your instructor. Identify each as to its use or type (if possible), and record this information in Table 18.3-1.
2. Measure the resistance of each coil and record in Table 18.3-1.
3. Record the rated resistance of each coil in Table 18.3-1.
4. Compare the rated resistance and the measured resistance of each coil and determine whether the coil is normal or has an open, or possible shorted turns. Record your results in Table 18.3-1.

TABLE 18.3-1. Inductor Resistance Measurement

No.	*Type Inductor*	*Rated R*	*Measured R*	*Normal, Short, Open?*
1				
2				
3				
4				
5	Power transformer	Primary		
		Secondary 1		
		Secondary 2		
		Secondary 3		

TABLE 18.3-2. Power Transformer Voltage Measurements

Test Method	V_{in}	*Secondary 1*	*Secondary 2*	*Secondary 3*
Signal generator		Measured *V*	Measured *V*	Measured *V*
AC line		Calculated *V*	Calculated *V*	Calculated *V*
		Measured *V*	Measured *V*	Measured *V*

Power Transformer Measurements

5. Identify the primary winding of the power transformer. Use the signal generator method for determining the output voltage of the power transformer. Record the signal generator output voltage in column V_{in} in Table 18.3-2.
6. Measure and record in the block labeled "Measured *V*" the voltage of each secondary winding of the power transformer.
7. Using the values measured in step 6, calculate the output voltage for each secondary winding when the transformer is connected to the ac line. Measure the ac line voltage and record it before making the calculations. Record the calculated values in the block labeled "Calculated *V*" in Table 18.3-2.
8. Connect the primary of the power transformer to the ac line and measure the output voltage of each secondary winding. Record these values in Table 18.3-2.

QUESTIONS

1. How closely did the rated values of inductor resistance compare with the measured values?
2. Could you easily identify the various windings of the power transformer?
3. Which inductor was open? ________. How could you tell?
4. In your own words explain why the high-voltage secondary winding of a power transformer has the most

turns and the highest resistance of any winding in the transformer.

5. What would cause any variations in the rated inductor resistance from the measured inductor resistance?
6. Refer to the data from Table 18.3-2. How did the experimental results compare with the actual voltage output for the windings in the transformer? Account for any discrepancies.

Answers to Self-Test

1. infinity
2. more
3. small
4. true
5. lower
6. false
7. High-voltage secondary

CHAPTER 19 IMPEDANCE: SERIES *RL* AND *RC* CIRCUITS

EXPERIMENT 19.1. Impedance of a Series *RL* Circuit

OBJECTIVE

To verify that the impedance Z of a series RL circuit may be calculated from the formula $Z = \sqrt{R^2 + X_L^2}$

INTRODUCTORY INFORMATION

Impedance of a Series *RL* Circuit

The total opposition to alternating current in an ac circuit is called the impedance of the circuit. Consider the circuit of Fig. 19.1-1. If it is assumed that the choke coil L, through which alternating current flows, has zero resistance, the current is impeded only by the X_L of the choke. That is, $Z = X_L$. In this case, if $L = 8$ H and $f = 60$ Hz, $X_L = 2\pi fL = (6.28)(60)(8) = 3015\ \Omega$. How much current will there be in the circuit if $V = 6.3$ V? Apply Ohm's law:

$$I = \frac{V}{X_L} = \frac{6.3}{3015} = 2.09 \text{ mA}$$

If there is resistance associated with the inductance or if L is in series with a resistor of, say, 3000 Ω (Fig. 19.1-2), there will be less than 2.09 mA of current. How much current will flow, assuming the same X_L as previously computed? *Measurement* shows that there is 1.485 mA in the circuit. It is evident that the total impedance of the resistor and the inductor L connected in series is *not* simply the arithmetic sum of R and X_L. If Z were the simple sum of R and X_L,

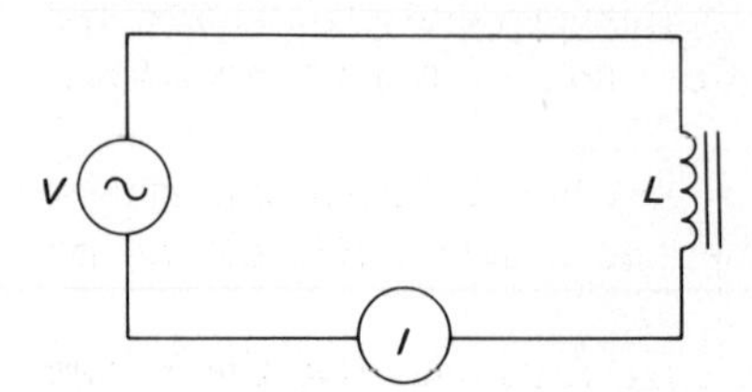

Fig. 19.1-1. The current I in a pure inductance is limited only by the inductive reactance X_L of the circuit.

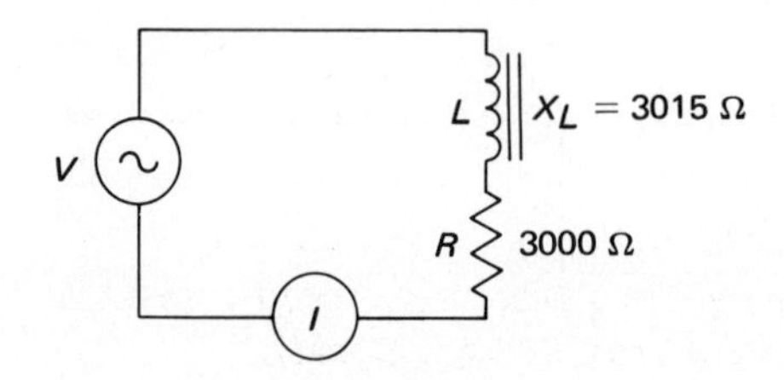

Fig. 19.1-2. The impedance Z of L in series with R is greater than that of L alone.

it would equal 6015 Ω. For this value, circuit current I, calculated by Ohm's law would be

$$I = \frac{V}{Z} = \frac{6.3}{6{,}015} = 1.047 \text{ mA}$$

which is incorrect according to our earlier measurement.

Vector Sum of *R* and X_L

We have shown in previous experiments that V and I are in phase in a pure resistance and 90° out of phase in capacitors and inductors. With some thought it can be reasoned that a reactance is also 90° out of phase with resistance; that is, R and X_L are 90° out of phase with each other in an RL circuit, while R and X_C are also 90° out of phase in an RC circuit.

Since X_L and R are 90° out of phase, they cannot be added directly. They must be added vectorially.

The vector diagram of Fig. 19.1-3 shows that R and X_L are at right angles to each other and that X_L leads R by 90°. Vector addition is achieved as follows: Line AB is drawn to scale to represent X_L, and line BC is drawn to scale to represent R, and at right angles to X_L. Line AC is then drawn, connecting points A and C. The total opposition Z is found by measuring the length of the resultant line AC and converting it to ohms by the scale factor.

Mathematically line AC is the hypotenuse of the right triangle ABC. Applying the Pythagorean theorem to this right triangle, we note that the length of the

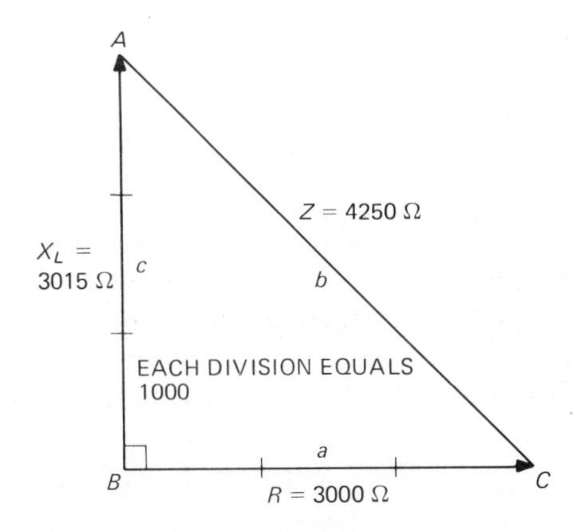

Fig. 19.1-3. Vector relations in a series RL circuit. Impedance is a vector quantity.

hypotenuse (b) is equal to the square root of the sum of the squares of the other two sides (a and c).

$$b = \sqrt{a^2 + c^2} \qquad (19.1\text{-}1)$$

Applying this theorem to the right triangle created by the lines drawn in scale length to represent X_L and R, we see that the equation for finding the length of the line representing Z is

$$Z = \sqrt{X_L^2 + R^2} \qquad (19.1\text{-}2)$$

This equation makes it possible to determine the impedance of an RL circuit if R and X_L are known.

One of the two methods for finding Z discussed to this point uses graph paper, the other uses mathematics.

1. Using graph paper, draw a line to represent the value of X_L and another line at right angles to represent the value of R. Draw these lines to scale. See Fig. 19.1-3. Now, measure the hypotenuse. The measure of the length of the hypotenuse is the scale value of Z. Note that in Fig. 19.1-3 each line division equals 1000 Ω, and R and X_L are each approximately 3 divisions long. The hypotenuse is 4.25 divisions long, or 4250 Ω.
2. For the purely mathematical method, use Eq. (19.1-2).

$$Z = \sqrt{X_L^2 + R^2} = \sqrt{(3015)^2 + (3000)^2}$$
$$= \sqrt{9.09 \times 10^6 + 9 \times 10^6}$$
$$= \sqrt{18.09 \times 10^6} = 4253\ \Omega$$

The second method points up a very important difference between the mathematics of ac and dc circuits. In a dc circuit consisting of series-connected resistances, the total opposition to current R_T is simply the sum of all the resistances. In an ac circuit consisting of a series-connected resistor and inductance, the total opposition to current, the impedance Z, is *not* the arithmetic sum of R and X_L, but is the vector sum of R and X_L.

Ohm's Law Applied to a Series *RL* Circuit

Ohm's law extended to ac circuits states that the impedance in a circuit equals the ratio of the applied voltage and the circuit current.

$$Z = \frac{V}{I} \qquad (19.1\text{-}3)$$

Thus, in the circuit of Fig. 19.1-2 we can measure V and I and compute Z by substituting the measured values of V and I in Eq. (19.1-3). We can also calculate Z by the equation given earlier [Eq. (19.1-2)]. If the results agree, that is, if the value of Z is substantially the same for each of these methods, then we have verification of the formula for Z which states that

$$Z = \sqrt{X_L^2 + R^2}$$

SUMMARY

1. The total opposition to current in an ac circuit is called the impedance Z of the circuit.
2. Reactance and resistance cannot be added arithmetically, but must be added vectorially to arrive at the circuit impedance.
3. Impedance can be calculated by Ohm's law for ac circuits using the formula $Z = V/I$.
4. Impedance can also be calculated by the formula $Z = \sqrt{X_L^2 + R^2}$.
5. In a third method, impedance can be found by laying off X_L and R, to scale, on graph paper at right angles (Fig. 19.1-3) and finding the vector sum by measuring the length of the line (hypotenuse) joining the ends of X_L and R. The measured length is the scale value of impedance.

SELF-TEST

Check your understanding by answering these questions.

1. The symbol for impedance is ________.
2. In an ac circuit containing only an inductor and an ac source, if the inductor has no resistance, X_L is the circuit ________.
3. The voltage drop across the resistance V_R lags the inductor voltage V_L in a series RL circuit by ________°.
4. In an ac RL circuit the resistance and the X_L are added directly to give the circuit impedance. ________ (true/false)

5. Ohm's law for dc circuits states that $R = V/I$; the same formula is used for ac circuits except that R is replaced by ________.

MATERIALS REQUIRED

- Power supply: Isolation transformer and variable autotransformer operating from the 120-V 60-Hz line
- Equipment: EVM; 0- to 25-mA ac meter
- Resistors: ½-W 2700-Ω; 5-W 5000-Ω
- Inductor: 8 H at 50 mA direct current, with resistance of 250 Ω
- Miscellaneous: SPST switch; fused line cord

PROCEDURE

1. Connect the circuit of Fig. 19.1-4. S is open. R is a 5-W, 5000-Ω resistor. L is an iron-core choke rated 8 H at 50 mA direct current. V_A is a variable autotransformer plugged into T, a 1:1 isolation transformer, which derives its power from the 120-V 60-Hz line. M is a 0- to 25-mA ac meter, and V is an EVM on the ac voltage range.
2. Close S. **Power on.** Adjust the output of the ac supply until the meter measures 15 mA (0.015 A). If an alternating current meter is not available to determine the circuit current, measure the ac voltage drop across the resistor and by Ohm's law calculate the circuit current. Measure the rms voltage V_A applied to the circuit and record it in Table 19.1-1. Also measure and record V_R, the voltage across R, and V_L, the voltage across L. Compute and record X_L and Z. ($X_L = V_L/I$; $Z = V_A/I$.)
3. Substitute the computed values of X_L and R (5 kΩ) in the equation $Z = \sqrt{R^2 + X_L^2}$. Find Z and record it in Table 19.1-1. Show your computations.
   Confirm the measured value of I by computing I from the equation $I = V_R/R$. Record it in Table 19.1-1. **Power off.**
4. Replace R with a 2700-Ω resistor. **Power on.** Repeat steps 2 and 3 and record your results in Table 19.1-1.
5. Using X_L and R from above, find Z by the scale-drawing method for the circuit employing the 5000-Ω resistor. Show your work.

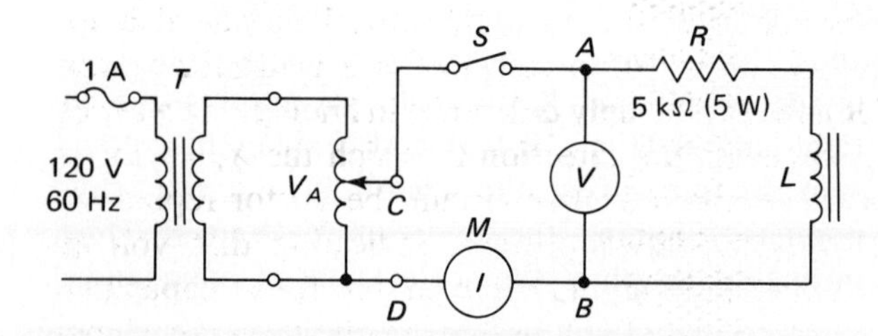

Fig. 19.1-4. Determining the impedance Z of an RL circuit.

QUESTIONS

1. Did the values for Z found in steps 2 and 3 agree? Give reasons for any discrepancies.
2. Refer to Table 19.1-1. Is the measured value of I the same as the calculated value of I? Explain why it should be the same.
3. Define Z.
4. What are the possible sources of error in this experiment?
5. Did the value for Z found by scale drawing agree substantially with the calculated values found in steps 2 and 3?
6. Why would the scale-drawing method for determining Z not be as accurate as the other methods?
7. Did you confirm the formula $\mathbf{Z} = \sqrt{X_L^2 + R^2}$? How?

Answers to Self-Test

1. Z
2. impedance or Z
3. 90
4. false
5. Z

TABLE 19.1-1. Impedance of a Series RL Circuit

R, Ω	I, A	V_A, V	V_R, V	V_L, V	$X_L = \frac{V_L}{I}$, Ω	$Z = \frac{V_{AB}}{I}$, Ω	$Z = \sqrt{R^2 + X_L^2}$, Ω	$I = \frac{V_R}{R}$, A
5000	0.015							
2700	0.015							

EXPERIMENT 19.2. Impedance of a Series *RC* Circuit

OBJECTIVE

To verify that the impedance Z of a series RC circuit may be calculated from the formula

$$Z = \sqrt{X_c^2 + R^2}$$

INTRODUCTORY INFORMATION

Impedance of a Series *RC* Circuit

The total opposition to alternating current in an ac circuit is called the impedance of the circuit. Consider the circuit of Fig. 19.2-1. If it is assumed that the capacitive circuit through which alternating current flows has zero resistance, the current is impeded only by the reactance X_c of the capacitor. That is, where $R = 0, Z = X_c$. In this case, if $C = 0.5\ \mu\text{F}$ and $f = 60$ Hz,

$$X_c = \frac{1}{2\pi fC} = \frac{10^6}{6.28(60)(0.5)} = 5305\ \Omega$$

How much current will there be in the circuit if $V =$ 12.6 V? Apply Ohm's law:

$$I = \frac{V}{X_c} = \frac{12.6}{5305} = 2.375\ \text{mA}$$

If there is resistance associated with the capacitor (leakage) or if C is in series with a resistor of, say, 5000 Ω (Fig. 19.2-2), there will be *less* than 2.375 mA in the circuit. How much current will there be in the circuit, assuming the same X_c as previously computed? *Measurement* shows that there is 1.73 mA in the circuit. It is evident that the total impedance of the resistor and the capacitive reactance, connected in series, is *not* simply the arithmetic sum of R and X_c. If it were the simple sum of the R and X_c,

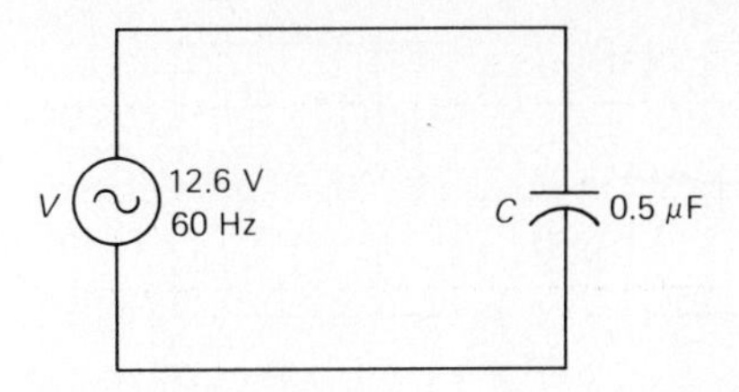

Fig. 19.2-1. Current in a pure capacitive circuit is limited only by the capacitive reactance X_c of the circuit.

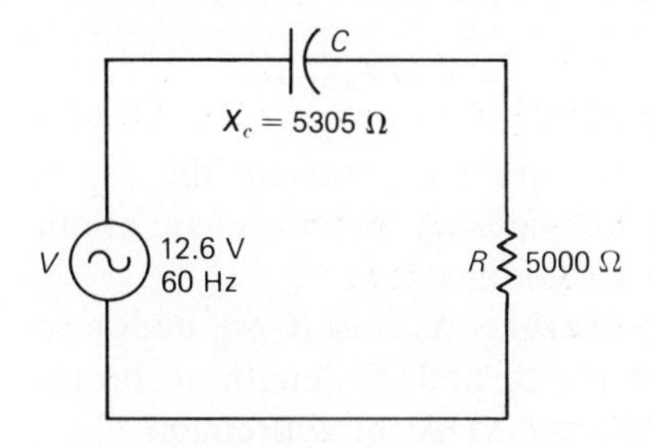

Fig. 19.2-2. The impedance Z of C in series with R is greater than that of C alone.

Z would be 10,305 Ω, and circuit current when calculated by Ohm's law would be

$$I = \frac{V}{Z} = \frac{12.6}{10{,}305} = 1.22\ \text{mA}$$

The current 1.22 mA is incorrect, according to our earlier measurement.

Determining *Z* in a Series *RC* Circuit

We have learned previously that V and I are in phase in a pure resistance, and 90° out of phase in capacitors and inductors. In Exp. 19.1 it was shown that in a circuit where a resistor and an inductor are in series, the impedance must be found by vector addition, that is, by adding R and X_L vectorially. This rule also applies to a series RC circuit; that is, Z is the vector sum of R and X_c. The only difference in finding Z in a series RC circuit is the direction in which the X_c vector is drawn. In the inductive circuit the vector representing X_L was drawn upward, indicating that voltage leads current in an inductive circuit. In the capacitive circuit the vector for X_c is drawn pointing down, showing that current leads voltage by 90°. See Fig. 19.2-3.

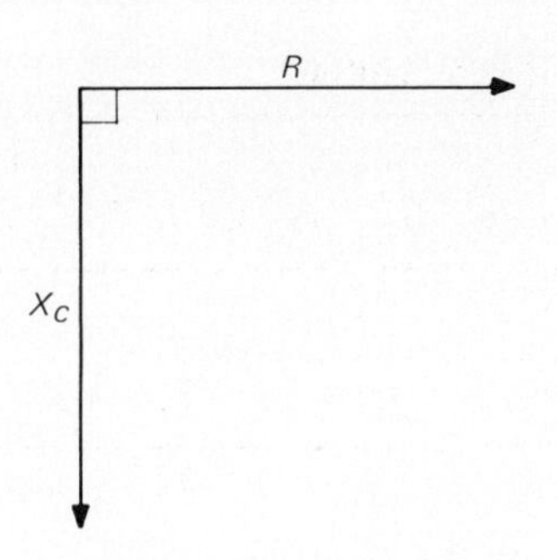

Fig. 19.2-3. The vector for X_c is drawn pointing down to show that I leads V in a capacitive circuit.

In the example given in Fig. 19.2-2, using the formula $Z = \sqrt{X_c^2 + R^2}$, we find that $Z = 7290\ \Omega$. Likewise, by knowing the measured current and the applied voltage, Z can be calculated by Ohm's law for ac circuits.

$$Z = \frac{V}{I} = \frac{12.6}{0.00173} = 7283\ \Omega$$

Though the two figures are not precisely the same, they are quite close considering meter measurement error and decimal round-off error.

To find Z by scale drawing, X_c and R are drawn to scale length on graph paper, and the length of the resultant vector is measured. This measurement (BC) will give the scale value of Z. See Fig. 19.2-4. This method of finding Z is the least accurate of the three, but if care is given to measurement and the scale drawing is made as large as possible, the result will be surprisingly close.

It can be seen that the calculation of Z in an RC circuit is the same as the calculation of Z in an RL circuit. It must be remembered, however, that X_c and X_L are 180° out of phase. Also, in an inductive circuit the voltage leads the current, and in the capacitive circuit the current leads the voltage.

SUMMARY

1. The total opposition to current in an ac circuit is called the impedance of the circuit.
2. Reactance and resistance cannot be added arithmetically but must be added vectorially to determine circuit impedance.
3. Impedance can be calculated by Ohm's law for ac circuits using the formula $Z = V/I$.
4. Impedance of a series-connected RC circuit can be calculated using the formula $Z = \sqrt{X_c^2 + R^2}$.
5. Impedance can be found in a series RC circuit by drawing to scale X_c and R, 90° out of phase, and then finding their vector sum. This is done by measuring the length of the *resultant* line, which is the scale value of Z.

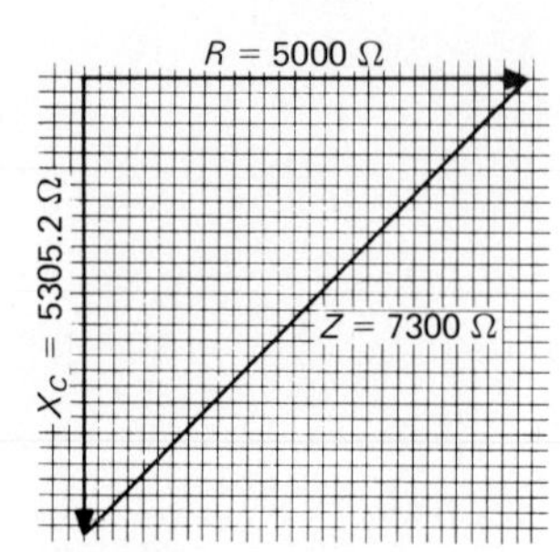

Fig. 19.2-4. Vector relations in a series RC circuit. Impedance is a vector quantity.

SELF-TEST

Check your understanding by answering these questions.

1. If a capacitive circuit has no leakage resistance and there is no resistor connected in the circuit, the X_c is the circuit ________.
2. The voltage drops across the resistance V_R and the capacitor V_c in a series RC circuit are ________° out of phase.
3. In an RC circuit the resistance and X_c are added arithmetically to give the circuit impedance. ________ (true/false)
4. The current leads the voltage in a series RC circuit. ________ (true/false)

MATERIALS REQUIRED

- Power supply: Isolation transformer and variable autotransformer operating from the 120-V, 60-Hz line
- Equipment: EVM; 0- to 25-mA ac meter
- Resistors: 1-W 5000-Ω; ½-W 22,000-Ω
- Capacitors: 0.5-μF; 0.1-μF
- Miscellaneous: SPST switch; fused line cord

PROCEDURE

1. Connect the circuit of Fig. 19.2-5. S is open. R is a 5000-Ω resistor; C is a 0.5-μF capacitor. V_A is a variable autotransformer plugged into T, a 1:1 isolation transformer. M is an ac milliammeter set on the 0- to 25-mA range. V is an EVM set to the 100-V ac range.
2. Close S. Adjust the output of the autotransformer until M measures 10 mA. With the voltmeter, measure the voltage V across the RC circuit and record it in Table 19.2-1. Compute Z by the formula $Z = V/I$ and record it in the column V/I.

   NOTE: If an ac ammeter is not available, measure the voltage drop across R and calculate the circuit current by Ohm's law.

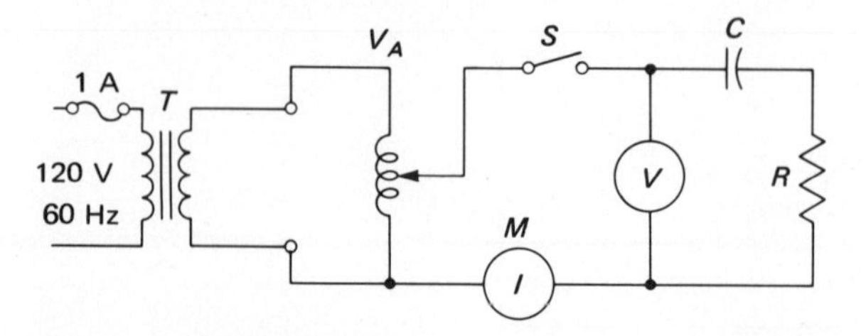

Fig. 19.2-5. Experimental circuit for determining the impedance Z of a series RC circuit.

TABLE 19.2-1. Impedance of a Series *RC* Circuit

R, Ω	C, μF	I, A	V, V	$Z = \frac{V}{I}$, Ω	X_c, Ω	$Z = \sqrt{X_c^2 + R^2}$, Ω	V_R, V	$I = \frac{V_R}{R}$, A
5 kΩ	0.5	0.010						
22 kΩ	0.1	0.002						

3. Compute X_c using the formula $X_c = 1/(2\pi fC)$ where f is the power-line frequency. Record X_c in Table 19.2-1. Calculate and record Z using the formula $Z = \sqrt{X_c^2 + R^2}$.
4. Measure and record V_R, the voltage across R. Compute I, using the formula $I = V_R/R$. Record I in the column V_R/R.
5. **Power off.** Replace C by a 0.1-μF capacitor and R by a 22,000-Ω resistor. Switch the milliammeter to the 0- to 5-mA range. Repeat steps 2, 3, and 4, but set the output of the variable autotransformer for 2 mA of current.

QUESTIONS

1. What are the possible sources of error in the data in Table 19.2-1?
2. Did the measured value of I compare favorably with the calculated value of I? Explain any discrepancies.
3. Did the two methods of calculating Z compare favorably? Explain any discrepancies.
4. Define Z.
5. In an RC circuit what other way besides making R larger can the circuit Z be made greater?
6. Did the experiment confirm that Z in a series-connected RC circuit may be calculated using the formula $Z = \sqrt{R^2 + X_c^2}$? Explain how.

Answers to Self-Test

1. Z
2. 90
3. false
4. true

CHAPTER 20 SERIES RESONANCE

EXPERIMENT 20.1. Frequency Response of a Series-Resonant Circuit

OBJECTIVE

To determine by experiment (*a*) the resonant frequency and (*b*) the frequency-response characteristic of a series-resonant circuit

INTRODUCTORY INFORMATION

Characteristics of a Series-Resonant Circuit

Some characteristics of a series-connected *RLC* circuit are:

1. The voltage drop across a reactive component is equal to the product of the current in the circuit and the reactance of the component. $V = IX_L$, and $V = IX_c$.
2. Capacitive reactance X_c tends to cancel the effects of inductive reactance X_L, and vice versa.
3. The impedance of an *RLC* circuit is given by the formula

$$Z = \sqrt{R^2 + (X_L - X_c)^2}$$

   Note that $(X_L - X_c)$ in the above formula illustrates the cancellation effect of X_c and X_L.
4. Because X_c and X_L cancel completely when they are of equal value, the impedance of the *RLC* circuit is minimum when $X_c = X_L$. When this condition exists, current in the circuit is maximum, and the circuit is said to be resonant. This occurs at a particular frequency f_R, called the *resonant frequency*. The resonant frequency can be calculated using the formula

$$f_R = \frac{1}{2\pi\sqrt{LC}}$$

Consider the circuit in Fig. 20.1-1. Assume V is an ac generator whose frequency and output voltage are manually adjustable. For each frequency there will be a current I, such that $I = V/Z$. The voltage drops across R, L, and C can be calculated by the expressions IR, IX_L, and IX_c, respectively.

If the generator frequency is changed but V remains constant, the current will change, and the voltage drops across R, L, and C will change. The reason, as we have studied previously, is that both X_L and X_c are dependent on frequency. The higher the frequency, the greater the X_L but the lower the X_c. The lower the frequency, the lower the X_L but the higher the X_c. So, as the frequency is increased, X_L increases, and so does the voltage drop across X_L. At the same time, X_c decreases, as does the voltage drop across X_c.

Resonant Frequency

There is a frequency f_R, called the *resonant frequency* of the series *RLC* circuit, at which $X_L = X_c$. The resonant frequency can be determined using the formula $f_R = 1/(2\pi\sqrt{LC})$, where f is in Hz, L in henrys, and C in farads. At f_R,

$$Z = \sqrt{R^2 + (0)^2} = R$$

Therefore, a minimum impedance Z exists at f_R, and maximum current will flow in the circuit.

At f_R the applied voltage appears across R and the current may be computed from the formula

$$I = \frac{V}{R}$$

Since *at resonance* I is limited only by the value of R, the circuit is said to be *resistive*. For all frequencies higher than f_R, X_L is larger than X_c. The net reactance $X_L - X_c$ is inductive, and the circuit is *inductive*. For all frequencies lower than f_R, X_c is larger than X_L, and the circuit is capacitive.

At resonance the voltage across L and the voltage

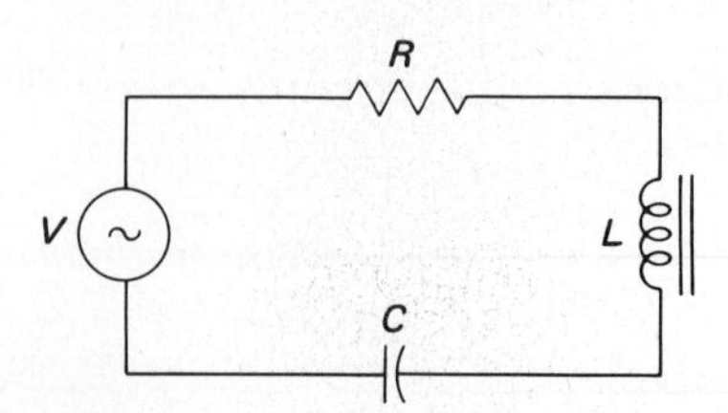

Fig. 20.1-1. A series *RLC* circuit.

across C are maximum and equal. Theoretically, at the resonant frequency, when the reactance becomes zero, the current becomes infinite (assuming no circuit R) and the voltages across L and C become infinitely large. Practically, this condition is never reached because there will be some resistance in the circuit due to circuit wiring and coil resistance.

The amount of coil resistance therefore determines the current flow through the circuit at resonance if there is no external resistance.

Applications of Series-Resonant Circuits

Series-resonant circuits are used extensively in communications electronics. For example, the series-resonant circuit connected from an antenna across the input to a radio receiver, in Fig. 20.1-2, acts as a trap to exclude an unwanted frequency, f_R. At the resonant frequency of the LC circuit in Fig. 20.1-2, Z is minimum and I is maximum. This means that at f_R the incoming signal on the antenna has an easy path to ground through the resonant circuit consisting of C and L and is bypassed around the radio receiver. This trap circuit can be used to "tune out" an unwanted radio station or interference.

In Fig. 20.1-3, the LC circuit is in series with the signal path from the antenna to the radio receiver. Here, at the resonant frequency f_R, when Z is minimum, the signal finds little opposition to entering the receiver. But what of other frequencies? We found earlier that at any frequency except f_R the impedance of the circuit increases. This would mean that frequencies other than the one at which the LC circuit is resonant would find more opposition to entering the receiver. They would not get through to the receiver, and therefore would not be heard. In effect our radio receiver has been "tuned" to a single frequency f_R (or a narrow band of frequencies on either side of f_R) by the series-resonant circuit. This radio receiver is similar to the single-band receivers sold as weather receivers, aircraft receivers, etc.

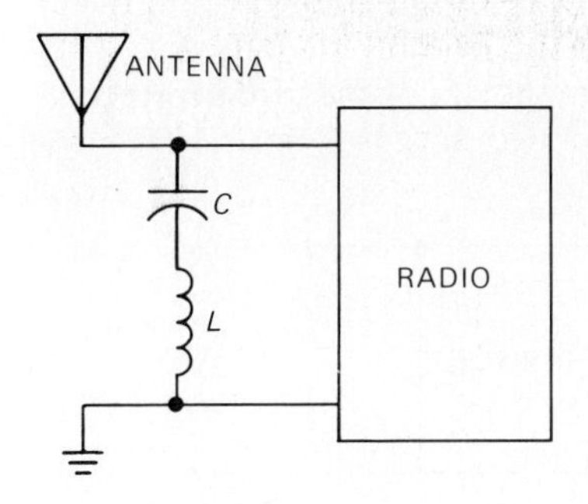

Fig. 20.1-2. A series-resonant circuit used to "trap out" unwanted radio signals.

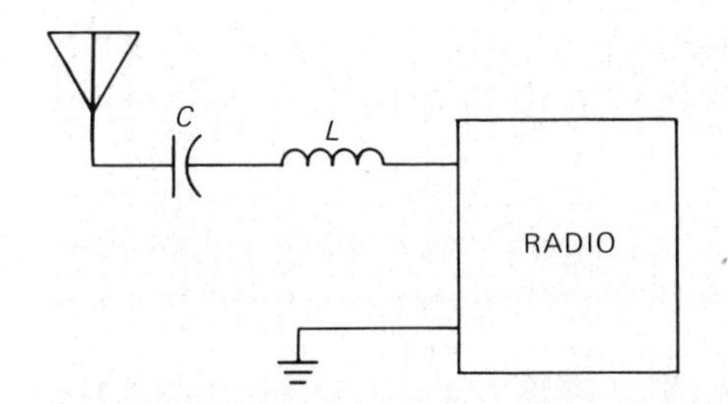

Fig. 20.1-3. A series-resonant circuit used to "tune in" a narrow band of frequencies.

Frequency Response of a Series-Resonant Circuit

Series-resonant circuits respond differently to the frequencies above and below f_R. A sharply tuned resonant circuit will "tune out" almost all frequencies except f_R. A broadly tuned resonant circuit may actually be resonant over a wide band of frequencies. This can be shown by a graph. When the amplitude of the circuit current (or the voltage across L or C) is plotted against f, the resulting graph is called a *frequency-response curve.* Figure 20.1-4 shows a frequency-response curve for (*a*) a sharply tuned and (*b*) a broadly tuned resonant circuit. Here, the response curve shows how V_C (which equals $I \times X_c$) varies with frequency. Note that in Fig. 20.1-4*a* the response falls off rapidly on each side of the resonant frequency. In Fig. 20.1-4*b*, however, the circuit is broadly tuned, and several frequencies on either side of the resonant frequency have the same amplitude as f_R. In each response curve the voltage across one of the reactive components, in this case C, is plotted against frequency.

SUMMARY

1. In a series RLC circuit capacitive reactance tends to cancel the effects of inductive reactance, and vice versa.
2. A series-connected RLC circuit will be resonant at some frequency f_R.

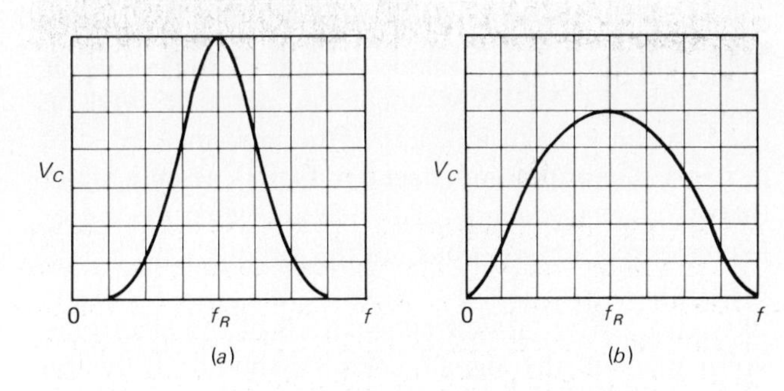

Fig. 20.1-4. Frequency-response curves for a series-resonant circuit: (*a*) sharply tuned curve; (*b*) broadly tuned curve.

3. At resonance $X_c = X_L$.
4. At resonance, circuit current is maximum and circuit impedance is minimum.
5. At resonance, the circuit impedance is equal to R since X_c and X_L cancel each other completely.
6. The resonant frequency of an LC circuit is given by the formula $f_R = 1/(2\pi\sqrt{LC})$.
7. The frequency response of a series RLC circuit is a graph which shows the relationship between circuit current and frequency as f is varied on either side of resonance.

SELF-TEST

Check your understanding by answering these questions. (All questions pertain to a series RLC circuit.)

1. When a resonant circuit is connected in series with a signal source, it allows the resonant frequency through. ________ (true/false)
2. At frequencies above resonance, X_c is smaller than X_L. ________ (true/false)
3. At frequencies below resonance, X_L is smaller than X_c. ________ (true/false)
4. When are X_c and X_L equal? ________
5. At resonance Z is ________ (maximum/minimum)?

MATERIALS REQUIRED

- Equipment: AF sine-wave generator; oscilloscope; EVM
- Resistor: ½-W 100-Ω
- Capacitors: 0.001-, 0.01-, 0.05-μF
- Miscellaneous: 30-mH (approximately) choke

NOTES ON PROCEDURE

In one part of this experiment the student will determine the resonant frequency f_R of a series LC circuit. There will be no resistance in the circuit other than the internal resistance of the coil. Resonance will therefore be determined by tuning the signal-generator frequency for maximum voltage V_L across L, or V_C across C. An EVM connected across C will be used as the voltage indicator. One fact must be kept in mind. The input impedance of the EVM may affect the resonant frequency of the circuit when the meter is connected across L or C in the circuit.

CAUTION: The ground connections of the electronic voltmeter and the signal generator should always be at the same common point, to avoid confusing effects which may arise from internal connections within the instruments. The component across which a voltage is to be measured is connected so that one leg is grounded. The order in which the components are connected in the series circuit will not affect circuit operation.

When loaded by a frequency-selective circuit, the voltage output of most signal generators will not remain constant as the frequency is varied. In order to compensate for this characteristic of the generator, the generator-output level must be monitored and maintained at the same voltage output, if necessary, at each new frequency setting of the instrument. Thus, a constant signal input to the circuit under test will be maintained. In this experiment the generator-output level will be monitored with an oscilloscope, and the generator-output cables will be terminated by a 100-Ω carbon resistor to minimize generator-signal voltage variations.

PROCEDURE

Resonant Frequency

1. Adjust the oscilloscope for proper viewing, and calibrate it for voltage measurement.
2. Set the frequency of the AF generator at 30 kHz, and connect the oscilloscope across the output of the signal generator. Terminate the generator-output cable with a 100-Ω carbon resistor.
3. Connect the circuit of Fig. 20.1-5. Leave the scope connected across the generator output. Set the output of the generator at 2 V p-p. *Maintain generator-signal output voltage at this level (2 V p-p) for the remainder of the experiment.*
4. Now connect the electronic voltmeter across C as in Fig. 20.1-5 and observe the rms voltage V_C across C. Vary the generator frequency first on one side of 30 kHz and then on the other side of 30 kHz until the maximum voltage V_C appears across C. Now read the generator frequency. This is the resonant frequency f_R of the circuit. Record

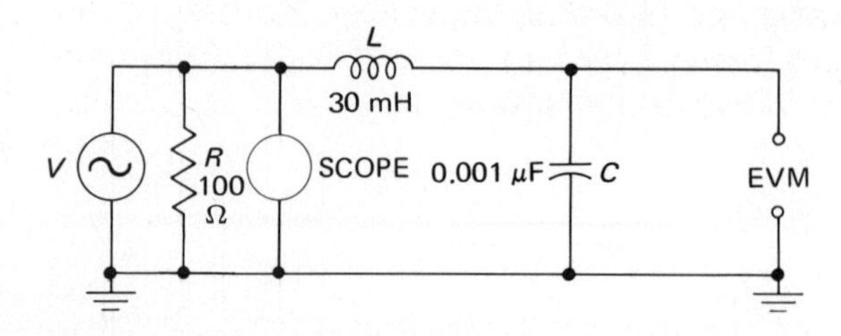

Fig. 20.1-5. Experimental circuit.

TABLE 20.1-1. Resonant Frequency of a Series *LC* Circuit

			Resonant Frequency, Hz	
Step	L, mH	C, μF	Measured	Calculated $f_R = \frac{1}{2\pi\sqrt{LC}}$
4 to 7	30	0.001		
5 to 7	30	0.05		
6 to 7	30	0.01		

this frequency in Table 20.1-1 in the column headed "Measured."

5. Replace the 0.001-μF capacitor in Fig. 20.1-5 with a 0.05-μF capacitor. Following the method in steps 3 and 4, find the new resonant frequency of the circuit and record it in Table 20.1-1.
6. Repeat step 5, using a 0.01-μF capacitor.
7. For each value of L and C used, compute the resonant frequency of the circuit using the formula $f_R = 1/(2\pi\sqrt{LC})$. Record each value of f_R in the appropriate column labeled "Computed."

Frequency Response

8. Leave the circuit connected as in step 6 and find the resonant frequency f_R. It should be the same frequency as in step 6. Record it in Table 20.1-2.
9. With the oscilloscope measure the generator (still connected across the circuit) output and readjust the generator level, if necessary, for 2 V p-p. *Maintain generator output at this level throughout.*
10. Now measure the voltage across C and record it in Table 20.1-2.
11. Fill in the "Frequency" column in Table 20.1-2, performing the required frequency additions and subtractions.
12. Make a frequency-response check of the circuit by setting the signal-generator frequency where possible to the frequencies listed in Table 20.1-2, noting and recording the voltage V_c across C for each frequency setting. *Be certain that the generator output is 2 V p-p at each frequency setting.*

TABLE 20.1-2. Frequency Response of a Series *LC* Circuit

Frequency Deviation, Hz	Frequency, Hz	V_C, V rms
$f_R - 5000$		
$f_R - 4000$		
$f_R - 3000$		
$f_R - 2000$		
$f_R - 1000$		
$f_R - 500$		
Resonance, f_R		
$f_R + 500$		
$f_R + 1000$		
$f_R + 2000$		
$f_R + 3000$		
$f_R + 4000$		
$f_R + 5000$		

QUESTIONS

1. In a series RLC circuit, is it possible for the voltages V_L and V_C to be higher than the applied generator voltage? If so, under what conditions?
2. What is the impedance of a series LC circuit at resonance?
3. In this experiment you measured only V_C. How did you know that f_R had been reached when V_C was maximum?
4. Would the same results have been reached if you had measured the voltage across L instead of across C? Explain.
5. Using the data in Table 20.1-2, draw a frequency response curve showing V_C as a function of f.
6. Did your experiment confirm the formula $f_R = 1/(2\pi\sqrt{LC})$? Explain.

Answers to Self-Test

1. true
2. true
3. true
4. At resonance
5. minimum

CHAPTER 21 PARALLEL RESONANCE

EXPERIMENT 21.1. Resonant Frequency of a Parallel-Resonant Circuit

OBJECTIVE

To determine by experiment the resonant frequency of a parallel LC circuit

INTRODUCTORY INFORMATION

Characteristics of a Parallel-Resonant Circuit

Resonant Frequency of a High-Q Circuit

Consider the circuit of Fig. 21.1-1*a* consisting of C and L in parallel. This theoretical circuit must be modified as in Fig. 21.1-1*b* to show the resistance R_L of coil L. It is assumed here that the Q of this circuit is high (that is, R_L is small compared with X_L) and that the resistance of C and the wiring resistance of the circuit are very low and may be ignored.

There is a particular frequency at which $X_L = X_C$. This frequency may be defined as the condition for parallel resonance in a high-Q circuit and is similar to the condition for series resonance.

There are other definitions for parallel resonance. Thus parallel resonance may be considered as the frequency at which the impedance of the parallel circuit is *maximum*. Also, parallel resonance may be considered as the frequency at which the parallel impedance of the circuit has unity power factor. These three definitions may lead to three different frequencies, each of which may be considered as the resonant frequency. In circuits whose Q is greater than 10, however, the three conditions lead to the same resonant frequency.

In a high-Q circuit, the formula for the resonant frequency f_R is the same as in the case of series resonance and is given by

$$f_R = \frac{1}{2\pi\sqrt{LC}} \tag{21.1-1}$$

Line Current

In the series-resonant LC circuit the current was the same throughout the circuit, but the voltage drops across L and C were 180° out of phase with each other. In a parallel LC circuit there are two paths for current, one through the inductor and the other through the capacitor. If the resistance R_L of the inductor is small compared with X_L, then at resonance the impedance X_C of the capacitive branch circuit is practically equal to the impedance $(\sqrt{X_L^2 + R_L^2})$ of the inductive branch. At resonance the current in each branch may therefore be considered equal. These currents together form the "line" current. However, since they are practically 180° out of phase, it is not their sum but their difference which is the line current. Another way of saying this is that line current in a parallel circuit is the vector sum of the individual branch currents. Since they are equal but opposite at resonance, their vector sum is practically zero.

The impedance which a parallel circuit presents to the circuit is defined as $Z = V/I_{\text{line}}$. At resonance, since line current is low, the impedance is very high.

As opposed to line current, however, the *circulating* current *inside* the parallel circuit is high at resonance. This circulating current is the current which "flip-flops" back and forth between the capacitor and the inductor, alternately charging and discharging each.

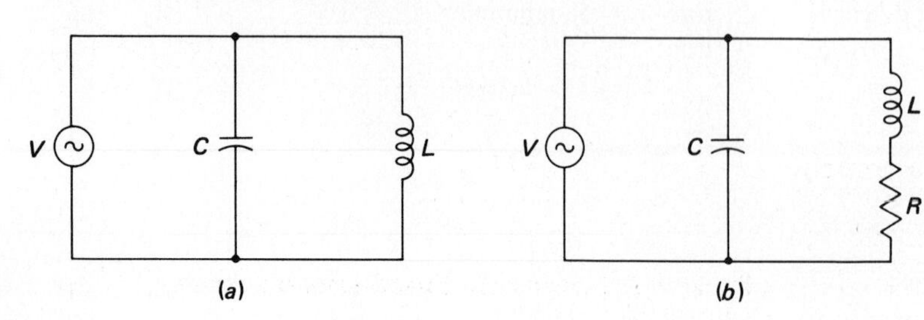

Fig. 21.1-1. Parallel-resonant circuits.

When a capacitor or an inductor is charged, it stores energy; when discharged, it gives up the energy stored in it. The current circulating inside the LC circuit switches the stored energy back and forth between L and C. The only energy used in this *oscillation* of current is that taken for overcoming any resistance in the circuit. This "lost" energy is the only energy the *source* must supply to the parallel LC circuit in the form of electric current. It is this storage action of the parallel-resonant circuit which gives rise to the term "tank circuit," often used in describing a parallel-resonant LC circuit.

SUMMARY

1. The formula for the resonant frequency of a high-Q parallel-resonant circuit is

$$f_R = \frac{1}{2\pi\sqrt{L \times C}}$$

2. The currents in the two legs of a parallel LC circuit are 180° out of phase, if there is no resistance in either leg.
3. The "circulating" current in the parallel resonant "tank" circuit is high at resonance.
4. The impedance of a parallel LC circuit is maximum at resonance.

SELF-TEST

Check your understanding by answering these questions.

1. In a parallel LC circuit at frequencies below resonance, current in the capacitive branch is higher than current in the inductive branch. ________ (true/false)
2. The impedance of a parallel LC circuit at resonance is ________. (maximum/minimum)
3. At resonance the line current in a parallel LC circuit is minimum. ________ (true/false)
4. At resonance, the branch currents of a parallel LC circuit are not equal. ________ (true/false)
5. The phase angle between the two branch currents in a parallel LC circuit is ________°.
6. The formula for the resonant frequency of a high-Q parallel LC circuit is the same as for a series-resonant circuit. ________ (true/false)

MATERIALS REQUIRED

- Equipment: AF sine-wave generator; oscilloscope
- Resistor: ½-W 10,000 Ω
- Capacitor: 0.01-μF
- Miscellaneous: 30-mH (approximately) choke

PROCEDURE

1. Connect the circuit of Fig. 21.1-2. Adjust the scope controls for proper viewing and the AF sine-wave generator controls for a 4-V p-p signal output.
2. Connect the scope across the 10,000-Ω resistor with the ground of the scope going to ground of the circuit.
3. Set generator frequency at 10,000 Hz and observe the response on the scope.
4. Increase or decrease the generator frequency until a frequency f_R is reached where *minimum* voltage V_R appears across R. This point may be identified by the fact that V_R increases on either side of f_R. The frequency f_R is the resonant frequency of the circuit. Record here for future use: $f_R =$ ________ Hz (measured).
5. Compute, by formula, the f_R of the experimental circuit, and record here: $f_R =$ ________ Hz (computed).

QUESTIONS

1. Compare the procedure for finding f_R in this experiment with that for finding f_R in a series LC circuit.
2. Why is V_R minimum at resonance in the parallel LC circuit in this experiment?
3. In your own words, explain "circulating" current in a parallel-resonant LC circuit.
4. Why is circulating current also called *oscillating* current?
5. Why are there "losses" in a parallel-resonant circuit?
6. How do the measured and computed values of f_R compare? Discuss any difference in values.

Answers to Self-Test

1. false
2. maximum
3. true
4. false
5. 180
6. true

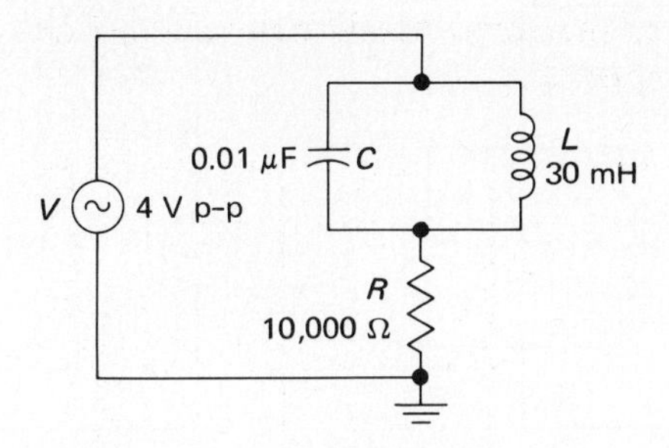

Fig. 21.1-2. Experimental parallel-resonant circuit.

EXPERIMENT 21.2. Impedance of a Parallel-Resonant Circuit

OBJECTIVE

To measure the effect of frequency on impedance of a parallel LC circuit

INTRODUCTORY INFORMATION

Impedance

Figure 21.2-1 is a graph of impedance versus frequency in a parallel-resonant circuit. This graph resembles the frequency-response characteristic of a series-resonant circuit. It shows that circuit impedance is maximum at resonance and falls off on either side of resonance.

The impedance of a parallel LC circuit can be determined mathematically. Impedance is calculated using the formula

$$Z = \frac{V_T}{I_{\text{line}}} \qquad (21.2\text{-}1)$$

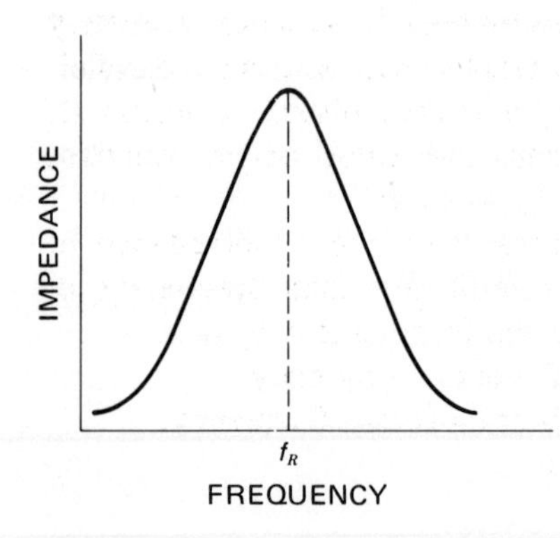

Fig. 21.2-1. Impedance versus frequency in a parallel-resonant circuit.

where Z is the impedance in ohms, V_T is the signal in volts measured across the parallel circuit, and I_{line} is the line current in amperes. This formula is another application of Ohm's law for ac circuits. Here, the voltage V_T across the LC circuit is divided by the line current coming into and out of the LC circuit.

Line current can be measured experimentally by inserting an ac ammeter into the line as in Fig. 21.2-2*a*. Or it may be calculated by measuring the voltage V_R across a series-connected resistor R as in Fig. 21.2-2*b* and using Ohm's law. Using the second method, we have

$$I_{\text{line}} = \frac{V_R}{R} \qquad (21.2\text{-}2)$$

Let us consider the characteristics of the currents in the two branches of a parallel LC circuit. These currents, I_L and I_C, are 180° out of phase. The vectors for the two branch currents at resonance are shown in Fig. 21.2-3, equal in amplitude but 180° out of phase. They cancel each other; that is, their vector sum is zero. Therefore the LC circuit impedance at resonance is theoretically infinite. Practically, there is never complete cancellation, but line current is very low, and therefore impedance at resonance is very high.

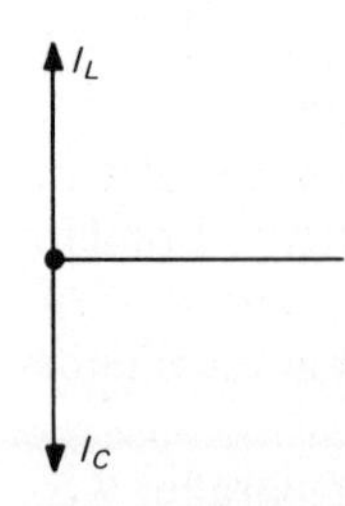

Fig. 21.2-3. Vector relationship between the branch currents in a parallel *RLC* circuit at resonance.

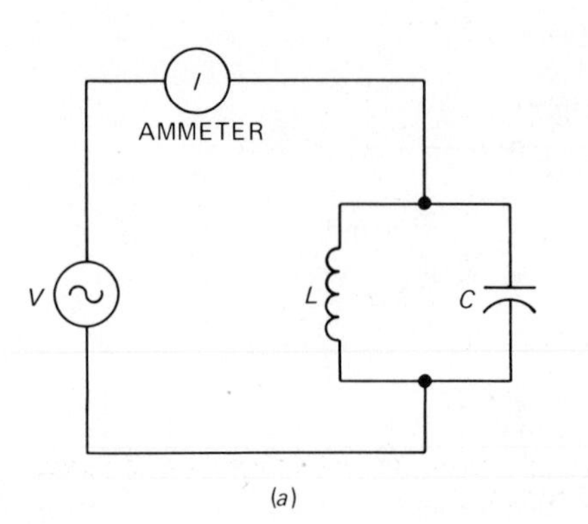

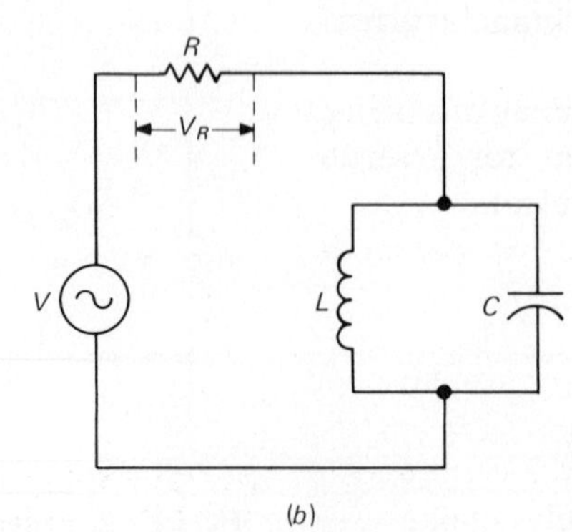

Fig. 21.2-2. Two methods used to determine line current for a parallel-resonant circuit.

For frequencies below resonance, X_L is lower than X_C. Therefore inductive current is higher than capacitive current. Line current, which equals $I_L - I_C$ in this case, is inductive and higher than the current was at resonance. Therefore circuit impedance is lower. The greater the frequency deviation from f_R, the higher the line current, and the lower the circuit impedance.

For frequencies higher than f_R, the circuit is capacitive. Again, the greater the frequency deviation from f_R, the lower the LC circuit impedance.

In this experiment you will calculate the value of impedance Z of the circuit from the relationship

$$Z = \frac{V_T}{I_{\text{line}}}$$

and will plot a graph for Z of the circuit from the calculated values of Z.

SUMMARY

1. The circuit impedance of a parallel LC circuit is maximum at f_R.
2. The circuit impedance of a parallel LC circuit falls off on either side of f_R.
3. Impedance of a parallel LC circuit is calculated by the formula $Z = V_T/I_{\text{line}}$.
4. The line current can be found by measuring the voltage V_R across a series-connected resistor R and using the formula $I_{\text{line}} = V_R/R$.
5. The two branch currents of a parallel LC circuit are 180° out of phase.

SELF-TEST

Check your understanding by answering these questions.

1. Currents 180° out of phase can be said to be flowing in ________ (the same/opposite) direction(s).
2. When two currents 180° out of phase are added vectorially, the result is the ________ between the two currents.
3. If a parallel LC circuit is operating at a frequency lower than the resonant frequency f_R, it would behave as a(an) ________ circuit.
4. If the parallel LC circuit in question 3 were to have its currents drawn in vector form, which vector would be longer? ________
5. What is the formula for finding the impedance of a parallel LC circuit? $Z =$ ________

MATERIALS REQUIRED

- Equipment: Audio sine-wave generator; oscilloscope
- Resistor: ½-W 10,000-Ω
- Capacitor: 0.01-μF
- Miscellaneous: 30-mH (approximately) choke

PROCEDURE

1. Connect the circuit of Fig. 21.2-4. Adjust the scope controls for proper viewing and the AF generator controls for a 4-V p-p signal output.
2. Connect the scope across the 10,000-Ω resistor with the ground of the scope going to ground of the circuit.
3. Set generator frequency at 10,000 Hz and observe the waveform on the scope.
4. Increase or decrease the generator frequency until a frequency is reached where minimum voltage v_R appears across R. This point may be identified by the fact that v_R increases on either side of f_R. The resonant frequency of the circuit is f_R. Record it in Table 21.2-1. This frequency should be the same as that found in Exp. 21.1.
5. Check the generator output and maintain it at 4 V p-p during the remainder of this experiment.
6. Adjust the generator frequency in turn to each of the frequencies listed in Table 21.2-1, maintaining a 4-V p-p output signal. At each frequency measure and record the voltage across the parallel-resonant circuit v_T and the voltage v_R across R.
7. Compute and record the value of line current, using the formula $i_{\text{line}} = v_R/R$.
8. Compute and record the value of impedance of the parallel LC circuit for each frequency in Table 21.2-1. Use the formula $Z = v_T/i_{\text{line}}$.
9. Draw a graph of Z versus frequency.
10. Also draw a graph of v_T versus frequency.

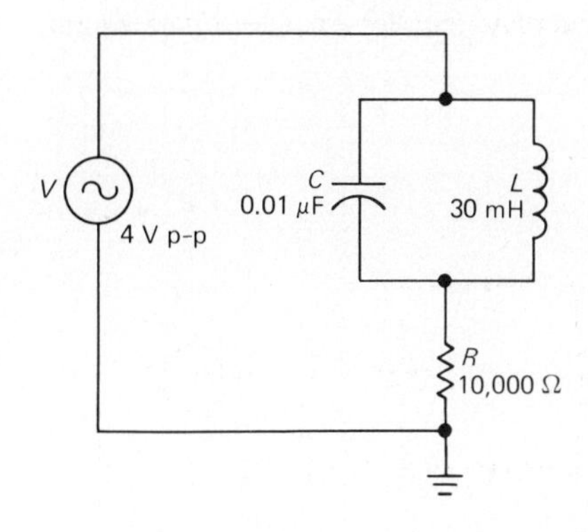

Fig. 21.2-4. Experimental circuit to determine how Z varies with f in a parallel LC circuit.

TABLE 21.2-1. Impedance of a Parallel-Resonant Circuit

Frequency Deviation, Hz	*Frequency,* Hz	v_T, V p-p	v_R, V p-p	$i_{line} = \frac{v_R}{R}$	*Z Computed* = $\frac{v_T}{i_{line}}$
$f_R - 6000$					
$f_R - 5000$					
$f_R - 4000$					
$f_R - 3000$					
$f_R - 2000$					
$f_R - 1000$					
$f_R - 500$					
f_R					
$f_R + 500$					
$f_R + 1000$					
$f_R + 2000$					
$f_R + 3000$					
$f_R + 4000$					
$f_R + 5000$					
$f_R + 6000$					

QUESTIONS

1. What are the sources of error for this experiment?
2. What would happen to the graph of step 9 if a low-valued resistance were placed in parallel with the *LC* circuit?
3. What does the graph drawn in step 9 tell about the impedance of the *LC* circuit (*a*) below f_R, (*b*) at f_R, (*c*) above f_R?
4. Can the impedance of a parallel *LC* circuit be found vectorially? How?
5. What does the frequency-response curve drawn in step 10 indicate about the signal voltage developed across a parallel *LC* circuit?

Answers to Self-Test

1. opposite
2. difference
3. inductive
4. I_L
5. v_T/i_{line}

CHAPTER 22 NONLINEAR RESISTORS: VDRs AND LDRs

EXPERIMENT 22.1. Voltage-Dependent Resistor (VDR)

OBJECTIVE

To determine by experiment the relationship between resistance and voltage across a VDR

INTRODUCTORY INFORMATION

Nature of Varistors

In an ordinary resistor, the relationship between current, voltage, and resistance obeys Ohm's law. That is, if the voltage applied doubles, so does the current. The resistance remains the same regardless of the amount of current flowing through it or the voltage drop across it, except at extremely high frequencies or if operated at extremes of heat or cold.

The voltage-dependent resistor, known also as a VDR or a varistor, unlike the conventional resistor, is "nonohmic." This means that it does *not* obey Ohm's law. The current does not increase in the same ratio as the applied voltage but increases by a much larger factor. For example, in a specific VDR, the current might increase five times for a doubling of the applied voltage. The resistance of the varistor, therefore, decreases at a nonlinear rate, inversely with voltage.

The resistive material of the varistor is what gives it this characteristic. A specially processed semiconductor, silicon carbide, is the material used in the manufacture of a varistor. In the manufacturing process, silicon carbide is combined with a binder, usually a ceramic material, and formed under high pressure and temperature into a variety of shapes, sizes, and ratings.

The type of varistor most often seen by the service technician is the disk type (Fig. 22.1-1), though they are available in the form of rods, washers, and stacked disks and washers. Varistors may be supplied with or without leads or cooling fins, or they may be stacked to handle higher voltages and have screw connections.

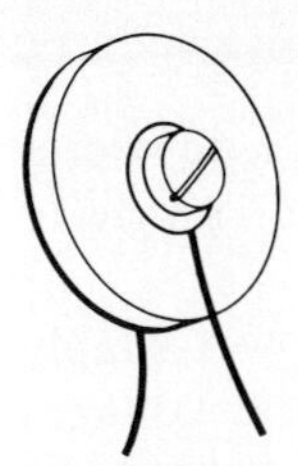

Fig. 22.1-1. Disk varistor.

Varistor Performance

Figure 22.1-2 shows the current-voltage characteristic of a conventional resistor and a varistor. Notice that the characteristic of the conventional resistor is a straight line (linear), while that of the varistor is nonlinear.

Standard varistors are supplied with maximum ratings within the following ranges: power, 0.25 to 0.5 W per square in of exposed surface (up to 12 W per varistor in standard off-the-shelf varieties); current, 1 μA to 1 A and more; voltage, 1 V to more than 10,000 V. Varistors with special ratings are also available from the manufacturers.

Varistor Uses

Varistors have long been used for protection against lightning. When the voltage across a line increases

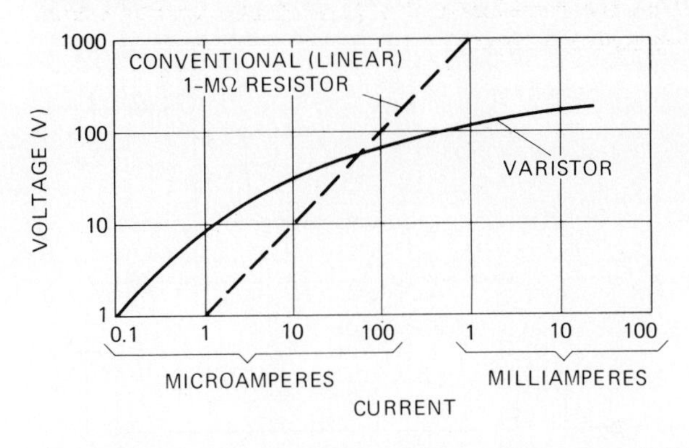

Fig. 22.1-2. Graph of current versus voltage characteristics of a conventional resistor and varistor.

sharply, as it would if lightning were to strike, the resistance of the varistor decreases to practically a dead short. The excess current caused by the lightning is then shunted to ground by the varistor. Once the lightning pulse is over, the varistor resistance returns to normal, keeping the regular circuit current within its normal path.

The technician will encounter varistors in color television receivers. In the degaussing system of a TV receiver, a varistor is connected in series with the degaussing coil along with a thermistor as in Fig. 22.1-3. The purpose is to permit an alternating current of short duration to flow through the degaussing coil when the TV receiver is turned on. The alternating current is used to "degauss" or demagnetize the color picture tube. If the picture tube becomes magnetized, impure color will result. Operation of the system, shown in simplified form in Fig. 22.1-3, is as follows. As alternating current begins to flow, the *thermistor* offers a high resistance, and most of the current is shunted through the varistor and the degaussing coil. As the voltage increases across the varistor, its resistance decreases, permitting more current through the degaussing coil, degaussing the CRT. Within a few seconds, as the *thermistor* heats due to the current through it, its resistance decreases. More of the alternating current flows through the thermistor instead of the varistor-degaussing coil combination. Now, since there is little voltage drop across the low-resistance thermistor, the degaussing coil and varistor (which are in parallel with the thermistor) have little voltage across them, and so the varistor's resistance becomes high enough to reduce the degaussing current to a level which will not affect CRT performance. If the alternating current were allowed to stay on the degaussing coil, ac interference in the color picture would be extremely objectionable.

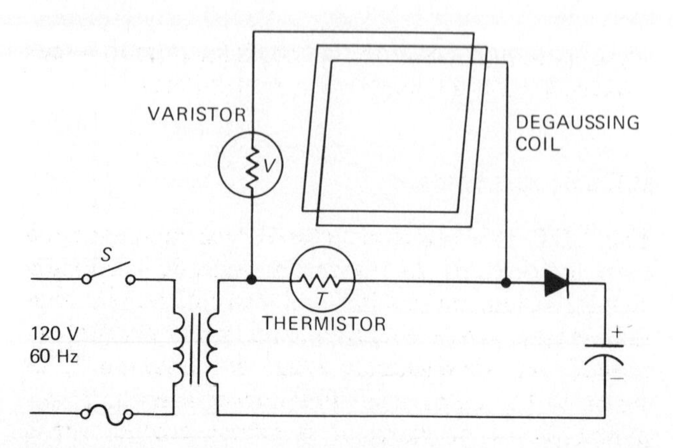

Fig. 22.1-3. Simplified TV power supply with degaussing circuit.

SUMMARY

1. The varistor is a special type of resistor which is nonohmic.
2. As the voltage increases across a varistor, resistance decreases; when the voltage decreases, resistance increases.
3. The resistance decrease in a varistor varies nonlinearly with the increase in voltage across it.
4. The varistor is a common element in color TV systems.

SELF-TEST

Check your understanding by answering these questions.

1. The varistor is made of ________.
2. If the voltage across a varistor doubles, its resistance halves. ________ (true/false)
3. The varistor does not follow Ohm's law and is said to be non ________.
4. The varistor is a type of semiconductor. ________ (true/false)
5. Varistors are commonly used in what consumer product? ________

MATERIALS REQUIRED

- Power supply: Variable regulated dc source
- Equipment: Multirange dc microammeter (the current ranges of a 20,000-Ω/V VOM will suffice); EVM
- Resistor: ½-W 5600-Ω
- Miscellaneous: Varistor (Carborundum, Electronics Division, Globar Plant, Niagara Falls, N.Y.) Part #333 BNR-4, or the equivalent; SPST switch

NOTE: If a high-voltage variable regulated dc source is not available, a variable ac source may be used. If an ac supply is used, an ac multirange microammeter will be required.

PROCEDURE

1. Connect the circuit of Fig. 22.1-4. M is a dc microammeter, and V_1 is an EVM set on dc voltage to measure V_{AB}. V is set to zero V.
2. Close S. Increase the output of the power supply so that V_{AB} takes, in turn, each of the voltages shown in Table 22.1-1. Measure and record the current at each of these voltages.

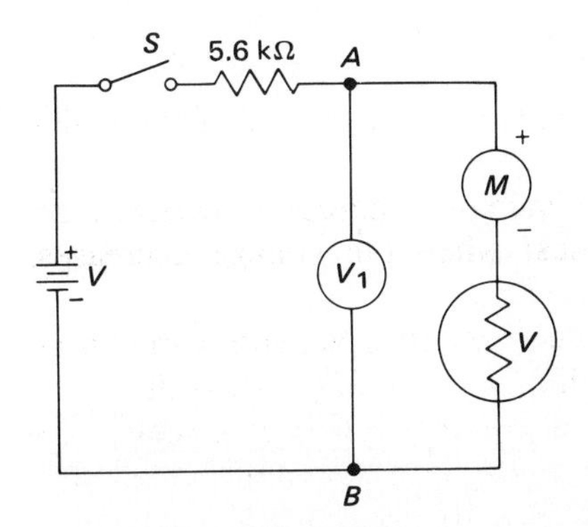

Fig. 22.1-4. Experimental circuit to determine resistance-voltage characteristic of a varistor.

3. Compute and record the resistance of the varistor at each voltage setting, using Ohm's law.

QUESTIONS

1. What is the difference between a linear and nonlinear resistor?
2. Explain how a varistor of the proper value can protect a circuit against a voltage surge.
3. From the data in Table 22.1-1 is it clearly shown that the varistor is a nonlinear resistor? Explain.
4. From your data on voltage and resistance of the varistor, comment on the relationship between them.

Answers to Self-Test

1. silicon carbide
2. false
3. ohmic
4. true
5. TV receivers

TABLE 22.1-1. VDR Characteristics

V_{AB}, V	I, A	$R = \frac{V}{I}$, Ω
0		
10		
20		
30		
40		
50		
60		
70		
80		
90		
100		
110		
120		
130		
140		
150		
160		

EXPERIMENT 22.2. Light-Dependent Resistor (LDR)

OBJECTIVE

To determine by experiment the relationship between resistance and the intensity of light applied to an LDR

INTRODUCTORY INFORMATION

The light-dependent resistor is known by several other names. It is sometimes called a photoconductive cell, photocell, or photoresistor. Each of these names can be used accurately except "photocell," which is often confused with the photovoltaic cell, a cell which produces a voltage when light strikes it.

LDR Characteristics

The LDR is a semiconductor device manufactured from compounds of cadmium selenide, cadmium sulfide, indium antimonide, and lead sulfide. In manufacture the active semiconductor materials are deposited on an insulating base of glass, mica, or ceramic. The element is often encapsulated in transparent plastic or glass, or is merely coated with a protective layer of plastic film. See Fig. 22.2-1. When light strikes the conductive surface of the LDR, its resistance decreases. The decrease in resistance depends on the intensity of light falling on the conductive surface. Figure 22.2-2 is a graph of resistance-versus-light intensity for a particular LDR. Light intensity is given in footcandle units.

LDRs operate with either alternating or direct current. Like an ordinary resistor, the cell has no polarity, and so a dc source can be connected to it without regard to polarity. Cell voltage and current levels have been improved greatly in the past few years, and depending on type, voltage ranges may exceed 600 V direct or peak alternating current. The maximum current may exceed 0.5 A, and the light-to-dark resistance ratio may be greater than 10,000 to 1.

If all the specifications of an LDR were considered, it would be found that there are nine different listings. These are maximum voltage, maximum current, maximum wattage, dark current versus voltage rating, light-to-dark resistance ratio, time constant, temperature range, light sensitivity, and light frequency response.

The service technician normally needs to know only three or four of these:

Maximum Voltage. Just as with a diode, there is a maximum voltage which can be applied without damaging the LDR. This voltage must not be exceeded.

Maximum Current. As with all semiconductors, the LDR has a limit as to the amount of current it can conduct without being damaged. This rating must not be exceeded.

Dark Resistance. This specification gives the resistance of the device when no light is striking its conductive surface.

Light-to-Dark Resistance Ratio. Here, the ratio of light resistance is given with respect to the dark resistance of the LDR. For example, if the resistance of the device is given as 100 Ω in the dark, with a light-to-dark resistance ratio of 10, the resistance in the specified light would be 10 Ω.

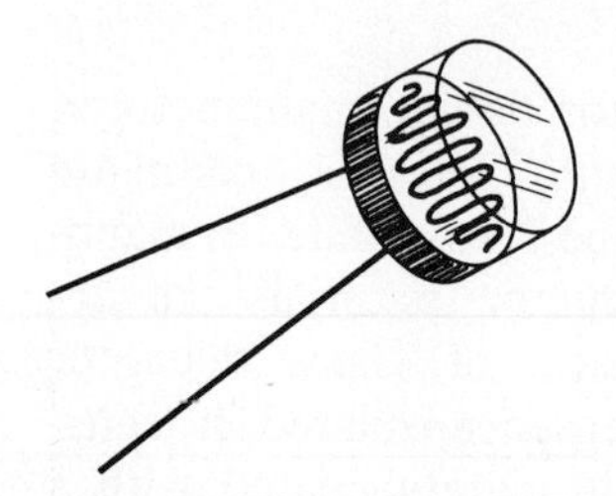

Fig. 22.2-1. Typical LDR.

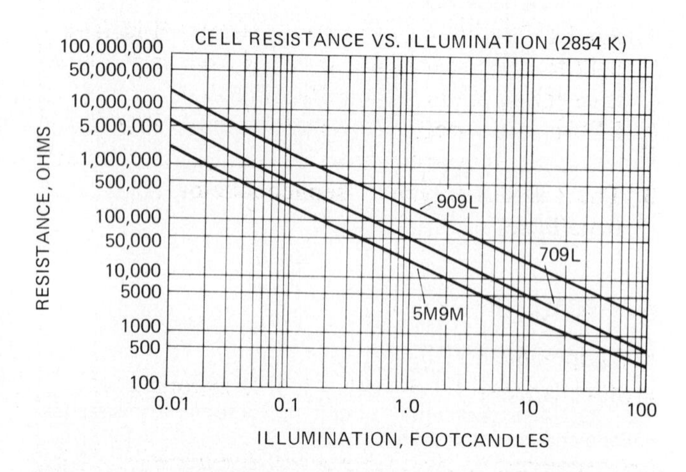

Fig. 22.2-2. Graph of light versus resistance in an LDR. (*Clairex Corp.*)

LDR Used in TV Receivers

Many television receivers use the LDR as a light sensor for automatic control of TV brightness and contrast as changes occur in room light level. Normally, when room brightness increases, the contrast of the television picture must be increased for best picture viewing. Likewise, when the room light level decreases, the picture contrast must be decreased. By sensing the room's ambient light level, the LDR provides information to the appropriate television circuitry, which then automatically controls both picture brightness and contrast. At least one TV manufacturer is currently using the LDR to control not only brightness and contrast, but also color and tint levels.

SUMMARY

1. The LDR is a light-sensitive resistor whose resistance changes when light strikes its light-sensitive element.
2. The resistance of the LDR decreases when the light level shining on it increases.
3. LDRs are commonly specified by their maximum voltage and current ratings, light-to-dark resistance ratio, and dark resistance.
4. LDRs are semiconductors manufactured from compounds including cadmium selenide, cadmium sulfide, indium antimonide, and lead sulfide.

SELF-TEST

Check your understanding by answering these questions.

1. An LDR with a dark resistance of 150 Ω has a light-to-dark resistance ratio of 100 to 1. What is its resistance in the specified light?
2. LDRs are used in automatic brightness and ________ control circuits of ________ receivers.
3. The LDR is a type of semiconductor. ________ (true/false)

MATERIALS REQUIRED

- Equipment: EVM
- Miscellaneous: CLAIREX CL 5M2 or equivalent

PROCEDURE

NOTE: In this experiment no attempt will be made to make precise light-sensitivity measurements, because the equipment required—including a controlled light chamber—is too costly. Our concern here is to acquaint the technician with a rough measure of resistance-versus-light intensity of an LDR.

1. Measure the resistance of the LDR supplied **(a)** in total darkness, and **(b)** in bright light. Record both values in Table 22.2-1.
2. Calculate the light-to-dark resistance ratio of the LDR and record in Table 22.2-1.
3. Measure the resistance of the LDR in various light levels. Record these readings in Table 22.2-1.
4. Draw a graph of the LDR resistance plotted against the light level.

TABLE 22.2-1. LDR Characteristics

LDR Resistance, Ω	*Light-to-Dark Resistance Ratio*
Total darkness ________ Ω	
Bright light ________ Ω	
Low light ________ Ω	X
Medium light ________ Ω	
Bright light ________ Ω	

QUESTIONS

1. Was the calculated light-to-dark resistance ratio comparable with the manufacturer's specifications?
2. What would be necessary in order to produce a technically accurate graph of light versus resistance of the LDR?
3. What uses, other than those described, can you think of for the LDR?

Answers to Self-Test

1. 1.5 Ω
2. contrast; television
3. true

CHAPTER 23 SEMICONDUCTOR DIODES: CHARACTERISTICS AND TESTING

EXPERIMENT 23.1. Semiconductor-Diode Biasing

OBJECTIVES

To measure the effects of forward and reverse bias on current in a junction diode

INTRODUCTORY INFORMATION

Semiconductors and Vacuum Tubes

Electronics is concerned with the theory and application of devices which control current. Included among these are semiconductors and vacuum tubes.

Semiconductors are solids whose resistivity lies between that of electrical conductors and insulators. Transistors, junction diodes, zener diodes, tunnel diodes, integrated circuits, and metallic rectifiers are examples of semiconductors. These are used in computers, in radio and television receivers, and in other electronic products.

Vacuum tubes, invented before transistors, perform similar control functions. But semiconductors are fast replacing vacuum tubes. However, because vacuum tubes are still found in some electronic products, the technician should understand the principles and operation of tubes. In this book we shall be interested mainly in semiconductors, though some tube experiments are included. In this experiment we shall begin with electronics by studying the operation of semiconductor junction diodes.

Semiconductor devices are used in electronics in many ways. They may be used as rectifiers, amplifiers, detectors, oscillators, and switching elements. Some of the characteristics which make the semiconductor so popular are:

1. The semiconductor is small and light in weight, permitting miniaturization of electronic equipment. Figure 23.1-1 shows the size of an early transistor.
2. Recent developments in solid-state technology have led to *microminiaturization*. Figure 23.1-2 shows an integrated circuit (IC) containing circuits made up of transistors, resistors, capacitors, and interconnections. The work done by this tiny device compares favorably with that previously performed by a vacuum-tube device whose dimensions were about 100 times as large!

   Advances in the technology of integrated circuits have brought the art of ICs far beyond the early device in Fig. 23.1-2. There are now large scale integrated circuits (LSICs) which hold on one small chip thousands of transistors, diodes, and resistors. And the end of microminiaturization is not yet in sight.
3. Semiconductors are solids. They therefore have no elements which can vibrate. Element vibration in vacuum tubes was one cause of tube troubles.
4. Semiconductors use little power and give off less heat than tubes in similar applications. They do not need warm-up time and will operate as soon as power is applied.
5. Semiconductors are rugged.

Semiconductor Materials and Impurities

Silicon is the material from which most semiconductor devices are presently constructed. Germanium was first used in the manufacture of transistors and junction diodes. But now germanium devices are but a small percentage of the semiconductor output. Silicon is less heat-sensitive and therefore more desirable.

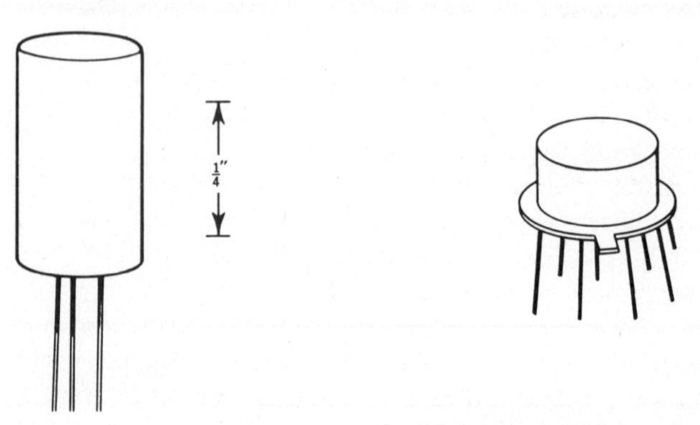

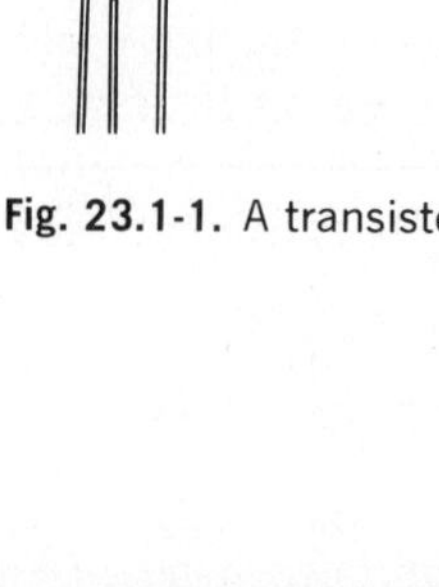

Fig. 23.1-1. A transistor.

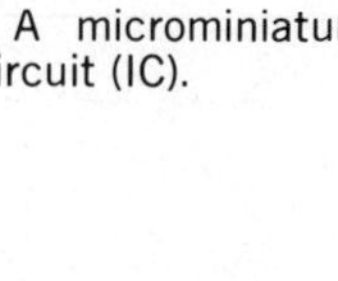

Fig. 23.1-2. A microminiature integrated circuit (IC).

Germanium and silicon must be highly purified before they can be made into effective semiconductor materials. In their pure state these semiconductors have a very low conductivity; that is, their resistivity is high. The conductivity of germanium and silicon may be increased by the addition of very small quantities of certain "impurities." The addition of controlled quantities and types of impurities, called *doping*, changes the electron-bond structure within the atoms of these elements and provides them with current carriers, increasing their conductivity.

Current Carriers in a Semiconductor

In a vacuum tube, negatively charged electrons are considered the current *carriers*. To this concept of negative-charge carriers must be added positive-charge carriers to explain current flow in semiconductor diodes and transistors. Positive-charge carriers are called *holes*. They are considered to have mass, mobility, and velocity. Current in semiconductors is carried on by the movement of negative charges (free electrons) and positive charges (holes). It is beyond the scope of this material to consider the physics of holes. We shall, however, be concerned with the current resulting from the movement of free electrons and holes and with the control of this current.

Impurities such as arsenic and antimony increase the conductivity of silicon by increasing the number of negative (N) charge carriers (free electrons). For this reason, silicon which has been doped with arsenic or antimony is called *N type*. Some holes exist in N-type silicon, but these are in the minority and so are called *minority carriers*. Current in N-type silicon may be considered as carried on by free electrons, which are the *majority carriers*.

Impurities such as indium and gallium increase the conductivity of silicon by increasing the number of positive (P) charge carriers (holes). Silicon which has been doped with indium or gallium is called *P type*. Some free electrons exist in P-type silicon, but these are the minority carriers. Current in P-type silicon may be considered as carried on by holes, which are the majority carriers.

Holes have an attraction for free electrons. When a free electron and a hole "meet," the free electron "fills" the hole, neutralizing its charge. The free electron is said to combine with the hole. In this process, both the hole and free electron are lost as current carriers. While this action is taking place, however, new current carriers are being formed at other points in the semiconductor.

The movement of current carriers may be controlled by applying an external battery voltage V_{AA} across the semiconductor (see Fig. 23.1-3). Holes in the P-type silicon are repelled by the positive terminal of V_{AA} and move toward the negative terminal. Free electrons enter the silicon from the negative terminal of V_{AA} and move toward the holes. Combinations of free electrons and holes take place. While these combinations are being formed, additional mobile electrons and holes are released in the silicon from an electron-hole pair. The electrons move toward the positive battery terminal and the holes toward the negative battery terminal. Electrons and holes continue to recombine and to tear away, so a constant current in the *external* circuit is maintained.

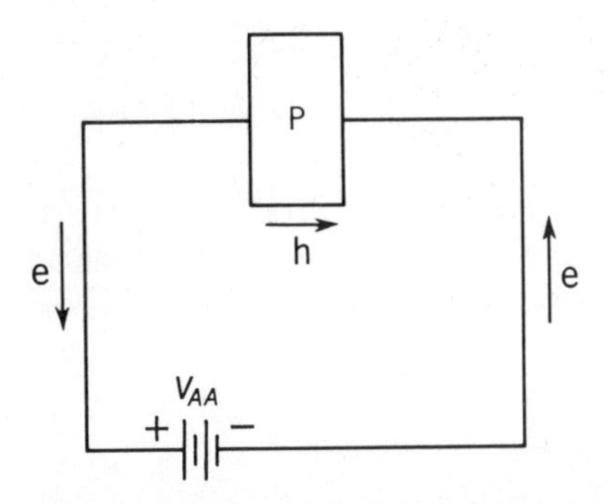

Fig. 23.1-3. Movement of free electrons and holes is controlled by connecting a battery to a doped silicon crystal.

Operation of a Semiconductor Junction Diode

When P- and N-type silicon are joined as in Fig. 23.1-4, a junction diode is created. This two-element device is unique in that it can pass current readily in one direction but not in the other.

The theory of negative and positive carriers can be used to explain this characteristic. Consider first the effect of connecting a battery V_{AA} across this diode with the polarity shown in Fig. 23.1-5. Free electrons enter the N-type silicon at the negative terminal of V_{AA}. These in turn repel the free electrons in the N-type silicon, and these free electrons move toward the PN junction. The holes in the P-type silicon are repelled by the positive terminal of the battery and also move toward the PN junction, where combination of free electrons and holes takes place. The current carriers lost in these combinations are re-

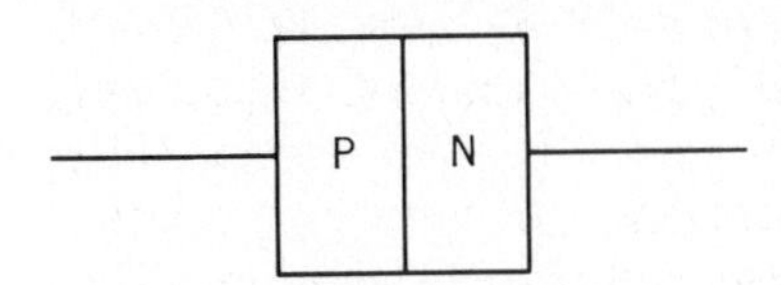

Fig. 23.1-4. Junction diode.

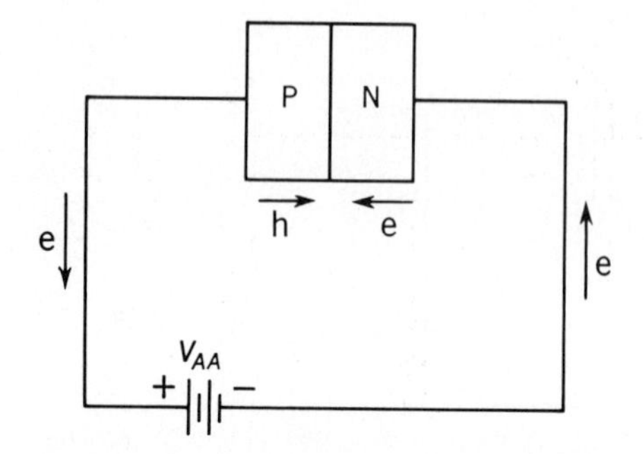

Fig. 23.1-5. Current flow in a junction diode, forward bias.

placed by new current carriers resulting from separation of electron-hole pairs. The free electrons created in the P-type silicon are attracted to the positive terminal and move in the external circuit as shown. The process is a continuous one, and current is maintained. Moreover, if V_{AA} is increased, current in the diode will increase.

This manner of connecting the negative battery terminal to the N-type and the positive battery terminal to the P-type silicon results in current and is called *forward bias*. Because there is current flow in this connection, the diode is said to have a low forward resistance.

The reverse-bias connection is shown in Fig. 23.1-6. The positive terminal of the battery attracts free electrons in the N-type silicon away from the PN junction. The negative terminal of the battery attracts the holes in the P type away from the PN junction. Therefore there are no combinations of free electrons and holes, and the majority current carriers in the diode do not support current. In this *reverse-bias* connection, there is a very small current in the diode. This current is due to the minority carriers, that is, the holes in the N type and free electrons in the P type. For the minority carriers, battery polarity is correct to support current. Only a few microamperes of current are a result of the minority carriers. This is shown by the dotted arrows in Fig.

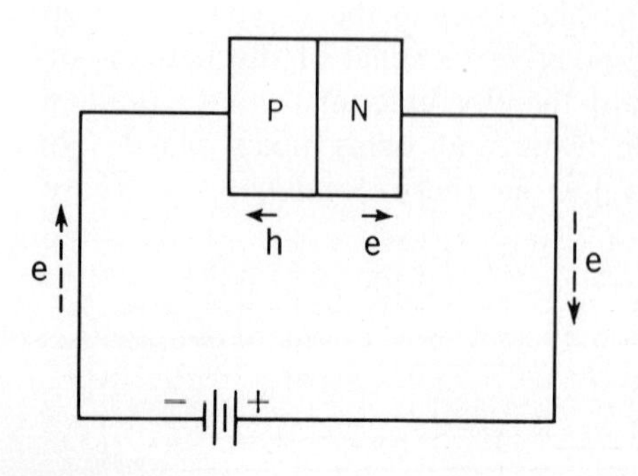

Fig. 23.1-6. Effect of reverse bias on junction diode.

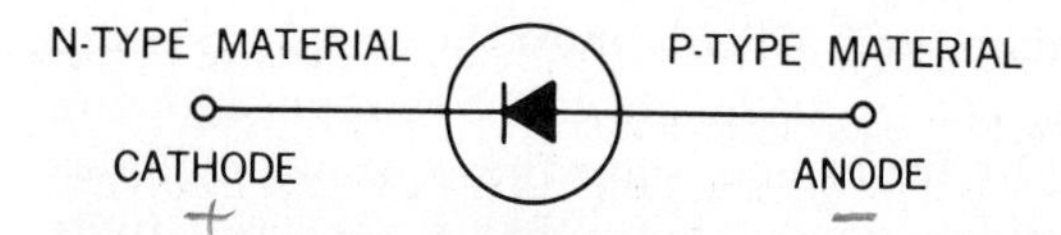

Fig. 23.1-7. Circuit symbol for a semiconductor diode.

23.1-6. The reverse-bias connection results in a high resistance in the diode.

There is a limit not only to the forward-bias but also to the reverse-bias voltage which may be placed across the diode. If the forward or reverse bias is increased beyond its limiting value, there is a sharp increase in forward or reverse current, respectively, which may permanently damage the diode.

Figure 23.1-7 is the circuit symbol for a semiconductor diode. The terminal marked "anode" (identified by the arrowhead) is connected to P-type material, while that marked "cathode" is connected to N-type material. Physically, the cathode is identified by a circular band or by a plus (+) sign at the cathode end of the diode as in Fig. 23.1-8. Reference to Fig. 23.1-5 shows that to cause current in this diode, the positive terminal of a battery must be brought to the anode, the negative terminal to the cathode, in a forward-bias arrangement.

SUMMARY

1. Semiconductors are used to control current in electronics. They are preferred to vacuum tubes because they are much smaller, use less power, and permit microminiaturization of electronic products.
2. Basic semiconductor materials are silicon and germanium. Silicon is much more widely used than germanium.
3. In their *pure* state silicon and germanium are insulators.
4. When silicon and germanium are *doped* with

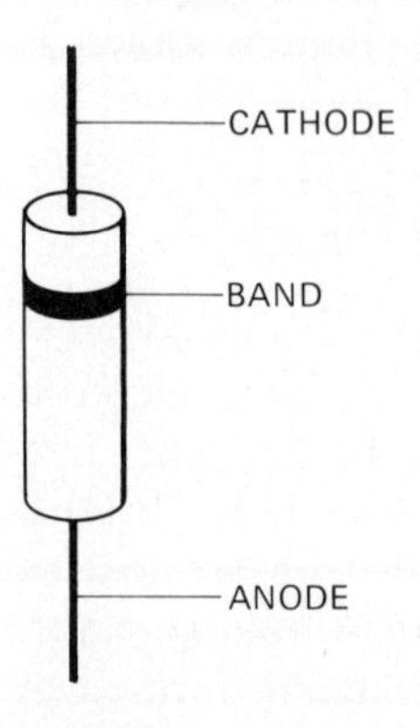

Fig. 23.1-8. Band on a junction diode is near the cathode terminal.

certain impurities, their resistivity decreases and they become *semiconductors*.

5. After doping, silicon and germanium become either N-type (negative current carriers—electrons) semiconductors or P-type (positive current carriers—holes) semiconductors. Whether they become N or P depends on the kind of impurity added by doping.
6. When a piece of N-type and a piece of P-type silicon are joined, as in Fig. 23.1-4, we have a *junction diode*.
7. A junction diode has unidirectional current characteristics; that is, it will permit current to flow through it in one direction but not in the other.
8. A junction diode must be forward-biased to permit current flow through it. To forward-bias a PN junction, the positive terminal of a power source must be connected to the P-type, the negative terminal of the source to the N-type semiconductor.
9. When the power terminals are reversed, that is, when the positive power terminal is connected to the N-type and the negative power terminal is connected to the P-type, the junction diode is reverse-biased and will not permit current flow.
10. Current flow in a junction diode is supported by the movement of electrons and holes.

SELF-TEST

Check your understanding by answering these questions.

1. The most popularly used semiconductor material is SILICON.
2. Germanium and silicon in their pure form are ________ (conductors/insulators).
3. In silicon which has been doped with impurities such as arsenic there is an increased number of ________ (negative/positive) charge carriers and the material is ________ (N/P) type.
4. A junction diode may be compared with a resistor because it permits current to flow through it in either direction. ________ (true/false)
5. To forward-bias a junction diode, connect the ________ (positive/negative) lead of a battery to the P-type terminal of the diode and the ________ (positive/negative) lead of the battery to the N-type terminal of the diode.

MATERIALS REQUIRED

- Power supply: Variable, regulated, low-voltage dc source
- Equipment: Electronic voltmeter; VOM (or a 0–10-mA, and a 0–100-mA meter)

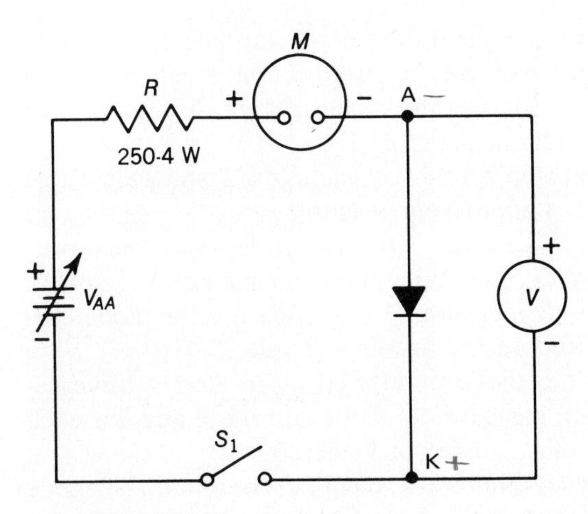

Fig. 23.1-9. Measuring the effect of forward bias on current flow in a diode.

- Resistors: 4-W 250-Ω
- Semiconductors: 1N5625 (silicon) or equivalent
- Miscellaneous: SPST switch

PROCEDURE

1. Examine the 1N5625 silicon diode assigned to you, and identify the anode and cathode terminals.
2. Connect the circuit of Fig. 23.1-9 with the diode forward-biased. V is an electronic voltmeter set to measure dc voltage. M is a VOM set to measure

E/I = R

TABLE 23.1-1. Forward and Reverse Junction Diode Biasing

Bias	V_{AK}, V	*I*, mA	*R (diode)*, Ω
Forward	0.1	.0004	250
	0.3	.0450	6.67
	0.5	3.0	.167
	0.6	14.6	.041
	0.65	27.5	.024
*	0.7	49.1	.014
Reverse	0.5	0	∞
	1.0	0	∞
	1.5	.0001	15K
	2.0	.0002	10K
	2.5	.0002	12.5k
	3.0	.0003	10K

current. Set the output of the variable dc supply so that the voltage V_{AK} *across the diode* measures 0.1 V. Measure the diode current, if any. Record the results in Table 23.1-1.

3. Repeat step 2 for every value of V_{AK} shown in Table 23.1-1. Record your results.
4. Reverse the diode in the circuit. Readjust the power supply, if necessary, until the voltage V_{AK} across the diode measures 0.5 V. Measure the diode current. Record the results in Table 23.1-1.
5. With the diode connected as in step 4 (reverse-biased), measure the diode current, if any, for each value of V_{AK} listed in Table 23.1-1.
6. Applying Ohm's law, compute the diode resistance for each value of V_{AK}, when it is forward- and reverse-biased. Record the results in Table 23.1-1.

QUESTIONS

1. Under what conditions will a silicon junction diode turn on? Refer to your measurements in Table 23.1-1 and explain.

Answers to Self-Test

1. silicon
2. insulators
3. negative; N
4. false
5. positive; negative

EXPERIMENT 23.2. Voltampere Characteristic of a Junction Diode

OBJECTIVE

To determine and graph the voltampere characteristic of a junction diode, using a curve tracer or point-by-point measurement

INTRODUCTORY INFORMATION

Voltampere Characteristic by Curve Tracer

The voltampere characteristic of a diode is a graph which shows how current in that diode varies with the voltage applied across it. An instrument called a curve tracer is used to display, automatically, the VA characteristic. The curve tracer may be self-contained, including a cathode-ray oscilloscope display. Or the curve tracer may simply generate the ac voltage required for the diode. It is then necessary to connect the curve tracer to an ordinary lab oscilloscope. Simply follow the manufacturer's instructions for the test setup.

Voltampere Characteristic (Point-by-Point Method)

The voltampere characteristic of a diode may also be obtained using the point-by-point method. Experimentally, this is done by measuring the current in the diode for a successive number of higher applied voltages, and plotting a graph of current versus voltage. The student will experimentally determine the forward voltampere characteristic of a silicon junction diode in this experiment, and will discover some interesting facts about the diode used. The student will note that very little current flows in the diode for low levels of applied voltage. Thus below approximately 0.7 V forward bias, a *silicon* diode draws little current. For forward-bias voltages equal to or higher than 0.7 V, the diode is turned ON and permits current to flow. The second fact which will become very evident is that beyond 0.7 V, very slight increases in forward-bias voltage result in large increases of current in the diode. A typical forward voltampere characteristic for a silicon diode is shown in Fig. 23.2-1.

The turn-on forward-bias voltage for silicon diodes is typically 0.7 V. For germanium diodes it is 0.3 V.

Maximum Current and Voltage Levels

Current, in a semiconductor diode which is forward-biased, increases with an increase in anode-to-cathode voltage. But there is a limit to the amount of current which can be permitted to flow safely through a diode. Beyond this limit, the diode will overheat and will be destroyed.

When the diode is reverse-biased, the small current due to minority carriers remains relatively constant, that is, independent of the bias voltage, up to a certain voltage. Beyond this safe level of reverse bias, a phenomenon called "avalanche breakdown" takes

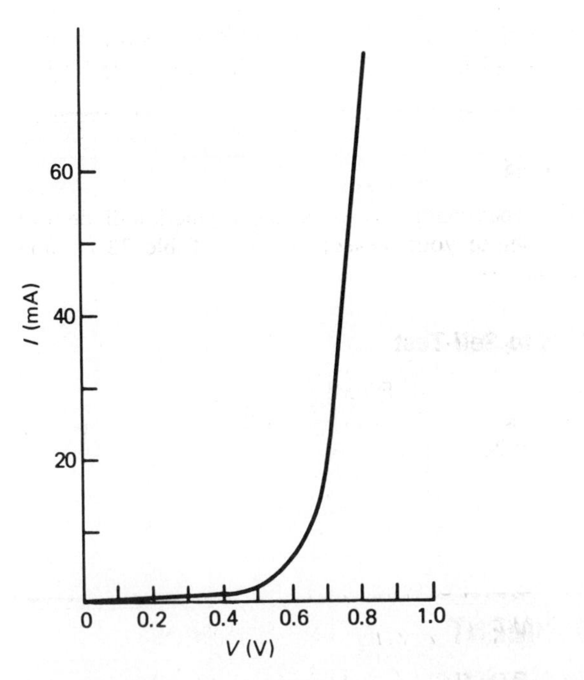

Fig. 23.2-1. Forward voltampere characteristic of a silicon junction diode.

place when a heavy surge of current occurs, which may again destroy the diode. The diode must therefore be operated within these two safe limits. The limits of safe operation will normally be specified by the manufacturer under the headings, "maximum forward voltage (V_{FM})" and "maximum reverse voltage (V_{RM})." Peak forward current (I_{FM}) may also be specified.

Junction Diode as a Switch

Once a junction diode is turned on, it appears to act like a *closed* switch. Current in the circuit containing the diode then seems to be limited only by the external resistance in the circuit. A reverse-biased diode will not permit significant current flow, and acts like an *open* switch.

The switch analogy is only approximately true. Consider a closed switch. The resistance measured across the contacts of a closed switch is zero, whereas a turned-on diode does have a measurable forward resistance R_F. It is true that R_F is small and may be neglected in many applications, but it does exist.

Now consider an open switch. The resistance measured across the contacts of an open switch is infinitely high, and an open switch will not permit current flow through it. A reverse-biased junction diode, however, does permit some current through it. Hence, though its reverse resistance R_R is very high, it is not infinite.

However, as an approximation, it is frequently useful to compare the action of a junction diode to the operation of a switch.

SUMMARY

1. The voltampere (VA) characteristic of a diode graphically shows how current in the diode varies with a change in voltage across the diode.
2. The VA characteristic may be determined automatically by using a curve tracer with an oscilloscope as the display unit.
3. A point-by-point method may also be used to measure diode currents for varying levels of voltage across the diode. A graph is then made of measured current versus applied voltage.
4. There is a limit to the maximum forward and maximum reverse voltage which may be placed across a junction diode. Operation beyond these limits may destroy the diode by overheating, resulting from excessive current.
5. The turn-on point for a silicon diode is approximately 0.7 V; that is, the diode must be forward-biased 0.7 V, before it will conduct appreciably.
6. The turn-on point for a germanium diode is 0.3 V.
7. Current, in a forward-biased diode, increases with an increase in voltage across it.
8. A junction diode acts like a *closed* switch *after* it is turned on; like an *open* switch *before* it is turned on. However, this analogy is only approximate, because a junction diode does have a measurable forward resistance and a high but not infinite reverse resistance.

SELF-TEST

Check your understanding by answering these questions.

1. The voltampere (VA) characteristic of a junction diode is a GRAPH of voltage versus current.
2. The VA characteristic may be shown automatically using a CURVE TRACER, or it may be determined by the POINT-by-POINT method.
3. After a diode is turned on, an increase in forward voltage across the diode will result in a(n) ________ (increase/decrease) of current through the diode.
4. A diode acts like a(n) CLOSED switch when it is on.

MATERIALS REQUIRED

- Power Supply: Variable, regulated, low-voltage dc source

- Equipment: EVM; VOM; curve tracer
- Resistor: 4-W 250-Ω
- Semiconductors: 1N5625 or equivalent silicon; 1N34A or equivalent germanium
- Miscellaneous: SPST switch

PROCEDURE

VA Characteristic (Point-by-Point Method)

1. Connect the circuit of Fig. 23.2-2. When power is applied, the diode will be forward-biased.
2. **Power on.** Set the output of the regulated power supply so that the voltage V_{AK} from anode to cathode is 0 V. Measure the current, if any, and record in Table 23.2-1.

   Increase the voltage in Table 23.2-1 in 0.1- and 0.05-V steps as in the table to a maximum of 0.75 V. Measure the current. Record the results. For each condition compute and record the forward resistance of the diode.
3. Reverse the diode in the circuit. The diode is now reverse-biased. Measure the current, if any, as the voltage of the supply is varied in 5-V steps from 0 to 40 V, and record the results in Table 23.2-1. For each condition, compute and record the reverse resistance of the diode.
4. Plot a graph of V versus I for both bias conditions.

Germanium Diode

5. **Power off.** Substitute a 1N34A germanium diode for the silicon diode in Fig. 23.2-2. The diode is connected in the forward-bias position.

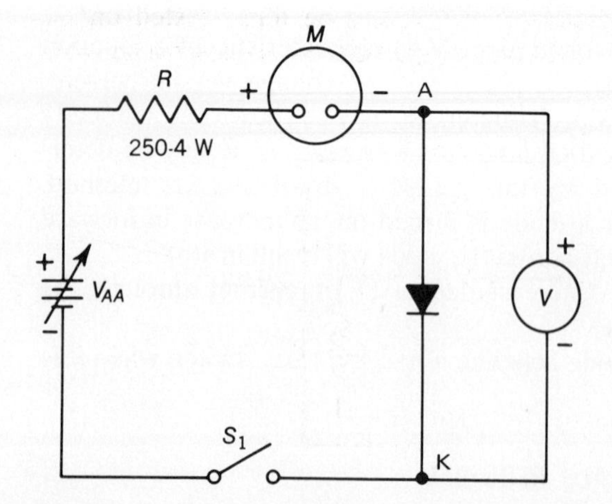

Fig. 23.2-2. Experimental circuit to determine voltampere characteristic of a diode.

TABLE 23.2-1. Voltampere Characteristics

Forward			Reverse		
V_{AK}, V	I, mA	R, Ω	V_{AK}, V	I, mA	R, Ω
0	0	∞	0	0	∞
0.1	.0004	250	5	.0005	10K
0.2	.0055	36.36	10	.0015	6.7K
0.30	.0570	5.26	15	.0016	9.4K
0.40	.5000	.80	20	.0025	8K
0.50	3.15	.159	25	.0027	9.3K
0.55	6.75	.081	30	.0032	9.4K
0.60	13.28	.045	35	.0037	9.5K
0.65	28.40	.023	40	.0042	9.5K
* 0.70	49.80	.014			
0.75	78.30	.009			

6. Set the output of the supply at 0 V. **Power on.**
7. Determine the voltage across the diode V_{AK} at which the diode turns on. V_{AK} = 0.3 V.

Curve Tracer

8. Display the characteristic curves of the silicon and germanium diodes using a curve tracer. Photograph or draw these curves for comparison with the plotted curves.

QUESTIONS

1. What are the limitations, if any, on (*a*) forward bias, (*b*) reverse bias of a junction diode? Were the limitations exceeded in this experiment? How do you know?
2. What portion, if any, of the voltampere characteristic of the forward-biased diode in this experiment is linear? Mark it on your graph.
3. What is the turn-on voltage for (*a*) 1N5625? (*b*) 1N34A?
4. What is the relationship between diode current and diode resistance?

Answers to Self-Test

1. graph
2. curve tracer; point, point
3. increase
4. closed

EXPERIMENT 23.3. Resistance Testing a Semiconductor Diode

OBJECTIVE

To test a junction diode with an ohmmeter and identify the anode and cathode of the diode

INTRODUCTORY INFORMATION

Testing a Semiconductor Diode with an Ohmmeter

CAUTION: Some ohmmeters use high-voltage batteries which may destroy a transistor junction.

A resistance check may be used as an approximate test of a semiconductor diode's operation. Recall that the polarity of the terminals of the battery contained within an ohmmeter appears at the leads of the ohmmeter. In Fig. 23.3-1, lead *A* is positive, lead *B* negative. An ohmmeter test of a diode which is operating normally will show that the diode has a low forward resistance and a high back (reverse) resistance. If the positive ohmmeter lead (*A* in Fig. 23.3-1) is connected to the anode of a diode and the negative lead (*B*) to the cathode, the diode will be forward-biased. Current will flow, and the diode will measure low resistance. On the other hand, if the ohmmeter leads are reversed, the diode will be reverse-biased. Very little current will flow, and the diode will measure a very high resistance. If a semiconductor diode measures a very low forward *and* a low reverse resistance, it is probably damaged (shorted). On the other hand, an open diode is indicated if the forward resistance is unusually high or infinite.

Identifying the Anode and Cathode of a Diode

The cathode end of a semiconductor diode is usually marked by a circular band or by a plus (+) sign. If the diode is unmarked, it is simple to determine by a resistance check which is the anode, which the cathode. First, the polarity of the ohmmeter leads is determined by checking with a voltmeter across the ohmmeter terminals. Then the ohmmeter-lead position which measures the forward resistance of the diode is determined. In this position, the positive ohmmeter lead is connected to the anode, the negative lead to the cathode.

Low-Power Ohms Function of an Ohmmeter

The battery in a nonelectronic ohmmeter, like that in Fig. 23.3-1, may be 1.5 V or higher. If it is, it can forward-bias a silicon junction diode beyond the 0.7 V required for conduction. Similarly it can forward-bias a germanium junction diode beyond the 0.3 V needed for conduction. That is why it is possible to make the previously discussed ohmmeter checks of semiconductor diodes. In troubleshooting some semiconductor circuits, however, low-power (LP) electronic ohmmeters are used whose open circuit lead voltage is lower than the 0.7 or 0.3 V. The low-power ohms (LP Ω) function of this type of ohmmeter *cannot* be used to measure the forward resistance of a diode, nor can it be used to identify the anode and cathode of a diode. Fortunately, some manufacturers have provided, in addition to the *low-power ohms* function, a *normal ohms* function. Resistance tests of a semiconductor diode may be made by using the normal ohms function of an ohmmeter whose terminal voltage is higher than 0.7 V.

SUMMARY

1. A junction diode may be *ohms* tested by an ohmmeter. When the leads of the meter are connected across the diode so that it is forward-biased, the meter will show the *low* forward resistance (R_F) of the diode. When the leads are reversed, the meter will measure the high reverse resistance (R_R).
2. A junction diode *cannot* be ohms tested on the *low-power* ohms function of an electronic voltmeter.

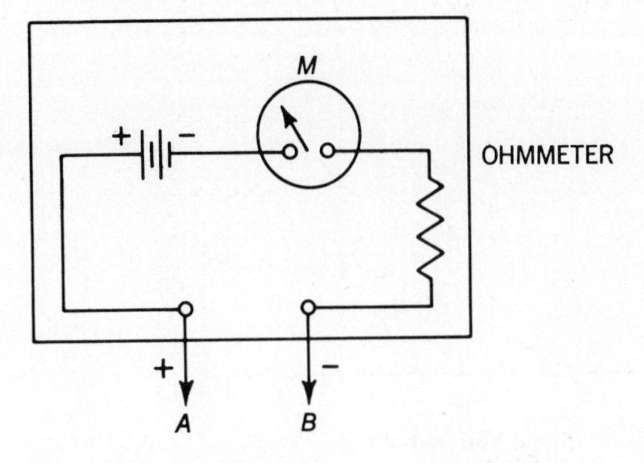

Fig. 23.3-1. Polarity of ohmmeter leads.

3. The anode and cathode of an unmarked semiconductor diode can be identified by an ohmmeter. First, the polarity of the voltage at the terminals of the ohmmeter must be determined. Then the diode's forward resistance is checked. The low-resistance reading occurs when the diode is forward-biased. For this condition the positive lead of the ohmmeter is connected to the *anode*, the negative lead to the cathode.

SELF-TEST

Check your understanding by answering these questions.

1. The forward resistance of a silicon diode is LOW, the reverse resistance HIGH.
2. The forward resistance of a diode may be checked, approximately, with a(n) OHMMETER.
3. The polarity of the voltage at the terminals of an ohmmeter may be checked with a(n) VOLTMETER.

MATERIALS REQUIRED

- Equipment: EVM; VOM
- Semiconductors: 1N5625; 1N34A; unmarked diode

PROCEDURE

1. Measure the forward and reverse resistance of the 1N5625 silicon diode. Record the results in Table 23.3-1.

TABLE 23.3-1. Diode Resistance Measurements

Diode	R_F *(forward)*, Ω	R_R *(reverse)*, Ω	$\frac{R_R}{R_F} = r$
1N5625	2.9 K	30 M	10344.9
1N34A	1 K	170 K	170

2. Compute the back-to-forward resistance ratio r of the diode and record the result.
3. Repeat steps 1 and 2 for a 1N34A germanium diode and record the results in Table 23.3-1.
4. Your instructor will give you an unmarked diode. Identify the anode and cathode.

QUESTIONS

1. How did you find which lead of an ohmmeter is positive, which negative?
2. How did you identify the anode of an unmarked diode?
3. Does the ratio $r = R_R/R_F$ (see Table 23.3-1) give any clue as to whether a semiconductor diode is good or bad? Explain.
4. Which of the diodes used in this experiment is silicon? Which germanium? How do you know?

Answers to Self-Test

1. low; high
2. ohmmeter
3. voltmeter

CHAPTER 24 VACUUM-TUBE DIODES

EXPERIMENT 24.1. Vacuum Tubes: Diode Characteristics

OBJECTIVE

To become familiar with the physical structure of vacuum tubes

INTRODUCTORY INFORMATION

Physical Structure

Vacuum tubes ushered in the age of electronics. Though vacuum tubes are being quickly replaced by solid-state devices, such as semiconductor diodes, transistors, integrated circuits, and other types, vacuum tubes are still manufactured and sold for the replacement market, and in some instances for the manufacture of new equipment. For the present, at least, it is still important for the technician to have some knowledge of tubes, their operation, and applications. In this experiment you will become familiar with some of the physical properties of vacuum tubes.

Vacuum tubes have gone through many changes since the introduction of the Fleming valve. Physically, the early tubes were large, cumbersome laboratory devices. Modern production and know-how have changed these into the small, sturdy, efficient tubes of today.

A vacuum tube contains a cathode, which is the electron emitter; a plate or anode, which is the electron collector; and any additional electrodes required to serve a specific electron-control application. The electrodes are rigidly mounted and housed in a highly evacuated (vacuum) glass, ceramic, or metal envelope. Connections from each electrode within the tube are brought out through airtight seals to pins at the tube base. Originally the base was separate and was made of metal or phenolic material. An example of this type of construction is the "octal-based" tube, which has a metal or glass envelope. Today modern tubes have glass bases which are merely an extension of their glass envelopes. The 7- and 9-pin *miniature* tubes are examples of this type, as are also the 9-pin "novar" and the 12-pin "duo-decar" types. There is another type of tube, called the *nuvistor,* which is housed in a ceramic-metal envelope. Its design is different from that of conventional tubes. There are also subminiature, glass-envelope, glass-based tubes which were used in hearing aids before the transistor replaced them.

There are sockets for each tube type. However, for special applications some subminiature tubes are not socket-mounted.

Electrically, receiving tubes have evolved from the simple diode, or two-element tube, to complex, multielement tubes. A single envelope frequently houses two or more tubes.

Tubes are identified by a fairly standard numbering system. However, there are exceptions to the system which is described below.

1. The first part of a tube designation is a number or numbers which tell the approximate voltage required for the filament of the tube. For example, the filament of a 6J5 requires 6.3 V, the filament of a 5U4 requires 5 V, and the filament of a 50L6 requires 50 V.
2. The second part of a tube designation is a code letter or group of letters for identifying the tube types and functions.
3. The third part of a tube designation is a number which sometimes indicates the quantity of useful elements brought out to the base terminals. For example, a 6SN7 has seven elements, if we consider the filaments as one element.
4. The fourth part of a tube designation, if included, consists of a letter or letters describing some particular characteristic of the tube. For example, a 5Y3-G has a glass (G) envelope. A 5Y3-GT is the same as a 5Y3-G, except that the glass envelope is smaller (GT) than that of the G. Suffix letters A, B, etc., may also be used to indicate improved construction or special characteristics. In general any tube with a suffix may be used to replace an earlier one with the same type number but without a suffix. The reverse of this rule may or may not be true,

depending upon application and the type of change indicated by the suffix.

It must be emphasized, however, that this description is only general and that there are many exceptions to it. The only method for securing accurate information about the tube is to refer to a tube manual.

Each tube has a basing diagram which shows how the elements in the tube are brought to the base pins. The pins on a tube base are numbered in order, starting with 1 and reading in a clockwise direction from the key of the tube, looking at the pin end of the tube. Figure 24.1-1 shows an octal tube base illustrating this system. The *bottom* view of a tube socket (Fig. 24.1-2) is identically numbered.

Tubes with glass bases (such as the 7- and 9-pin miniatures) do not have a separate keyway. Extra spacing between the first and last pins serves as the guide to the identification of pin numbers. This spacing also serves as the key to seating the tube in its socket.

Vacuum-Tube Diode

A symbolic diagram of the tube which you will be using in this experiment is shown in Fig. 24.1-3. It is a twin diode. Note that the symbols for the elements in a vacuum-tube diode are different from those of a semiconductor diode. We shall call the righthand tube diode 1 and the lefthand tube diode 2. Pin 1 is the cathode of diode 1 (K_{D1}), and pin 7 its plate (P_{D1}). Similarly pins 5 and 2 are, in that order, the cathode (K_{D2}) and plate (P_{D2}) of diode 2. The filament (H), pins 3 and 4, heats the cathodes of both tubes. The only connection shown between elements is for the filàment, pins 3 and 4. So we can expect to measure resistance between pins 3 and 4. There is no physical connection between any other pins in the tube, and thus a resistance check between any other two pins in the tube should show infinite resistance (an open).

Before a vacuum tube can start working, its fila-

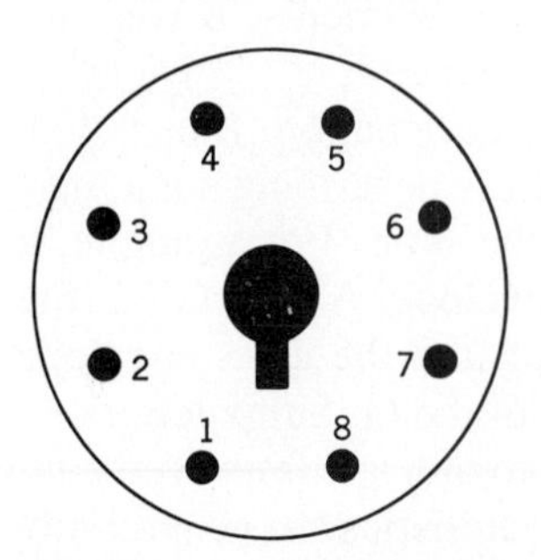

Fig. 24.1-1. Numbering system for pins on tube base.

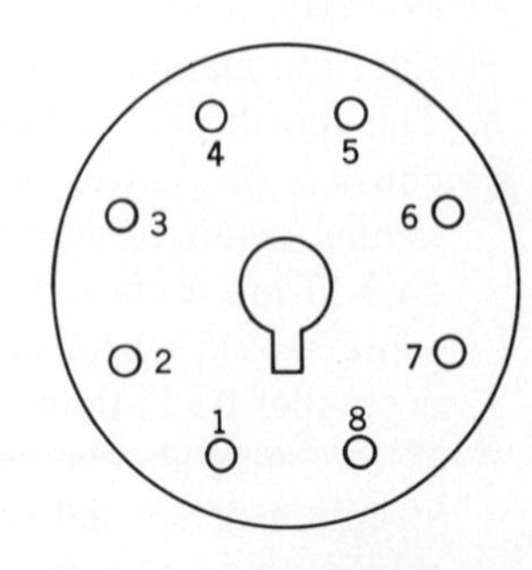

Fig. 24.1-2. Tube-socket (bottom view) numbering.

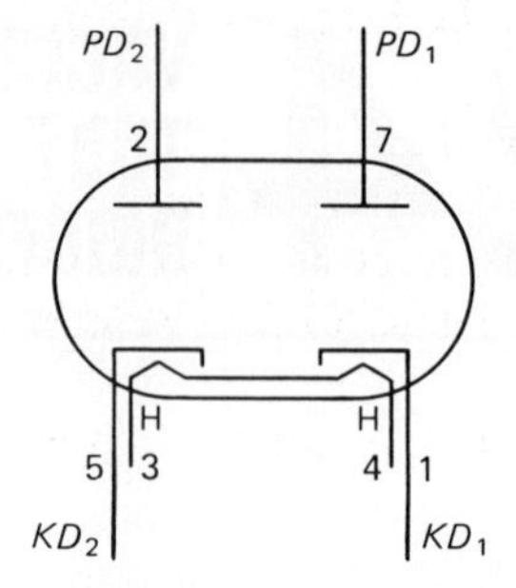

Fig. 24.1-3. Schematic diagram for a duo-diode (6AL5).

ments must heat the cathode sufficiently to emit or give off electrons. Heating time may be 10 s or longer. You can see the glow of the filaments through the envelope of a glass tube. A semiconductor diode has no filament, needs no warm-up time, and starts to operate as soon as power is applied.

The envelope of a glass tube is fragile and should not be struck or scratched. Once the glass is cracked or broken, the vacuum seal is broken and the tube is useless.

SUMMARY

1. Vacuum tubes are highly evacuated glass, metal, or ceramic envelopes which contain tube elements, rigidly mounted within.
2. These elements are a cathode, the electron emitter; a plate or anode, the electron collector; and others required by the function of the tube.
3. Connections from the elements inside the tube are brought to pins at the base of the tube. These pins plug into tube sockets.
4. Manufacturers provide tube-basing diagrams identifying the tube elements, so that circuit connections may be made to them, by way of tube sockets. These diagrams are found in a tube manual.
5. Tubes are identified by a standard numbering system.
6. Tube characteristics are fully described in tube manuals published by tube manufacturers.
7. The cathodes of present-day tubes are *indirectly* heated by filaments inside the cathode housing. The filament is electrically insulated from the cathode.

SELF-TEST

Check your understanding by answering these questions.

1. The space inside the envelope of a tube is a _________.

2. Current in the filament of a tube heats the filament and causes it to ________.
3. The heated filament heats the ________.
4. The ________, when heated, emits electrons.
5. A 6AL5 tube has ________ active elements.
6. The voltage required to heat a 6AL5 tube is ________ V rms.
7. The two active elements of a vacuum-tube diode are the ________ and ________.

MATERIALS REQUIRED

- Power supply: 6.3 V rms source for tube filament
- Tube: 6AL5
- Miscellaneous: Tube socket

PROCEDURE

(Fill in the steps, where required.)

1. Examine the 6AL5. Its envelope is made of ________.

2. There are ________ (how many) pins extending from its base.

3. Inside the glass envelope can be seen metallic ________ which are connected to the ________ pins.

4. The wide space between two pins on the tube base acts as the ________. It identifies pin no. ________ and pin no. ________ on the 6AL5.

5. Draw a bottom view of the tube pins looking directly down on them. Label each of the pin numbers.

6. Refer to a tube manual. Identify the element corresponding to each pin number.

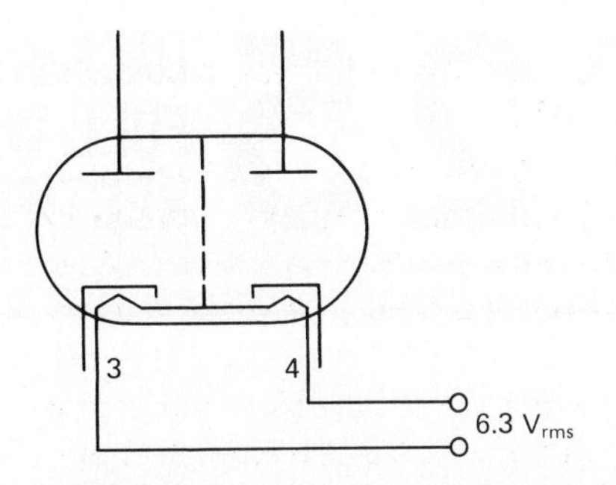

Fig. 24.1-4. Connecting filaments of a 6AL5 to 6.3-V rms source.

7. From the manual determine the **a.** heater voltage: ________ V; **b.** heater current: ________ A.

8. Plug the tube into its socket. Connect the filaments to a 6.3-V rms source, as in Fig. 24.1-4. What happens as the filaments heat up? ________.

9. Touch the tube while the filaments are hot. Has the tube envelope become warm? ________.

QUESTIONS

1. Can the two diodes in a 6AL5 tube be used independently of each other? Refer to the tube manual for your answer.
2. Which of the seven pins of a 6AL5 is not used (that is, not connected to any element within)?
3. Draw a bottom view of a 6AL5 tube and identify each of the pins.

Answers to Self-Test

1. vacuum
2. glow
3. cathode
4. cathode
5. five
6. 6.3
7. cathode; plate

CHAPTER 25 HALF-WAVE AND FULL-WAVE RECTIFIERS

EXPERIMENT 25.1. Half-Wave Rectifier

OBJECTIVE

To observe and measure the input and output waveforms of a half-wave rectifier

INTRODUCTORY INFORMATION

DC and ac voltages and currents serve the power requirements of the wide variety of electronic devices. Thus we have seen how alternating current is used to heat the filaments of a vacuum tube. On the other hand, in studying the static voltampere characteristics of both solid-state and vacuum-tube diodes, we saw that a dc supply served as the power source. Both dc and ac power sources then must be available.

Because it is more efficient and economical to transmit, ac power is generally distributed by utility companies. This necessitates the rectification (changing) of ac into dc voltages and currents. In this experiment we shall be concerned with electronic means of achieving rectification.

Direct current is current which flows in only one direction. The diode is well suited to accomplish rectification, since it permits current to flow in only one direction. Either vacuum-tube or solid-state diodes may be used as rectifiers. In this experiment we shall be concerned with solid-state-diode rectification.

Silicon, selenium, germanium, and copper-oxide rectifiers are solid-state devices which serve as power rectifiers. In this experiment we shall use a silicon rectifier, the type most widely used in electronics today.

Diffused-Junction Silicon Rectifier

These rectifiers are made by diffusing controlled amounts of "impurities" into thin wafers of silicon. The result is a highly reliable rectifier. The silicon rectifier has several advantages over the selenium, germanium, and copper-oxide rectifiers. Thus silicon can operate at higher temperatures (175°C) than the other solid-state devices. It can be constructed to withstand higher reverse-bias breakdown voltages and to exhibit low reverse-bias current. Silicon rectifiers can handle high forward currents.

An ideal rectifier is one which acts as a zero-resistance closed switch when it is forward-biased and an infinite-resistance open switch when it is reverse-biased. That is, it is ON when its anode is positive relative to the cathode, OFF when the anode is negative relative to the cathode. Though this ideal is never realized, the silicon rectifier approaches it.

Figure 25.1-1 shows the voltampere characteristic of a silicon rectifier. When it is forward-biased, the rectifier exhibits extremely low forward resistance R_F. The graph shows that when there is 0.6 V across the diode, it permits 0.2 A of current for a forward resistance $R_F = 0.6/0.2 = 3\ \Omega$. When there is 0.8 V across it, the current is 0.8 A for an $R_F = 0.8/0.8 = 1\ \Omega$. The forward resistance decreases as the current through the diode increases.

The reverse-bias characteristic is equally revealing. Now the current axis is in microamperes and the reverse-bias axis is in 100-V divisions. At 300 V there is approximately 0.4 μA of current, for a reverse resistance

$$R_R = \frac{300}{0.4 \times 10^{-6}} = 750\ \text{M}\Omega$$

At 500 V,

$$R_R = \frac{500}{8 \times 10^{-6}} = 62.5\ \text{M}\Omega$$

Though the silicon rectifier is more tolerant of heat than semiconductors made of other materials, it is still heat-sensitive. The graph in Fig. 25.1-1 is the characteristic of a silicon rectifier at 100°C. If the junction temperature is increased, the forward voltage decreases. If the junction temperature is decreased, the forward voltage increases. If the maximum operating temperature, usually 175°C, is exceeded, the rectifier will fail. To avoid excessive heating of the rectifier junction, heat sinks are used.

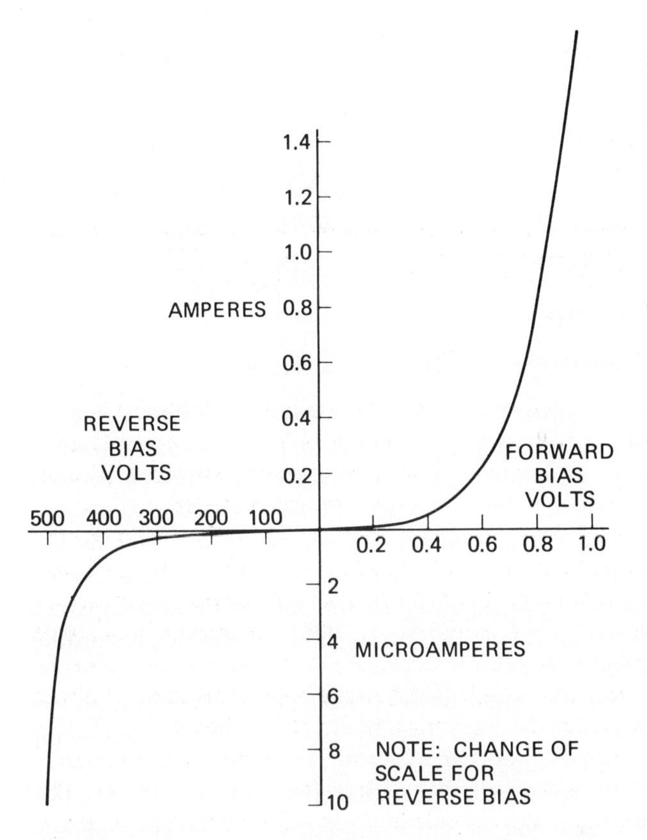

Fig. 25.1-1. Voltampere characteristic of a silicon rectifier.

They dissipate the heat developed and ensure trouble-free operation.

Silicon-Rectifier Ratings

Rectifier characteristics usually supplied by the manufacturer include:

1. The *peak inverse voltage* (PIV), which is the maximum reverse bias that can be applied across a rectifier without having the rectifier break down
2. Maximum sine-wave voltage input (rms)
3. Average half-wave rectified forward current with resistive load, at a specified temperature
4. Peak-recurrent forward current at a specified temperature
5. Maximum forward voltage at specified values of current and temperature
6. Maximum reverse current at maximum reverse voltage
7. Operating and storage temperatures
8. A derating factor, to determine the amount of current through a rectifier for a higher-than-rated temperature

The characteristics of a silicon rectifier, such as the maximum current it can safely handle, the maximum reverse bias (PIV), and the maximum input (rms) voltage, are determined by its construction and size. A wide range of silicon rectifiers exist which can deliver load currents from 200 mA to 1000 A, whose PIV ratings vary from 100 to more than 1000 V. These rectifiers can be connected in parallel for increased load-current requirements, or in series to increase the PIV capabilities of the stack.

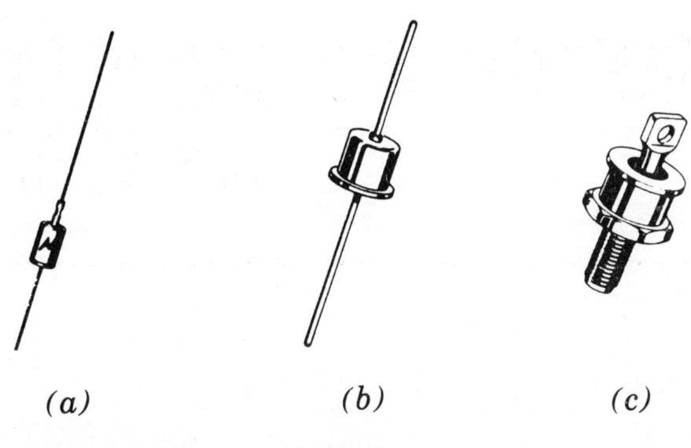

Fig. 25.1-2. Silicon rectifier types.

Silicon rectifiers come in various shapes and sizes including the small flangeless type with two axial leads (like the germanium diodes), Fig. 25.1-2*a*, the top-hat single-ended type, Fig. 25.1-2*b*, and the stud-mounted type, Fig. 25.1-2*c*. Types *a* and *b* are connected in the circuit like a resistor or capacitor. Type *c* is screwed onto a metal chassis, making the chassis serve as a heat sink for the rectifier.

Half-Wave Rectification

Consider the circuit of Fig. 25.1-3. A sinusoidal 6.3-V rms voltage is applied across the series-connected diode D_1 and the load resistor R_L. The input voltage v_{in} is an ac voltage which changes in polarity every $^1/_{120}$ s. During the positive alternation the anode is positive with respect to the cathode, and current flows. During the negative alternation there is no current, because the anode is negative with respect to the cathode.

It is apparent that current through the diode will result in a voltage drop across R_L, the series-con-

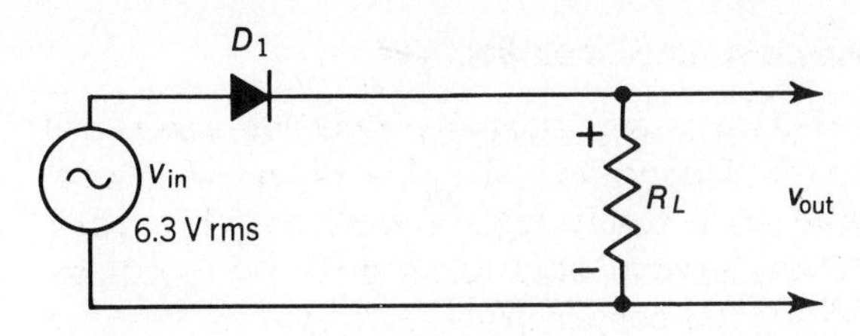

Fig. 25.1-3. Diode half-wave rectifier.

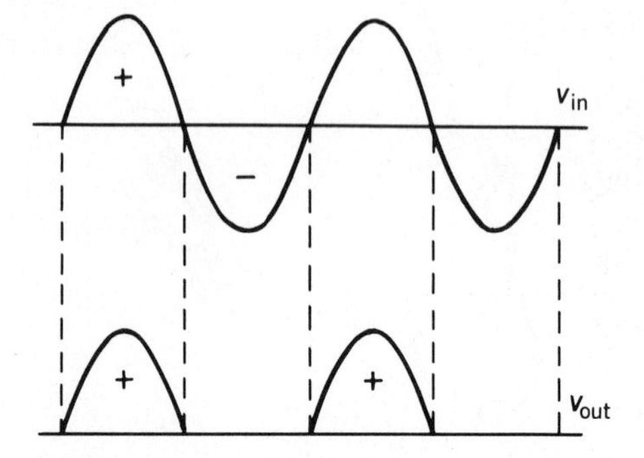

Fig. 25.1-4. Half-wave rectifier waveforms.

nected load resistor. Moreover, since the variation of current will follow the variation of input voltage, the output voltage v_{out} across R_L should follow the positive alternation which causes current. Figure 25.1-4 shows the waveforms v_{in} and v_{out}. It should be noted that v_{out} is no longer an ac voltage, but rather a pulsating dc voltage.

The diode may therefore be compared to a switch which closes only when its anode is positive with respect to cathode. The diode has a certain internal resistance (its forward resistance) which is in series with the load resistor R_L. Hence the diode can be replaced by an equivalent resistance R_F, and the effective input voltage by a generator putting out positive alternations periodically (see Fig. 25.1-5).

The voltage v_{out} across R_L will therefore be a positive alternation, like the input voltage v_{in}, but smaller than v_{in} (see Fig. 25.1-6).

The internal forward resistance R_F of the diode rectifier should be small for maximum output v_{out} across R_L. This resistance depends on the design of the type of diode used. The higher the current rating of the solid-state diode, the lower will be the internal resistance R_F of the rectifier diode. The voltage drop across the rectifier will be 1 V or less.

The process which takes place when a diode conducts during one alternation of the input cycle is called *half-wave rectification.*

The student may be puzzled by the fact that the output of a half-wave rectifier is pulsating direct current rather than constant level, unvarying dc

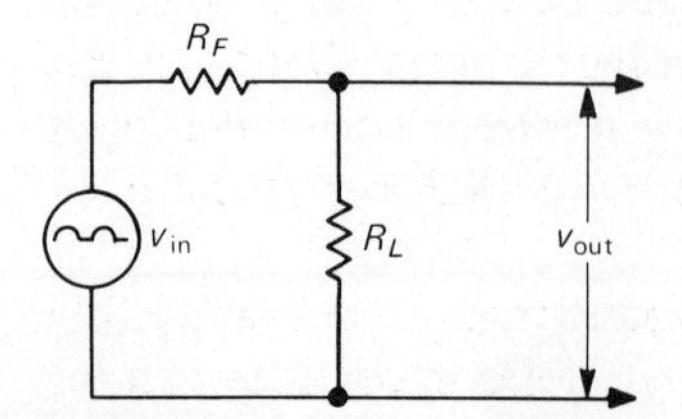

Fig. 25.1-5. Equivalent circuit for half-wave rectifier.

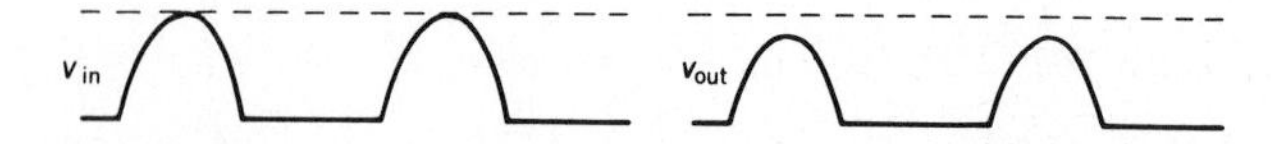

Fig. 25.1-6. Equivalent half-wave circuit waveforms.

voltage. Pulsating direct current is changed into constant dc voltages by filter networks. Power-supply filters will be considered in a future experiment.

Transformer-Fed Half-Wave Rectifier

The source for a rectifier power supply may be the ac line directly, or a rectifier may be transformer-fed. The advantage of a supply which operates directly from the line is its simplicity. The disadvantage is that the input voltage is fixed to the level of the line voltage (120 V rms/60 Hz), and one side is always at the same potential as the side of the line to which it is connected. This arrangement can create hazardous conditions.

Where higher or lower voltages are required, power transformers are usually used, as in the half-wave rectifier supply of Fig. 25.1-7.

Power transformers are built with a primary winding and one or more insulated secondary windings. The primary winding receives 120 V/60 Hz from the line. The secondary windings are either stepup or stepdown windings. Figure 25.1-8 is a diagram of the transformer T_1 which will be used in this experiment. It contains a 120-V primary and a 26-V center-tapped secondary. The center tap is not needed for the half-wave rectifier, but it is required for the full-wave rectifier which will be studied in the next experiment.

The secondary of transformer T_1 is rated at 1 A. This means that the maximum current which may be safely drawn from the secondary is 1 A. Transformers are rated for their input and output voltages *and* for the current they can deliver.

An advantage which a transformer-fed supply has over a line-derived transformerless supply is that the output voltage of the transformer supply is isolated from the power line. The reason is that the secondary

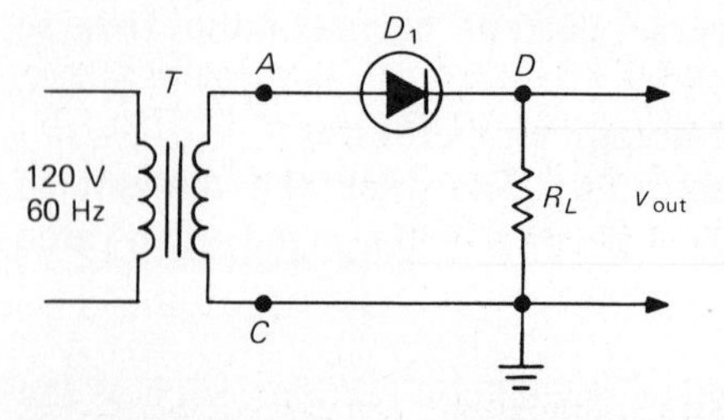

Fig. 25.1-7. Transformer connected to half-wave rectifier.

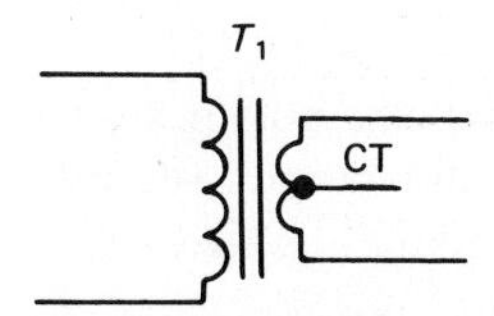

Fig. 25.1-8. Voltage stepdown power transformer with center-tapped secondary.

of the transformer serves as the input to the rectifier, and the secondary has no direct connection to the primary. The line, of course, is connected to the primary. This is an important safety feature.

Observing Phase Relations in a Circuit

Dual-trace oscilloscopes simplify the study of phase relations in an ac circuit. Where this type of instrument is not available, a single-trace scope may be used.

Oscilloscopes contain "line" sync or "line" triggering inputs. This makes it possible to observe phase relations in a circuit receiving its input from a line-derived (60-Hz) source, as in this experiment.

Suppose, for instance, in Fig. 25.1-7 it is required to observe the phase of the input waveform AC and the output waveform DC. The procedure is to set the oscilloscope on line sync or triggering. The vertical input of the oscilloscope is then connected to point A, the ground to point C. After the vertical-gain (volts per division) controls are adjusted for the desired waveform amplitude, the oscilloscope sweep (time per division) and sync (or triggering) controls are adjusted for a stable presentation of 1 or 2 cycles on the screen. The horizontal-gain controls (where available) are adjusted so that the waveform is approximately 4 in wide (if a 5-in oscilloscope is used). The vertical-centering controls are adjusted so that the waveform is centered with respect to the X axis. The horizontal-centering controls are then set so that the start of the positive alternation of the waveform on the left coincides with a major vertical graticule line, as in Fig. 25.1-9. This is the *reference* input waveform. The oscilloscope sweep, sync, horizontal-gain, and centering controls must not be varied until the required phase relations have been observed.

The vertical-input (hot) lead of the oscilloscope is now connected to the top of R_L, point D. The observed output waveform is then in proper time phase with the input reference waveform.

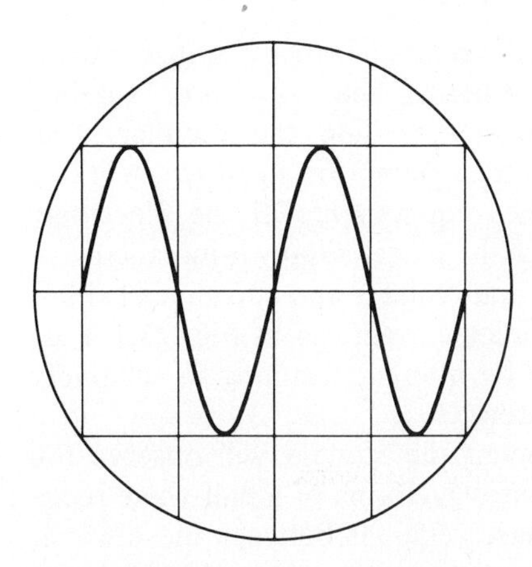

Fig. 25.1-9. Reference waveform properly centered for observing phase relations in rectifier circuit.

SUMMARY

1. Power companies distribute ac electrical power because this method of distribution is more efficient and economical than distribution of direct current.
2. Because *dc* voltage and current are required for electronic devices, it becomes necessary to convert alternating into direct current by a process called *rectification*.
3. Silicon rectifiers are in greater use in electronics than any other solid-state rectifier.
4. Silicon power rectifiers are rated for their peak inverse voltage (PIV); for the peak forward and reverse voltage they can withstand; for the average forward current they can supply and the peak forward current they can provide at a specified temperature; for their operating and storage temperature; and for the maximum rms sine-wave voltage input they can tolerate.
5. A wide range of silicon power rectifiers exists which can meet load current demands in the range of 200 mA to 1000 A.
6. A single rectifier diode, connected as in Fig. 25.1-3, serves as a half-wave rectifier, in which only *one-half* of the ac waveform is applied to the load.
7. The rectified output of a half-wave rectifier appears as one-directional current pulses (Fig. 25.1-4). The rectifier has transformed the ac waveform into pulsating direct current.
8. The rectifier diode accomplishes this process because a rectifier diode conducts only when its anode is positive relative to its cathode. In the half-wave rectifier with a 60-Hz sine-wave input this occurs during the $^1/_{120}$ s (8.3 ms) of the positive alternation.
9. A rectifier, when conducting, is not a perfect switch, but does exhibit some internal resistance R_F. Because of this resistance there is some voltage loss (drop) across the diode.

10. Power rectifier circuits normally utilize power transformers. Whether the secondary winding of the transformer feeding the rectifier, Fig. 25.1-7, is a voltage stepup or stepdown winding depends on the requirements of the electronic device for which the power supply is intended.
11. The rectifier output voltage and current can either be pulsating direct current, as in Fig. 25.1-4, or the pulses can be filtered, resulting in relatively pure direct current.
12. In this experiment the student will observe the input and output waveforms of a half-wave rectifier and the phase relations between these waveforms, using an oscilloscope set to "line" sync/triggering.

SELF-TEST

Check your understanding by answering these questions.

1. Because dc power is used mainly in electronics, power companies distribute direct current for this need. ________ (true/false)
2. RECTIFICATION is the process in which alternating is converted into direct current.
3. A half-wave rectifier, like that in Fig. 25.1-7, receives a 74-V p-p input. The output waveform across R_L, point D with respect to C, is: (*a*) ________ (positive/negative); (*b*) in phase with the ________ (positive/negative) alternation of the input voltage; (*c*) about 37 V p-p.
4. To reverse the polarity of the output waveform in a half-wave rectifier, it is necessary to reverse the DIODE in the circuit.
5. In a half-wave rectifier there is one output pulse for each alternation of the input sine wave. ________ (true/false)
6. Transformers are used in power supplies to isolate the output from the LINE.
7. When a rectifier is conducting, it acts like a ________ (closed or open) switch.
8. Phase relations may be observed between the input and output waveform in the circuit of Fig. 25.1-7 by setting the oscilloscope on LINE synchronization/triggering.
9. The most popular rectifier in electronics today is the ________ (germanium/silicon) rectifier.

MATERIALS REQUIRED

- Equipment: Oscilloscope (dual-trace, optional); EVM
- Resistor: ½-W 10,000-Ω
- Solid-state diode: 1N5625
- Miscellaneous: Two SPST switches; power transformer 120-V primary, 26-V center-tapped secondary at 1 A (Triad F40X or equivalent); fused line cord

PROCEDURE

1. Connect the circuit of Fig. 25.1-10 and set the oscilloscope on *line* sync or triggering. *Have an instructor check your circuit before proceeding.*
2. Connect the vertical input lead of the oscilloscope to the anode of D_1, the ground lead to point C. Calibrate the vertical amplifiers of the oscilloscope for voltage measurement. Close switch S_1. **Power on.** Close switch S_2.
3. Adjust the vertical attenuator, horizontal-gain, sweep, and sync/triggering controls for viewing the reference waveform V_{AC}, as described in the section headed Observing Phase Relations in a Circuit. The waveform viewed should be identical with the reference waveform in Table 25.1-1.

   With the oscilloscope, measure the peak-to-peak voltage v_{in} (v_{AC}). Record the results in Table 25.1-1. With an EVM measure the dc voltage, if any, across points AC. Record the results in Table 25.1-1.
4. Connect the vertical input lead of the oscilloscope to point D. Draw the waveform v_{out} observed across R_L, in proper time phase with the reference waveform. Measure and record the peak-to-peak voltage of the waveform, and the dc voltage, if any, across R_L.
5. Open S_1 and S_2. Break the connection between S_2 and point A, and connect S_2 between the center tap, point B, and the anode of D_1. *When power is applied, the half-wave rectifier will receive only half the voltage of the secondary winding.*
6. Close S_1 and S_2. Observe, measure, and record in Table 25.1-1, v_{in} (v_{BC}) and v_{out} (v_{DC}). Also measure and record the dc voltage, if any, across BC and DC.

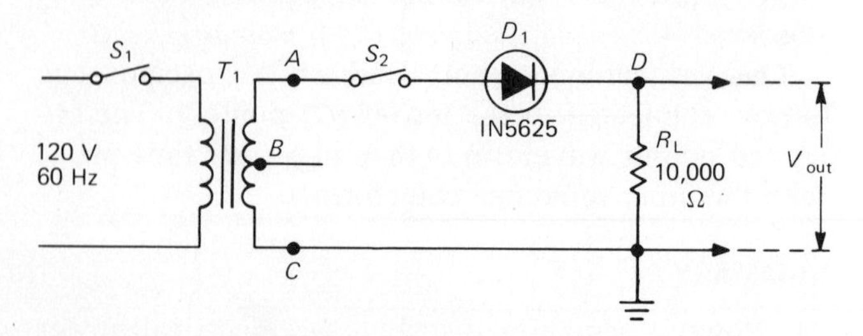

Fig. 25.1-10. Experimental half-wave rectifier circuit.

TABLE 25.1-1. Half-Wave Rectifier Waveforms and Voltages

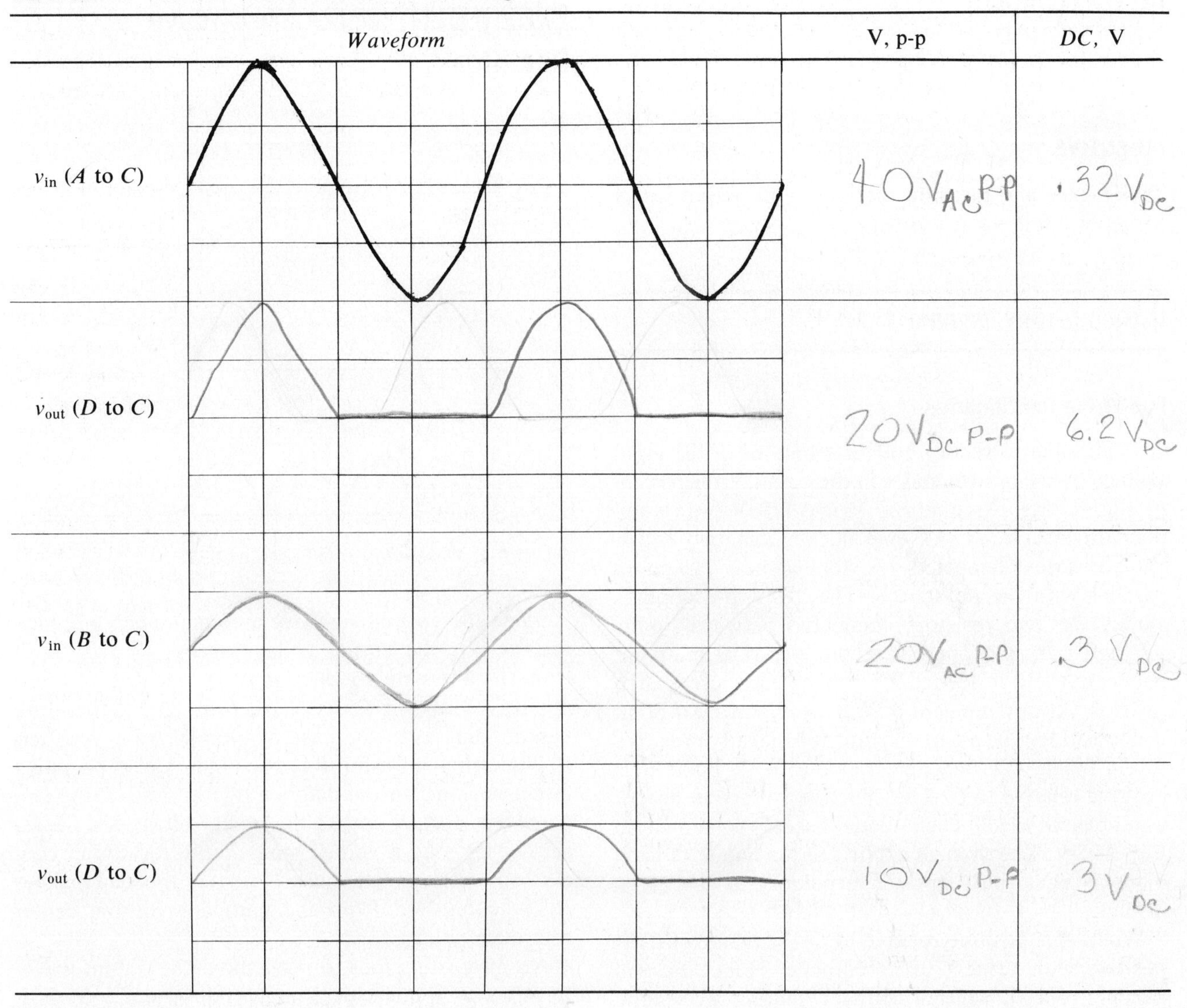

	Waveform	V, p-p	*DC*, V
v_{in} (*A* to *C*)		40 V AC P-P	.32 V DC
v_{out} (*D* to *C*)		20 V DC P-P	6.2 V DC
v_{in} (*B* to *C*)		20V AC P-P	.3 V DC
v_{out} (*D* to *C*)		10 V DC P-P	3 V DC

QUESTIONS

1. Explain what is meant by rectification.
2. Describe the nature of the output voltage v_{out} of a half-wave rectifier.
3. Compare with relation to the input frequency the frequency of the output voltage of a half-wave rectifier. Refer to the waveforms in Table 25.1-1.
4. From your data, what conclusion can you draw about the relationship between the dc voltage across R_L and the peak input voltage in a half-wave rectifier?
5. Refer to your data and explain the operation of a half-wave rectifier.

Answers to Self-Test

1. false
2. Rectification
3. (*a*) positive; (*b*) positive; (*c*) 37
4. rectifier or diode
5. false
6. line
7. closed
8. external or line
9. silicon

EXPERIMENT 25.2. Full-Wave Rectifier

OBJECTIVE

To observe and measure the input and output waveforms of a full-wave rectifier

INTRODUCTORY INFORMATION

Full-Wave Rectification

It is possible to rectify both alternations of the input voltage by using two diodes in the circuit arrangement of Fig. 25.2-1. Assume 6.3 V rms (18 V p-p) is applied to the circuit. Assume further that two equal-valued series-connected resistors R are placed in parallel with the ac source. The 18 V p-p appears across the two resistors connected between points AC and CB, and point C is the electrical midpoint between A and B. Hence 9 V p-p appears across each resistor. At any moment during a cycle of v_{in} if point A is positive relative to C, point B is negative relative to C. When A is negative relative to C, point B is positive relative to C. The effective voltage in proper time phase which each diode "sees" is shown in Fig. 25.2-2. The voltage applied to the anode of each diode is equal but opposite in polarity at any given instant.

When A is positive relative to C, the anode of D_1 is positive with respect to its cathode. Hence D_1 will conduct, but D_2 will not conduct. During the second alternation, B is positive relative to C. The anode of D_2 is therefore positive with respect to its cathode, and D_2 conducts while D_1 is cut off.

There is conduction then by either D_1 or D_2 during the entire input-voltage cycle.

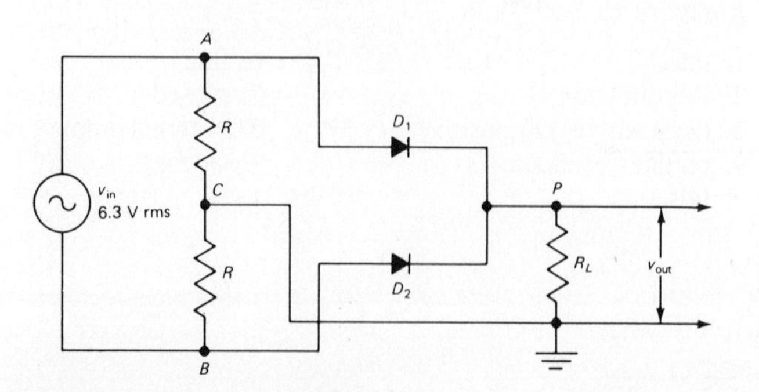

Fig. 25.2-1. Full-wave rectifier.

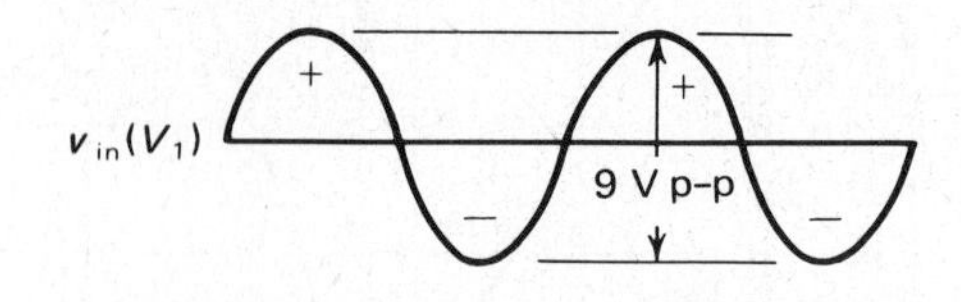

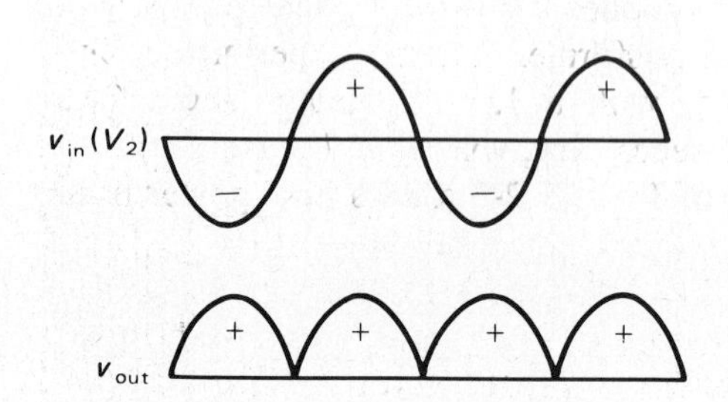

Fig. 25.2-2. Waveforms of full-wave rectifier.

Since the two diodes have a common-cathode load resistor R_L, the output voltage across R_L will result from the alternate conduction of D_1 and D_2. The output waveform v_{out} across R_L in Fig. 25.2-2 therefore has no gaps as in the case of the half-wave rectifier.

The output of a full-wave rectifier is also pulsating direct current. In the diagram of Fig. 25.2-1, the two equal resistors R across the input voltage are necessary to provide a voltage midpoint C for circuit connection and zero reference. Note that the load resistor R_L is connected from the cathodes to this center reference point C.

An interesting fact to note about the output waveform v_{out} is that its peak amplitude is not 9 V as it was in the case of the half-wave rectifier using the same power source, but is less than $4\frac{1}{2}$ V. The reason, of course, is that the peak positive voltage of A relative to C is $4\frac{1}{2}$, not 9 V, and part of the $4\frac{1}{2}$ is lost across R.

Though the full-wave rectifier of Fig. 25.2-1 fills in the conduction gaps, it delivers less than half the peak output voltage that results from half-wave rectification.

Transformer-Fed Full-Wave Rectifier

Figure 25.2-1 is not a practical full-wave power supply rectification circuit because of the use of the two center-tapped resistors R. The voltage drop across R, when its respective diode is conducting, subtracts from the voltage V_{out} and reduces the output voltage. Moreover the dc voltage and current requirements of electronic circuits vary, depending on the nature of the devices used in the circuit, on the power consumed, and on other factors. The voltage and current requirements of a specific electronic device will therefore determine the design of the power supply for that device.

Solid-state devices generally require low-voltage, high-current supplies. Vacuum-tube circuits, on the other hand, require high-voltage, low-current supplies. The component which makes it possible to design unique power supplies for these varied requirements is the power transformer. In this experiment power transformer T_1, with a 120-V primary and a 26-V center-tapped secondary, will again be used.

The circuit of Fig. 25.2-3 shows how power transformer T_1 is connected in a full-wave rectifier circuit. The anodes of rectifier diodes D_1 and D_2 are fed by the secondary voltages AC and BC, respectively. Since C is the center tap, each diode anode receives 13 V rms. The load resistor R_L is connected from the junction of the cathodes of D_1 and D_2, point D, to the center tap on the secondary winding, point C. The output voltage appears across R_L.

When the power is applied to the primary of T_1, D_1 and D_2 operate as a full-wave rectifier. Each diode "sees" only half the voltage appearing across the secondary, and each diode conducts alternately. When the connection between B and the anode of D_2 is open, D_1 acts as a half-wave rectifier. When the connection between A and the anode of D_1 is open, D_2 acts as a half-wave rectifier.

Power rectifiers D_1 and D_2 are rated for the current they must deliver to a circuit and for the peak forward and peak inverse voltages they can withstand.

One of the disadvantages of the circuit of Fig. 25.2-1 was the voltage loss across R. The need for R in Fig. 25.2-3 is eliminated by the center tap on the transformer secondary. The dc resistance between the center tap and either end of the secondary winding is very low, and hence the voltage drop across this resistance is negligibly low.

You will recall from the last experiment that another advantage which a transformer power supply has over a transformerless circuit is that the output voltage V_{out} of a transformer-fed supply is *line-isolated,* since there is no direct connection between the primary (line) winding and the secondary winding.

SUMMARY

1. A full-wave rectifier rectifies *both* alternations of the input ac voltage.
2. Two diodes are required to operate a full-wave rectifier, connected as in Fig. 25.2-1.
3. A practical full-wave rectifier power supply uses a center-tapped power transformer, as in Fig. 25.2-3. In such a circuit, each diode receives one-half of the voltage across the secondary winding.
4. Each of the rectifier diodes in a full-wave rectifier receives voltages equal in amplitude but 180° out of phase. As a result the diodes conduct alternately. That is, when one is ON, the other is OFF, and vice versa.
5. There are no gaps in the output voltage waveform (see Fig. 25.2-2).
6. In a transformer-fed full-wave rectifier, the resistance of the secondary windings must be very low to reduce power losses in the output.

SELF-TEST

Check your understanding by answering these questions.

1. In the full-wave rectifier of Fig. 25.2-3 diode D_1 conducts when ___D2___ is cut off, and vice versa.
2. The output voltage in the full-wave rectifier of Fig. 25.2-3 is ___PULSATING DC___.
3. If the frequency of the source from which transformer T_1 in Fig. 25.2-3 receives its power is 60 Hz, the frequency of the output waveform is ___120___ Hz.
4. In Fig. 25.2-3 diode D_1 is open. The circuit operates like a ___1/2 WAVE___ rectifier.

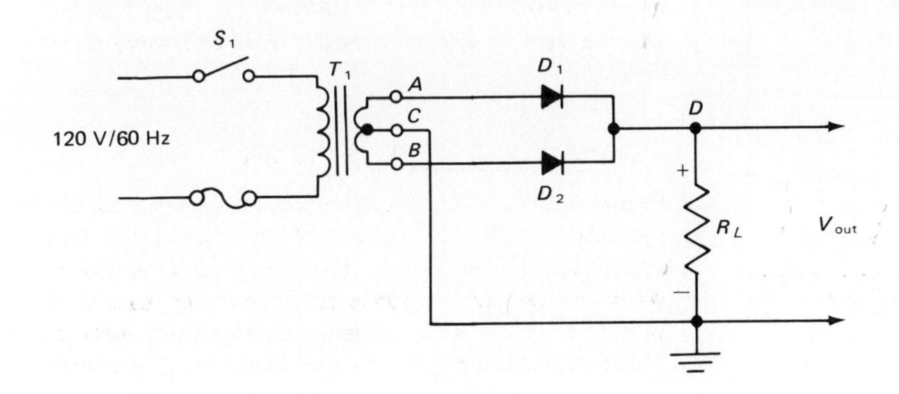

Fig. 25.2-3. Transformer-fed full-wave voltage rectifier.

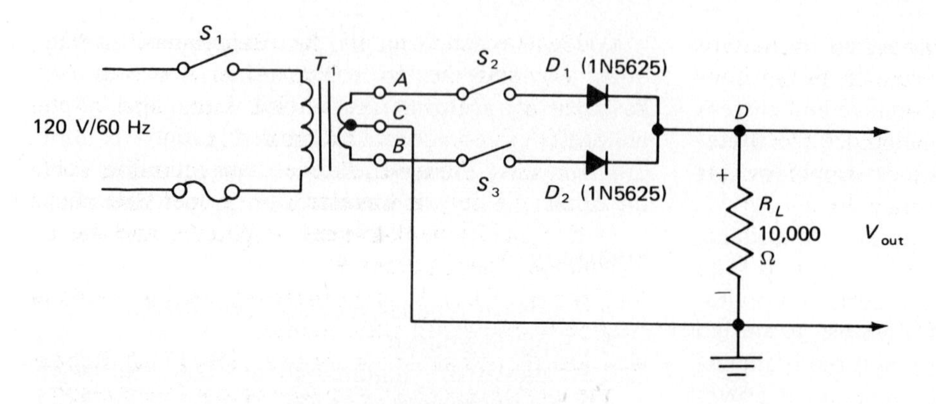

Fig. 25.2-4. Experimental full-wave voltage rectifier.

TABLE 25.2-1. Measurements in Full-Wave Rectifier

	Waveform	V, p-p	*DC*, V
v_{in} (A to C)		20 V_{AC} P-P	.325 V_{DC}
v_{in} (B to C)		20 V_{AC} P-P	.325 V_{DC}
v_{out} (D_1) S_2 closed, S_3 open		10 V_{DC} P-P	3 V_{DC}
v_{out} (D_2) s_2 open, s_3 closed		10 V_{DC} P-P	3 V_{DC}
v_{out} (full wave) S_2, S_3 closed		10 V_{DC} P-P	6 V_{DC}

5. For the same conditions as in question 4, there is one output pulse for each alternation of the input sine wave. ________ (true/false)
6. Phase relations between the input and output waveforms in Fig. 25.2-3 may be observed by an oscilloscope using LINE synchronization.

MATERIALS REQUIRED

- Equipment: Oscilloscope; EVM
- Resistor: ½-W 10,000-Ω
- Solid-state diodes: Two 1N5625
- Miscellaneous: Three SPST switches; power transformer 120-V primary, 26-V center-tapped secondary at 1 A (Triad F40X or the equivalent); fused line cord

PROCEDURE

1. Connect the circuit of Fig. 25.2-4. *Have an instructor check your circuit before proceeding.*
2. Connect the vertical input lead of the oscilloscope to the anode of D_1, the ground lead to point C. Calibrate the vertical amplifiers of the oscilloscope for voltage measurement. Set sync/triggering control on *line*.
3. Close switch S_1. **Power on.** Close switch S_2, but keep S_3 open.
4. Adjust the oscilloscope controls for viewing the reference waveform v_{AC} (see Exp. 25.1), which should be the same as that in Table 25.2-1.

   With the oscilloscope, measure the peak-to-peak voltage v_{AC} and record in Table 25.2-1. With an EVM, measure and record the dc voltage, if any, across points AC.
5. Open switch S_2. Connect the vertical input lead of the oscilloscope to the anode of D_2. Close S_3. Draw v_{BC}, in Table 25.2-1, in proper time phase with the reference waveform.

   Measure and record the peak-to-peak voltage of v_{BC} and the dc voltage, if any, across BC.
6. Open S_3. Connect the vertical input lead of the oscilloscope to point D (across R_L).

   Close S_2. Observe, measure, and record in Table 25.2-1 the output waveform (in proper time phase with v_{AC}), its peak-to-peak amplitude, and the dc voltage, if any, across R_L.
7. Open S_2. Close S_3 and repeat the measurements in step 6 and record your results.
8. Close S_2. All switches are now closed **(on)**. Repeat the measurements in step 6 and record your results.

QUESTIONS

1. How does the output of a full-wave rectifier differ from that of a half-wave rectifier regarding: (*a*) waveform? (*b*) dc voltage?
2. What is the frequency of the output waveform of the experimental full-wave rectifier? Of the input?
3. From your data, what conclusion can you draw about the relationship between the dc voltage across R_L and the peak input voltage in: (*a*) A half-wave rectifier? (*b*) A full-wave rectifier?
4. In which steps of the procedure was the circuit a half-wave rectifier (*a*) for D_1? (*b*) for D_2?
5. In which step of the procedure was the circuit a full-wave rectifier?
6. Would switches S_2 and S_3 normally be used in an industrial full-wave rectifier circuit?

Answers to Self-Test

1. D_2
2. pulsating direct current
3. 120
4. half-wave
5. false
6. line

CHAPTER 26 POWER SUPPLIES: FILTERING AND REGULATION

EXPERIMENT 26.1. Power-Supply Filter Elements

OBJECTIVE

To measure the effects of filter elements on dc output voltage and ripple

INTRODUCTORY INFORMATION

The full-wave rectifier, Fig. 26.1-1*a*, changes alternating current to pulsating direct current, Fig. 26.1-1*b*. The pulses can be smoothed out by filters. In this experiment we shall study filter elements and measure their effect on the output dc voltage level and on the ripple (unsmoothed voltage variations).

Capacitors and chokes are used as passive filter elements. Resistors are sometimes also found in filter circuits.

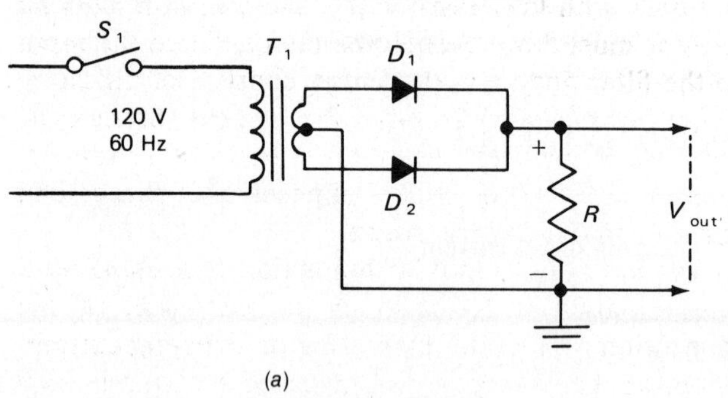

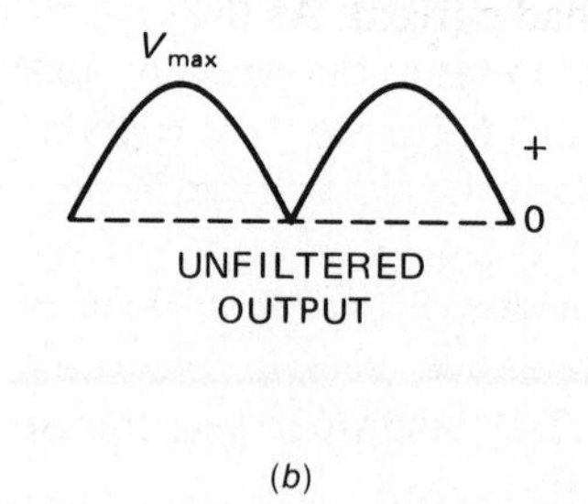

Fig. 26.1-1. Unfiltered output of full-wave rectifier.

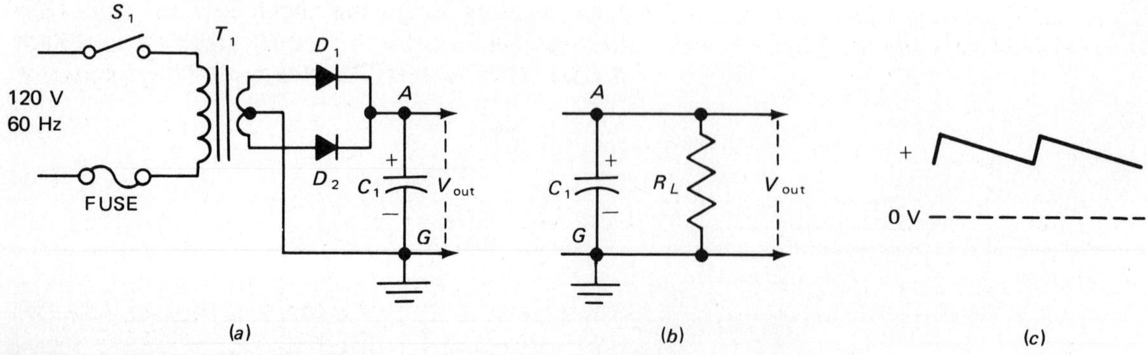

Fig. 26.1-2. Effect of filter capacitor on output of full-wave rectifier.

Capacitors as Filters

The capacitance of a capacitor and its reactance determine how well it will work as a filter. The larger the capacitance, the more effective it will be in smoothing out pulsating direct current and in keeping a low ripple level.

Consider the operation of the circuit in Fig. 26.1-2*a*. Here capacitor C_1 replaces resistor R in the preceding circuit. You will recall from the experiment before that each diode conducts alternately; that is, first one diode conducts and then the other. The capacitor charges through each diode to the peak of the positive alternation that appears across one-half the secondary winding. The polarity of the voltage developed across C_1 makes the cathode of each diode positive with respect to ground. C_1 has no path through which it can discharge except its own parallel leakage resistance, which is ordinarily very high. So C_1 holds its charge, and the dc voltage across C_1 biases the rectifiers D_1 and D_2 to cut off.

The rectifier diodes conduct only during the peaks of the positive alternations of input ac voltage, replacing the small charge that C_1 has lost while discharging. An EVM connected across C_1 will measure a *dc* voltage, approximately equal to the peak of the input *ac* voltage to each diode. An oscilloscope, connected across C_1, will show on its most sensitive range a very small ripple. The ripple represents the charge and discharge of C_1.

If a load resistor R_L is connected across C_1, Fig. 26.1-2*b*, the rectified, filtered dc voltage appears across R_L, and the load resistor draws current from the supply. The resistance of R_L determines how much current is drawn. The lower the resistance, the higher the current. Since the rectifiers are still cut off during a large portion of the input cycle, the current drawn by the load is actually supplied by C_1, which discharges through R_L. If the load current is high, that is, if the resistance of R_L is low, the dc output voltage drops appreciably during the discharge cycle and rises during the interval that C_1 is charging through the rectifiers. The output voltage V_{out} is no longer a steady voltage, but varies between some maximum and minimum value in the manner shown in Fig. 26.1-2*c*. This variation in capacitor charge is the ripple that is observed with an oscilloscope across C_1. If C_1 is replaced by a capacitor of higher capacitance, the ripple is decreased. The dc voltage measured with an EVM depends on the value of C_1 and on the load current. As the capacitance is made larger, the dc voltage increases. As the load current increases, the dc voltage decreases for the same size capacitor.

Why does a larger-valued capacitor act as a better filter element? One reason is that it can store a larger charge than a smaller capacitor. It can therefore supply more current to a load than a capacitor of smaller value. In this respect it is like a water reservoir. A half-barrel of water drawn from a large reservoir will not noticeably lower the water level of the reservoir. On the other hand, drawing half a barrel from a *barrel* of water will lower the water level in the barrel to one-half.

A capacitor acts as a filter also because of its reactance in an *ac* circuit. Capacitive reactance X_C varies inversely with capacitance and with frequency. The larger the capacitance, the lower the X_C. A low value of X_C, in parallel with R_L in Fig. 26.1-2*b*, offers an easier path for ripple current than does R_L. That is, the capacitor acts as a good bypass for ripple current.

The higher the ripple frequency to be filtered, the smaller the value of capacitance required to filter it. In the low-voltage power supplies of TV receivers, where line frequency ripple is low (60 or 120 Hz), large electrolytic capacitors are used as filter elements. By contrast, the input frequency of the high-voltage supply of a TV receiver is 15,750 Hz. As a result capacitors of very much lower value are needed to filter the TV high-voltage supply.

Choke as Filter

A choke is also an effective filter element. For power supplies operating from the line (60 Hz), iron-core chokes are used. Iron-core chokes have a high inductance. The larger the inductance, the more effective the choke is in smoothing out power-supply ripple. An inductor opposes a change in current. This characteristic tends to keep a relatively constant current in a fixed load.

Choke L is connected as a filter element in the full-wave rectifier circuit, Fig. 26.1-3*a*. Observe that it is in series with load resistor R_L, and therefore all load current must flow through the choke. Since the input to the filter and load at point A consists of pulsating direct current, Fig. 26.1-3*b*, the current through the load would tend to follow the variations in the unfiltered input. The choke opposes the changes in current, smoothing the output.

Another way to look at the action of a choke as a filter element is to consider its reactance, or the opposition it offers to alternating or changing current. Suppose the inductive reactance of a choke is high for the supply ripple, and the load resistance is low compared with the inductive reactance. Most of the ripple appears across the choke, and less across the load resistor, because X_L and R_L act as an ac voltage divider. This suggests that the power-line frequency,

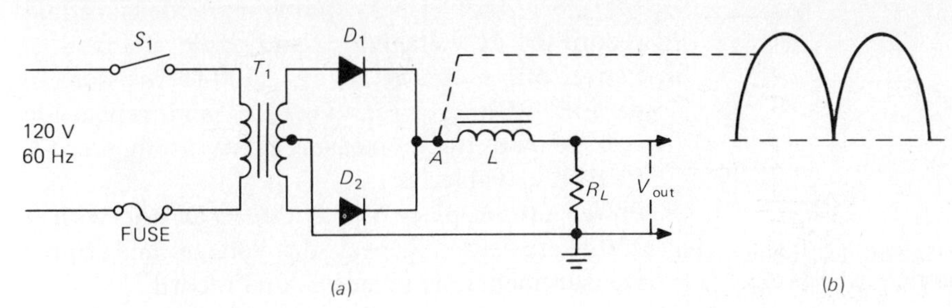

Fig. 26.1-3. (*a*) Iron-core choke used to filter; (*b*) unfiltered output of full-wave rectifier.

which determines the ripple frequency in the output of a rectifier, also determines the size of choke which will be effective as a ripple filter. Because X_L increases with an increase in frequency, and becomes smaller with a decrease in frequency, ($X_L = 2\pi fL$) line frequencies call for large chokes. Higher frequencies permit smaller chokes to be used as filter elements.

The windings of a choke have resistance which must be considered when the choke acts as a filter element. The resistance affects the level of *dc* voltage which appears across the load. If we replace the choke by a resistor R_C equal to the resistance of the windings, as in Fig. 26.1-4, the *dc* voltage which the load resistor R_L receives is reduced, owing to the voltage-divider action of R_C and R_L.

Resistors as Filter Elements

Resistors sometimes replace chokes as filter elements. Resistors do not store a charge. They offer the same opposition to *alternating* as they do to *direct* currents. Their only filter action is the voltage-divider action they produce in combination with other filter elements.

SUMMARY

1. Filters are used to smooth the pulsating dc output of rectifiers.
2. Passive filter elements are capacitors, chokes, and resistors.
3. The larger the value of capacitance, the more effective is its filtering action.
4. The larger the inductance value of a choke, the more effective is its filtering action.
5. A filter capacitor is connected in parallel with the load, Fig. 26.1-2*b*.
6. A filter choke is connected in series with the load, Fig. 26.1-3*a*.
7. For rectifiers operating from the 60-Hz power line, filters use large electrolytic capacitors. If chokes are used in these filters, they are the iron-core type.

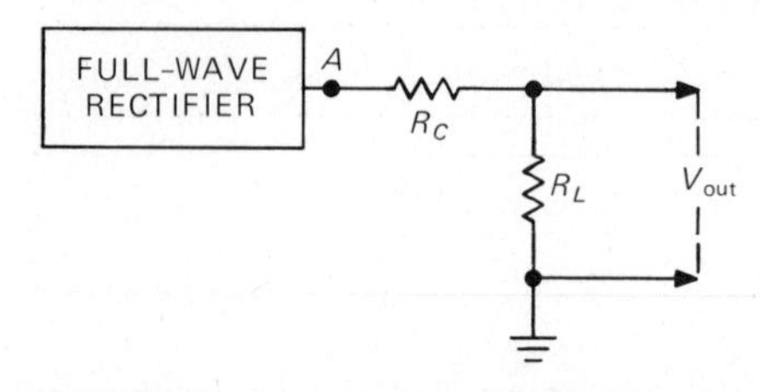

Fig. 26.1-4. Equivalent circuit shows dc voltage-divider action of resistance of choke windings R_C and load resistance R_L.

SELF-TEST

Check your understanding by answering these questions.

1. The unfiltered output of the full-wave rectifier in Fig. 26.1-1 is ________ (positive/negative).
2. With no load, Fig. 26.1-2*a*, the dc output is approximately equal to the ________ (average/rms/peak) value of input voltage across each rectifier.
3. ________ capacitors are used as filter elements in line-derived power-supply filters.
4. A capacitor opposes a change in ________.
5. A choke opposes a change in ________.
6. Resistors are ________ (good/poor) filter elements.

MATERIALS REQUIRED

- Equipment: Oscilloscope; EVM or VOM
- Resistors: ½-W 2700-Ω; two 2-W 250-Ω
- Capacitors: 100 μF/50V; 25 μF/50V
- Solid-state rectifiers: Two 1N5625 or equivalent
- Miscellaneous: Power transformer, T_1, 120-V primary, 26-V 1-A center-tapped secondary; SPST switch

PROCEDURE

Unfiltered Output

1. Connect the circuit of Fig. 26.1-1. $R = 2700\ \Omega$, $D_1 = D_2 =$ 1N5625.
2. Close S_1. **Power on.** With an oscilloscope observe the voltage waveform across R, v_{out}, and measure its peak-to-peak voltage. Record in Table 26.1-1. Also measure and record the *dc* voltage across R.

Output Filtered by Capacitor

3. Open S_1 **(power off).** Remove R from the circuit and substitute C_1, a 25-μF capacitor for R, as in Fig. 26.1-2*a*. Close S_1 **(power on).** Measure and record the dc voltage V_{out}, and ripple, if any.
4. **Power off.** Connect R_L, a 2700-Ω resistor, in parallel with C_1, Fig. 26.1-2*b*, and repeat dc voltage and ripple measurements as in step 3. Record in Table 26.1-1.
5. **Power off.** Replace the 2700-Ω resistor with a 250-Ω resistor. Repeat dc voltage and ripple measurements, as in step 3, and record.

TABLE 26.1-1. Capacitor as a Filter Element

Step	*Conditions*	*Waveform*	V p-p	V dc
2	Fig. 26.1-1 $R = 2700\ \Omega$ No C		10	5.9
3	Fig. 26.1-2*a* $C_1 = 25\ \mu F$ no R		0	9.7
4	Fig. 26.1-2*b* $C_1 = 25\ \mu F$ $R_L = 2700\ \Omega$		1	8.9
5	Fig. 26.1-2*b* $C_1 = 25\ \mu F$ $R_L = 250\ \Omega$		5	7.0
6	Fig. 26.1-2*b* $C_1 = 100\ \mu F$ $R_L = 250\ \Omega$		2	8.5

6. **Power off.** Replace C_1 with a 100-μF capacitor. Repeat dc voltage and ripple measurements as in step 3 and record.

Choke as a Filter Element

7. **Power off.** Connect the circuit of Fig. 26.1-3*a*. $L = 8$ H, $R_L = 250\ \Omega/2$ W.
8. **Power on.** Measure and record in Table 26.1-2 the input waveform to the filter, point A, and its peak-to-peak amplitude. Also measure and record the *dc* voltage at point A.
9. Measure and record the waveform across R_L, and the dc voltage across R_L.

TABLE 26.1-2. Choke as a Filter Element

Step	*Conditions*	*Waveform*	V p-p	V_{out}, dc
7,8	Fig. 26.1-3 $L = 8$ H $R_L = 250\ \Omega$		10	5.9
9	Fig. 26.1-3 $L = 8$ H $R_L = 250\ \Omega$		1	3.0
10	Fig. 26.1-3 $L = 8$ H $R_L = 2700\ \Omega$		3	5.4

10. **Power off.** Replace R_L with a 2700-Ω resistor. **Power on.** Repeat the ripple voltage and dc voltage measurements across R_L and record them.

Resistor as a Filter Element

11. **Power off.** Replace the choke in Fig. 26.1-3 with a 250-Ω resistor. $R_L = 2700\ \Omega$.
12. **Power on.** Measure and record in Table 26.1-3 the ripple and dc voltage, V_{out}.
13. **Power off.** Replace R_L with a 250-Ω load resistor. The 250-Ω filter resistor R_C is still connected.
14. **Power on.** Measure and record in Table 26.1-3 the ripple and dc voltage, V_{out}.

QUESTIONS

Refer to your experimental data for the answers to these questions.

1. Which of the filter elements provided the highest filtered *dc* output voltage V_{out}? Why?
2. In the circuit of Fig. 26.1-2*b*, is the dc output voltage affected by the size of the load resistor? How?
3. In the circuit of Fig. 26.1-2*b*, is the ripple voltage affected by the size of R_L? How?
4. In the circuit of Fig. 26.1-2*b*, are (*a*) V_{out}, dc and (*b*) ripple voltage affected by the size of C_1? How?
5. In the circuit of Fig. 26.1-3, what determines the level of (*a*) ripple voltage in V_{out}? (*b*) dc voltage V_{out}?
6. How does the resistor R_C, Fig. 26.1-4, compare with the choke as a filter element? with the capacitor?

Answers to Self-Test

1. positive
2. peak
3. Electrolytic
4. voltage
5. current
6. poor

TABLE 26.1-3. Resistor as a Filter Element

Step	*Conditions*	*Waveform*	V p-p	V_{out}, dc
11, 12	Fig. 26.1-4 $R_C = 250\ \Omega$ $R_L = 2700\ \Omega$		9	5.3
13, 14	Fig. 26.1-4 $R_C = 250\ \Omega$ $R_L = 250\ \Omega$		5	3.0

EXPERIMENT 26.2. Power Supply π-type Filter

OBJECTIVE

In a transformer power supply to determine experimentally and compare the effectiveness of a *CLC* π (pi)-type filter with a *CRC* filter

INTRODUCTORY INFORMATION

π-type *CLC* Filter

In Exp. 26.1 we studied the action and effectiveness of capacitors (C), chokes (L), and resistors (R) as filter elements. In this experiment we shall see how these elements are combined in filter networks to provide better filtering.

For low load-current applications, a capacitive filter, that is, a capacitor in parallel with the load resistor, may be adequate to maintain a relatively constant *dc* level. For higher load currents such as those used in radio, hi-fi, and television receivers, more effective filters are necessary to ensure low-ripple output voltages.

Figure 26.2-1 shows a filter network which is very frequently used in a television receiver. The pulsating dc output of the full-wave rectifier is filtered by electrolytic capacitor C_1, choke L, and electrolytic capacitor C_2. The output *dc* voltage V_{out} across C_2 is applied to the load resistor R_L.

The effect of C_1, L, and C_2 is to improve filtering action by increasing the charge stored in these reactive components. Load current drawn from this supply will cause less ripple in the output than the same load would cause in the output of a power supply using, say, simply a capacitor in parallel with the load, or a choke in series with the load. The arrangement of filter elements in Fig. 26.2-1 is called a *pi* filter, so-called because of the Greek letter π which the circuit resembles.

Because the first element in the filter is capacitor C_1, it is called a *capacitor input filter*. Since the input capacitor charges to the peak of the *ac* voltage supplied by the transformer, a capacitor input filter provides the maximum dc output voltage for a load. Large capacitors are needed for line-derived power supplies. Therefore C_1 and C_2 are electrolytics.

The maximum current which the power-supply rectifiers must pass is the surge current which results when a device, such as a television receiver, is first turned on. Large-valued electrolytic capacitors cause high surge currents. So there is a practical limit to the size of input capacitance which may be used with a particular rectifier. Rectifier manuals specify the maximum value of input capacitance which may be used with a specific rectifier.

The resistance R_C of the choke in Fig. 26.2-1 and the load resistor R_L make up a voltage divider. The highest dc voltage, call it V_1, appears across the input capacitor C_1. Direct current (I_L) flows in R_L and in the choke, resulting in the voltage drops, $V_C = I_L \times R_C$, across the resistance R_C of the choke, and $V_L = I_L \times R_L$, across the load. By Kirchhoff's voltage law,

$$V_1 = V_C + V_L$$

Therefore

$$V_L = V_1 - V_C$$

So it is clear that the filtered voltage V_L which the load receives is less than the unfiltered voltage V_1 at the cathode of the rectifier. For large load currents a choke with high inductance and low internal resistance is required.

The dc voltage across the output filter capacitor C_2 is termed $V+$. The value of $V+$ depends on the ac

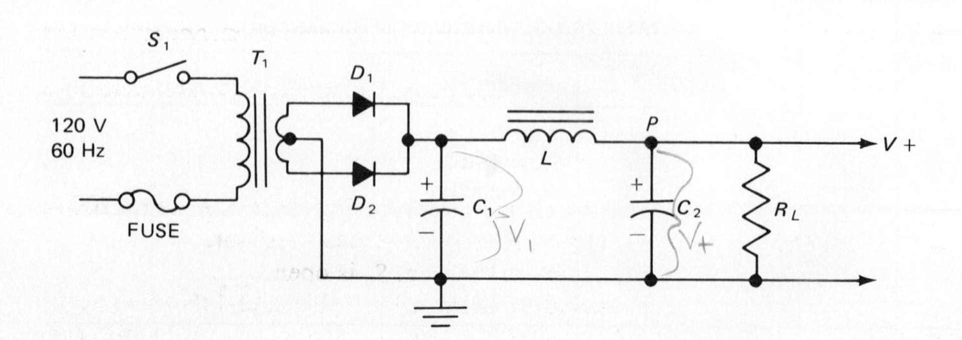

Fig. 26.2-1. Rectifier with π-type *CLC* filter.

voltage across the secondary of the power transformer, the size of the filter capacitors and choke, and the value of load current.

π-type *CRC* Filters

Frequently a resistor R is used to replace the choke as a filter element. Figure 26.2-2 illustrates this type of filter. However, a resistor is not as effective a filtering component as a choke. Therefore, if circuit requirements or cost suggest the use of a resistor rather than a choke, larger-valued filter capacitors C_1 and C_2 are required to compensate for lack of the choke.

For solid-state devices, requiring low voltages, either regulated supplies or power supplies using π-type *CRC* filters, such as that shown in Fig. 26.2-2, are employed. For the *CRC*-filtered supplies it is not unusual to find capacitors being used whose values are 500 to 1000 μF. For vacuum-tube devices requiring much higher voltages than solid-state circuits, power supplies with π-type *CLC* filters are normally used. Capacitor values in *CLC* filters are about 50 to 100 μF. The student may wonder why *CRC* filters are not designed for high-voltage supplies. The reason is that large-valued high-voltage electrolytic capacitors are exceedingly bulky and expensive. Low-voltage electrolytic capacitors, on the other hand, are much less bulky and also less costly.

SUMMARY

1. Combinations of capacitors, chokes, and resistors make better filters than individual filter elements acting alone.
2. The π filter is a popular-type filter used to smooth out pulsating direct current in the output of a power supply.
3. A *CLC* π filter, as in Fig. 26.2-1, uses two electrolytic capacitors, C_1 and C_2, and a choke to clip the pulse peaks and fill in the valleys in the output of a power-supply rectifier.

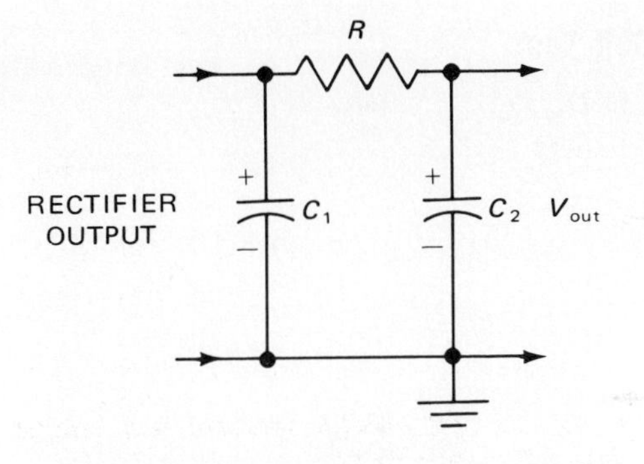

Fig. 26.2-2. Resistor R used in π-type *CRC* filter.

4. The filter in Fig. 26.2-1 is called a capacitor input filter.
5. A resistor in a *CRC* arrangement is sometimes used to replace the choke in a π filter, as in Fig. 26.2-2.
6. The resistor is not as effective a filtering element as a choke, but it is smaller and cheaper.
7. When *CRC* π filters are used, oversized electrolytic capacitors make up for the poor filtering action of the resistor.

SELF-TEST

Check your understanding by answering these questions.

1. π-type filters are used to supply low-ripple voltages for ________ (high/low) current loads.
2. The dc voltage V_1 across C_1, Fig. 26.2-1, is ________ (higher/lower) than the voltage across the load resistor.
3. The limit on the maximum capacitance value of C_1 in Fig. 26.2-1 is determined by the characteristics of D1 and D2.
4. If C_1 were eliminated in Fig. 26 2-1, and just the choke and C_2 were left, the dc output voltage would be unchanged, but the ripple would increase. ________ (true/false)
5. Assume that the resistance R of the *CRC* filter in Fig. 26.2-2 is the same as the resistance R_C of the choke windings in Fig. 26.2-1, and the values of C_1 and C_2 are the same in both filters. The two filters are then equally effective. ________ (true/false)

MATERIALS REQUIRED

- Equipment: Oscilloscope; EVM or VOM
- Resistors: ½-W 2700-Ω; 2-W 250-Ω
- Capacitors: Two 100-μF/50V
- Solid-state Rectifiers: Two 1N5625 or equivalent
- Miscellaneous: Power transformer T_1, 120-V primary, 26-V 1-A center-tapped secondary; resistance decade box; two SPST switches, 8-H 250-Ω choke

PROCEDURE

CLC π-type Filter

1. Connect the circuit of Fig. 26.2-3. R_1 is a 2700-Ω resistor. R_L is a 250-Ω load resistor.
2. Close S_1. **Power on.** S_2 is open.
3. Measure and record in Table 26.2-1 the dc voltages V_1 across C_1 and V_2 across C_2. Also observe,

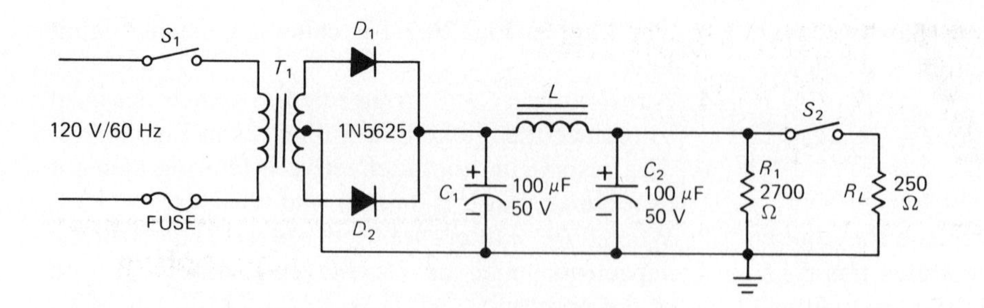

Fig. 26.2-3. Experimental rectifier and filter.

TABLE 26.2-1. ***CLC*** **π-type Filter**

Conditions	Waveform		V p-p (RIPPLE)	V, dc	
No load S_2 open	v_1		1	V_1	9.2
	v_2		0	V_2	8.4
(270) 250-Ω load S_2 closed	v_1		2	V_1	8.6
	v_2		0	V_2	4.5

measure, and record the ripple voltage waveforms v_1 across C_1 and v_2 across C_2.

4. Close S_2 and repeat your dc and ripple voltage measurements in step 3 and record.

CRC π-type Filter

5. **Power off.** Remove the choke from the circuit. Measure and record the resistance R_C of the choke. 250 Ω.
6. Set the resistance decade box for a terminal resistance equal to the value of R_C measured in step 5. Connect the resistance decade box set to the value of R_C, in place of the choke. You now have a *CRC* filter.
7. Close S_1, **power on.** S_2 is open.
8. Repeat step 3. Record your results in Table 26.2-2.
9. Close S_2 and repeat step 8.

Table 26.2-2. ***CRC*** **π-type Filter**

Conditions	Waveform		V p-p (RIPPLE)	V, dc	
No load S_2 open	v_1		1	V_1	9.2
	v_2		0	V_2	8.3
(270) 250-Ω load S_2 closed	v_1		2	V_1	8.6
	v_2		0	V_2	4.2

QUESTIONS

1. Which filter, if any, provided the highest dc load voltage, under 250-Ω load?
2. Which filter, if any, provided the lowest output ripple, under 250-Ω load?
3. What conclusions, if any, can you draw from your answers to questions 1 and 2?
4. How did the dc output voltage and ripple for the *CLC* filter compare with the dc output voltage and ripple for the *CRC* filter, when the load resistance was 2700 Ω? What conclusions can you draw from your answer to question 3?
5. If you had measured the *dc* voltage across the choke, under 250 Ω load, what would it have been? Why?
6. How effective was the *CLC* filter under 250-Ω load, as compared with the capacitor filter in Exp. 26.1-1.?

Answers to Self-Test

1. high
2. higher
3. D_1; D_2
4. false
5. false

CHAPTER 27 TROUBLESHOOTING A POWER SUPPLY

EXPERIMENT 27.1. Troubleshooting a Power Supply I: Voltage and Resistance Measurements

OBJECTIVE

To measure the value of dc and ac voltages and resistances in a power supply which is operating normally

INTRODUCTORY INFORMATION

DC Voltage and Ripple

In troubleshooting defective sections of electronic equipment, such as television receivers, the technician makes dc and ac voltage measurements and resistance checks at test points in the circuit. To have a standard for comparison, the technician should know the values at these test points when the section is operating normally. This is as true of a power supply as of any other circuit.

The television manufacturer usually supplies normal voltages and resistance values in a technical manual. When this information is not available, it is still possible to make a fairly close estimate of what these values should be.

Consider the full-wave rectifier power-supply circuit in Fig. 27.1-1. This has a π-type filter, consisting of an input capacitor C_1, a filter resistor R_C, an output filter capacitor C_2, and a bleeder resistor R_B. Because it is a capacitor input filter, the dc voltage V, across C_1, is approximately equal to one-half the peak-to-peak ac voltage across the secondary winding of T_1. So if the peak-to-peak voltage across the secondary of T_1 is, say, 50 V, each rectifier "sees" half of that, or 25 V. The input capacitor C_1 charges to 25 V at the peak of the positive alternation, through whichever rectifier is ON at that time.

The output voltage $V+$ across C_2 (and therefore also across R_B and the load) is lower than 25 V because of the voltage drop across R_C due to the bleeder and load currents I_T. You will recall from a previous experiment that $V+$ equals V_1 minus $I_T \times R_C$; that is, $V+ = V_1 - I_T \times R_C$. Moreover, the greater the load current, the lower is $V+$.

If we know the total bleeder and load current I_T and the value of R_C, we can easily calculate $V+$. For example, let $I_T = 40$ mA, $R_C = 100\ \Omega$, and $V_1 = 25$ V. Then $V+ = 25 - 0.04 \times 100 = 21$ V. Generally, we can say that

$$V+ = V_1 - I_T \times R_C \qquad (27.1\text{-}1)$$

It is clear then that the dc output voltage $V+$ in a power supply like that in Fig. 27.1-1 is lower than the

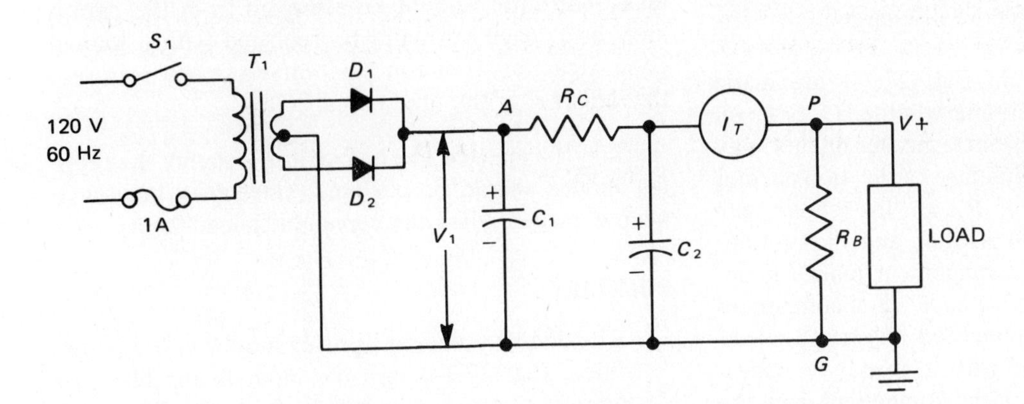

Fig. 27.1-1. Power supply connected to furnish $V+$ with respect to common return G.

dc voltage V_1 across the input filter capacitor. How much lower it will be depends on the circuit conditions.

What about ripple? The ripple at $V+$ is definitely lower than the ripple at the input filter capacitor C_1. How much lower depends on the size of the filter capacitors and on the amount of load current. Generally, the larger the capacitors, the lower the ripple. That is, ripple varies *inversely* with the size of the capacitor. However, ripple varies *directly* with load current; thus, the higher the load current, the higher the ripple.

The effect of a *leaky* capacitor C_1 or C_2 is to reduce the dc output voltage $V+$. Also, ripple increases. An *open* C_1 reduces $V+$ and also causes the ripple to increase. An open C_2 increases ripple very much but does not affect $V+$ appreciably.

AC Voltage

Under normal operation the line voltage (120 V/60 Hz in Fig. 27.1-1) is present across the primary of transformer T_1. The voltage specified for the secondary of the transformer should be measured across the secondary winding. Line voltage and voltages across the one or more secondary windings of a transformer are given in rms values. Voltages measured from the anodes of D_1 and D_2, respectively, to the center tap on the secondary should be approximately equal to one-half the measured value across the secondary winding. AC voltage measurements in a transformer power supply such as that in Fig. 27.1-1 can be made of the voltage across the primary winding (line voltage), the voltages across each of the secondary windings, and the voltages from each end of a secondary winding to any tap on that winding.

Resistance

NOTE: Before resistance measurements may be made in a power supply, power should be turned off and all electrolytic capacitors should be discharged.

Resistance measurements in Fig. 27.1-1 are made from $V+$ to ground to determine whether C_1 or C_2 is short-circuited. Other resistance measurements may be made to determine continuity of the transformer windings, the resistance of R_C or the choke, the bleeder resistor, the ON-OFF switch, and the rectifier diodes. Where the circuit arrangement makes it impossible to measure the resistance of a component without measuring the combination resistance of another component in parallel with it, it may be necessary to disconnect *one end* of the component from the circuit and then measure its resistance. If manufacturers have supplied resistance values on the equipment diagrams, it may not be necessary to disconnect any component.

Specific resistance values are related to the parts used. The following are suggested as guidelines in the circuit of Fig. 27.1-1, where C_1 and C_2 are 100-μF/50-V capacitors, $R_C = 100\ \Omega$, $R_B = 5000\ \Omega$, the load is *disconnected,* and T_1 is a 4:1 stepdown transformer, whose secondary winding is rated at 1 A:

P to ground, 5000 Ω
A to ground, 5100 Ω
R_C(*A* to *P*), 100 Ω
Resistance of secondary winding of $T_1 = 0.2\ \Omega$
Resistance of primary winding of $T_1 = 1.2\ \Omega$

When the bleeder and load are both disconnected from the circuit, capacitors C_1 and C_2 will give a charging indication when an ohmmeter is first connected from point *P* or *A* to point *G*. The meter will first read close to zero resistance. Then if the meter leads are left connected, the capacitors will slowly charge toward the supply voltage in the ohmmeter. The charging time constant is *long* because of the large values of C_1 and C_2. Therefore the measured resistance will *gradually* increase. If the leads are left on long enough, the resistance may measure 1 MΩ or higher. It is not necessary to leave the meter leads connected for any great length of time. A capacitor charging indication is evidence that C_1 and C_2 are not shorted.

NOTE: An ohmmeter test of a capacitor is not a conclusive test of a leaky capacitor because the supply voltage of the ohmmeter is relatively low. The capacitor may register good on an ohmmeter test, but may break down when its rated voltage is placed across the capacitor. A better test of a capacitor is to check for dc voltage at *P* and at *A*. If the voltage is lower than normal and the ripple is excessive, one or both capacitors may be defective. One lead of a suspected leaky capacitor should be removed from the circuit and the capacitor should be replaced with a known good capacitor. If circuit function is properly restored, the defect is in the original capacitor.

NOTE: If a shorted filter capacitor is found, the rectifier diodes should be checked, since they may have been damaged when the capacitor failed.

SUMMARY

1. The dc test points in a power supply with a π-type filter, Fig. 27.1-1, are the input to the filter, V_1 across C_1, and the output from the filter, $V+$ across C_2.

2. In a filtered dc supply with load, the dc input to the filter, V_1, is higher than $V+$, the dc output from the filter.
3. The relationship between $V+$, V_1, and V_C (the voltage across the filter resistor) is,

$$V+ = V_1 - V_C$$

4. The dc voltage at the input to a π-type filter in a full-wave rectifier supply is approximately one-half the peak-to-peak voltage across the secondary of the transformer.
5. The dc voltage at the output of a filter ($V+$) depends on the load current, on the resistance of the filter resistor or choke, and on the size of the filter capacitors. The larger the capacitors, the higher is $V+$. The higher the load current, the lower is $V+$.
6. In a power supply with π-type filter, Fig. 27.1-1, the ripple voltage at $V+$ is lower than the ripple voltage across C_1.
7. The ripple voltage on $V+$ under load is higher than without load. A supply which is operating properly will furnish a low ripple voltage on $V+$ under load.
8. An open input filter capacitor C_1, Fig. 27.1-1, will cause a lower $V+$ and an increased ripple.
9. An open output filter capacitor C_2 will cause a higher ripple but will not appreciably affect $V+$.
10. Leaky filter capacitors cause reduced $V+$ and higher ripple.

SELF-TEST

Check your understanding by answering these questions. Refer to the circuit in Fig. 27.1-1.

1. In the circuit the *dc* voltage V_1 across $C_1 = 50$ V. If $R_C = 100$ Ω and $I_T = 0.2$ A, then $V+ =$ ________ V.
2. If $V+$ drops and there is an appreciable increase in ripple, a likely cause of trouble is an open ________ (C_1/C_2).
3. The bleeder resistor $R_B = 10{,}000$ Ω. $R_C = 500$ Ω. The load resistance = 500 Ω.
   a. The resistance measured across $C_1 =$ ________ Ω.
   b. The resistance measured from $V+$ to ground = ________ Ω.
4. The primary voltage is 120 V. T_1 is a 2:1 step-down transformer. The ac voltage measured from either end of the secondary winding to the center tap is ________ V rms.
5. The resistance of the secondary winding in question 4 is ________ (higher/lower) than the resistance of the primary.

MATERIALS REQUIRED

- Equipment: Oscilloscope; EVM or VOM
- Resistors: ½-W 100-, 2700-Ω; 2-W 250-Ω
- Solid-state rectifiers: Two 1N5625 or the equivalent
- Capacitors: Two 100 μF/50 V
- Miscellaneous: Power transformer T_1, 120-V primary, 26-V 1-A center-tapped secondary; SPST switch; fused line cord

PROCEDURE

Voltage Measurements and Ripple

1. Connect the circuit of Fig. 27.1-2. Close S_1, **power on.**
2. Measure and record in Table 27.1-1 the dc voltages V_1 and $V+$.
3. Measure and record the *ac* voltages across: **a.** Secondary; **b.** anode of D_1 to the center tap (ground); **c.** anode of D_2 to the center tap.
4. Measure and record the ripple of **a.** V_1; **b.** $V+$.

Resistance Measurements

5. Open S_1, **power off.** Remove the line plug from the power outlet. Discharge capacitors C_1 and C_2.
6. Measure and record in Table 27.1-2 the resistance

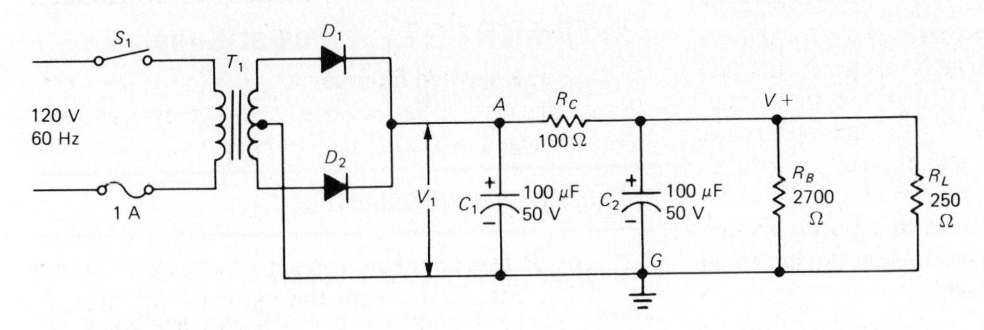

Fig. 27.1-2. Experimental power supply.

TABLE 27.1-1. Voltage Measurements and Ripple

Where Measured	V dc
V_1	
$V+$	
	V ac, rms
Secondary T_1	
Anode D_1 to center tap	
Anode D_2 to center tap	
	V p-p
Ripple on V_1	
Ripple on $V+$	

of: **a.** Primary of T_1; **b.** secondary of T_1; **c.** anode of D_1 to center tap; **d.** anode of D_2 to center tap; **e.** cathodes of D_1 and D_2 to ground; **f.** $V+$ to ground.

TABLE 27.1-2. Resistance Measurements

Where Measured	*Resistance,* Ω
Primary of T_1	
Secondary of T_1	
Anode of D_1 to center tap	
Anode of D_2 to center tap	
Cathodes of D_1 and D_2 to ground	
$V+$ to ground	

TABLE 27.1-3. Open Filter Capacitors

Open	V+, dc	*Ripple,* V p-p
C_1		
C_2		

Open Filter Capacitors

7. Disconnect filter capacitor C_1. **Power on.** Measure, and record in Table 27.1-3, $V+$ and the ripple on $V+$.
8. **Power off.** Restore C_1 to the circuit. Open C_2. **Power on.** Measure, and record in Table 27.1-3, $V+$ and ripple on $V+$.

QUESTIONS

1. How does the measured value of $V+$ compare with that of V_1 in the experimental circuit?
2. What is the voltage drop across the 100-Ω filter resistor?
3. What is the total load and bleeder current in the experimental circuit?
4. What is the peak-to-peak voltage across the secondary of T_1?
5. What percentage of the peak-to-peak voltage across T_1 (question 4) is the dc voltage V_1 measured across C_1?
6. Which of the filter capacitors affects $V+$ more, C_1 or C_2?
7. If the resistance measured across C_2, the output filter capacitor, is zero, what is the most likely cause of trouble?
8. If the resistance measured across the secondary of T_1 is infinite, what is the most likely cause of trouble?

Answers to Self-Test

1. 30
2. C_1
3. (*a*) 1000; (*b*) 500
4. 30
5. lower

EXPERIMENT 27.2. Power-Supply Troubleshooting II

OBJECTIVE

To locate and correct the defect in a power supply which has no output dc voltage or low $V+$ and high ripple

INTRODUCTORY INFORMATION

Troubleshooting a Power Supply

In troubleshooting an electronic device such as a TV receiver or radio, defects may sometimes be traced to the power supply. Thus, in Fig. 27.2-1, if the dc output voltage $V+$ is lower than normal, or if the ripple voltage at $V+$ or at A is higher than normal, the trouble may be in the power supply circuit *or* in the load, for a large increase in load current may give these indications.

A first step in the troubleshooting process is to isolate the trouble to the load or to the supply by disconnecting the load from the supply. If the measured dc voltage at P is now higher than the rated $V+$ voltage under load, and if the ripple voltage at P is appreciably lower than the rated ripple under load, the trouble must be in the load circuit. However, if the $V+$ voltage without load is still low, and/or the ripple is still higher than normal, the trouble is in the supply.

Troubleshooting the supply requires measuring the dc voltages at P and at A with respect to ground and measuring also the ripple voltages at these points. The results of these measurements may give us a clue to the trouble. We shall consider several possibilities to illustrate the troubleshooting process.

No $V+$ Voltage

If $V+ = 0$ V, the trouble may be due to any of the following defects: (*a*) open line cord or defective power outlet, (*b*) open fuse, (*c*) open switch, (*d*) open or shorted transformer winding, (*e*) defective D_1 *and* D_2, (*f*) shorted C_1, (*g*) shorted C_2, (*h*) open R_C, (*i*) shorted R_B (very unlikely), or (*j*) short-circuited load. The problem then is to find the defective component and replace it with a known good one. This may involve one or all of the following: voltage measurements, resistance measurements, parts substitution.

Before going further, check the power outlet for line voltage. If this is normal and if $V+$ measures 0 V, the next check is to measure the *dc* voltage V_{AG}. If this is normal, the trouble is a shorted C_2, an open R_C, a shorted load, or a shorted R_B (very unlikely). A resistance check of these components will determine the trouble.

If V_1 across C_1 also measures 0 V direct current, the trouble may be a shorted C_1, an open line cord or defective plug, open fuse, open switch, open or shorted transformer winding, defective D_1 *and* D_2. However, before resistance-checking these components, another voltage check is indicated. Measure the ac voltage across the secondary of T_1.

It is possible to eliminate four of the suspected components with this one check. If the ac voltage across the secondary is normal, then the line cord, the fuse, the switch, and the power transformer are okay.

If there is no ac voltage across the secondary, remove the power plug from the ac outlet and proceed as follows: Connect an ohmmeter across the two *hot* prongs of the power plug. Close the switch. If the meter indicates continuity in the circuit (about 1 to 10 Ω), then the power plug, the line cord, the switch, the fuse, and the primary of the transformer are okay. If the meter shows infinite resistance, then it will be necessary to make a continuity check of the plug and each *hot* wire in the line cord, of the fuse, the switch, and the primary of T_1. One of these components will be open and should be replaced.

If the fuse is open, the cause may be a temporary overload. The simplest test is to replace the fuse and apply power to the circuit. If the circuit operates, the trouble has been corrected. If the fuse blows again, the trouble must still be found.

If all components from the line cord through the primary of T_1 are okay, the trouble must be an open secondary in T_1. Check for continuity in the secondary to confirm this conclusion.

NOTE: A shorted transformer winding would also give no ac voltage across the secondary, but in that case there would be such an increase in alternating current that the fuse would also blow.

Now assume that there is ac voltage across the secondary winding but that $V+$ *and* V_1 both measure 0 V dc. The trouble then may be a shorted C_1 or open rectifiers D_1 *and* D_2.

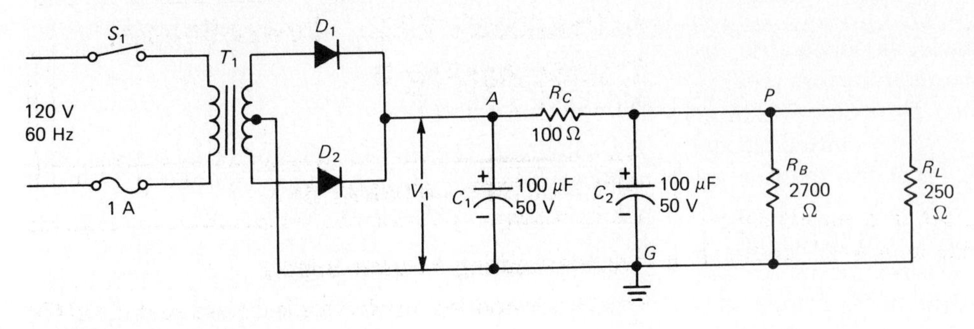

Fig. 27.2-1. Experimental power supply.

NOTE: If only one rectifier were open, there would still be dc voltages V_1 and $V+$, though the voltages would be lower than normal. One other possibility is an open center tap on the secondary of the transformer or open wiring from the center tap to point G. The defective component or the open lead from the center tap to point G may be found by resistance measurements.

NOTE: In the preceding discussion the assumption was made that the wiring between components was okay. This may not be so. As a final check the *connections and wiring* between components must be checked for continuity.

Low V+, High Ripple

An increase in load current or leaky electrolytic capacitors are the usual reasons for low $V+$ and high ripple. Of course, as we noted previously, an open D_1 *or* D_2 can also cause this problem.

An oscilloscope check of $V+$, Fig. 27.2-1, will indicate not only the amplitude of ripple voltage but the *frequency* of the ripple. The frequency should be 120 Hz for a full-wave rectifier. If it is 60 Hz, then either D_1 *or* D_2 is defective. These can be checked by determining their forward resistance. When the defective rectifier is replaced, the dc voltage and ripple levels should return to normal.

The usual cause for low $V+$ *and* high ripple is an open or leaky C_1 or C_2. The simplest procedure is to replace these capacitors, one at a time, with known good capacitors. When the defective capacitor is replaced, both dc voltage and ripple levels should return to normal.

SUMMARY

1. In troubleshooting it is always desirable to isolate the trouble to part of a circuit, thus eliminating those components in that part of the circuit which are good.
2. If there is no $V+$ in the output of the power supply, then trouble may be due to: (*a*) open line cord or defective electrical receptacle, (*b*) open fuse, (*c*) open switch, (*d*) open or shorted transformer winding, (*e*) *both* rectifiers D_1 *and* D_2 open, (*f*) shorted C_1, (*g*) shorted C_2, (*h*) open R_C, (*i*) shorted bleeder resistor R_B (very unlikely), (*j*) short circuit in the load.
3. To determine whether the trouble is to the right or left of C_1, Fig. 27.2-1, measure the dc voltage across C_1. If it is normal, then all the components to the left of C_1, including C_1, are good. In this case make a resistance check of C_2, R_C, R_B, and the load. Either R_C is open or C_2, R_B, and the load are shorted. Replace the defective part with a known good part.
4. If there is no dc voltage across C_1, the trouble may be a shorted C_1 or a defective component to the left of C_1. In that case check the resistance of C_1. If it is okay, proceed as follows:
5. Measure the *ac* voltage across the secondary of T_1. If it is normal, then the ac input circuit components are good. If there is 0 V across the secondary, resistance-check the fuse, line cord, and primary and secondary of T_1. If these are all good, resistance-check the center tap to the top and bottom of the secondary of T_1. If this is normal, then either the wiring between components is causing the trouble or *both* D_1 and D_2 are open. Check wiring and D_1 and D_2.
6. All resistance checks are made with power OFF and the line cord removed from the power receptacle.
7. Low $V+$ and high ripple usually indicate leaky filter capacitors. These should be replaced, one at a time, with known good capacitors and then measured for $V+$ and ripple after each replacement.

SELF-TEST

Check your understanding by answering these questions. Refer to Fig. 27.2-1.

1. If there is no $V+$, a possible cause may be an open C_1. ________ (true/false)
2. If there is no $V+$, a possible cause may be an open load resistor R_L. ________ (true/false)
3. A shorted C_2 will cause no $V+$. ________ (true/false)
4. A shorted secondary winding in T_1 will most likely cause the fuse to blow. ________ (true/false)
5. If there is no ac voltage across the secondary of T_1, you would check the transformer and the components to the right of the transformer. ________ (true/false)
6. Low $V+$ may be due to an open rectifier, such as D_1. ________ (true/false)

MATERIALS REQUIRED

- Equipment: Oscilloscope; EVM or VOM
- Resistors: ½-W 100-, 2700-Ω; 2-W 250-Ω
- Capacitors: Two 100-μF/50-V
- Solid-state rectifiers: Two 1N5625 or equivalent
- Miscellaneous: Power transformer T_1, 120-V primary, 26-V 1-A center-tapped secondary; SPST switch; *defective components* for *troubleshooting*

PROCEDURE

Troubleshooting

Your instructor will insert trouble into the power supply of Fig. 27.2-1. Troubleshoot the circuit, keeping a step-by-step record of each check that you make in the order performed. Record these checks in the standard troubleshooting report which will be used for all servicing procedures. When you have found the trouble, correct it and notify your instructor, who will put another trouble into the circuit. Again service it, using a *separate* troubleshooting report to

TROUBLESHOOTING REPORT

Student's Name(s) ______________________ Experiment # ________

Class ______________________ Date ________

1. Describe trouble. ______________________

____

2. Preliminary inspection: describe troubles found, if any. ______________________

____

3. Describe procedure (list steps in order performed).

Test Point	*Voltage, ac or dc*	*Resistance, Ω*	*Waveforms and V p-p*	*Other*

4. Describe trouble found and list parts used for repair: ______________________

____

5. Was circuit operation normal after repair? ______________________

____

6. Instructor's signature: ______________________

record your results. Service as many troubles as time will permit.

QUESTIONS

1. Trouble in an electronic device has been isolated to the dc power supply. Assume it is the supply of Fig. 27.2-1. $V+$ voltage is lower than normal, and the ripple on $V+$ is much higher than normal. What are the troubles which could give these effects?
2. Explain the procedure you would follow to find the defective component for the conditions in question 1.
3. In Fig. 27.2-1 the ac voltage measured from the anode of D_1 to the anode of D_2 is 60 V rms. The ac voltage measured from either anode to the center-tap connection on the secondary is zero. The output of the dc supply ($V+$) is zero. What is the trouble in the circuit?

Answers to Self-Test

1. false
2. false
3. true
4. true
5. false
6. true

CHAPTER 28 VOLTAGE-DOUBLER POWER SUPPLIES AND TROUBLESHOOTING

EXPERIMENT 28.1. The Half-Wave Voltage Doubler

OBJECTIVE

To measure the dc voltage and ripple at key test points in a half-wave voltage doubler

INTRODUCTORY INFORMATION

Line-Operated Half-Wave Power Supply

In vacuum-tube television receivers, power supplies operating directly from the power line are often used. These make costly power transformers unnecessary.

A typical silicon-rectifier half-wave supply operating directly from the line is shown in Fig. 28.1-1. No transformer is used. The ground return of the supply is connected directly to the line. Therefore there is no isolation from the line. Lack of line isolation is the main disadvantage of this type of supply.

Refer to Fig. 28.1-1. The rectifier D_1 is connected to deliver a positive dc voltage, $V+$, to the load. C_1, R_1, and C_2 make up a π-type filter. R_2, a surge-current-limiting resistor, protects the rectifier against overload caused by the high charging currents of the electrolytics when power is first turned on. Thermistors which have a high cold resistance and very low hot resistance are sometimes used as protective resistors.

The rectifier conducts on the positive alternation and is cut off on the negative. The resultant 60-Hz pulsating current is filtered and appears as a relatively ripple-free dc voltage $V+$. This type of supply under normal load puts out about 100 V, $V+$. R_3 is a bleeder resistor.

Half-Wave (Cascade) Voltage Doubler

A voltage doubler circuit is frequently used to supply a higher $V+$ output. There are two types of doublers, the half-wave (cascade) doubler and the full-wave doubler. In this experiment we shall work with the half-wave doubler.

Figure 28.1-2 is the circuit of a practical, transformerless half-wave voltage doubler, used with vacuum-tube series-filament television receivers. This circuit operates directly from the ac line. C_1, C_2, and C_3 are electrolytic capacitors. D_1 and D_2 are silicon rectifiers. R_1 is a surge-current-limiting resistor, and R_B is a bleeder resistor.

To understand how this circuit operates, assume that when power is applied, D_1 is not in the circuit. On the negative alternation of the line voltage at R_1, D_2 conducts and charges C_1 to the peak line voltage (169 V) with the polarity shown.

Next consider that D_1 is also in the circuit. C_1, just charged, now acts as a battery in series with the power source. On the positive alternation, therefore, D_1 "sees" a positive voltage equal to twice the line-voltage peak (see Fig. 28.1-3). D_1 now conducts and charges C_2 with the polarity shown. The voltage across C_2 is therefore equal to approximately twice the peak line voltage, or 338 V. The purpose of D_2,

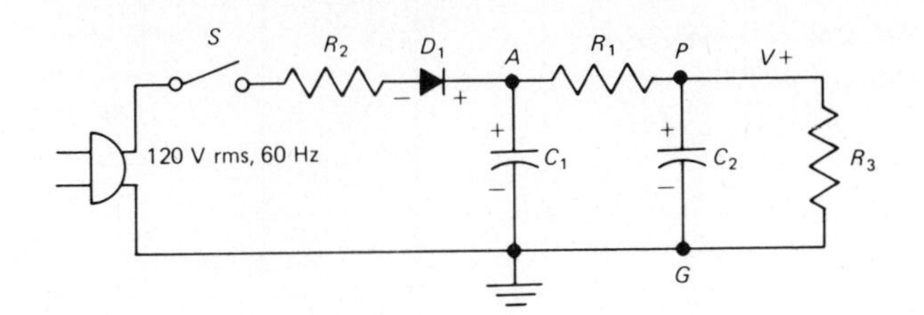

Fig. 28.1-1. Line-derived half-wave power supply.

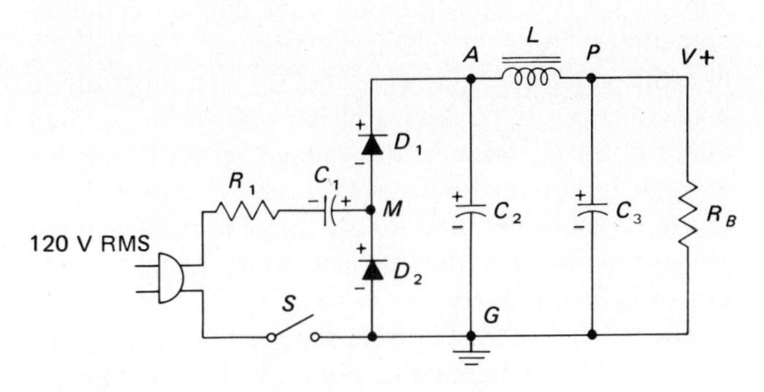

Fig. 28.1-2. Cascade voltage doubler (half-wave).

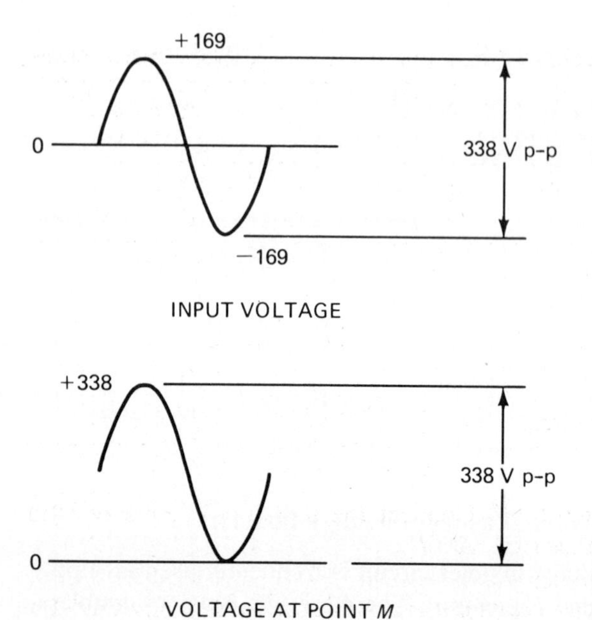

Fig. 28.1-3. Effect of voltage-doubler input circuits.

therefore, is to charge C_1 so that rectifier D_1 can receive twice the peak positive line voltage. The dc voltage from A to G results from conduction of D_1, and D_1 conducts only on positive alternations. Therefore this is a half-wave rectifier, and the ripple frequency is 60 Hz.

If this supply is to be used with a load of 300 to 400 mA, fairly large-size capacitors are needed with values from 150 to 300 μF.

The filter is the π-type. Note that electrolytic capacitors C_2 and C_3 must have twice (approximately) the voltage rating of C_1.

This supply can deliver 250 to 300 V, depending on the size of the filter components and on the load current.

Full-Wave Voltage Doubler

In addition to half-wave voltage doublers, full-wave doublers are also used in electronics. The circuit of Fig. 28.1-4 is a solid-state full-wave doubler and filter sometimes found in television receivers. D_1 and D_2 are the rectifiers. C_1, C_2, and C_3 are electrolytic capacitors. C_3 must have a higher voltage rating than either C_1 or C_2 because the voltage across it is twice as high (approximately) as that across either C_1 or C_2. R_B is a bleeder, and R_1 is a surge-current-limiting resistor to protect the rectifier from damage when power is first applied.

The amount of load current required determines the size of filter components. For low-current applications, the choke L can be replaced by a resistor.

SUMMARY

1. Transformerless, line-derived power supplies are sometimes used in vacuum-tube circuits.
2. Their advantage is that they eliminate the need for heavy, costly power transformers.
3. Their disadvantage is that they are not line-isolated, but are directly connected to the line.
4. One type of line-derived power supply is the half-wave rectifier circuit shown in Fig. 28.1-1. This circuit develops about 100 V dc with 60-Hz ripple.
5. If a higher voltage (higher than 100 V) is required, a half-wave voltage doubler, such as that in Fig. 28.1-2, may be used. The dc output ($V+$) of this supply depends on the size of the filter components and on the load current. The ripple frequency is 60 Hz.
6. In Fig. 28.1-2, the voltage rating of C_2 and C_3 must be approximately 350 V, somewhat higher than the peak-to-peak value of the line voltage. C_1 is rated at about 175 V.

SELF-TEST

Check your understanding by answering these questions.

1. The circuit of Fig. 28.1-1 ________ (is/is not) isolated from the power line.
2. For an input of 120 V rms, the output of the half-wave supply of Fig. 28.1-1 is about ________ V.
3. For a 120-V rms input, the output of the half-wave doubler in Fig. 28.1-2 is about ________ V.
4. The ripple frequency in the output of Fig. 28.1-1 is ________ Hz.
5. The ripple frequency in the output of Fig. 28.1-2 is ________ Hz.

MATERIALS REQUIRED

- Equipment: Oscilloscope; EVM or VOM
- Resistors: ½-W 5600-Ω; 1-W 100-Ω; 5-W 500-Ω

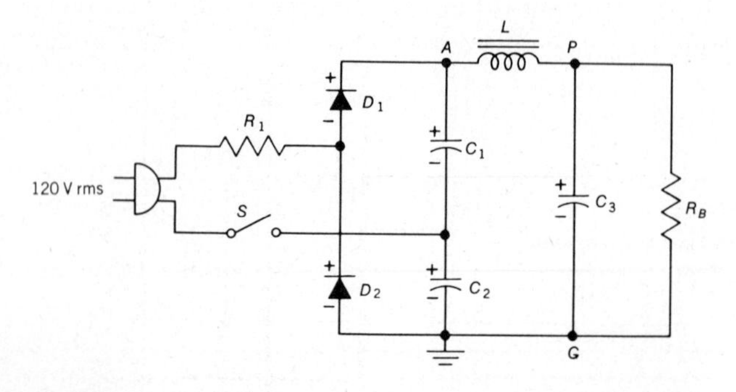

Fig. 28.1-4. Full-wave voltage doubler using silicon rectifiers.

- Capacitors: Three 100-μF/50-V
- Solid-state rectifiers: Two 1N5625 or the equivalent
- Miscellaneous: Power transformer T_1, 120-V primary, 26-V 1-A center-tapped secondary; SPST switch; fused line cord

PROCEDURE

NOTE: For safety, to isolate the dc output voltage from the input power line, a transformer is used in this experiment. Though a center-tapped transformer is not needed for a half-wave rectifier, to keep the number of parts to a minimum, the same center-tapped transformer T_1 is used as in Exp. 27.

The full secondary of T_1 is not connected, but just half of the secondary, as in Fig. 28.1-5. If the full secondary were used, we could get approximately double the *dc* output voltage. But for our purpose, we can learn the operation of a voltage doubler just as easily by using only half the secondary voltage. An advantage to us, in using half the secondary voltage, is that the dc voltage ratings of the filter capacitors are one-half what would otherwise be needed if the full secondary were connected.

1. **Connect** the circuit of Fig. 28.1-5. *Note again that only one-half of the secondary of T_1 is connected. There is no connection (NC) to the other half.*
2. **Power off.** Discharge all capacitors. Measure the resistance across R_B, the bleeder resistor. It should be approximately 5600 Ω. If it is appreciably less than 5600 Ω, **do not** apply power, but check the circuit to find out why the resistance is low.
3. If the circuit is normal, turn **power on.** Measure in turn, and record in Table 28.1-1, the dc no-load voltage across C_1, C_2, and C_3.
4. Observe the waveforms and measure, with an oscilloscope, the no-load voltage across C_1, C_2, and C_3. Record the results in Table 28.1-1.

TABLE 28.1-1. Voltages and Ripple in Half-Wave Doubler Supply

	No Load		*With Load (500 Ω)*	
Measured Across	*DC* V	V p-p	*DC* V	V p-p
C_1				
C_2				
C_3				

Ripple frequency: ______________ Hz

5. **Power off.** Connect the 500-Ω 5-W load resistor in parallel with R_B.
6. Repeat steps 3 and 4.
7. Measure and record the ripple frequency of $V+$ across C_3.

QUESTIONS

1. How does the *dc* output of the half-wave rectifier power supply in this experiment compare with that of the half-wave rectifier in Exp. 26.3 under load? Explain the difference, if any.
2. What is the percentage of regulation of this supply?
3. How much load current is drawn in the experimental circuit when a 500-Ω load is connected?
4. Are the voltage ratings of C_1, C_2, and C_3 in this experiment sufficient for the job? Explain.
5. Approximately what would be the dc output voltage $V+$ if the entire secondary of T_1 had been used in Fig. 28.1-5?

Answers to Self-Test

1. is not
2. 100
3. 200
4. 60
5. 60

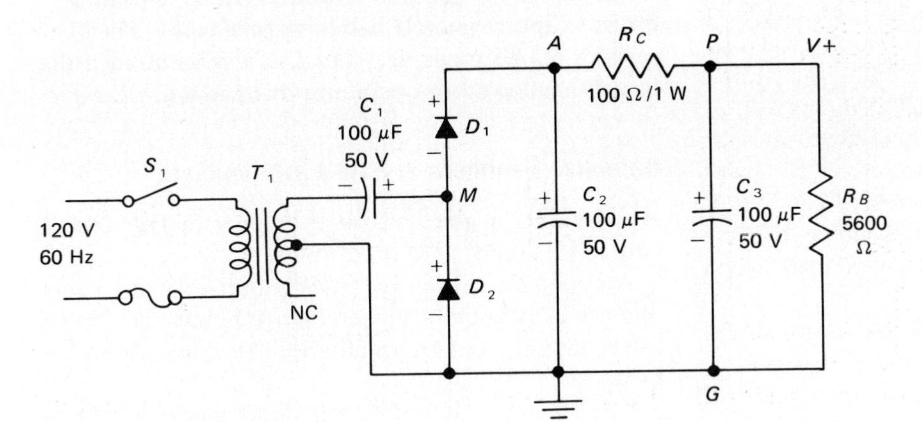

Fig. 28.1-5. Experimental cascade (half-wave) voltage doubler.

EXPERIMENT 28.2. Troubleshooting Voltage-Doubler Power Supplies

OBJECTIVES

1. To measure the effects of defective components in doubler supplies
2. To troubleshoot voltage-doubler supplies

INTRODUCTORY INFORMATION

Defective Capacitors in Experimental Full-Wave Doubler

The experimental full-wave doubler in Fig. 28.2-1 has three filter capacitors. C_1 and C_2 are the input, C_3 is the output capacitor. The effect of C_1 on $V+$ and ripple is about the same as that of C_2. If either C_1 or C_2 is open, $V+$ is reduced, and the ripple on $V+$ is very much increased. If C_3 is open, $V+$ remains the same, but the ripple on $V+$ is very much higher.

A shorted C_3 will result in 0 $V+$. The input *dc* voltage across *AG* will drop, and the ripple on V_{AG} will increase. R_C, the filter resistor, will become very hot because the entire dc voltage V_{AG} appears across it. If power remains on for any length of time with C_3 shorted, R_C will burn.

A shorted C_1 causes $V+$ voltage to drop to about one-half its normal value. Rectifier D_1 becomes very hot and will burn out if the circuit is permitted to remain on for any length of time. The ripple on $V+$ does not increase appreciably, but its frequency changes from 120 to 60 Hz.

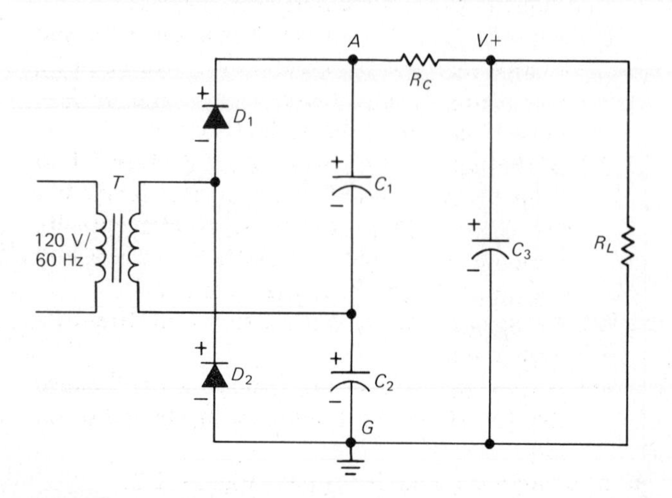

Fig. 28.2-1. Full-wave voltage doubler.

A shorted C_2 causes the same effects as a shorted C_1, with the exception that rectifier D_2 overheats.

NOTE: In this experiment we shall measure the effects of open but not shorted components to prevent damage to circuit components.

Open Rectifiers in Experimental Full-Wave Doubler

An open rectifier, D_1 or D_2 in Fig. 28.2-1, will cause $V+$ to drop to a very low value, 5 to 6 V, without affecting the ripple voltage to any extent. Ripple frequency will change to 60 Hz.

The effects of defective capacitors and rectifiers on $V+$, V_{AG}, and ripple are summarized in Table 28.2-1.

Defective Capacitors in Experimental Half-Wave Doubler

In the half-wave doubler of Fig. 28.2-2, an open C_1 will completely disable the circuit. There will be no $V+$ and *no* V_{AG}.

An open C_2 will cause $V+$ to drop and the ripple voltage on $V+$ to increase. V_{AG} will also be reduced, and the ripple voltage on V_{AG} will become very high.

An open C_3 will not affect $V+$ but will cause V_{AG} to drop, although not as much as an open C_2. The ripple voltage on $V+$ will become higher than normal.

A shorted capacitor C_1 will cause $V+$ to drop to approximately half its normal value. Ripple will increase, and rectifier D_2 will become very hot. If operation continues for any length of time, D_2 will burn out.

A shorted C_2 causes $V+$ and V_{AG} to drop to 0 V.

A shorted C_3 causes $V+$ to drop to 0 V, while V_{AG} drops to approximately half its normal value. The filter resistor R_C becomes very hot and will burn out if the circuit is allowed to operate in this fashion.

Defective Rectifiers in Half-Wave Doubler

An open D_1 in the half-wave doubler of Fig. 28.2-2 causes $V+$ and V_{AG} to drop to 0 V.

An open D_2 causes $V+$ to drop to a very low value while V_{AG} is slightly higher than $V+$, but very much lower than it would normally be. The ripple on $V+$ increases.

The effects of defective capacitors and rectifiers on

TABLE 28.2-1. Effects of Defective Components on Experimental Full-Wave Doubler Operation

Defective Component	*V+*, V dc	*Ripple on V+*	*V_{AG}*, V dc	*Ripple on V_{AG}*	*Ripple Frequency,* Hz	*Effect on Other Components*
Open C_1	Reduced	Large increase	Reduced	Large increase	60	
Open C_2	Reduced	Large increase	Reduced	Large increase	60	
Open C_3	Remains same	Large increase	Remains same	Remains same	120	
Shorted C_1	Drops ½ value	No effect			60	D_1 overheats
Shorted C_2	Drops ½ value	No effect			60	D_2 overheats
Shorted C_3	0		Drops	Increases	120	R_C overheats
Open D_1	Drops to 5 V	No effect			60	
Open D_2	Drops to 5 V	No effect			60	

the operation of the half-wave doubler of Fig. 28.2-2 are summarized in Table 28.2-2.

Troubleshooting Voltage-Doubler Circuits

The procedure for troubleshooting voltage-doubler supplies is the same as that for troubleshooting simple half-wave and full-wave rectifier supplies. An attempt is made to isolate the trouble to a defect in the circuit by voltage, ripple, and resistance measurements. These are compared with known standard values measured under normal operation. An analysis of the measurements should lead to a suspected component, which is then replaced with a part known to be good. New measurements are then made to determine whether circuit operation is normal.

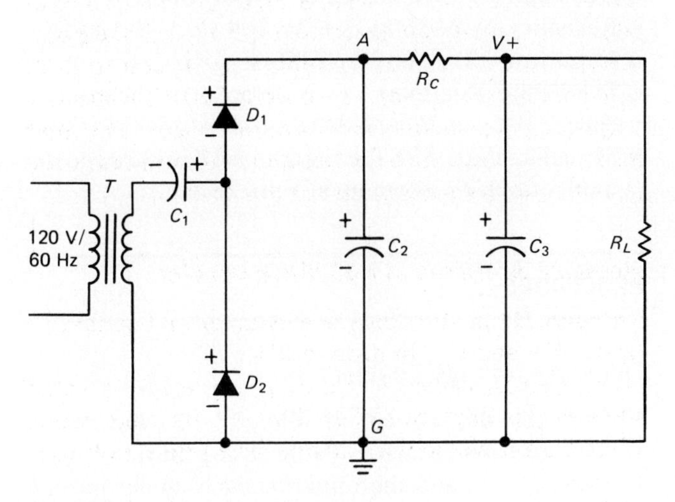

Fig. 28.2-2. Half-wave voltage doubler.

SUMMARY

1. Defective capacitors in a full-wave or half-wave voltage doubler will result in changed direct current and/or ripple voltages in the circuit.
2. In a full-wave doubler circuit, Fig. 28.2-1:
   (*a*) An open C_1 or C_2 will result in a lower $V+$ and a large increase in ripple on $V+$. An open C_3 will cause a large increase in ripple on $V+$, but will not affect the *dc* level of $V+$.
   (*b*) A shorted C_1 or C_2 will not only cause $V+$ to drop to about one-half its normal level, but will cause D_1 or D_2 to overheat and eventually burn out. A shorted C_3 reduces $V+$ to zero and causes R_C to overheat and burn out.
   (*c*) An open D_1 or D_2 will cause $V+$ to drop to a very low level.
   (*d*) Any of the defects (*a*) through (*c*) will cause the 120-Hz ripple frequency to change to 60 Hz.
3. In a half-wave doubler circuit, Fig. 28.2-2,
   (*a*) An open C_1 causes $V+$ to drop to 0 V.

TABLE 28.2-2. Effects of Defective Components on Half-Wave Doubler Operation

Defective Component	*V+, V dc*	*Ripple on V+*	*V_{AG}, V dc*	*Ripple on V_{AG}*	*Effect on Other Components*
Open C_1	0		0		
Open C_2	Drops	Large increase	Drops	Large increase	
Open C_3	Drops	Large increase	Drops	No effect	
Shorted C_1	Drops ½ value	Increases			D_2 overheats
Shorted C_2	0		0		
Shorted C_3	0		Drops ½ value		R_C overheats
Open D_1	0		0		
Open D_2	Very low	Increases	Very low		

(*b*) An open C_2 or C_3 causes $V+$ to drop and ripple on $V+$ to increase.
(*c*) A shorted C_1 causes $V+$ to drop to about one-half its normal value, while increasing the ripple on $V+$. Rectifier D_2 overheats and will burn out after continued operation.
(*d*) A shorted C_2 or C_3 causes $V+$ to drop to 0 V. In the case of a shorted C_3, R_C overheats and will eventually burn out.
(*e*) An open D_1 or D_2 causes $V+$ to drop to 0 V or a very low value, respectively.

4. In troubleshooting voltage-doubler circuits, voltage, ripple, and resistance measurements are made in the circuit, and the results are compared with the known good values. The comparison usually suggests which part is defective.

SELF-TEST

Check your understanding by answering these questions.

NOTE: For questions 1, 2, and 3 normal voltages are $V+ = 26$ V, ripple on $V+ = 0.4$ V p-p, $V_{AG} = 32$ V, ripple on $V_{AG} = 5$ V p-p.

1. In the full-wave voltage-doubler circuit of Fig. 28.2-1, the following voltages are measured: $V+ = 19$ V, ripple on $V+ = 4$ V p-p. Ripple frequency = 60 Hz. Probable cause of trouble is (*a*) open C_1, (*b*) open C_2, (*c*) open C_3, (*d*) open C_1 or C_2.
2. In Fig. 28.2-1, $V_{AG} = 23$ V, $V+ = 0$ V, and R_C becomes very hot. The most probable cause of trouble is ________ ________.
3. In Fig. 28.2-1, $V+ = 15$ V, and D_2 becomes very hot. The most probable cause is ________ ________ ____.

NOTE: Normal voltages for questions 4 and 5 are $V+ = 24$ V, ripple on $V+ = 0.6$ V p-p, $V_{AG} = 29.5$ V, ripple on $V_{AG} = 4.0$ V p-p.

4. In Fig. 28.2-2, $V+ = 0$ V. Nothing overheats. A possible cause of trouble is a shorted ________ ($C_1/C_2/C_3$).
5. In Fig. 28.2-2, $V+ = 0$ V, $V_{AG} = 0$ V. Nothing overheats. A possible cause of trouble is (*a*) C_1 open, (*b*) C_2 open, (*c*) D_1 open, (*d*) either C_1 or D_1 open.

MATERIALS REQUIRED

- Equipment: Oscilloscope; EVM or VOM
- Resistors: ½-W 5600-Ω; 1-W 100-Ω; 5-W 500-Ω
- Capacitors: Three 100-μF/50-V
- Solid-state Rectifiers: Two 1N5625 or the equivalent
- Miscellaneous: SPST switch; fused line cord; power transformer T_1, same as in Exp. 28.1

PROCEDURE

Full-Wave Voltage Doubler

1. Connect the circuit of Fig. 28.2-3. **Power on.** Measure and record in Table 28.2-3 the dc voltages V_{AG} and $V+$, and the ripple on each, under normal operation.

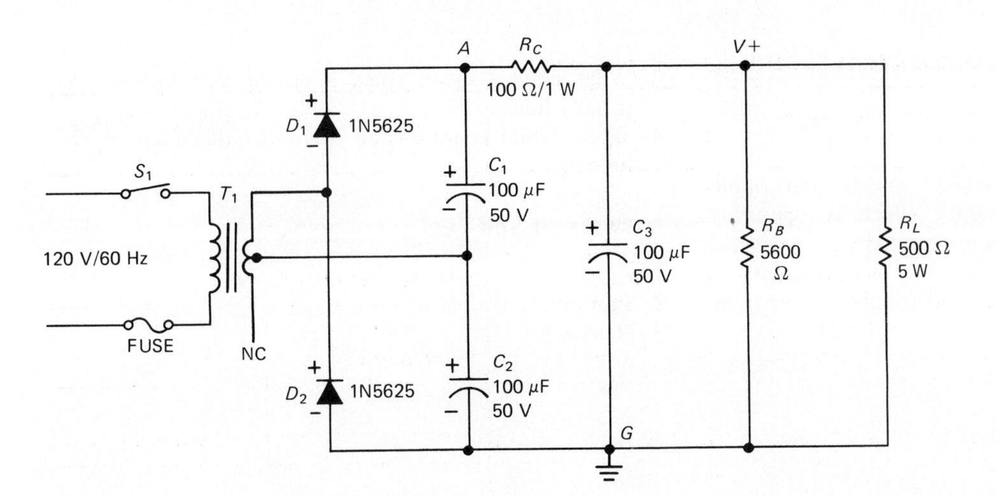

Fig. 28.2-3. Experimental full-wave voltage doubler.

TABLE 28.2-3. Full-Wave Voltage-Doubler Defects

Condition	DC V		Ripple (V p-p) On	
	$V+$	V_{AG}	$V+$	V_{AG}
Normal				
Open C_1				
Open C_2				
Open C_3				
Open D_1				
Open D_2				

TABLE 28.2-4. Half-Wave Voltage-Doubler Defects

Condition	DC V		Ripple (V p-p) On	
	$V+$	V_{AG}	$V+$	V_{AG}
Normal				
Open C_1				
Open C_2				
Open C_3				
Open D_1				
Open D_2				

2. **Power off.** Open C_1 and repeat step 1. Check if any component overheats. Reconnect C_1 in the circuit.
3. Repeat step 2 for every condition in Table 28.2-3.

Half-Wave Voltage Doubler

4. Connect the circuit of Fig. 28.2-4. Repeat steps 1, 2, and 3 for this circuit and record the results in Table 28.2-4.

Troubleshooting

5. Your instructor will insert trouble into one of the rectifier circuits in this experiment. Troubleshoot the circuit and correct the trouble. Keep a record of your steps, as you make them, and record them and all measurements in the standard troubleshooting report.

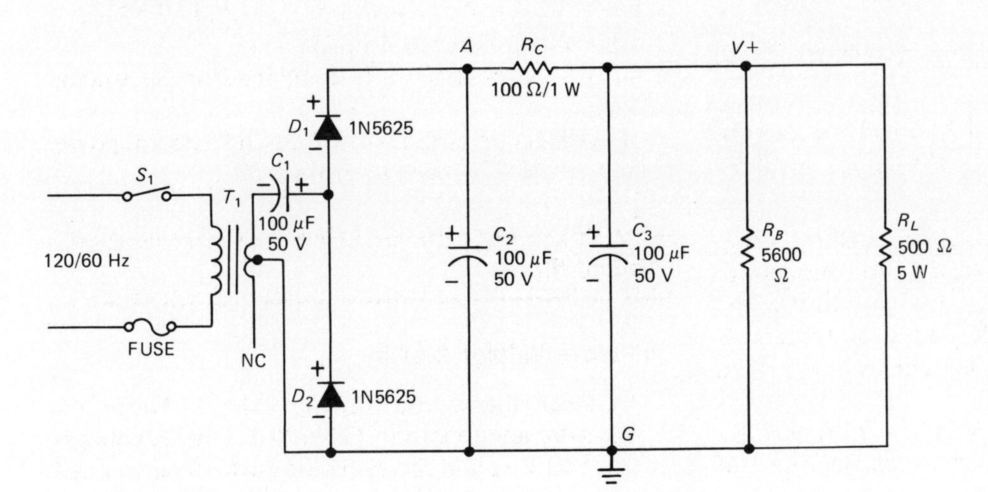

Fig. 28.2-4. Experimental half-wave voltage doubler.

QUESTIONS

1. What is the purpose of making voltage and ripple measurements in: (*a*) A supply which is operating normally? (*b*) A supply which is defective?
2. What is the purpose of making the measurements in this experiment of voltages and ripple for specific defective components?
3. What does it mean when a circuit part becomes unusually hot?
4. What should you do when you notice that a part is overheating?

Answers to Self-Test

1. *d*
2. shorted C_3
3. shorted C_2
4. C_2
5. *d*

CHAPTER 29 JUNCTION TRANSISTORS

EXPERIMENT 29.1. Junction Transistor Familiarization

OBJECTIVE

To identify the elements of a junction transistor and learn rules for working safely with transistors

INTRODUCTORY INFORMATION

The Three-Element Transistor

Our experiments to date have dealt with diodes, two-element devices. In this experiment we shall become familiar with bipolar transistors. These are three-element devices. The third element in the transistor extends its characteristics and applications far beyond those of the diode. These transistors are called *bipolar* because their operation depends on negative- and positive-charge carriers.

The study of semiconductors resulted in the discovery and development of the transistor. This device can handle many electronic jobs, including amplification, oscillation, and detection. Before the transistor the vacuum tube performed this work. The transistor has taken over because of the many advantages it has over the tube. You will recall some of these advantages from your introduction to semiconductors. The transistor is small, allowing miniaturization of electronic equipment. It is solid, so its elements cannot vibrate. The transistor operates with low power-supply voltages and does not require any warm-up period. It does not need filaments, and so it has fewer circuit connections than a tube and does not use as much power as a tube. It develops less heat than tubes, but it is more sensitive to heat.

There are many types of transistors. These may be classified according to the material from which they are made, either silicon or germanium. Most transistors now are made of silicon. Another classification is the process by which they are made. Here we find many types of junction transistors: grown-junction, alloy-junction, drift-field, mesa, epitaxial mesa, planar, and the now obsolete point-contact transistors. Transistors are classified according to the maximum power they use, from low power (50 mW) to medium (2 to 10 W) and higher.

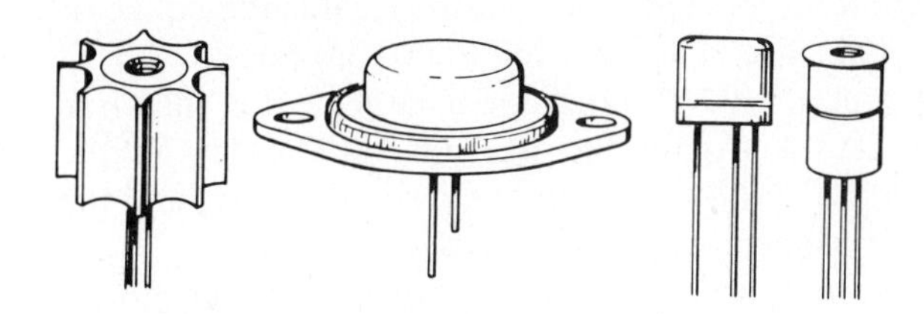

Fig. 29.1-1. Some transistor shapes. (*Sylvania Electric Products, Inc.*)

Transistors come in different shapes and sizes, as shown in Fig. 29.1-1. They differ also in the basing arrangement and in the way they are mounted in the circuit. Some are socket-mounted like tubes. The sockets are designed to follow the physical basing arrangement of the transistor. Some transistors have leads for direct soldering into a circuit.

Junction Transistors

Junction transistors are an extension of semiconductor diodes. They come in two arrangements, NPN and PNP. Figure 29.1-2 is a representation of an NPN junction transistor. It is formed by sandwiching a very thin strip of P-type silicon between two wider strips of N-type silicon. Three leads are brought out from the individual semiconductor elements. The entire assembly is enclosed in a moisture-proof housing. The size of the elements and their physical shape will differ depending on how the junction is constructed. The characteristics of various types will also differ.

In Fig. 29.1-2, the wafer on the left is called the

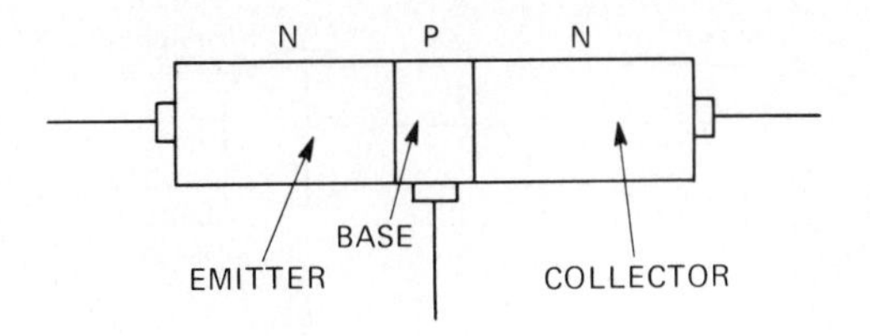

Fig. 29.1-2. NPN junction transistor.

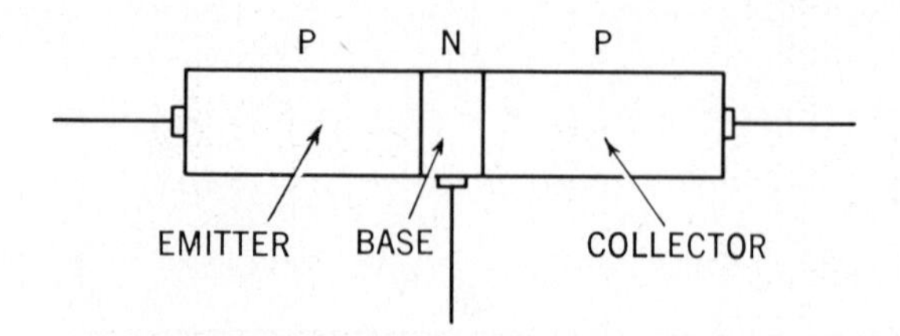

Fig. 29.1-3. PNP junction transistor.

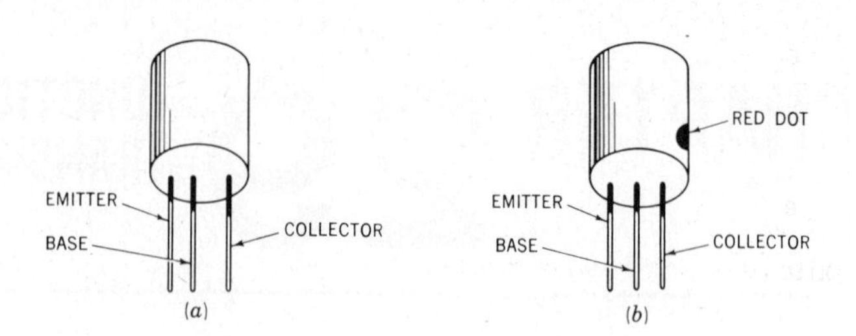

Fig. 29.1-5. Methods of identifying transistor leads.

emitter (E), the P-type wafer in the middle is the base (B), and the N-type wafer on the right is the collector (C). The base is about 1 mil (0.001 in) thick.

A junction PNP transistor arrangement is shown in Fig. 29.1-3. Again the element on the left is the emitter. The N-type center wafer is the base, and the P-type element on the right is the collector.

Electrons are the majority current carriers in an NPN transistor, holes in a PNP type.

Transistor Symbols, Basing, and Mounting

The schematic symbols for PNP and NPN transistors are shown in Fig. 29.1-4*a* and *b*. The element with the arrowhead is the emitter. In an NPN transistor the arrow points *away* from the base. In a PNP type the arrow points toward the base.

Two methods are generally used for bringing transistor connections out of the envelope. The first employs long leads. An example of this is the transistor shown in Fig. 29.1-5*a*. The three leads of *this* transistor are all in line and may be identified as emitter, base, and collector in the order shown. The collector spacing is greater than that of the other two elements. A variation of this system also uses three in-line leads. The collector is identified by a red dot which is painted close to it on the transistor case, as in Fig. 29.1-5*b*. The usual method for connecting the transistors of Fig. 29.1-5 into the circuit is to solder the flexible leads to the proper terminals in the circuit.

A second method uses rigid pins for bringing transistor connections out of the envelope, as in Fig. 29.1-6*a*. This is a plug-in type of transistor and requires a matching socket to hold it. Another basing arrangement permits the selection of either the first or second mounting method. In this case, the transistor is supplied with flexible leads welded to rigid base pins. The flexible leads are used if the transistor is to be soldered into the circuit. If a socket is to be used, the flexible leads are easily removed with a pair of cutters. A triangular type of socket and basing arrangement is also used on other socket-mounted transistors (Fig. 29.1-6*b* and *c*).

Other typical arrangements of transistor basing and socket arrangements are shown in Fig. 29.1-7.

Transistors are rated according to their ability to dissipate power. Thus a transistor used as a low-level audio amplifier may have a power rating which is low, say 50 mW. A transistor used as an output amplifier must have a higher wattage rating. The casings of power transistors are especially designed to permit cooling. For example, some power transistors use radial fins for conducting the heat away (see Fig. 29.1-1). Other types use a metal shell which mounts onto the metal chassis of the equipment where it is employed. The collector of this type of transistor is connected to the transistor housing. The chassis then conducts the heat away. This type of transistor has a higher power rating when physically clamped onto the metal chassis as described, and a lower power rating when mounted off the chassis. Good design requires that transistors be mounted in the coolest part of the chassis. Fans are sometimes installed inside equipment to cool off the transistors and other components.

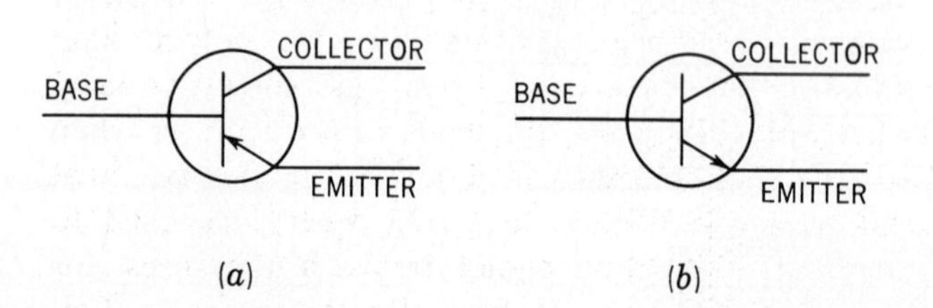

Fig. 29.1-4. Schematic symbol for (*a*) PNP; (*b*) NPN transistor.

Rules for Working with Transistor Circuits and for Making Measurements in Them

Tools

Small tools are required for use in transistor circuits because of the small size of the transistors and their

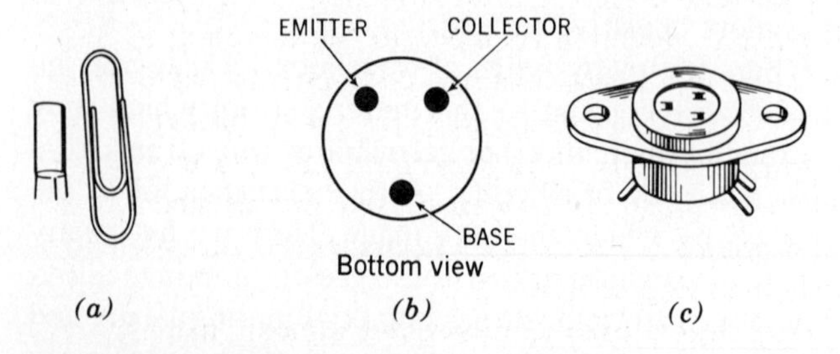

Fig. 29.1-6. (*a*) Plug-in transistor. Triangular basing identification. (*b*) Basing diagram (bottom view). (*c*) Socket. (*CBS*)

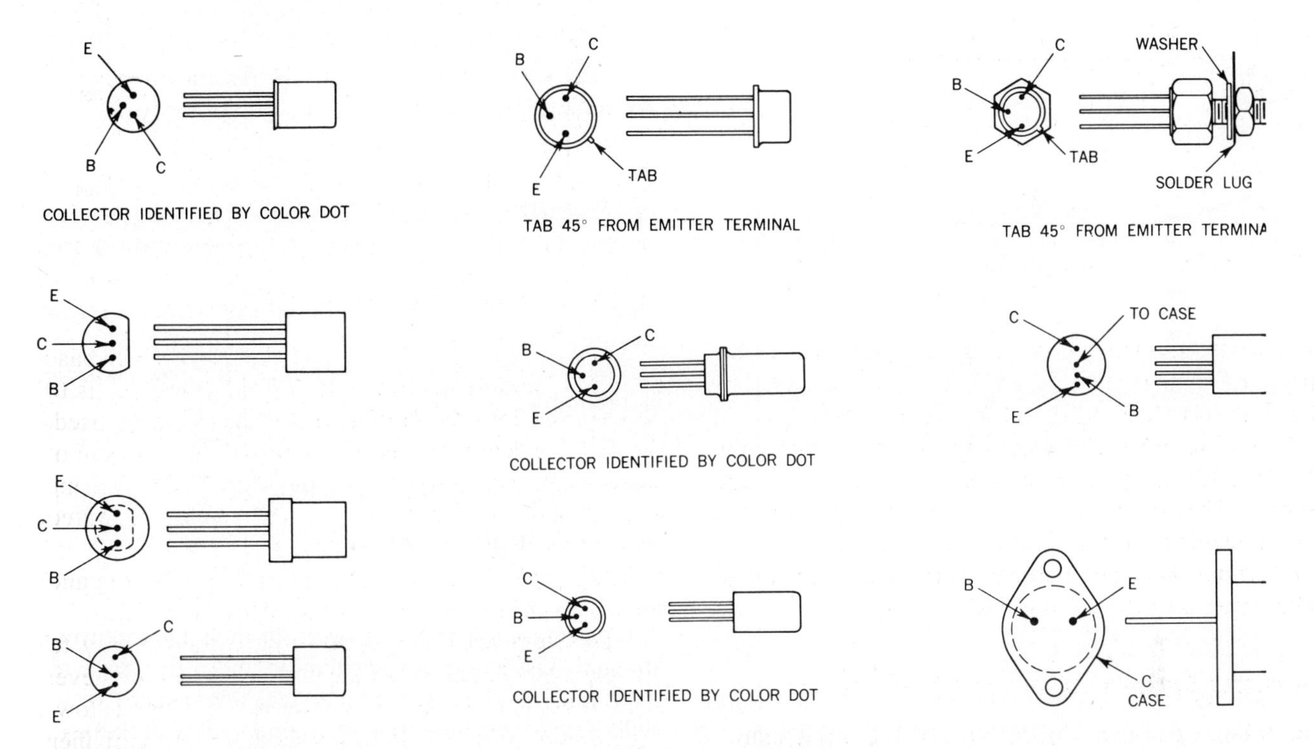

Fig. 29.1-7. Typical transistor basing and socket arrangements. (*Symphonic Radio Corp.*)

associated components. Moreover, transistorized equipment is usually miniaturized, and working with ordinary tools would be difficult. A listing of tools includes:

- Miniature cutters
- Miniature needle-nose pliers
- An assortment of service-type tweezers
- A 25-W pencil-type soldering iron
- A soldering aid
- A jeweler's eyepiece or a magnifying lens (stand-mounted)

Precautions in Physical Handling of Transistors

Though they are sturdy, transistors may be easily damaged if improperly handled. You should exercise great care in working with low-power transistors with flexible leads, since these leads may be easily broken. The same precaution applies to the handling of transistor components. These components have been miniaturized, and leads may be easily broken if not handled carefully. For maximum life, the transistors and associated components used in this and future experiments should be permanently mounted on breadboarding devices.

Transistors can be destroyed almost instantly. A short circuit from base to collector, in an operating circuit, will almost always destroy a transistor. Short-circuiting a pair of terminals with a meter probe or a tool may destroy a transistor.

At this point you must be cautioned against inserting transistors into a circuit, or removing transistors from a circuit, with power ON. This practice may permanently damage a transistor because of high transient currents which may develop. A good rule to follow is to *be certain that power is off before inserting a transistor or other component or removing it from the circuit.*

Other Precautions

Check of Circuit Connections. All connections should be checked against the circuit diagram before power is applied. Transistors should not be connected to a voltage source without some limiting resistance in the circuit.

Transistor Soldering. Soldering of transistors where necessary should be accomplished quickly. Low-wattage irons (25 W) are recommended. Transistor pigtails should be kept as long as possible consistent with circuit design considerations to reduce heat transfer. The same type of heat sink should be used when soldering transistor leads in a circuit as when soldering semiconductor diodes. An effective heat sink is created when long-nose pliers are used to grasp the transistor pigtail between the transistor body and the point of heat application, as in Fig. 29.1-8.

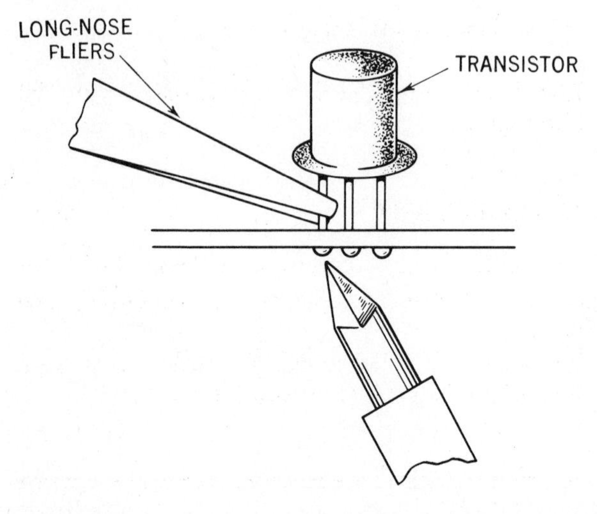

Fig. 29.1-8. Pliers act as heat sink when transistors are soldered in circuit.

Voltage Measurements. All test equipment should be isolated from the power line. If the equipment is not isolated, an isolation transformer should be used. In making voltage measurements in transistor circuits, care should be taken to minimize the possibility of accidental short circuits between closely spaced terminals. Accidental short circuits may apply improper or excessive voltage to the transistor elements and may destroy the transistor.

Resistance Measurements. Transistors may be damaged during resistance measurements. If a shunt-type ohmmeter is used on the low-resistance ranges, it may supply excessive current to the transistor and destroy it. On higher ranges some meters use batteries up to 15 V. You must therefore be cautious in making ohmmeter checks in transistor circuits. A good rule to follow is to use a "low-power" ohmmeter, described below. If it is necessary to check the resistance of components in a transistor circuit, using an ordinary ohmmeter, the transistor should be removed first if it is a plug-in type. When resistance measurements with the transistor are made in-circuit, allowance must be made for conduction through the transistor. A recommended method is to make two sets of resistance measurements, reversing the ohmmeter leads for the required reading. The higher reading is more correct because on the higher reading the transistor was reverse-biased.

Some VOMs and many EVMs now contain a *low-power ohms* (LPΩ) function, specifically designed for making resistance measurements in transistor circuits. On the LPΩ function not enough voltage is developed to forward-bias transistor junctions. Therefore the LPΩ function is used for resistance measurements in circuits containing solid-state devices.

Use of a Signal Generator as a Signal Source in Transistor Circuits. Excessive signal-generator output may destroy a transistor. Generator output should therefore be set at minimum at the start of signal injection. As an additional safety measure, the generator should not be coupled directly into the circuit. Loose capacitive coupling, wherever possible, is recommended.

Identifying Transistor Elements

In working with a transistor, it is necessary to know whether it is NPN or PNP and which is the collector terminal, which the emitter, and which the base. Identifying information is given in the manufacturer's manual. If a manual is not handy, and if the diagrams in Fig. 29.1-7 are not conclusive, a transistor tester will help determine the elements and will indicate whether the transistor is NPN or PNP. In this experiment you will have the opportunity to use both a transistor manual and a transistor tester in identifying transistor elements and types.

SUMMARY

1. Transistors are solid-state devices which extend the range and application of the two-element diode.
2. Silicon is the element from which most transistors and other semiconductors are made today.
3. Junction transistors consist of a very thin element called the *base,* sandwiched between two elements called the *emitter* and *collector,* Figs. 29.1-2 and 29.1-3.
4. Bipolar transistors are either PNP or NPN types, where the first letter designates the emitter material type, the middle letter, the base, and the last letter, the collector material type. They therefore contain two *junctions,* the emitter-base junction and the collector-base junction.
5. The characteristics of junction transistors depend on the extent of doping of the emitter, base, and collector; on the geometry of the transistor; and on the method of manufacture.
6. Transistors are heat sensitive, and the safe temperatures within which they may be operated are specified by the manufacturer.
7. Transistors may be of the solder-in or plug-in type. The plug-in type use sockets. These sockets vary in shape and size to match the transistor.
8. Transistors are rated according to their ability to

dissipate power, varying from very low (50 mW) to very high.
9. In soldering transistors, low-wattage soldering irons are used to prevent overheating.
10. Transistors should be handled carefully to prevent damage to their fragile leads.
11. Always turn power OFF before inserting a transistor in, or removing it from, a circuit.
12. Transistors may be destroyed accidentally when measuring voltages at their terminals. Care must therefore be exercised to prevent short-circuiting terminal leads during measurement in the circuit.
13. Transistors may be damaged in making ohmmeter checks of their elements if the ohmmeter supplies excessive current to the transistor. Therefore, the high-ohms (low-current) ranges of an ohmmeter should be used.
14. In making resistance checks in a circuit containing a transistor or other solid-state device, the *low-power ohms* function of the meter should be used. This will prevent the transistor elements from being forward-biased, and will ensure proper resistance readings.

SELF-TEST

Check your understanding by answering these questions.

1. Present-day transistors are usually made of ________ (germanium/silicon).
2. Three-element transistors contain ________ (two/three) junctions.
3. A dot on the transistor case, or the spacing between elements, is frequently used to identify transistor elements. ________ (true/false)
4. Transistors are sensitive to ________. Therefore ________ sinks should be used in soldering transistor terminals in a circuit.
5. ________ should be exercised in making voltage measurements at transistor elements in a circuit, to prevent ________ to the transistor.
6. Transistors are hardy devices and therefore little care need be exercised in handling them. ________ (true/false)
7. In making resistance tests in a transistor circuit, the low-power ohms function of a meter should be used. ________ (true/false)
8. In a ________-type transistor the arrow points to the base.
9. In a junction transistor the ________ lies between the ________ and ________.
10. The basing diagram for a transistor may be found in a ________ ________.

MATERIALS REQUIRED

- Equipment: Transistor tester with facilities for identifying transistor type (PNP or NPN) and elements
- Transistors: An assortment of transistors at least one of which is not identified (by number)

PROCEDURE

1. Examine the transistors you have received. One will not be identified. List the others by number.
2. Draw a basing diagram for each.
3. Refer to a transistor manual and identify on the basing diagram each element. Indicate also whether it is NPN or PNP.
4. Read the instructions for the use of the transistor tester in identifying the elements of a transistor. Follow these instructions and use the tester to determine the E, B, and C of the unknown transistor. Identify these on the basing diagram for the transistor. Indicate also whether it is PNP or NPN.

QUESTIONS

1. Were you able to find all the transistors in one transistor manual? If not, where did you obtain the data for the others?
2. What difficulty, if any, did you have in identifying the elements of the unknown transistor, using a transistor tester?
3. How did your results (step 4) compare with the basing diagrams in Fig. 29.1-7?

Answers to Self-Test

1. silicon
2. two
3. true
4. heat; heat
5. Care; damage
6. false
7. true
8. PNP
9. base; emitter; collector
10. transistor manual

EXPERIMENT 29.2. Transistor Biasing

OBJECTIVES

1. To measure the effect on collector current of forward- and reverse-biasing the emitter-to-base of an NPN transistor
2. To measure the effect on collector current of forward- and reverse-biasing the emitter-to-base of a PNP transistor

INTRODUCTORY INFORMATION

NPN Transistor Biasing

A junction transistor can be considered as consisting of two diodes. The emitter-to-base junction is considered as one diode, the base-to-collector, the other (see Fig. 29.2-1). Because this transistor operates with negative or positive charge carriers, it is called *bipolar*. In an earlier experiment we saw that current flow in a semiconductor diode took place when the diode was forward-biased, that is, when the positive terminal of a battery was connected to the P-type anode and the negative terminal to the N-type cathode. We observe that the emitter-to-base diode in Fig. 29.2-1 is *forward*-biased, but the base-to-collector diode is *reverse*-biased. Why? And what effect do these bias connections have on the operation of a transistor?

In the emitter-to-base diode of the NPN transistor, electrons are the *majority current* carriers, and they originate in the N-type emitter. If the collector were not present, current flow in the emitter-base section would take place in the manner described in an earlier experiment dealing with semiconductor diodes. However, the presence of the N-type collector and its connection to the *positive terminal* of V_{CC} completely changes the path of electron-current flow. Because the base is so thin, only a *small percentage* of the electrons emitted by the emitter combine with the holes in the P-type base. About 95 percent of the electrons pass through the very thin base and are attracted to the positive battery terminal in the collector. The emitter-collector circuit is externally completed through the two batteries V_{EE} and V_{CC} connected in series-aiding.

From the foregoing description of current flow in a transistor, it is evident that:

1. The emitter is the source of current carriers.
2. The emitter-base current is very small.
3. The emitter-collector current is high.

It can also be seen that changes in emitter-base bias will cause changes in emitter current. As an example, an increase in forward (emitter-base) bias will result in increased emitter current, and therefore will also cause increased collector current. Base current will increase or decrease very little when emitter current increases or decreases. It is clear, therefore, that collector current can be easily controlled by bias changes in the emitter-base section of a transistor, biased as in Fig. 29.2-1. The names *emitter* and *collector* in a transistor are associated with the job that these elements perform.

Figure 29.2-2 is a simplified diagram showing the direction of electron current in the *external* circuit of an NPN transistor. The current shown as I_{CBO} is a very small leakage current which will be discussed later. The current in the *emitter* is the *total* current. This current divides into base and collector currents. The sum of the base and collector currents equals the emitter current. The emitter-base current (determined by base-emitter bias voltage) causes and controls the emitter-collector current.

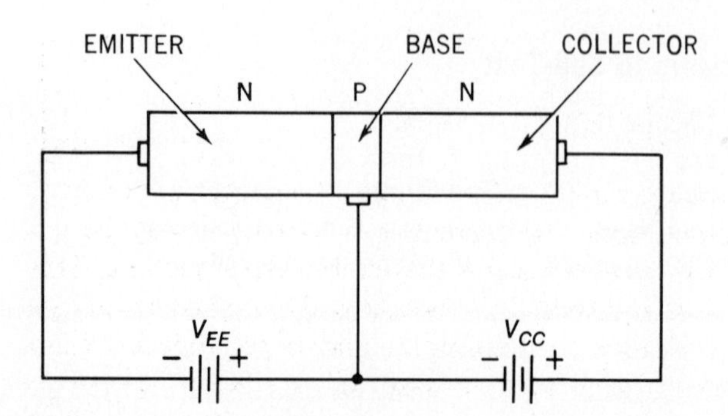

Fig. 29.2-1. Biasing an NPN transistor.

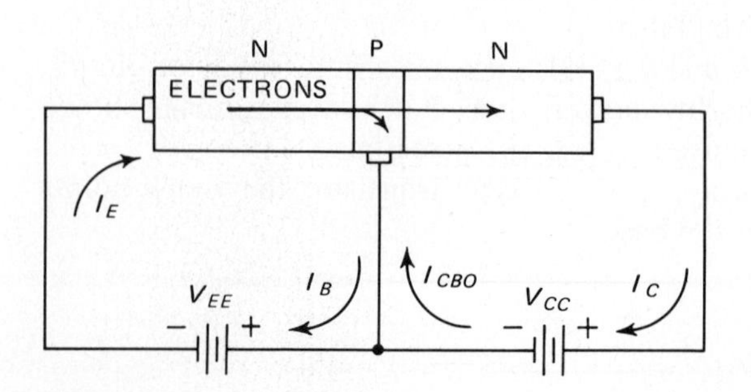

Fig. 29.2-2. Electron flow in external circuit of NPN transistor.

The bias connections in Figs. 29.2-1 and 29.2-2 are the battery connections to a transistor operating as an amplifier. The base-emitter is forward-biased; the base-collector is reverse-biased. A simple scheme to remember battery connections to a transistor is the following: *The battery connections to the emitter and base are the same as the letter of the impurity symbols of the emitter and base.* For example, in the NPN transistor, the emitter (on the left) is N and the base is P. The N-type emitter therefore receives the N (negative) terminal of the battery, and the P-type base, the P (positive) terminal. *The collector receives a battery polarity opposite to its impurity type.* In an NPN transistor the collector is N-type. Therefore it receives a *positive* battery terminal.

PNP Biasing

The same rules for battery connections apply to a PNP transistor, but because the impurities in a PNP type are opposite to those in an NPN transistor, battery polarities are naturally reversed.

Figure 29.2-3 is the bias arrangement for a PNP transistor. The emitter-base is forward-biased by V_{EE}. The base-collector is reverse-biased by V_{CC}. Holes are the current carriers inside the transistor. Electron-current flow in the *external* circuit is shown in Fig. 29.2-4. As you can see battery polarities and current flow in a PNP transistor are opposite to those of an NPN transistor. Again emitter current is the *total* current. The sum of collector and base currents equals the emitter current.

The rules for biasing hold for the PNP transistor as well as for the NPN type. The P-type emitter receives the P (positive) terminal of V_{EE}. The N-type base receives the N (negative) terminal of V_{EE}. The P-type collector receives the N terminal (opposite polarity) of V_{CC}.

Bias control of emitter-base current also controls collector current.

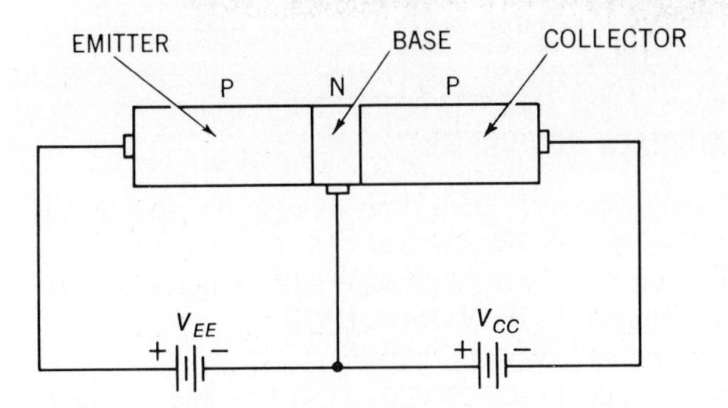

Fig. 29.2-3. Biasing a PNP transistor.

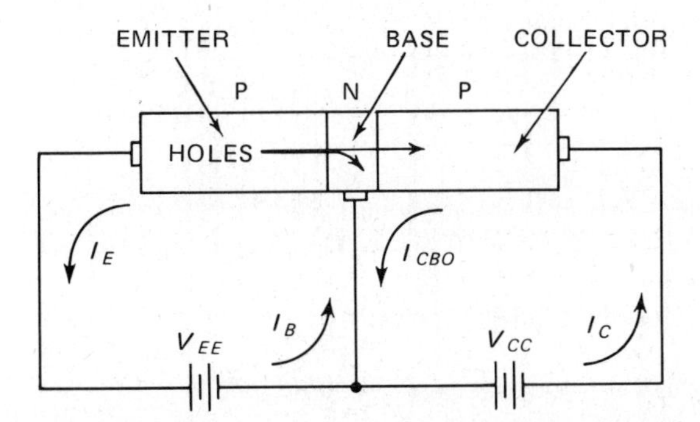

Fig. 29.2-4. Electron-current flow in external circuit of PNP transistor.

I_{CBO} and Thermal Runaway

An important characteristic of the collector-base circuit should be noted. This is the collector current flowing with the collector-base junction biased in the reverse direction and with the emitter-base open-circuited (termed I_{CBO} or I_{CO}). This leakage current I_{CBO} is due to minority carriers in the collector and base. I_{CBO} is in the range of a few microamperes for germanium and a few nanoamperes for silicon, and it increases with an increase in temperature.

An important factor which affects the operation of a transistor is its operating temperature. A transistor is very sensitive to temperature changes. Increased temperature results in increased current. This in turn leads to added heat and more current. If this chain reaction, which is called "runaway," is uninterrupted, it may result in complete destruction of the transistor because of excessive heat. In the design of transistor circuits, thermal runaway is prevented by negative-feedback arrangements which you will study later. The normal range of temperature within which a transistor may be operated safely is specified by the manufacturer. Silicon transistors are more tolerant of heat than are germanium transistors, and their temperature operating range is therefore much wider than germanium.

Cutting Off Collector Current

Since the emitter-base diode is the source of collector current, collector current in a transistor may be cut off by reverse-biasing the emitter-base section. This is accomplished by reversing the polarities of the bias battery to the emitter and base. Figure 29.2-5 shows the emitter-to-base of an NPN transistor reverse-biased. There will be no collector current for this connection.

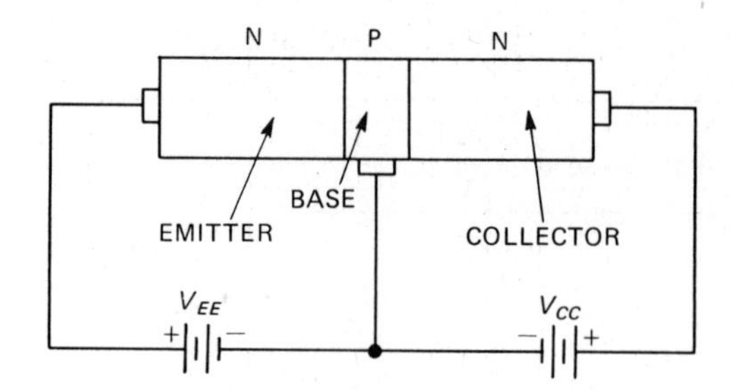

Fig. 29.2-5. Reverse-biased base-to-emitter junction.

Transistor Parameter Symbols

The following list gives the symbols used to identify transistor parameters.

Symbol	*Meaning*
I_C	Collector current
I_E	Emitter current
I_B	Base current
V_C	Voltage at the collector
V_E	Voltage at the emitter
V_B	Voltage at the base
V_{CC}	Supply voltage to the collector
V_{EE}	Supply voltage to the emitter
V_{BB}	Supply voltage to the base
I_{CB}	Collector-to-base current (the second subscript may be used in any of the above to avoid confusion)
V_{KJ}	Circuit voltage between elements, for example, elements K and J
V_{CB}	Voltage between collector and base

SUMMARY

1. For purposes of biasing, the transistor may be considered as consisting of two diodes, the emitter-base diode and the collector-base diode.
2. In most applications the emitter-base diode is forward-biased, the collector-base diode is reverse-biased. To bias a PNP transistor, therefore, requires that the positive (P) terminal of a battery be connected to the emitter, the negative terminal to the base; the negative terminal of a battery to the collector, the positive to the base (see Fig. 29.2-3).
3. To bias an NPN transistor requires reversing the polarities of both batteries, so that the emitter is negative relative to the base, and the collector is positive relative to the base, as in Fig. 29.2-1.
4. Adopting the convention that current flow is the same as electron flow helps determine the direction of current flow in the *external* circuit of a transistor (see Figs. 29.2-2 and 29.2-4).
5. The emitter *injects* (emits) or is the source of the current carriers *in* a transistor. The collector receives (collects) most of the current carriers. The base controls the collector current.
6. The emitter current is the total current, and it divides into collector current and base current.
7. Current flow inside a transistor is carried on by the majority current carriers; these are electrons in N-type material, holes in P-type material.
8. There are also *minority* carriers. These are electrons in P-type and holes in N-type material. Biasing of the collector-base junction, which is *reverse* for the majority carriers, is *forward* for the *minority* carriers. The small *minority*-carrier current in the collector-base junction with the emitter open is called leakage current or I_{CBO}.
9. Though I_{CBO} is very small in silicon transistors, it increases with heat and may destroy a transistor unless it is checked.
10. Reverse-biasing the base-emitter junction of a transistor will cut off collector current.

SELF-TEST

Check your understanding by answering these questions.

1. The emitter-base junction of a transistor is ________ biased; the collector-base ________ biased.
2. In a PNP transistor the emitter is ________ (positive/negative) relative to the base, while the collector is ________ (positive/negative) relative to the base.
3. Battery polarities in an NPN transistor are ________ (reversed/the same) as compared with those in a PNP transistor.
4. In a transistor biased as in Fig. 29.2-1, $I_E = 10$ mA and $I_B = 0.05$ mA. $I_C =$ ________ mA.
5. If the battery polarity of V_{EE} is reversed, Fig. 29.2-3, ________ current and ________ current will be ________ ________.
6. Thermal runaway results in uncontrolled ________ in a transistor, and will destroy the transistor because of the excessive ________ which is generated.

MATERIALS REQUIRED

- Power supply: Dual low-voltage dc source (or 1½-V and 6-V dc sources)
- Equipment: Two multirange milliammeters (or the current ranges on VOMs), EVM
- Resistors: ½-W 100-, 820-Ω
- Semiconductors: 2N6004 (NPN) and 2N6005 (PNP) properly mounted or their equivalent

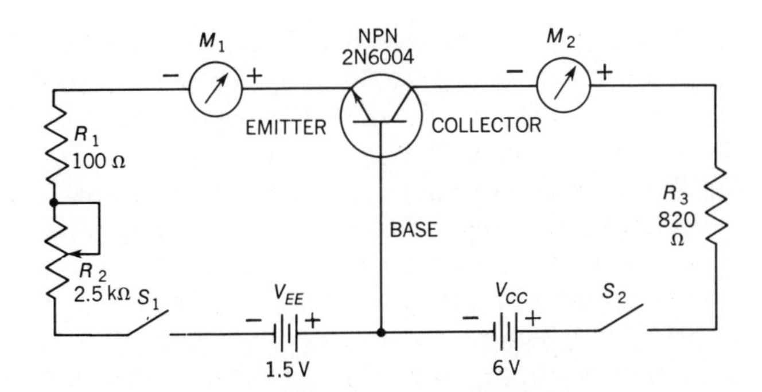

Fig. 29.2-6. Experimental NPN transistor circuit.

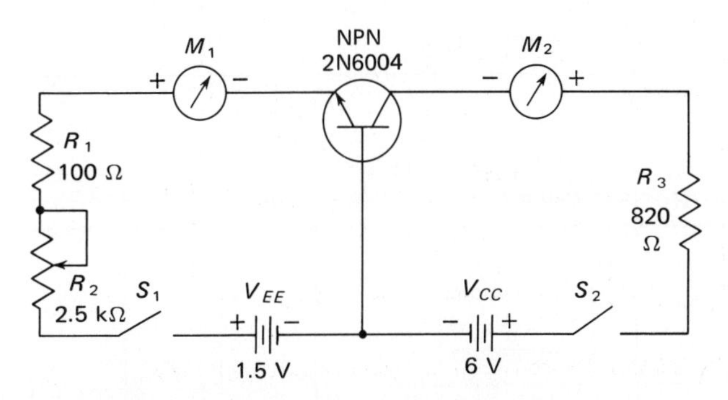

Fig. 29.2-7. Experimental NPN transistor circuit.

- Miscellaneous: 2-W 2500-Ω potentiometer; two SPST switches

PROCEDURE

NPN Biasing

1. Set the dual power-supply outputs so that $V_{EE} =$ 1.5 V and $V_{CC} = 6$ V.
2. Connect the circuit of Fig. 29.2-6. M_1 and M_2 are milliammeters. **Power on.** (Close S_1 *and* S_2.)

NOTE: In the experiment **power on** will mean *closing both* S_1 and S_2. **Power off** will mean *opening both* S_1 and S_2.

3. Set R_2 for maximum resistance, therefore *minimum* bias current. Set milliammeters to their proper range. Measure and record in Table 29.2-1 I_E and I_C. Also measure and record V_{EB} and V_{CB}. Indicate polarity of voltage.
4. Set R_2 for minimum resistance (maximum bias current) and repeat the measurements in step 2.
5. **Power off.** Reverse polarity of V_{EE} (emitter supply) and meter M_1 connections (Fig. 29.2-7).

TABLE 29.2-1. NPN Transistor Biasing

Step	Condition	*Current,* mA		V, *dc*	
		I_C	I_E	V_{EB}	V_{CB}
3	Fig. 29.2-6 R_2 maximum				
4	R_2 minimum				
5,6	Fig. 29.2-7				
7	$I_{CBO} =$	μA			

6. **Power on.** Vary R_2 over its entire range. Record I_E and I_C (if any). Measure and record V_{EB} and V_{CB}, showing polarities.

Measuring I_{CBO}

7. Open S_1. Set M_2 on its most sensitive range (50 μA or lower, if available). Measure and record I_C. This is the value of I_{CBO} for the conditions in the circuit of Fig. 29.2-7.
8. **Power off.**

PNP Biasing

9. Connect the circuit of Fig. 29.2-8. Repeat steps 2 through 4 for the PNP transistor. Record your results in Table 29.2-2.
10. **Power off.** Reverse polarity of V_{EE} and meter M_1 connections, Fig. 29.2-9, and repeat step 6, recording your results in Table 29.2-2.
11. Following the procedure in step 7, measure and record I_{CBO} for the PNP transistor.

QUESTIONS

1. Which of the experimental circuits shows proper biasing of the emitter and collector circuit for (*a*) an NPN

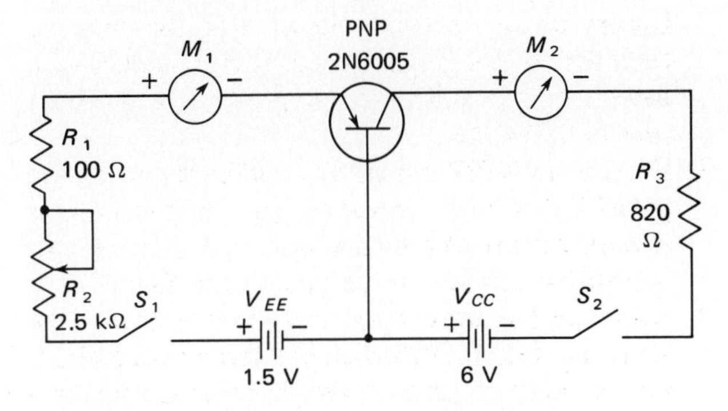

Fig. 29.2-8. Experimental PNP transistor circuit.

TABLE 29.2-2. PNP Biasing

Step	Condition	Current, mA		V, dc	
		I_C	I_E	V_{EB}	V_{CB}
9	Fig. 29.2-8 R_2 maximum				
	R_2 minimum				
10	Fig. 29.2-9				
11	I_{CBO} = μA				

transistor? (*b*) a PNP transistor? Explain your answer by referring to the test data.

2. What is the effect on I_C of increasing emitter bias? Refer to your measured data to prove your answer.
3. What are the effects on I_C of reverse-bias on the emitter-base circuit? Refer to your measured data to prove your answer.
4. What is I_{CBO}? Explain how I_{CBO} was measured.

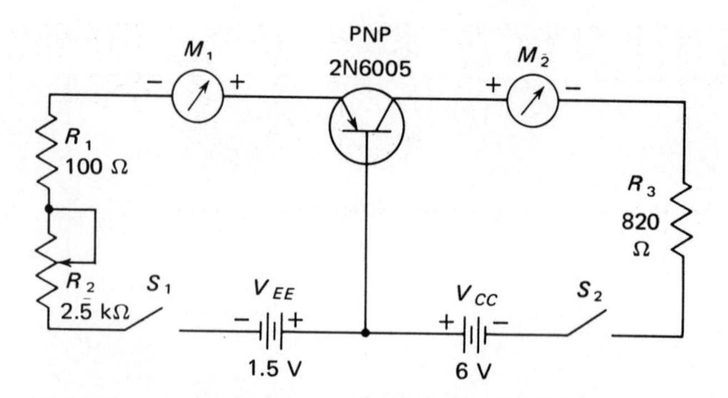

Fig. 29.2-9. Experimental PNP transistor circuit.

5. What is the effect on I_C of (*a*) increasing I_E? (*b*) decreasing I_E? Refer to your data to verify your answer.

Answers to Self-Test

1. forward; reverse
2. positive; negative
3. reversed
4. 9.95
5. emitter; collector; cut off
6. current; heat

CHAPTER 30 TRANSISTOR GAIN

EXPERIMENT 30.1. Current Gain (β) in a Common-Emitter Configuration

OBJECTIVES

1. To measure the effects on I_C of varying I_B
2. To determine beta (β)

INTRODUCTORY INFORMATION

Beta (β)

The grounded emitter is the most frequently used arrangement in transistor circuits. We should therefore understand the meaning of beta (β), the current gain of a grounded-emitter transistor.

In the grounded-emitter configuration, the input signal is applied to the base. Current gain is now computed as follows:

$$\beta = \frac{\Delta I_C}{\Delta I_B} \quad (V_{CE} \text{ constant}) \tag{30.1-1}$$

Equation (30.1-1) says that β is the ratio of the change in collector current ΔI_C caused by a change in base current ΔI_B with collector voltage V_{CE} held at a constant value. β, then, is the current amplification factor in a grounded-emitter amplifier. Another symbol for β is h_{fe}.

In a junction transistor, values of β are always greater than 1. The reason becomes clear when we examine the base circuit of a transistor connected as a grounded-emitter amplifier (Fig. 30.1-1). The directions of emitter current and collector current are opposite in the base circuit as shown here. Base current I_B is the difference between I_E and I_C. Emitter and collector current are almost equal in value. Therefore base current is small. Small changes in base current must therefore produce large changes in collector current.

Determining Beta Experimentally

A test setup for measuring β in an NPN transistor is shown in Fig. 30.1-2. Base current is varied by adjustment of R_2. M_1 is used for measuring base current. Collector current is measured by M_2. R_4 is used for maintaining a constant collector voltage V_{CE} measured by EVM, M_3.

To determine β, R_2 is first set for some current reading, I_{B1}. R_4 is adjusted to maintain a given value of collector voltage V_{CE}. Collector current I_{C1} is measured by M_2. R_2 is then varied for a slightly higher (or lower) value of base current I_{B2}. R_4 is readjusted to maintain the same value V_{CE} as previously measured. Collector current I_{C2} is now read. β may be computed as follows:

$$\beta = \frac{\Delta I_C}{\Delta I_B} = \frac{I_{C2} - I_{C1}}{I_{B2} - I_{B1}} \tag{30.1-2}$$

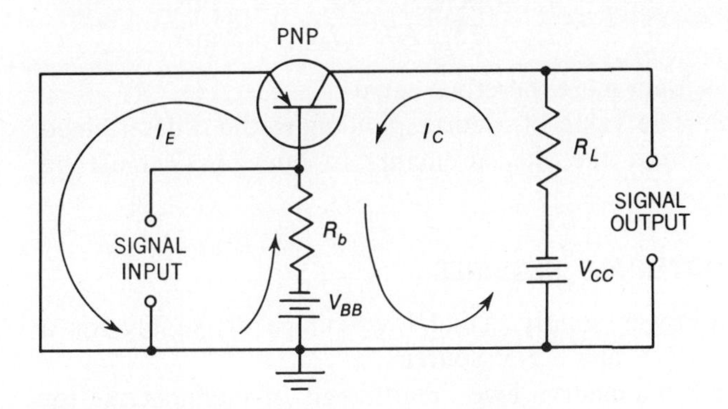

Fig. 30.1-1. Current in a grounded-emitter amplifier.

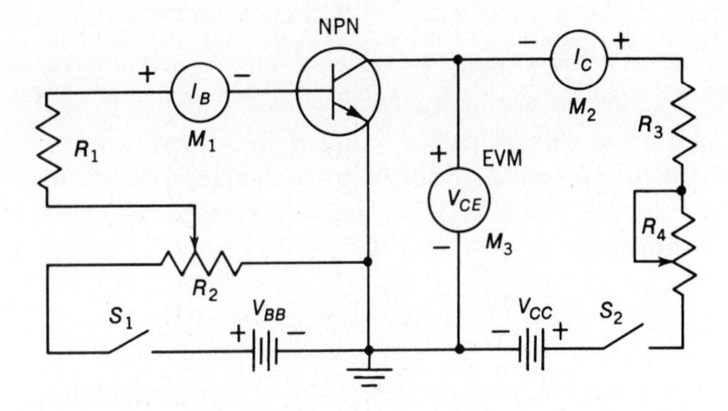

Fig. 30.1-2. Test circuit for measuring β in an NPN transistor.

An example will clarify how Eq. (30.1-2) is used. Suppose a base current I_{B2} of 0.1 mA causes a collector current I_{C2} of 4.5 mA, and a base current I_{B1} of 0.05 mA results in a collector current $I_{C1} = 4.0$ mA, with V_{CE} held constant at, say, 5 V. Then, substituting the values of I_B and I_C in our formula, we get

$$\beta = \frac{I_{C2} - I_{C1}}{I_{B2} - I_{B1}} = \frac{4.5 - 4.0}{0.1 - 0.05} = 100$$

This means that the change in collector current is 100 times greater than the change in base current.

Alpha

In a grounded- or common-base transistor circuit, emitter current I_E controls collector current I_C. This control characteristic of the grounded-base circuit is called alpha, and its symbol is the Greek letter α. Alpha is defined as the ratio of a change in collector current, ΔI_C, to the change in emitter current, ΔI_E, with collector voltage held constant. Thus

$$\alpha = \frac{\Delta I_C}{\Delta I_E} \qquad (V_{CB} \text{ constant})$$

Grounded-emitter circuits are more frequently used than grounded-base. Therefore the β of a transistor is a more commonly used figure of merit than α. Service-type transistor testers check the β of a transistor. Having found β, it is then possible to compute α, if that information is required, from the relationship

$$\alpha = \frac{\beta}{1 + \beta} \qquad (30.1\text{-}3)$$

An example will show how Eq. (30.1-3) is used. Suppose the β of a transistor is found to be 170. Substituting this value in our formula, we get

$$\alpha = \frac{170}{1 + 170} = \frac{170}{171} = 0.9942$$

SUMMARY

1. In a grounded-emitter (CE) configuration (Fig. 30.1-1), the input signal is applied to the base.
2. In this arrangement current gain is called beta (β).
3. Beta is defined as the ratio of the change in collector current ΔI_C caused by a change in base current ΔI_B, with collector-emitter voltage V_{CE} held constant.

$$\beta = \frac{\Delta I_C}{\Delta I_B} = h_{fe} \qquad (V_{CE} \text{ constant})$$

4. A practical way to determine β experimentally, using general-purpose laboratory meters, is to connect the circuit of Fig. 30.1-2, set the base current to a desired value I_{B1} and the voltage V_{CE} to a desired value, and measure I_{C1}. Collector current I_{C2} is next measured for a higher or lower value of base current I_{B2}, with V_{CE} set at the same level as in the first measurement. Beta is then computed as follows:

$$\beta = \frac{I_{C2} - I_{C1}}{I_{B2} - I_{B1}}$$

5. When V_{CE} and V_{CB} are both held constant, there is an important relationship between α and β which is given by this formula:

$$\alpha = \frac{\beta}{1 + \beta}$$

If we have measured β, we may find α by substituting in the above formula.

SELF-TEST

Check your understanding by answering these questions.

1. Current gain in a CE configuration is called _________.
2. Current gain in a CE configuration is always _________ than 1.
3. In a CE configuration the transistor element which is common to the input and output signal is the _________.
4. In a CE circuit the input signal is applied to the _________.
5. In determining β experimentally, the voltage V_{CE} must be _________ _________.
6. The formula which gives β in terms of collector and base current is:

   $\beta =$ _________ (   )
7. In the circuit of Fig. 30.1-2 the following measurements were made with $V_{CE} = 5$ V:

   $I_{B1} = 50\ \mu A \qquad I_{C1} = 4.5$ mA
   $I_{B2} = 75\ \mu A \qquad I_{C2} = 9.5$ mA

   Beta must therefore equal _________.
8. The value of α corresponding to the β determined from the measurements in question 7 must be _________.

MATERIALS REQUIRED

- Power supply: Dual low-voltage dc supply, or a $1\frac{1}{2}$-V and a 9-V source
- Equipment: Two multirange milliammeters (or VOMs); EVM

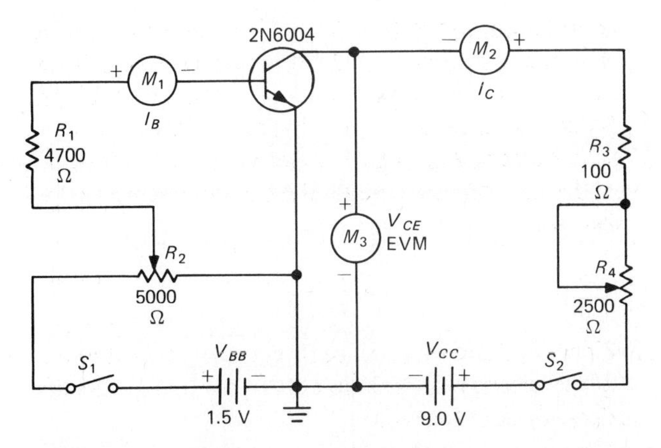

Fig. 30.1-3. Experimental circuit for measuring β in an NPN transistor.

- Resistors: ½-W 100- and 4700-Ω
- Semiconductors: 2N6004 with appropriate socket
- Miscellaneous: 2-W 2500- and 500-Ω potentiometers; two ON-OFF switches

PROCEDURE

Measuring Beta (h_{fe})

1. Connect the circuit of Fig. 30.1-3. Set $V_{EE} = 1.5$ V, and $V_{CC} = 9$ V. M_1 and M_2 are multirange milliammeters or equivalent ranges on 20,000-Ω/V VOMs. *R_4 must be set for maximum resistance before power is applied.*
2. Check circuit for proper connections.
3. **Power on.** Vary R_2 from maximum to minimum resistance, thus increasing base current I_B from its minimum to maximum value in this circuit. Observe the effect on I_C as I_B is increased. Record the results in Table 30.1-1.
4. Adjust R_2 for 10 μA base current. Adjust R_4 for 6 V at the collector. Measure and record I_C.
5. Adjust R_2 for maximum base current. Adjust R_4 to maintain 6 V. Measure and record I_C.
6. Adjust R_2 for 30 μA of I_B. Adjust R_4 for 6 V. Measure and record I_C.
7. Adjust R_2 for 40 μA of I_B. Adjust R_4 to maintain 6 V. Measure and record I_C.
8. **Power off.** Compute and record β. Show your computations.

QUESTIONS

1. Define β.
2. Explain in detail a procedure for measuring β.
3. Using the value of β determined in this experiment, find α. Show formula and all work.
4. A transistor has $\beta = 25$ and $I_C = 3$ mA when $I_B = 100$ μA. Assume linear operation.
   (*a*) What is the value of I_C when $I_B = 125$ μA?
   (*b*) What is the value of I_B when $I_C = 2$ mA?

Answers to Self-Test

1. beta (β)
2. greater
3. emitter
4. base
5. held constant
6. $\Delta I_C/\Delta I_B$ (V_{CE} constant)
7. 200
8. 0.995025–

TABLE 30.1-1. Test Data for Measuring β

Step	I_B, μA	I_C, mA	*Effect on I_C of Increasing I_B*	
4	10		Step 3	
5	Maximum:			
Step	I_B, μA	I_C, mA	*Collector volts*	
6	30		6	
7	40		6	
8	$\beta = \frac{\Delta I_C}{\Delta I_B} =$ ________ =			

CHAPTER 31 COMMON-EMITTER CHARACTERISTICS

EXPERIMENT 31.1. Transistor Data

OBJECTIVE

To become familiar with the nature of the data found in transistor manuals

INTRODUCTORY INFORMATION

Transistor Data

Transistors are designed with unique characteristics to meet certain application requirements. The manufacturer provides data sheets in which these characteristics are given. Data are furnished in both *tabular* and *graphical* form. It is important for the technician to understand the meaning of these data charts and graphs.

Data sheets are found in transistor manuals supplied by manufacturers of transistors or in transistor reference handbooks published commercially. The nature of the data depends on the source and on the intended use of the transistor. As a general rule the following specifications are available for each type:

1. *A brief description of the transistor and suggested applications.* As an example, the 2N6004 is listed by General Electric as an NPN silicon, planar, passivated epitaxial transistor for use in higher-voltage general-purpose amplifiers and switches in industrial applications. The PNP version of this unit is the 2N6005, JEDEC (Joint Electron Device Engineering Council) TO 18.
2. *Mechanical data, including dimensions and basing.* Figure 31.1-1*a* is an outline drawing (TO 18) of the 2N6004 transistor and its dimensions. The basing drawing, Fig. 31.1-1*b*, locates the positions of the emitter E, collector C, and base B. Figure 31.1-1*b* is a *bottom* view of the transistor, with the pins facing the viewer.
3. *Maximum ratings.* These ratings are usually based on the absolute maximum system defined by JEDEC and standardized by EIA (Electronic Industries Association) and NEMA (National Electrical Manufacturers Association). "*Absolute-maximum ratings* are limiting values of operating and environmental conditions which should *not* be exceeded . . . under any condition of operation." For example, for the 2N6004, maximum ratings are given as follows:

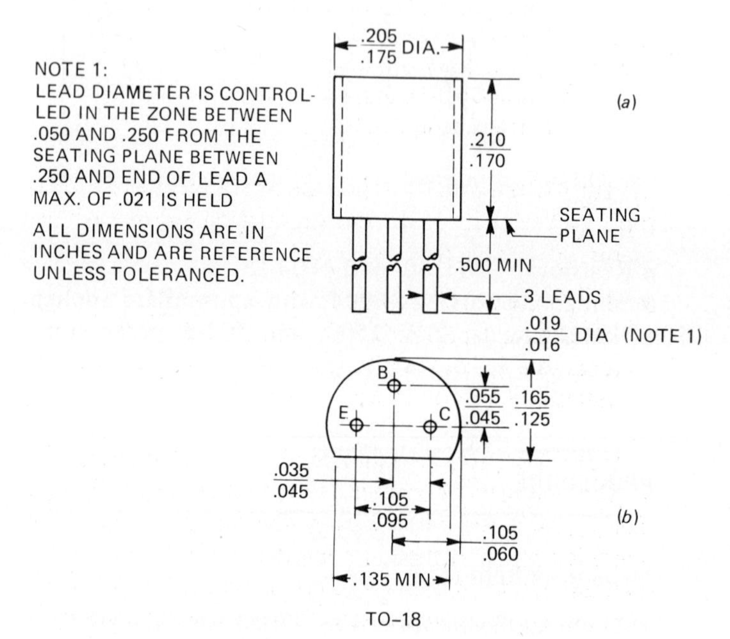

Fig. 31.1-1. (*a*) Outline drawing of a 2N6004; (*b*) basing diagram. (*General Electric Co.*)

Absolute Maximum Ratings: (T_A = 25°C, unless otherwise specified)

	Symbol	
Voltages		
Collector to emitter	V_{CEO}	40 V
Collector to emitter	V_{CES}	50 V
Emitter to base	V_{EBO}	5 V
Collector to base	V_{CBO}	50 V
Current		
Collector	I_C	500 mA
Dissipation		
Total power ($T_C \leqq 25$°C)	P_T	0.8 W
Total power ($T_A \leqq 25$°C)	P_T	0.4 W
Derate factor ($T_C > 25$°C)		8.0 mW/°C
Derate factor ($T_A > 25$°C)		4.0 mW/°C
Temperature		
Storage	T_{STG}	−65 to +150°C
Operating	T_J	−65 to +125°C

NOTE: The symbols used above have the following meanings:

Symbol	Meaning
V_{CEO}	Collector-to-emitter voltage with base open
V_{CES}	Collector-to-emitter voltage with base shorted to emitter
V_{EBO}	Emitter-to-base voltage with collector open
V_{CBO}	Collector-to-base voltage with emitter open
P_T	Average continuous total power dissipation
T_C	Case temperature
T_A	Ambient temperature
T_{STG}	Storage temperature
T_J	Junction temperature

NOTE: If the transistor is to be operated at a temperature higher than 25°C, the dissipation value must be derated by the amounts shown.

4. *Electrical characteristics:* ($T_A = 25$°C, unless otherwise specified). Figure 31.1-2 gives some characteristics for the 2N6004. Other characteristics are omitted here.

Not all transistor manuals list their characteristics in this manner. Moreover, the symbols used, though fairly standardized, are not always the same. The technician should become familiar with the transistor manual in use. Within its covers is a wealth of information on the many solid-state devices, an explanation of their operation, the symbols used, and application notes.

Transistor manuals frequently contain interchangeability charts in which the product numbers made by other manufacturers are listed, and suggested replacement numbers are given. This information is helpful where exact replacements cannot be obtained. Included also is a listing of obsolete and discontinued numbers, circuits of useful devices, and drawings and dimensions (outlines) of all transistors. Exact manufacturer replacements should always be used. In many cases transistors with closely controlled characteristics are required; these can be ensured only if exact replacements are used.

SUMMARY

1. Data of transistor characteristics are furnished by the manufacturer in tabular (chart) and graphical (curve) form.
2. Transistor data sheets include the following:
   (*a*) Transistor identifying number, for example, 2N6004
   (*b*) Descriptive information, such as NPN or PNP, silicon or germanium, how manufactured, and applications
   (*c*) Mechanical information, including dimensions, basing, and mounting

Static Characteristics		
Collector-to-emitter breakdown voltage ($I_C = 10$ mA, $I_B = 0$)	$V_{(BR)CEO}$	40 V
Emitter-to-base breakdown voltage ($I_E = 100$ μA, $I_C = 0$)	$V_{(BR)EBO}$	5 V
Collector-to-emitter saturation voltage ($I_C = 10$ mA, $I_B = 1$ mA)	$V_{CE(sat)}$	0.08 V
Base-to-emitter saturation voltage ($I_C = 10$ mA, $I_B = 1$ mA)	$V_{BE(sat)}$	0.6 to 0.8 V
Collector cutoff current ($V_{CB} = 25$ V, $I_E = 0$)	I_{CBO}	0.2 to 10 nA
Forward-current-transfer ratio ($V_{CE} = 1$ V, $I_C = 100$ μA)	h_{FE}	50
($V_{CE} = 1$ V, $I_C = 1$ mA)	h_{FE}	Min. 70, typical 160
Dynamic characteristics		
Forward-current-transfer ratio ($I_E = 1$ mA, $V_{CE} = 10$ V, $f = 1$ kHz)	h_{fe}	Min. 70, max. 450
Collector-base capacitance ($V_{CB} = 10$ V, $I_E = 0$, $f = 1$ MHz)	C_{CB}	Typ. 4.2 pF
Emitter-base capacitance ($V_{EB} = 0.5$ V, $I_C = 0$, $f = 1$ MHz)	C_{EB}	Typ. 15 pF
Gain-bandwidth product ($V_{CE} = 10$ V, $I_E = -10$ mA, $f = 100$ MHz)	f_T	Typ. 300 MHz
Input impedance ($I_E = -1$ mA, $V_{CE} = 10$ V, $f = 1$ kHz)	h_{ie}	Typ. 4 kΩ

Fig. 31.1-2. A partial list of characteristics for the 2N6004 at case temperature = 25°C.

(*d*) Maximum ratings, including voltage, current, and temperature
(*e*) Electrical characteristics, static and dynamic, including breakdown and saturation voltages, current gain (forward-current-transfer ratio), frequency (gain-bandwidth product), output and input capacitance, etc.

SELF-TEST

Check your understanding by answering these questions.

1. Maximum ratings are limiting values of ________ and ________ conditions which should not be exceeded.
2. The symbol V_{CEO} stands for ________ to ________ voltage with ________ open.
3. The three different temperature characteristics which must be considered in operating a transistor are:
   (*a*) T_C, ________ temperature
   (*b*) T_A, ________ temperature
   (*c*) T_J, ________ temperature
4. The symbol for average, continuous total power dissipation is ________.
5. For every degree Celsius that a 2N6004 is operated beyond $T_C = 25°C$, ________ mW must be subtracted from ________ W.

MATERIALS REQUIRED

- Equipment: Transistor tester capable of identifying E, C, and B
- Transistor: 2N6004 or the equivalent
- Miscellaneous: Data sheet for the transistor used

PROCEDURE

Refer to the data sheets of the transistor you are using and indicate the following:

1. JEDEC number: ________ (PNP or NPN)
2. Maximum power dissipation: ________ W at 25°C, T_C
3. Maximum I_C: ________ mA
4. Maximum V_{CBO}: ________ V
5. Maximum V_{EBO}: ________ V
6. Operating temperature (T_J): ________ °C
7. Static h_{FE}: ________ (V_{CE} = ________, I_C = ________)
8. Dynamic h_{fe}: ________ (V_{CE} = ________, I_E = ________, f = ________)
9. TO# ________
10. Basing diagram ________
11. Using a transistor tester, identify the emitter, base, and collector of the transistor and compare with the basing diagram in step 10.

QUESTIONS

1. When connecting a transistor in a circuit, why must a technician know whether a transistor is NPN or PNP?
2. How can a technician identify the elements of a transistor (E, B, C)? Give two ways.
3. What is the importance of absolute maximum ratings?
4. What may happen if a transistor is permitted to operate steadily beyond its rated temperature?
5. Where can a technician find technical information about a transistor to be used in a circuit?

Answers to Self-Test

1. operating; environmental
2. collector; emitter; base
3. (*a*) case; (*b*) ambient; (*c*) junction
4. P_T
5. 8; 0.8

EXPERIMENT 31.2. Common-Emitter Characteristic Curves I

OBJECTIVES

1. To use a transistor curve tracer to display the family of average collector characteristic curves (V_{CE} versus I_C)
2. To compute ac beta (β) from the experimental curves

INTRODUCTORY INFORMATION

Average Collector Characteristic Curves—CE Connection

Transistor characteristic curves give a good deal of information about a transistor. Characteristic curves for specific transistors are usually found in transistor manuals. When they are not available, they can be plotted by a technician in one of several ways. In this experiment we shall learn how to use a curve tracer to observe a family of curves.

Figure 31.2-1 shows the V_{CE} versus I_C family of average collector characteristics for the 2N6004, common-emitter connection. Each curve in the family is obtained by plotting I_C for increasing values of V_{CE}, while holding base current I_B constant. Observe that each curve shows an increasing collector-current level for an increased input base-current level. It is interesting to see that for base-current levels up to 25 μA, collector current is practically independent of collector voltage and is mainly dependent on base current. For example, at $V_{CE} = 5$ V, when $I_B = 15$ μA, $I_C = 3$ mA. For the same base current I_C increases to 3.6 mA when $V_{CE} = 30$ V, a variation of 0.6 mA for a 25-V change in collector voltage. On the other hand, at $V_{CE} = 5$ V, an increase in I_B from 15 to 20 μA causes an increase in I_C from 3 to 4 mA. Base current is evidently much more effective in controlling collector current than is collector voltage. The ac β of the transistor is the number which tells how much more effective I_B is than V_{CE} in controlling I_C.

Another point of interest in Fig. 31.2-1 is the end point of each curve. For example, the 15-μA curve ends at $V_{CE} = 42$ V. At that point $I_C = 10$ mA. Multiplying $I_C \times V_{CE}$ gives the power dissipated by the collector, which in this case is

$$P_T = 10 \times 10^{-3} \times 42 = 420 \text{ mW}$$

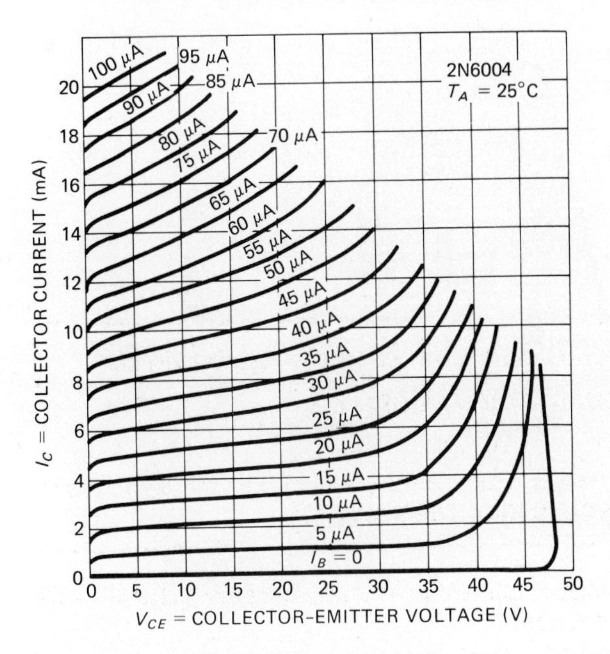

Fig. 31.2-1. Average collector characteristics (I_C versus V_{CE}) for the 2N6004. (*General Electric Co.*)

Now take the 20-μA curve. It ends at $V_{CE} = 41$ V, and at that point $I_C = 10.2$ mA (approx.). For this condition

$$P_T = 10.2 \times 10^{-3} \times 41 = 418 \text{ mW}$$

Similarly, the 50-μA curve gives $I_C = 14$ mA at $V_{CE} = 30$ V, and

$$P_T = 14 \times 10^{-3} \times 30 = 420 \text{ mW}$$

In each case the power dissipated is close to the 400-mW maximum power dissipation of the transistor operated at an ambient temperature T_A of 25°C. The end points of the curves are the limiting points at which the transistor can be operated safely and not exceed its maximum power dissipation rating.

The curves in Fig. 31.2-1 were experimentally determined with the transistor's ambient temperature T_A held at 25°C. If T_A had been higher, I_C would have been higher, and the permissible maximum dissipation would then have been lower. In other words, the curves could not have extended as far to the right as they do in the figure.

AC Beta

From the family of collector characteristics it is possible to compute β approximately. The formula for ac beta is

$$\beta_{ac} = \frac{\Delta I_C}{\Delta I_B} \qquad (V_{CE} \text{ constant})$$

AC beta is a dynamic characteristic. On the graph, the value of I_C is read for two different values of I_B, but the same value of V_{CE}. The values of I_C and I_B are then substituted in the formula and β computed. For example, when $V_{CE} = 5$ V and $I_B = 10$ μA, $I_C = 2$ mA. Also when $V_{CE} = 5$ V, and $I_B = 20$ μA, $I_C = 4$ mA. Substituting in our formula, we get

$$\beta = \frac{(4-2) \times 10^{-3}}{(20-10) \times 10^{-6}} = \frac{2}{10} \times 10^3 = 200$$

for the conditions specified. This result tells us that base current is 200 times more effective than collector voltage in controlling collector current for the given conditions.

Curve Tracer

It is possible to find, point by point, a series of points along any one of the curves, by a test circuit which will be studied in the next experiment. These points would then be joined to give one of the V_{CE} versus I_C characteristic curves. However, a more rapid method of plotting characteristic curves involves the use of an instrument called a *curve tracer*.

When an automatic curve tracer is used, the transistor is plugged into a receptacle on the instrument. The manual controls of the curve tracer are set to the required operating voltages and currents. The cathode-ray-tube (CRT) indicator of the curve tracer then displays a graph (curve) of the required transistor characteristics. It is possible to display a whole family of characteristic curves, as in Fig. 31.2-1, by proper settings of the curve-tracer controls.

There are several types of curve tracers. One is a completely self-contained unit, just described, which has both the CRT indicator and the circuitry required to generate the curves. One such instrument is shown in Fig. 31.2-2. Physically, this instrument resembles an oscilloscope, but it does not operate as an oscilloscope. That is, it will not show electrical waveforms. The curve tracer displays characteristic curves for transistors and other solid-state devices, such as diodes.

Another type is the curve-tracer unit which is used with a commercial oscilloscope. Such a unit also provides a satisfactory display of transistor characteristic curves. Figure 31.2-3 shows such a unit. Separate curve-tracer units are less expensive than the completely self-contained tracer in Fig. 31.2-2. They are widely used in servicing electronics equipment, such as TV receivers. With them, it is possible to obtain an in-circuit check of transistor operation. The curve seen on the oscilloscope, in servicing, is popularly known as the transistor "signature." Such signatures usually bear little resemblance to those in Fig. 31.2-1. Resistive and reactive circuit components create strange-looking traces. A practical way of using such traces is to have a record of standard signatures for particular makes and models of equipment and the specific circuits under test during *normal* operating conditions.

Oscilloscope cameras are available for photographing to obtain curves and families of characteristic curves for immediate use.

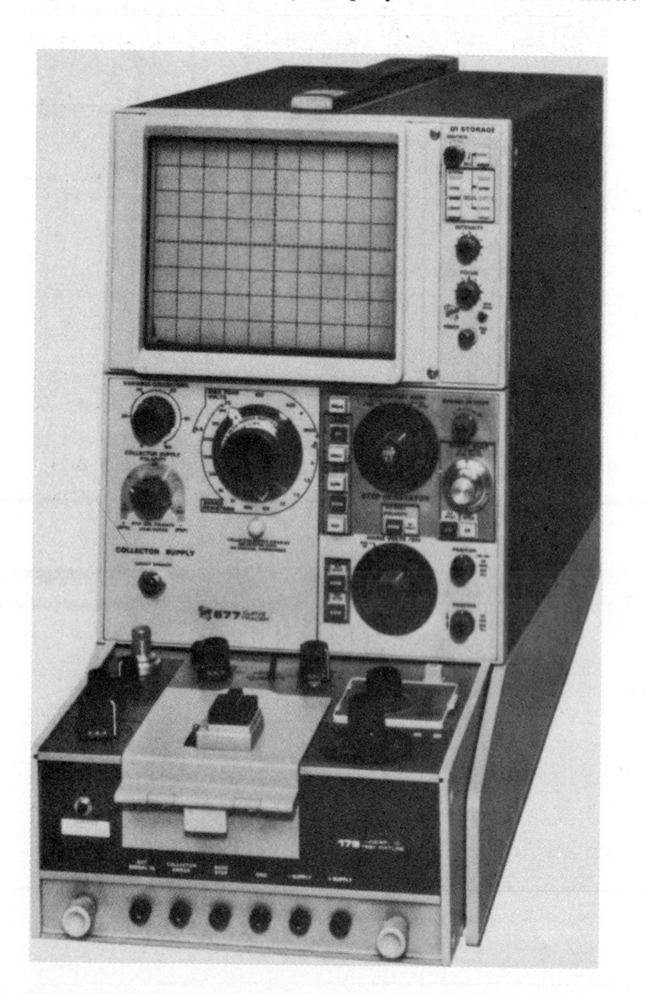

Fig. 31.2-2. Self-contained transistor curve tracer. (*Tektronix*)

SUMMARY

1. Transistor curves provide information about a transistor under operating conditions.
2. The curve which shows how I_C varies with changes in V_{CE} (I_B constant) is called the *average collector characteristic* for the CE configuration. A group of such curves for different values of I_B, Fig. 31.2-1, is called the *family of average collector characteristics*.
3. Transistors should not be operated beyond their maximum rated power. For the CE configuration

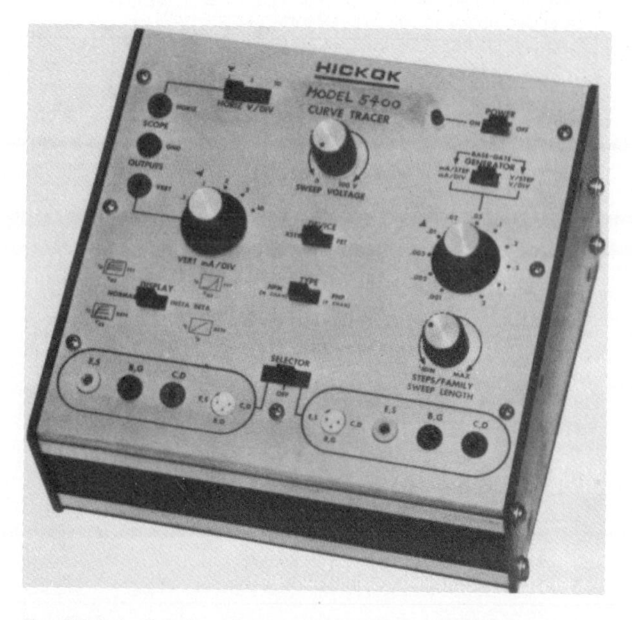

Fig. 31.2-3. Semiconductor curve tracer, used with an oscilloscope. (*Hickok Teaching Systems, Inc.*)

the product of V_{CE} and I_C is the collector dissipation of the transistor.

4. Transistor dissipation is specified for a specific operating temperature. If the temperature is higher than specified, transistor dissipation must be derated.
5. From the family of average collector characteristics it is possible to calculate the ac *beta* of a transistor. Beta is a number which tells how much more effective base current is than collector voltage in controlling collector current.
6. Average collector characteristics may be graphed using a point-by-point method.
7. Average collector characteristics may be displayed automatically on a CRT display using a curve tracer.
8. Curve tracers can be used in troubleshooting electronic equipment to test the operation of transistors in a circuit, but with certain restrictions.

SELF-TEST

Check your understanding by answering these questions.

1. The ________ of a transistor is the measure of the ability of the base to control collector current.
2. Current gain of a transistor may be determined from the family of ________ ________ characteristics.
3. The end points of the characteristic curve in Fig. 31.2-1 are the limiting points of collector current and voltage, beyond which the transistor should not be operated. ________ (true/false)
4. To determine the ac beta of a transistor, use the family of average collector characteristics of that transistor. For two different values of ________ but the same value of ________, read the values of I_C and substitute in the formula

   $\beta =$ ________

5. The family of average collector characteristics can most easily be plotted using a ________ ________ and an oscilloscope camera.

MATERIALS REQUIRED

- Equipment: Curve tracer; oscilloscope if required; oscilloscope camera, if available
- Transistors: 2N6004 or the equivalent

PROCEDURE

1. Read the operating instructions for the curve tracer assigned to you. Calibrate the vertical and horizontal amplifiers of your oscilloscope, if necessary, according to the instructions in the curve tracer and set the curve-tracer controls for observing the characteristic curves of a 2N6004 transistor.
2. Display the family of characteristic curves over the range V_{CE}-0 to 10 V with the curve tracer set for six I_B steps in increasing units of 10 μA, from 0 to 50 μA. Center the curves properly so that they are fully displayed on the CRT graticule.
3. Either photograph the curves or draw them on graph paper. Clearly mark the vertical (I_C) axis in proper 1- or 2-mA graduations, and the horizontal (V_{CE}) axis in 1-V graduations.
4. Compute the ac beta of the transistor at $V_{CE} = 6$ V. Show your computations.

QUESTIONS

1. How do the experimental V_{CE} versus I_C curves compare with the published curves for the transistor? Explain any discrepancies.
2. How does the value of ac beta, computed in step 4, compare with the value of h_{fe} in the published data for the 2N6004?
3. Refer to Fig. 31.2-1. What is the transistor dissipation when $V_{CE} = 35$ V and $I_B = 20$ μA? Show your computations.

Answers to Self-Test

1. beta
2. average collector
3. true
4. I_B; V_{EC}; $\dfrac{\Delta I_C}{\Delta I_B}$
5. curve tracer

CHAPTER 32 SOLID-STATE DIODE AND TRANSISTOR TESTING

EXPERIMENT 32.1. Solid-State Diode Testing

OBJECTIVES

1. To test semiconductor diodes with a commercial tester
2. To test semiconductor diodes with go–no go circuits

INTRODUCTORY INFORMATION

Ohmmeter Check of a Diode

In an earlier experiment we learned how semiconductor diodes could be tested with an ohmmeter. In one position of the ohmmeter leads the diode was forward-biased and measured a very low forward resistance R_F. In the reverse position of the ohmmeter leads, the diode was reverse-biased and measured very high reverse resistance R_R. If the ratio r of R_R to R_F was 100 or higher, that is, if $r = R_R/R_F \geqq 100$, the diode was assumed to be good.

There were two limitations on this test: (1) A low-power ohmmeter could not be used, and (2) the test was not always conclusive, because the diodes were not subjected to their normal operating voltages.

In this experiment we shall learn two dynamic tests of a diode which are more reliable than an ohmmeter check.

Checking Diodes with a Semiconductor Tester

A semiconductor tester will indicate whether a diode is good or bad. Since the voltage at the terminal leads of the tester is normally higher than that at the terminals of an ohmmeter, the semiconductor tester gives a more reliable indication of the diode's condition. However, even this is not a conclusive test, for though a diode may test okay, it still may break down under conditions of circuit operation.

Some semiconductor testers will also identify the anode and cathode terminals of the diode. They may also measure the reverse current I_R of a diode, for the terminal voltage output of the tester. It is also possible to get a relative indication of the forward current I_F of the diode. We say *relative* rather than *absolute* because in the forward position the meter needle will normally go off-scale.

Both in-circuit and out-of-circuit testers are available.

Another means of testing a diode is with a semiconductor curve tracer.

Diode Testing With Go-No Go Circuits

High-Voltage Rectifiers

A dynamic go–no go test for silicon rectifiers operating at 120 V or higher is possible with the circuit in Fig. 32.1-1. The rectifier to be tested, *D*, is connected between terminals *A* and *B* of the test jig. A *good* rectifier will give these test indications: With switch S_2 *open* and S_1 closed, the 25-W bulb will glow *dimly*, because it is operating only on the positive alternations of the 120-V input sine wave. (*D* acts as a *closed* switch on the *positive* and as an *open* switch on the *negative* alternation.) When S_2 is closed, the bulb will brighten.

An *open* rectifier will not permit the lamp to glow when S_2 is open.

A shorted rectifier will permit the bulb to glow just as brightly whether S_2 is open or closed.

Low-Voltage Rectifiers

The test circuit of Fig. 32.1-2 will check low-voltage rectifiers. A 6-V battery or power supply is the power

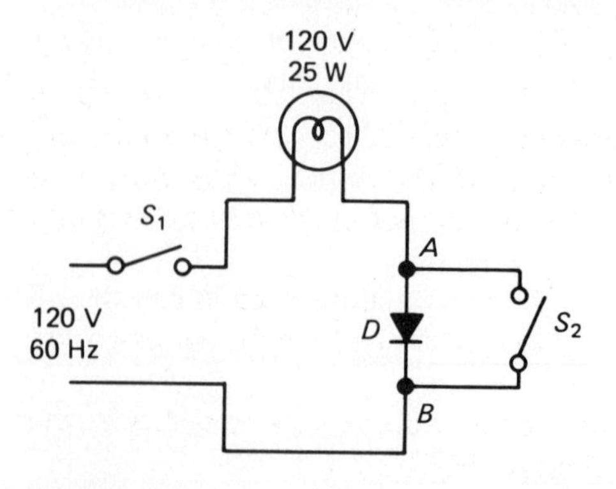

Fig. 32.1-1. Go–no go tests for 200-V silicon rectifiers.

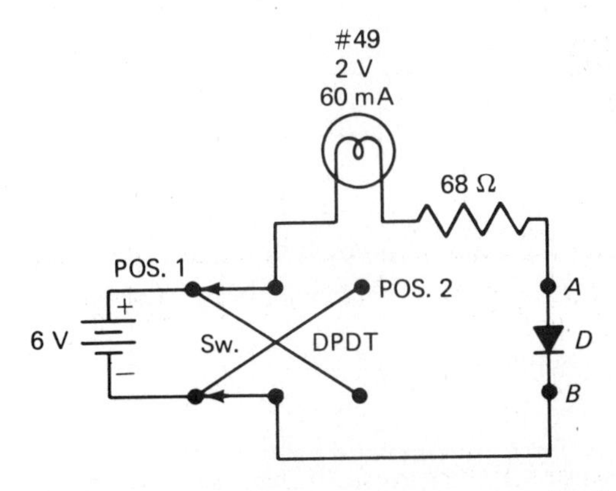

Fig. 32.1-2. Go–no go tests for low-current signal diodes.

source. A double-pole double-throw switch acts as a polarity-reversing switch for the circuit. In position 1 of the switch shown in the figure, the diode will be forward-biased. A good diode D will permit current flow in the circuit and the #49 pilot lamp will glow. If the lamp does not light in position 1 of the switch, the diode is *open*.

In position 2 of the switch, a *good* diode will not conduct and the lamp will *not* glow. If the lamp *does* glow in position 2 as well as in position 1, the diode is *shorted* and is defective.

NOTE: In the absence of a reversing switch, the test may still be made by reversing manually the polarity of the 6-V source in the test circuit.

SUMMARY

1. An ohmmeter may be used to make a continuity test of a semiconductor diode. A good diode will measure very low resistance in one position of the meter's leads, very high resistance in the other position of the meter's leads.
2. A semiconductor tester gives a more reliable indication of the condition of a solid-state diode than an ohmmeter. The tester will show whether the diode is good or bad. However, because a diode's circuit-operating voltages may be higher than those of the tester, a diode which checks okay on a tester may break down under circuit operation.
3. Some semiconductor testers identify the anode and cathode terminals of the diode. They may also measure the forward I_F and reverse I_R current of a diode.
4. Some testers will check a diode both in-circuit and out-of-circuit. Others will make only out-of-circuit measurements.
5. A high-voltage rectifier diode may be tested by the go–no go circuit of Fig. 32.1-1.
6. A low-voltage signal diode can be checked with the circuit of Fig. 32.1-2. If the lamp glows in position 1 of the switch, the diode is good. If it does not glow, the diode is bad. If the lamp glows in position 2 of the switch, the diode is defective, unless the diode polarity is reversed.
7. Semiconductor diodes may also be tested with a curve tracer.

SELF-TEST

Check your understanding by answering these questions.

1. In the circuit of Fig. 32.1-1, the lamp glows when S_1 *and* S_2 are closed. This indicates that: (*a*) D is good; (*b*) D is bad; (*c*) D may be good or bad.
2. In Fig. 32.1-1, with S_1 closed and S_2 open, the lamp does not glow. This indicates that D is: (*a*) open; (*b*) shorted; (*c*) good.
3. The test circuit of Fig. 32.1-1 ________ (is/is not) a conclusive test for 120-V silicon rectifiers.
4. In position 1 of Fig. 32.1-2 the lamp glows; in position 2 it does not. We can conclude that diode D ________ (is/is not) good.
5. If in Fig. 32.1-2 the lamp glows in position 2 of the switch but not in position 1, we can conclude that diode D is good but has been reversed in the circuit. ________ (true/false)

MATERIALS REQUIRED

- Power supply: Low-voltage dc source
- Equipment: Semiconductor tester
- Resistors: ½-W 68-Ω
- Semiconductors: Two low-current signal diodes, one of which is defective; two silicon 200-V rectifiers, one of which is defective
- Miscellaneous: Two SPST switches; 1 DPDT switch; 25-W light bulb and socket; #49 pilot light and socket; fused line cord

PROCEDURE

Semiconductor Diode Tester

1. Mark the low-current signal diodes 1 and 2; mark the 200-V rectifiers 3 and 4.
2. With the semiconductor tester determine whether the diodes are good or bad and record their condition in Table 32.1-1.

Diode Go–No Go Tests

3. Connect the test circuit of Fig. 32.1-1 and test the two 200-V silicon rectifiers. These are diodes 3 and

TABLE 32.1-1. Diode Tests

Diode Number		*Condition*
Low-current signal diode	1	
	2	
200-V silicon diode	3	
	4	

TABLE 32.1-2. Go–No Go Tests

Diode Number		*Condition*
200-V silicon diode	3	
	4	
Low-current signal diode	1	
	2	

4 from Table 32.1-1. Indicate in Table 32.1-2 whether the diodes are good or bad.

4. Connect the test circuit of Fig. 32.1-2 and test the two low-current signal diodes 1 and 2. Indicate in Table 32.1-2 whether they are good or bad.

QUESTIONS

1. Compare the results of the semiconductor tester, Table 32.1-1, with those of the go–no go tests. Comment on any unexpected results.
2. How does an in-circuit tester differ from an out-of-circuit tester?

Answers to Self-Test

1. *c*
2. *a*
3. is
4. is
5. true

EXPERIMENT 32.2. Transistor Testing

OBJECTIVES

1. To test transistors with an ohmmeter
2. To test transistors with a commercial semiconductor tester

INTRODUCTORY INFORMATION

Testing Transistors with an Ohmmeter

A conventional ohmmeter may frequently help the technician spot a defective transistor in an out-of-circuit test. Resistance tests are made between (1) emitter and base; (2) collector and base; and (3) collector and emitter. In testing resistance between any two terminals, the ohmmeter leads are first connected in *one* direction. Then they are reversed. In one of these lead positions the resistance between any two transistor terminals should be very high, about 10,000 Ω or more. In the other lead direction, the resistance between the two terminals should be low. Thus between emitter and base and collector and base the forward resistance should be about 100 Ω or less. Between emitter and collector the low-resistance reading should be about 1000 Ω.

NOTE: If a transistor *fails* this test, it is defective. If it passes, it may still be defective. A transistor tester, a curve tracer, or operation of a transistor in a live circuit are better means of testing a transistor than an ohmmeter.

CAUTION: As with diodes, the ohmmeter terminal voltage must not exceed the maximum voltage rating between any two terminals of the transistor, nor must the current delivered by the ohmmeter be higher than the maximum current rating of any junction.

Checking Transistors with a Transistor Tester

Transistor testers are instruments for checking transistors and solid-state diodes. There are three types of transistor testers: (1) quick-check in-circuit checker, (2) service-type tester, and (3) laboratory-standard tester. Each performs a unique function.

The in-circuit tester is primarily concerned with determining whether a transistor which has previously been performing properly in a circuit is still operational. The transistor's ability to amplify is taken as a rough index of its performance. Servicing with this type of tester will indicate to the technician whether the transistor is dead or is still operative. The advantage, of course, is that the transistor does not have to be removed from the circuit.

Service-type transistor testers usually perform three types of checks. They will test (1) the forward current gain, or beta, of a transistor, (2) the base-to-collector leakage current with emitter open (I_{CO}), and (3) short circuits from collector to emitter and base. Leakage tests cannot be made "in circuit" because of shunting by other components.

Measurements with the service-type tester are relative rather than absolute. However, they provide valuable information to the technician in the maintenance of transistorized equipment.

Laboratory-standard transistor testers, or analyzers, are used for measuring transistor parameters dynamically under operating conditions. The readings they give are absolute. Among the important characteristics measured are (1) I_{CBO}, collector current with emitter open (common base), (2) ac beta (common emitter), and (3) R_{in} (input resistance).

Transistor testers contain the necessary controls and switches for making the proper voltage, current, and signal settings. A meter with a calibrated "good" and "bad" scale is found on the front panel.

Transistor testers are designed to check solid-state diodes as well as transistors. There are also testers for checking high-power transistors and rectifiers.

SUMMARY

1. Transistor junctions may be tested with an ohmmeter. In testing resistance between any two terminals, the ohmmeter leads are first connected in one direction across the terminals and then reversed, just as in the case of a diode. In one of the leád positions (when the junction is reverse-biased) the resistance between the two terminals should be high, about 10,000 Ω or more. In the other position of the leads across the terminals, the resistance between the leads should be low.
2. In the forward-bias connection, resistance between emitter and base and collector and base should be about 100 Ω or less. Between emitter and collector the low-resistance reading should be about 1000 Ω.
3. In resistance-testing a diode or transistor, select the proper ranges of the ohmmeter so as not to exceed the maximum junction voltage or current.
4. Transistor testers will check both diodes and transistors. There are in-circuit testers, service-type testers, and laboratory-standard testers.
5. With in-circuit testers the transistor need not be removed from the circuit. The tester gives an indication of good or bad.
6. Service-type testers will, generally, make tests of (*a*) beta, (*b*) I_{CO} (base-to-collector leakage current with emitter open), and (*c*) short circuits from collector to emitter and base. Leakage cannot be measured "in circuit."
7. Laboratory-standard testers measure transistor performance dynamically. They give absolute readings of transistor parameters.
8. Curve tracers are also excellent devices for measuring out-of-circuit transistor performance.

SELF-TEST

Check your understanding by answering these questions.

1. The forward-resistance of the emitter-base junction of a transistor measures 10,000 Ω. The transistor is ________ (defective/nondefective).
2. Service-type transistor testers give relative rather than ________ values of transistor parameters.
3. If a transistor tester shows that a transistor is good, you may be certain that it is good. ________ (true/false)
4. In ohmmeter-testing a good transistor, the forward resistance between collector and base should be about ________ Ω.
5. A transistor tester will check the forward-current gain or ________ of a transistor.
6. Service-type testers usually check the base-to-collector leakage current with emitter open (I_{CO}). ________ (true/false)

MATERIALS REQUIRED

- Equipment: Transistor tester; VOM
- Semiconductors: Three or more transistors, of which one is defective

PROCEDURE

Resistance-Testing a Transistor

1. In Table 32.2-1, list each transistor type.

2. Measure and record in Table 32.2-1, for transistor

TABLE 32.2-1. Transistor Resistance Tests

Transistor			*Resistance, Ω*		
Type	*Number*	*Terminals*	R_R	F_R	*Condition, Good or Bad*
	1	E to B C to B C to E			
	2	E to B C to B C to E			
	3	E to B C to B C to E			

1, the R_R and F_R resistances of the **(a)** E to B (emitter-to-base) junction; **(b)** C to B (collector-to-base) junction; **(c)** E to C (emitter-to-collector).

3. Record in Table 32.2-1 whether your checks in step 1 show the transistor to be good or bad.
4. Repeat steps 1 and 2 for transistors 2 and 3.

Transistor-Tester Checks

5. Read the instructions for using the transistor tester assigned to you.
6. Test the three transistors you used previously and record in Table 32.2-2 whether they test good or bad.
7. Test also, if possible, and record transistor beta and I_{CO}.

TABLE 32.2-2. Transistor-Tester Checks

Transistor		*Condition*		
Type	*Number*	*Good or Bad*	*Beta*	I_{CO}, μA
	1			
	2			
	3			

QUESTIONS

1. How do the ohmmeter tests of the transistors compare with the transistor tester in determining whether a transistor is good or bad? Comment on any unexpected results.
2. Is the beta of a transistor a measure of the transistor's condition? Explain.
3. Is the I_{CO} of a transistor a measure of the transistor's condition? Explain.

Answers to Self-Test

1. defective
2. absolute
3. false
4. 100
5. beta
6. true

CHAPTER 33 COMMON-EMITTER AMPLIFIER CHARACTERISTICS

EXPERIMENT 33.1. Common-Emitter Amplifier Biasing

OBJECTIVES

1. To connect a transistor as a CE ac amplifier using voltage-divider bias
2. To observe the effect of bias stabilization on amplifier operation

INTRODUCTORY INFORMATION

The Transistor as an AC Amplifier

In Exp. 30.1 we learned that base current controlled the collector current in a transistor connected in a CE circuit. We also saw that the increase in collector current was much greater than the increase in base current. This *current gain* of the transistor in the CE connection was designated as beta (β), and β was defined as follows:

$$\beta = \frac{\Delta I_C}{\Delta I_B} \qquad (V_{CE} \text{ constant})$$

The bias conditions for which β was determined were as follows:

1. *The emitter-base junction was forward-biased* (and for this reason β was also defined as the *forward* current gain of the transistor in the CE connection).
2. *The collector-base junction was reverse-biased.*

These are the bias conditions that are used in amplifiers in audio, radio, and television circuits.

Transistor amplifiers are used to amplify direct or alternating currents and/or voltages. In this experiment we shall be concerned with alternating current *and* voltage amplifiers, for the CE amplifier is concerned with both.

Figure 33.1-1 is the circuit diagram of an NPN transistor connected as a grounded-emitter ac amplifier. Two batteries are used, V_{BB} for forward-biasing the base-emitter junction, and V_{CC} for reverse-biasing the collector-emitter junction. The input signal v_{in} is coupled by C_1 to the base. The output signal v_{out} is taken from the collector. R_1 limits the current in the base-emitter circuit. Together with V_{BB}, R_1 determines the base bias current or operating point. For large values of R_1 the base bias current I_B may be determined approximately from the formula

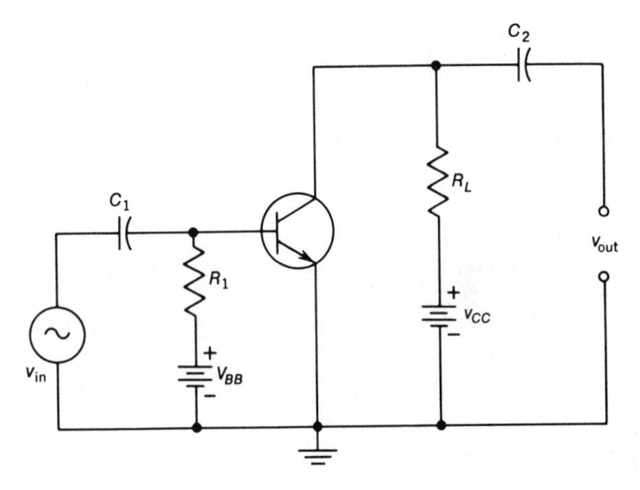

Fig. 33.1-1. NPN transistor connected as a grounded-emitter amplifier.

$$I_B = \frac{V_{BB}}{R_1} \qquad (33.1\text{-}1)$$

The small value of input resistance of the base was not considered in this approximation.

From our knowledge of transistor characteristics, we know that a small increase in base current will result in a large increase in collector current. A small decrease in base current will cause a large decrease in collector current. Currents in the base and collector circuits are therefore in phase. The extent of control of collector current depends on the β of the transistor in the common-emitter circuit.

The input-signal current (i_b) coupled into the base circuit by C_1 combines with the base bias current I_B and causes the base current to increase or decrease. The current waveforms in Fig. 33.1-2 illustrate this condition.

Assume the bias current I_B is $+100\ \mu$A and i_b is a sinusoidal, alternating-signal current. Assume that in

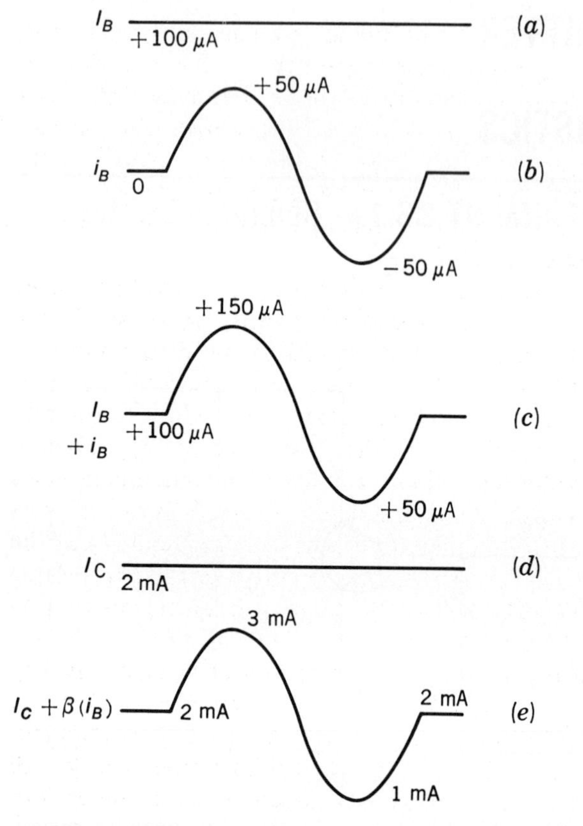

Fig. 33.1-2. Effect of signal current i_b on base current and on collector current in grounded-emitter amplifier.

one complete cycle i_b varies from 0 to $+50\ \mu A$, back to 0, then to $-50\ \mu A$, and back to 0. The result of coupling the signal into the base is to cause a sinusoidal base current to flow which is equal to the sum of the bias and signal currents ($I_B + i_b$). For the cycle of input signal described above, the base current will vary, respectively, from $+100\ \mu A$ to $+150\ \mu A$, back to $+100\ \mu A$, then to $+50\ \mu A$, and back to $+100\ \mu A$.

The resulting sinusoidal variations of current in the base circuit cause in-phase sinusoidal variations of collector current if the amplifier is operated over the linear portion of the transistor characteristic. A numerical example will illustrate this point. Assume that the collector current I_C is 2 mA when $I_B = 100\ \mu A$. For the cycle of signal current i_b in Fig. 33.1-2, collector current $I_C + \beta(i_b)$ will vary, respectively, say from 2 to 3 mA, back to 2 mA, then to 1 mA, and back to 2 mA. In this circuit then, a 100-μA change of signal current in the base has caused a 2-mA change of collector current, for a current gain β of 20.

What is the effect of operating the amplifier over the *nonlinear portion* of its characteristic? Assume that in the NPN transistor collector current I_C varies with base-to-emitter volts V_{BE} in the manner shown by the characteristic curve (Fig. 33.1-3). This graph

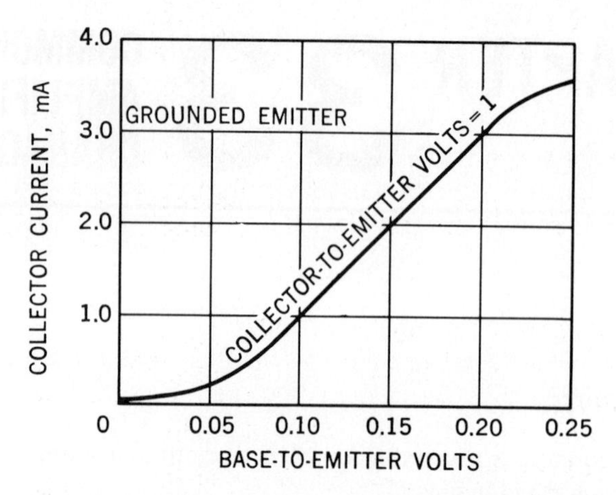

Fig. 33.1-3. Characteristic showing I_C as a function of V_{BE} for an NPN transistor.

shows that linear changes in collector current occur only in the range of 0.1 to 0.2 V base-to-emitter. If 0.15 V is maintained as the base-to-emitter voltage operating point, a signal swing of 0.1 V peak-to-peak in the base will be amplified without distortion. If a larger signal is coupled into the base (that is, if the amplifier is overdriven), nonlinear operation will occur, resulting in a flattening of the signal on the peaks, as in Fig. 33.1-4. In this case, the peak of the positive alternation drives the collector into the nonlinear portion of its characteristic. The peak of the negative alternation drives the collector to cutoff.

If a PNP transistor is used in this grounded-emitter configuration instead of the NPN, the effect of signal polarities is just the opposite. For the PNP transistor, both i_b and i_c would decrease on the positive alterna-

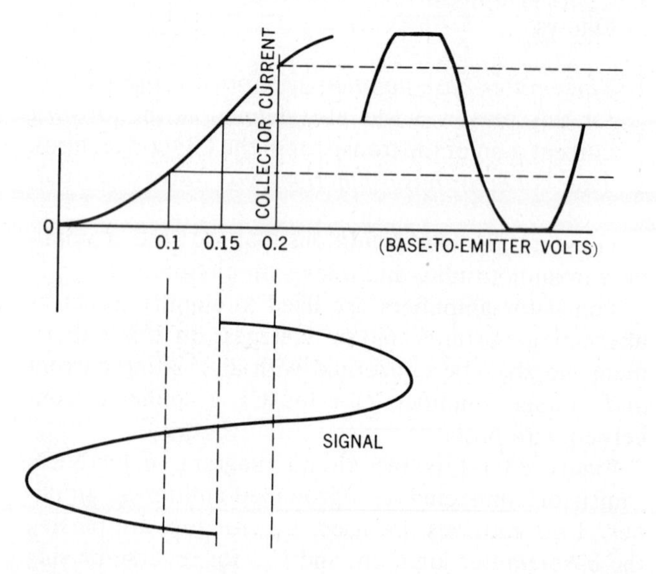

Fig. 33.1-4. Effect of overdrive on collector (output) current waveform.

tions of the input signal and increase on the negative alternations.

As a result of the input-signal current i_b, an amplified-signal current i_c flows in the collector circuit. The output-signal voltage v_{out} in the collector is equal to the product of i_c and R_L. That is,

$$v_{out} = i_c \times R_L$$

Here R_L is the collector load resistor.

If the base signal current i_b is sinusoidal and if the amplifier has been so designed that it will not distort the signal current in the output, the output voltage v_{out} is sinusoidal.

The transistor in Fig. 33.1-1 is a current amplifier. The signal *voltage gain* which results is dependent on the current gain of the transistor, on the circuit arrangement, and on the circuit parameters.

Biasing Methods and Stabilization

For an amplifier to provide *distortionless amplification,* the base must be biased properly so that the input signal operates over the linear portion of the transistor's characteristic. How a transistor is biased will therefore determine the output signal it will produce for a given level input signal.

The CE amplifier can be base-biased from a *single* power source, rather than from the two sources shown in Fig. 33.1-1. This characteristic—plus the fact that compared with a common-base (CB) amplifier, the CE amplifier has a higher input impedance, a lower output impedance, and provides current, voltage, *and* power gain—makes it the most widely used *junction* transistor amplifier.

The circuit of Fig. 33.1-5 illustrates one method of biasing a CE amplifier by using a common power supply V_{CC}. This arrangement is possible because both the base and collector of the PNP transistor must receive a dc voltage which is negative with respect to the emitter. The dc voltage divider consisting of R_1 and R_{BE} (the base-to-emitter resistance), drawing current from the battery V_{CC}, determines the base bias (current). A common battery may also be used with an NPN transistor, but here the polarity of the battery must be reversed, since the collector and base must receive a positive voltage with respect to the emitter.

The circuit of Fig. 33.1-5 is not practical for biasing the CE amplifier, and it is not used. The reason is that the *stability* of the grounded-emitter amplifier is greatly affected by temperature changes. Also, variations in transistor characteristics would affect amplifier performance. In the circuit shown, variations in transistor temperature and in transistor characteristics will change the operating point (the level of base-bias current), affecting in turn the gain and stability of the amplifier. Transistor runaway may occur from changes in operating temperature, resulting in the destruction of the transistor.

In order to protect the transistor against runaway and in order to stabilize the amplifier, a form of compensation is used which is known as *bias stabilization.* The circuit of Fig. 33.1-6 illustrates one type of stabilization. Resistor R_3 is connected in the emitter leg. The effect of this resistor is to offset or compensate for any slight increase in collector current due to an increase in operating temperature or to variations in transistor characteristics. Bias stabilization works in this manner. An increase in collector current will cause a larger voltage drop across R_3, with the polarity shown. This increased voltage developed across R_3 is in opposition to the forward bias in the base-emitter circuit and reduces base current, which in turn reduces collector current, balancing the tendency of the collector current to increase in value. The effect of R_3, therefore, is to provide negative dc feedback, or degeneration, to counteract the instability of the transistor. We shall study the effect of C_3 in the next experiment.

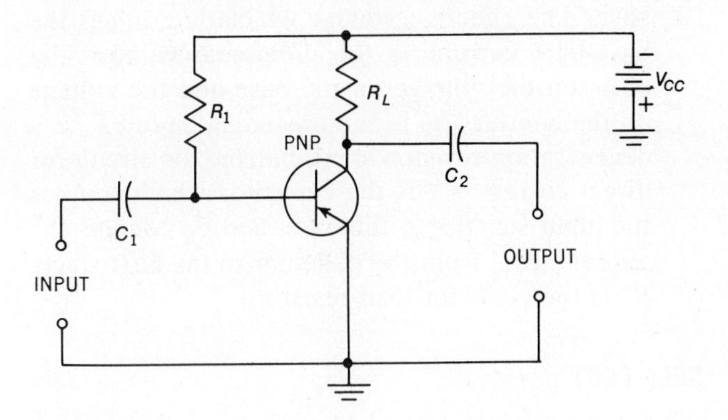

Fig. 33.1-5. Grounded-emitter amplifier using common power supply.

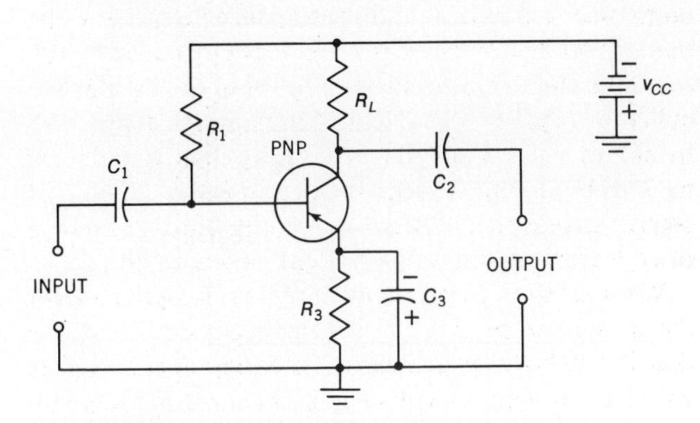

Fig. 33.1-6. Grounded-emitter amplifier with dc stabilization.

Voltage-Divider Bias

The most frequently used of all base-bias circuits is voltage-divider bias, illustrated in Fig. 33.1-7. Because an NPN transistor is connected here, a positive common supply V_{CC} is used. R_1 and R_2 make up a voltage divider on the base. Because the base is made *positive* by this voltage divider, there is emitter-base and emitter-collector current flow. The polarity of voltage developed across emitter resistor R_3 is shown in Fig. 33.1-7. The bias voltage, which determines the base-bias current, is the difference between the base and emitter voltages.

The other parts in Fig. 33.1-7 work in the same way as in Fig. 33.1-6. Thus C_1 couples the input signal to the base. C_2 connects the output signal to the stage which follows, not shown here. R_3 provides dc degeneration and compensates for current changes due to temperature and other changes. In a properly designed amplifier, the emitter resistor R_3 will compensate both for temperature changes and for variations in transistor β, for transistors with the same 2N number do have a range of β. C_3 bypasses R_3 for the ac frequencies which the circuit must handle. R_L is the collector load resistor.

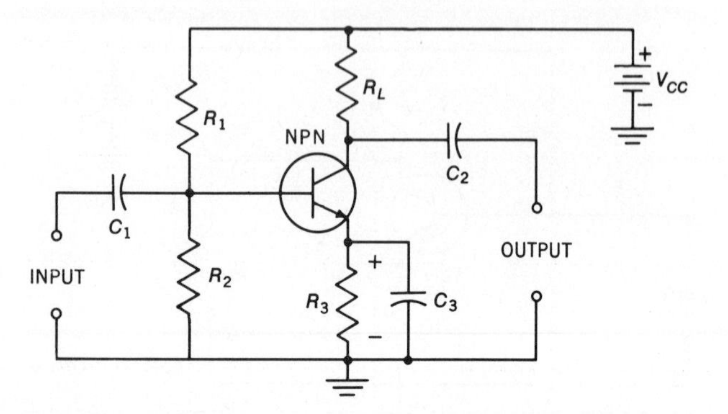

Fig. 33.1-7. Grounded-emitter amplifier using voltage-divider bias with dc stabilization.

SUMMARY

1. Transistors connected as CE amplifiers are used in audio, radio, and television, and in other applications.
2. The bias conditions for a CE distortionless amplifier are: (*a*) Emitter-base junction is forward-biased. (*b*) Collector-base junction is reverse-biased.
3. The base-bias current I_B in a CE amplifier is the direct current in the base which establishes the operating point for the circuit and determines the steady-state collector current I_C.
4. An ac signal i_b injected into the base alternately increases and decreases the base current, as in Fig. 33.1-2. That is, the base current is the algebraic sum of I_B and i_b.
5. The amplifier collector current corresponding to the base signal current is i_c, and the total collector current is $I_C + \beta \times i_b$ (Fig. 33.1-2).
6. As long as the amplifier is operated over the linear portion of its characteristic, 0.1 to 0.2 V V_{BE}, as in Fig. 33.1-3, the output signal is a faithful reproduction of the input. If the amplifier is overdriven by too high an input signal, as in Fig. 33.1-4, or is operated over the nonlinear portion of its characteristic (below 0.1 or above 0.2 V), the output signal is distorted; that is, its waveshape is different from the input.
7. A practical CE amplifier uses a *single power source* to bias the base-to-emitter junction and the base-to-collector junction. The circuit in Fig. 33.1-5 is not a practical biasing circuit because it does not compensate for the thermal (heat) sensitivity of a transistor which causes thermal runaway. Thermal runaway results when a transistor heats owing to current flow, increasing transistor current, causing more heat, more current, more heat, etc., until the transistor is destroyed.
8. A more practical base-bias circuit is that in Fig. 33.1-6 where emitter resistor R_3 provides negative feedback or degeneration to overcome the tendency of the transistor to increase its current as its heat increases.
9. R_3 in Fig. 33.1-6 also stabilizes the circuit for different values of transistor β. It should be understood that the β of transistors of the same type (that is, the same JEDEC number) may vary widely. R_3 helps make the gain of the circuit fairly constant over this range of variation of β.
10. An even more practical and more widely used base-biasing circuit is that in Fig. 33.1-7. R_1 and R_2 provide voltage-divider bias for the transistor. R_1, R_2, and V_{CC} determine the dc voltage on the base. The effective voltage which determines the base-bias current is the difference voltage V_{BE} between the voltage on the base and the voltage on the emitter. As in the preceding circuit R_3 is a degeneration resistor for stabilizing the circuit for direct current. C_1 is the capacitor which couples the input signal into the base, and C_2 couples the *output* signal from the collector to the next stage. R_L is the collector load resistor.

SELF-TEST

Check your understanding by answering these questions.

1. The B-to-E junction of a CE-connected transistor must be ________ biased and the C-to-E junction must be ________ biased if the transistor is to serve as a distortion-free amplifier.
2. In addition to the bias requirements, a CE-connected amplifier must be operated over the ________ ________ portion of its characteristic for distortion-free amplification.
3. The most popular bias arrangement for a CE amplifier is ________ ________ bias.
4. If R_1 and R_2 in Fig. 33.1-7 are, respectively, 22 kΩ and 4.7 kΩ and V_{CC} is +10 V, the dc voltage on the base will be ________ V approximately.
5. In an audio amplifier the collector to base must be ________ (forward-/reverse-) biased.

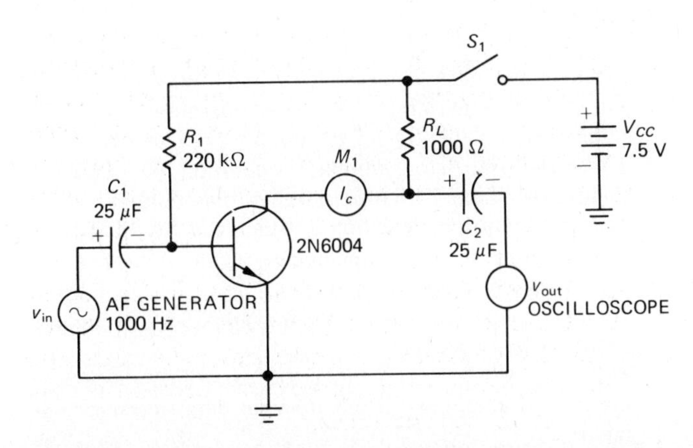

Fig. 33.1-8. Experimental grounded-emitter amplifier without bias stabilization.

MATERIALS REQUIRED

- Power supply: Variable low-voltage dc source
- Equipment: Oscilloscope; multirange milliammeter or VOM; EVM; AF sine-wave generator
- Resistors: ½-W 560-, 1000-Ω; 8.2-, 18-, 220-kΩ
- Capacitors: Two 25-μF/50-V, 100-μF/50-V
- Semiconductors: 2N6004 or the equivalent with appropriate mounting
- Miscellaneous: SPST switch; 120-V/60-W light bulb with socket and line cord

PROCEDURE

Transistor as CE AC Amplifier

1. Connect the circuit of Fig. 33.1-8. Set output of AF sine-wave generator, v_{in} at 10 mV p-p, 1000 Hz. M_1 is a dc milliammeter. The oscilloscope is connected to observe the amplified signal v_{out}.
2. Observe, measure, and record v_{in} and v_{out} in Table 33.1-1. Also draw their waveforms.
3. Measure and record the dc collector current I_C and the dc voltages V_{BE} and V_{CE}.

Bias Stabilization

4. Increase AF generator output until v_{out} distorts. Now reduce generator signal for the maximum undistorted output v_{out}.
5. Observe, measure, and record v_{in} and v_{out} in

TABLE 33.1-1. Transistor Connected as a CE AC Amplifier

Step	I_C, mA	Voltage, V		Waveform			
		V_{BE}	V_{CE}	v_{in}	V p-p	v_{out}	V p-p
1-3							
4-5							
6							
8							
9							

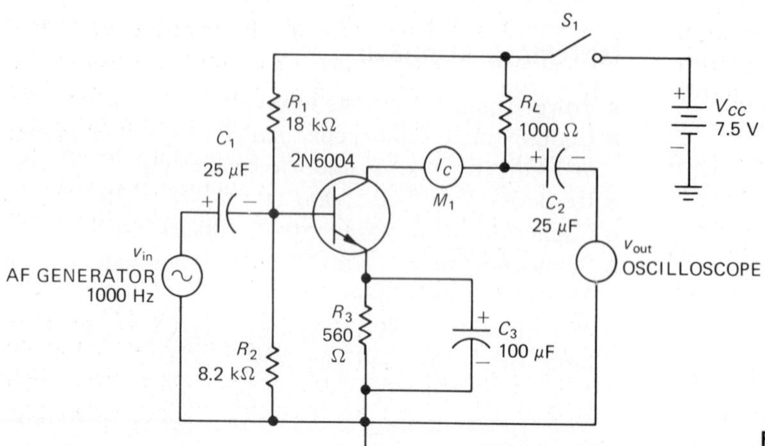

Fig. 33.1-9. Experimental grounded-emitter amplifier with bias stabilization and voltage-divider bias.

Table 33.1-1. Also draw their waveforms. Measure and record I_C, V_{BE}, and V_{CE}.

6. Heat the transistor by holding a lighted 60-W light bulb close to it for about 2 min (until waveform v_{out} distorts). Observe and record the waveform v_{out} and its peak-to-peak voltage. Measure also and record I_C, V_{BE}, and V_{CE}.
7. **Power off.** Change the circuit to that in Fig. 33.1-9. Check all circuit connections. Set AF generator to *minimum* output.
8. **Power on.** Adjust generator output v_{in} for the amplifier's *maximum undistorted* output v_{out}. In Table 33.1-1 draw the input and output waveforms. Measure and record their peak-to-peak amplitudes. Also measure and record I_C, V_{BE}, and V_{CE}.
9. Heat the transistor, as in step 6, for about 2 min. Observe, measure, and record v_{out} and draw its waveform. Measure and record I_C, V_{BE}, and V_{CE}.
10. **Power off.**

QUESTIONS

1. What is meant by bias stabilization? Why is it used?
2. How does heat affect transistor operation? Refer to your data in Table 33.1-1 to support your answer.
3. How can transistor amplifier operation be made independent of small changes in operating temperature?
4. Is your answer to question 3 supported by the data in this experiment? Refer specifically to the data in Table 33.1-1 to support your answer.
5. In this experiment how was the amplifier output distorted? List two ways.
6. How can we be certain that the experimental circuits are indeed amplifiers? Refer to your data to prove your answer.

Answers to Self-Test

1. forward; reverse
2. linear
3. voltage divider
4. 1.76
5. reverse

EXPERIMENT 33.2. Common-Emitter Amplifier Voltage Gain

OBJECTIVE

To measure the voltage gain of a CE amplifier (*a*) with an emitter bypass capacitor, and (*b*) with an unbypassed emitter resistor

INTRODUCTORY INFORMATION

Voltage Gain

The grounded-emitter amplifier is a current, voltage, and power amplifier. It is possible to determine experimentally the voltage gain of the amplifier by injecting a measured signal voltage into the input. The output signal voltage is then measured (an oscilloscope or ac voltmeter may be used), and the ratio of output signal to input signal is the required voltage-gain figure. That is,

$$A_v \text{ (voltage gain)} = \frac{v_{out}}{v_{in}} \qquad (33.2\text{-}1)$$

The amplifier must be operated over its linear region during this process.

It is possible to determine experimentally the range of linear operation of this amplifier. An audio signal, say a 1000-Hz sine wave, is injected into the input, and an oscilloscope is used to monitor the output at the collector. The attenuator on the signal generator is set at minimum at the start. The output of the generator is then increased, and the waveform is observed on the oscilloscope. The input signal is measured over the range where no output signal distortion is evident on the oscilloscope. This measurement will give the range of signal input over which no distortion occurs.

Emitter Bypass Capacitor

In Fig. 33.2-1 capacitor C_3, which is connected in parallel with emitter resistor R_3, prevents ac degeneration. If an ac signal were injected into the base of the transistor and C_3 were *not* present, the effect of R_3 would be to provide both dc *and* ac degeneration. For as collector current increased and decreased in step with the ac signal current in the base, an ac signal voltage would be developed across R_3, in phase with the signal voltage in the base. The effective base-to-emitter signal, which is the *difference* between the voltage on the base and the voltage on the emitter, would therefore become lower than the signal voltage on the base. Accordingly the transistor would "see" a *lower* input ac signal voltage, and the output signal in the collector would therefore be lower, thus *reducing the ac gain* of the amplifier.

C_3, connected across R_3, provides another path for ac signal current. If the capacitive reactance of C_3 is very much lower than the resistance of R_3 for the ac signal frequencies which the amplifier must handle, C_3 acts as a low-impedance path for the ac signal current. R_3 is then effectively bypassed, and the amount of ac degeneration is negligible. Signal gain is restored. R_3, however, still acts to provide dc bias stabilization.

A numerical example will illustrate this effect. Suppose at a certain frequency F, $X_{C3} = 100\ \Omega$ and $R_3 = 1000\ \Omega$. Then $^{10}/_{11}$ of the emitter ac signal current will flow through C_3 and only $^{1}/_{11}$ through R_3. The signal voltage developed across R_3 is therefore only $^{1}/_{11}$ of the voltage which would have been developed if C_3 were not present. Bypassing the emitter resistor with C_3, in this instance, appreciably reduces the effects of signal degeneration. However, the direct current through R_3 is not affected.

If an amplifier is required to handle more than one frequency, then the value of the emitter bypass capacitor must be chosen so that it provides adequate bypassing for the *lowest* of all the frequencies. Then it will also be a good bypass for all the higher frequencies.

To determine the size of C_3, it is therefore necessary to know the lowest frequency f_{min} it is required to bypass and the value of the emitter resistor R_3 in parallel with it. C_3 is considered a good bypass if at f_{min} the capacitive reactance of C_3 is one-tenth the resistance of R_3, that is, if

$$X_{C3} = \frac{R_3}{10} \qquad (33.2\text{-}2)$$

SUMMARY

1. The voltage gain A_v of a common-emitter amplifier is determined by measuring the output-signal voltage v_{out} at the collector and the input-signal voltage v_{in} at the base, then substituting the measured values in the formula

$$A_v = \frac{v_{out}}{v_{in}}$$

2. An unbypassed emitter resistor in a CE amplifier will cause both ac and dc degeneration. AC degeneration will result in lower *voltage* gain.
3. To prevent ac degeneration in a CE amplifier, the emitter resistor is bypassed by a capacitor C,

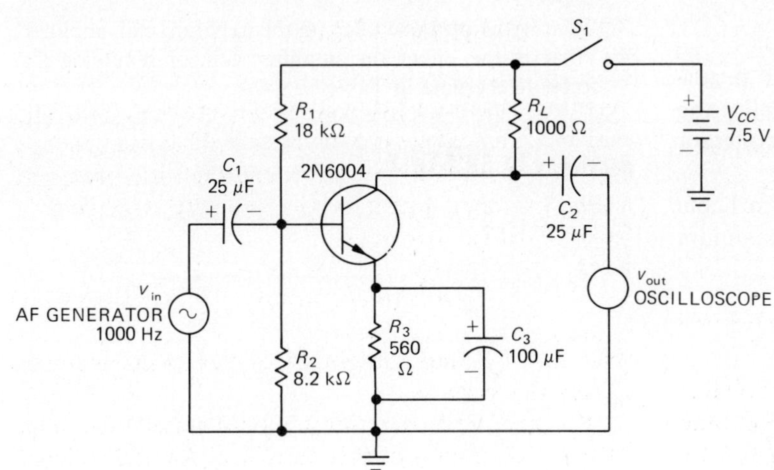

Fig. 33.2-1. Experimental grounded-emitter amplifier.

whose reactance X_C, at the lowest frequency the amplifier must handle, must be equal to or less than $\frac{1}{10}$ the value of the emitter resistor R. That is,

$$X_C \leqq \frac{R}{10}$$

4. The bypass capacitor does not affect the dc stabilization of the amplifier.

SELF-TEST

Check your understanding by answering these questions.

1. The ac signal voltage measured at the base of a CE amplifier is 0.1 V. The output signal measured at the collector is 10 V. The voltage gain of the amplifier is ________.
2. In the circuit of Fig. 33.2-1, what must be the lowest value of X_{C3} which will still permit C_3 to act as a good bypass? ________
3. What is the lowest frequency which meets the requirement in question 2? ________ Hz.

MATERIALS REQUIRED

- Power supply: Variable low-voltage dc source
- Equipment: Oscilloscope; EVM; AF sine-wave generator
- Resistors: ½-W 560-, 1000-Ω; 8.2, 18-, 220-kΩ
- Capacitors: Two 25-μF/50-V, 100-μF/50-V
- Semiconductors: 2N6004 or the equivalent, properly mounted
- Miscellaneous: SPST switch

PROCEDURE

Voltage Gain

1. Connect the circuit of Fig. 33.2-1. The AF generator, set at 1000 Hz, low-signal level, supplies the input signal v_{in}. An oscilloscope at the collector monitors the output v_{out}.
2. **Power on.** Determine, measure, and record in Table 33.2-1 the input-signal voltage for the maximum undistorted output signal.

TABLE 33.2-1. CE Amplifier Voltage Gain, Emitter Bypassed

Condition	$v_{in,p-p}$	$v_{out,p-p}$	*Gain*
Maximum undistorted signal			
One-half level in step 2			

TABLE 33.2-2. CE Amplifier Voltage Gain, Emitter Unbypassed

Condition	$v_{in,p-p}$	$v_{out,p-p}$	*Gain*
Maximum undistorted signal			

3. Measure also and record the maximum undistorted output-signal voltage.
4. Set the output of the signal generator to one-half the level in step 2. Measure and record v_{in} and v_{out}.
5. Compute and record the voltage gain of the amplifier for the signal levels in steps 2, 3, and 4.

Effect of Emitter Bypass Capacitor

6. Remove emitter bypass capacitor C_3. Determine, measure, and record in Table 33.2-2 the maximum input-signal voltage for which the amplifier output remains undistorted.
7. Compute and record the amplifier gain with the emitter unbypassed.

QUESTIONS

1. What is meant by voltage gain of an amplifier?
2. Is the gain of the CE amplifier constant in the range where the output signal remains undistorted? Refer specifically to the data in Table 33.2-1 to support your answer.
3. What is the purpose of C_3 in the experimental amplifier?
4. What is the effect on amplifier gain of removing C_3? Refer to the data in Table 33.2-2.

Answers to Self-Test

1. 100
2. 56
3. 28

CHAPTER 34 TROUBLESHOOTING A COMMON-EMITTER AMPLIFIER

EXPERIMENT 34.1. Troubleshooting a CE Amplifier I

OBJECTIVE

To make a dynamic test of an amplifier which will determine whether it is or is not operating normally

INTRODUCTORY INFORMATION

Troubleshooting Electronic Devices

Modern electronic devices are made up of arrangements of individual electronic circuits, called *stages*. There may be a few stages or very many, depending on how complex the device is. For example, a simple radio receiver may have from 7 to 10 stages. A color television receiver, on the other hand, may have from 50 to 100 stages, and each stage can consist of five or more components. It is clear then that troubleshooting a defective complex electronic device can be very time-consuming and difficult unless some logical procedure is used to localize the defective stage and then find the bad part in it.

The most suitable time to consider a complete, logical, troubleshooting procedure will come when you study a complete device. For the moment, let us assume that, following a logical troubleshooting procedure, the trouble has been isolated to a section of a receiver which has several stages, of which one is a common-emitter linear amplifier. How can the amplifier be tested to determine whether it is or is not operating normally? If it is normal, the trouble must lie in the other stages of the suspected section.

Dynamic Test of a CE Amplifier

As used in this experiment an ac *amplifier* will be considered a device which develops, *without distortion,* a larger signal voltage in the output than it receives in the input. This basic definition of an amplifier suggests a means of testing its performance dynamically. Consider the CE amplifier in Fig. 34.1-1. Assume that the voltage gain of this amplifier is 50 and that it operates as a linear amplifier (that is, it introduces no distortion) for a signal input in the range 10 to 100 mV. Then if a 50-mV sine wave is injected into the input terminals, a 2.5-V undistorted sine wave may be expected in the output. This is a dynamic test for the amplifier. If the amplifier passes this signal-injection and signal-tracing test, that is, if the undistorted output signal is 2.5 V for a 50-mV input, it may be assumed that every component in Fig. 34.1-1, including the power source, is operating satisfactorily.

If the exact voltage gain of the amplifier is not known, and if stage gain is not critical in trouble-

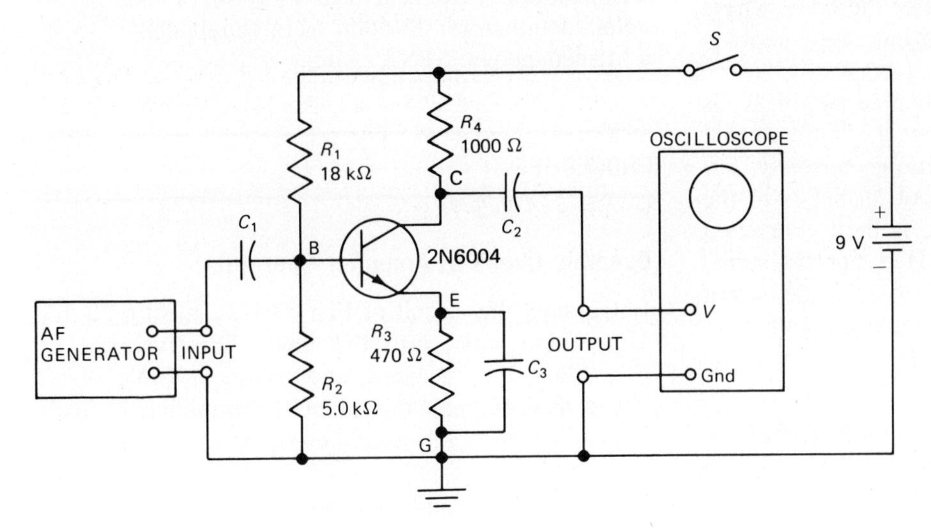

Fig. 34.1-1. Equipment hookup for dynamic test of a CE amplifier.

shooting a specific defect, the signal-injection and signal-tracing techniques can still be satisfactorily used. However all that the technician would be concerned with would be *distortionless amplification* of the stage, without specifying the stage gain.

The instruments used in this test are a sine-wave signal generator operating in the frequency range of the amplifier and an oscilloscope, used to observe the signal waveforms and to measure the input- and output-signal voltages.

Checking for an Open Input or Output Capacitor

In the example above, it was assumed that the stage checked out perfectly. Suppose, however, that there is no signal at the output terminals of the amplifier for a specified signal input. It is still possible, using the dynamic signal-tracing method, to determine in the case of Fig. 34.1-1 whether capacitors C_1 and C_2 are *open*. The procedure is as follows. A sine-wave signal, no larger than the amplifier can handle, is injected into the input terminals of the amplifier and observed at these terminals with an oscilloscope. If the observed signal is normal, the oscilloscope probe is moved to the base (point B) of the amplifier. The sine-wave signal at this point should be approximately the same as at the input terminals if the amplifier input is normal. If there is no signal at the base, two possibilities exist. The first is that capacitor C_1 is *open*. The second is that the base terminal is shorted to ground (because of a defective transistor or other component in the base circuit). An open capacitor may readily be found by connecting the "hot" generator lead directly to the base of the transistor and observing with an oscilloscope the output signal. If a normal output signal now appears, it indicates that capacitor C_1 is open. If no signal appears at the output terminals of the amplifier, the oscilloscope probe is connected to the base. An absence of signal at the base, where the generator is connected, would indicate a short in the base circuit.

We may also determine whether capacitor C_2 is open by signal tracing. Assume that the input circuit, including C_1, is found to be operating normally, but that there is no signal at the output terminals of the amplifier. The oscilloscope probe is then connected directly to the collector of Q. If a normal signal appears at the collector, but none exists at the output terminals, we have direct evidence that C_2 is open.

SUMMARY

1. In troubleshooting electronic devices the technician follows a logical procedure to isolate the trouble to a section of the receiver.
2. Then, using a signal-substitution and a signal-tracing method, he localizes the trouble to a stage and tests that stage to find the defective part.
3. A dynamic test of a CE amplifier is made by injecting a low-level sine wave into the input and observing the output waveform. If the output waveform is a sine wave, and the stage shows gain, the stage may be assumed to be working properly.
4. If in a dynamic test of a CE amplifier the output is distorted, or there is no output, there is trouble in that stage.
5. It is possible to find open coupling capacitors in the input and output of a CE amplifier by using the signal-substitution and signal-tracing methods.

SELF-TEST

Check your understanding by answering these questions.

1. A dynamic check of a CE amplifier requires these test instruments: (*a*) ________ (*b*) ________.
2. A CE amplifier has a voltage gain of 100. A 0.1-V p-p sine wave is brought into the input. The output signal should be a ________, ________ V p-p, if the amplifier is okay.
3. Assume that a dynamic test of the amplifier in Fig. 34.1-1 shows no output. A check at the collector of the transistor does show the proper waveform. We can conclude that ________ is defective.

MATERIALS REQUIRED

- Power supply: Variable, regulated low-voltage dc source
- Equipment: Oscilloscope, EVM, AF sine-wave generator
- Resistors: ½-W 560-, 1000-, 8200-, 18,000-Ω
- Capacitors: Two 25-μF/50-V; 100-μF/50-V
- Semiconductors: 2N6004 or the equivalent
- Miscellaneous: SPST switch

PROCEDURE

Dynamic Check of Amplifier Operation

1. Connect the circuit of Fig. 34.1-2. Inject a 20-mV p-p sine wave into the input. Observe with an oscilloscope, measure, and record in Table 34.1-1 the peak-to-peak output of the amplifier. Indicate also whether the amplifier operation is or is not normal.
2. Observe, measure, and record in Table 34.1-1 the

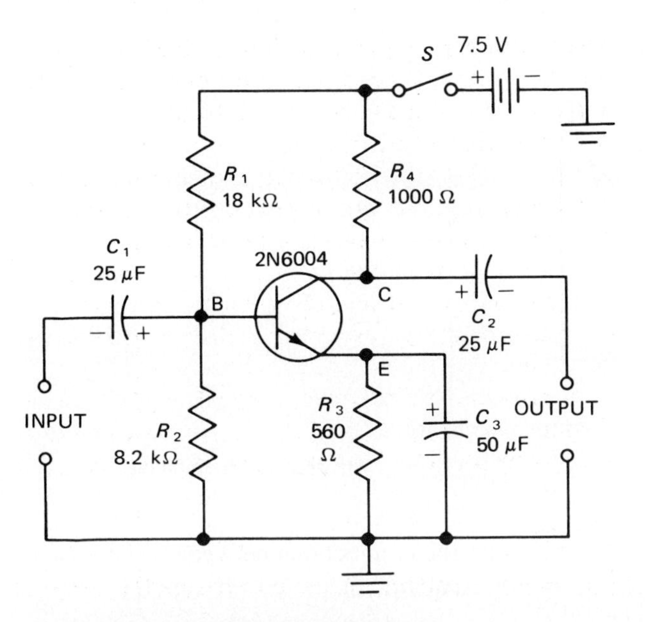

Fig. 34.1-2. Experimental CE amplifier.

peak-to-peak voltage of the signal at the base and collector of the amplifier.

3. With the generator still connected to the input of the amplifier, remove C_1. Observe, measure, and record in Table 34.1-1 the voltage of the waveform at the input, base, collector, and output of the amplifier. Is amplifier operation normal?
4. Replace C_1 and remove C_2 from the circuit. Observe, measure, and record the voltage of the waveform at the input, base, collector, and output. Is amplifier operation normal?

TABLE 34.1-1. Dynamic Checks

Step	*Condition*	V p-p				*Amplifier Operation*
		Input	*Base*	*Collector*	*Output*	
1,2	Normal					
3	C_1 open					
4	C_2 open					

QUESTIONS

1. What is the purpose of a dynamic test of an amplifier's operation?
2. Explain how signal tracing can be used to isolate trouble in a three-stage transistor amplifier to a specific stage. Assume the coupling between stages is capacitive.
3. What is the gain of the experimental amplifier?

Answers to Self-Test

1. AF generator; oscilloscope
2. sine wave; 10
3. C_2

EXPERIMENT 34.2. Troubleshooting a CE Amplifier II

OBJECTIVE

To consider dc voltage norms at test points in an amplifier which is operating normally and to draw inferences as to the nature of the trouble from voltage measurements in an amplifier which is defective

INTRODUCTORY INFORMATION

DC Voltage Norms

The dynamic tests of an ac amplifier, Fig. 34.2-1, check the ac operation of the amplifier and determine whether the amplifier is or is not operating properly. If no signal appears in the output, and the input and

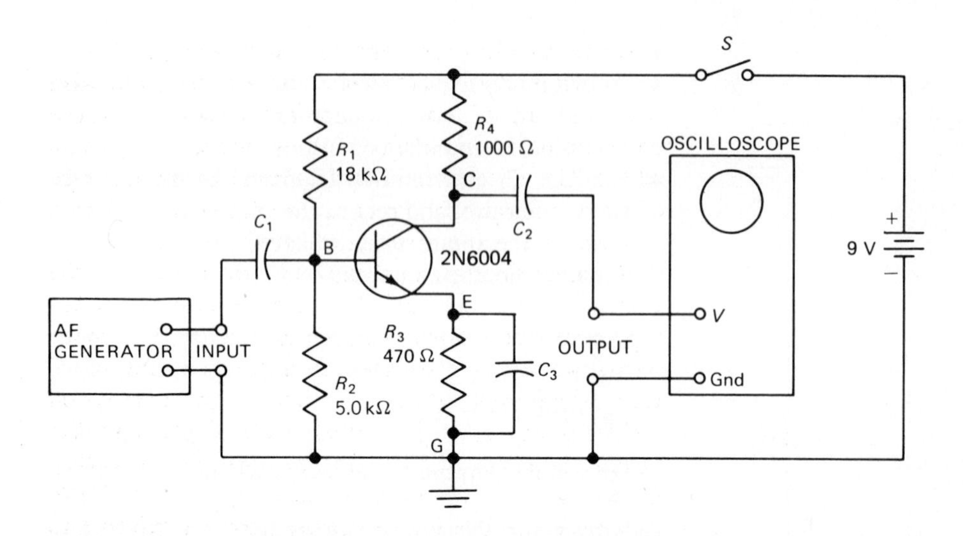

Fig. 34.2-1. CE amplifier.

output capacitors are found to be functioning properly, then the dc operation of the amplifier must be analyzed.

Figure 34.2-2 is the dc equivalent of the amplifier in Fig. 34.2-1. What steps should the technician take to determine which of the components or connections is defective? The first test is to determine whether the transistor is good. A transistor tester may be used to check the transistor. *Power is turned off,* and the suspected transistor is tested in the circuit or by removal from the circuit. If it is found defective, it must be replaced. If it is good, the next check is in order.

NOTE: If a transistor tester is not available, a substitution test is used. A known good transistor, with the same manufacturer's replacement number or JEDEC number, is substituted for the suspected one. A dynamic signal injection and oscilloscope test with power ON will now determine whether the transistor amplifier is operating normally. If it is, then the old transistor is defective and must be replaced. If the circuit is still not operating, the trouble points to one of the resistors, the connections between resistors and the transistor elements, or the power supply.

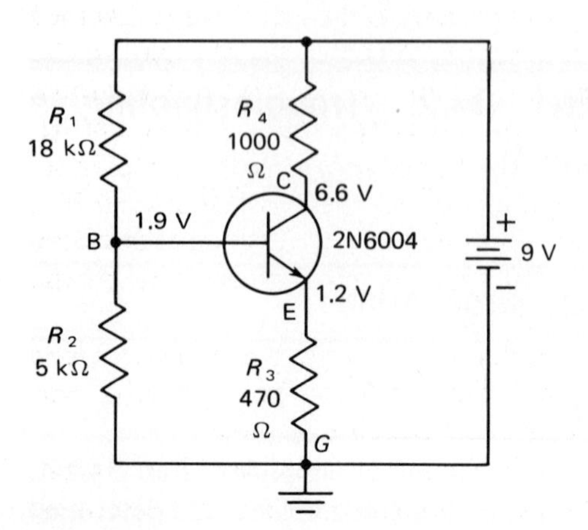

Fig. 34.2-2. DC equivalent of the amplifier in Fig. 34.2-1. Voltages shown are for normal operation

DC Voltage Measurements

The next series of checks involves dc measurements of the power supply and at the transistor elements. But for these tests to be meaningful, the technician must know what voltages to expect in an identical circuit whose operation is normal. The voltages shown in Fig. 34.2-2 are for normal operation of this amplifier. Thus the power supply is 9 V. There is 1.2 V at the emitter, 1.9 V at the base, and 6.6 V at the collector. These voltages are all with respect to common, G. What is significant about these voltages?

The fact that 1.2 V is developed across the emitter resistor shows that *there is current in the emitter.*

The fact that the 1.9 V at the base is *lower* than the voltage which might be expected from the voltage-divider action of R_1 and R_2 above shows that there is base current. If there were no base current, that is, if the connection from the junction of R_1 and R_2 to the base were effectively open, the voltage measured across R_2 (V_{BG}) would be:

$$V_{BG} = \frac{R_2}{R_1 + R_2} \times 9 = \frac{5000}{23{,}000} \times 9 = 1.96 \text{ V}$$

The fact that the measured voltage $V_{BG} = 1.9$ V suggests that the effective resistance from point B to G must be less than R_2. Therefore there must be a current path from base to ground other than R_2. This other path is the emitter-base resistance (since there *is* base current) R_{EB}, in series with R_3, as in Fig. 34.2-3. The branch $R_{EB} + R_3$ is in parallel with R_2, and that is why the measured voltage V_{BG} is lower than the 1.96 V calculated above.

The fact that 6.6 V is measured at the collector in Fig. 34.2-2 shows that there is enough collector cur-

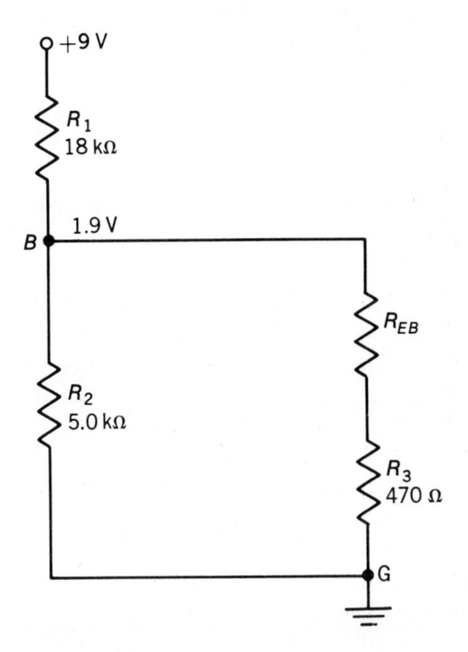

Fig. 34.2-3. Equivalent voltage-divider circuit which determines base voltage.

rent to cause a 2.4-V drop across R_4, the collector resistor.

The measured voltages in Fig. 34.2-2 therefore show that the supply voltage is normal, that there is emitter current, collector current, and base current. It is possible from the measured values and the known values of emitter and collector resistors to compute the direct currents in the emitter I_E, collector I_C, and base I_B. For

$$I_E = \frac{V_E}{R_E} = \frac{1.2}{470} = 2.55 \text{ mA}$$

$$I_C = \frac{V_C}{R_C} = \frac{2.4}{1000} = 2.40 \text{ mA}$$

and $I_B = I_E - I_C = 2.55 - 2.40 = 0.15$ mA

These results are consistent with our experience with transistor currents, for we demonstrated in a previous experiment that I_E is greater than I_C and that $I_B = I_E - I_C$ in a CE amplifier.

Let us analyze the voltages measured at the transistor elements in Fig. 34.2-2 from another point of view. We know that the emitter is 0.7 V (1.9 − 1.2) more negative than the base; that is, the *forward bias* on the emitter to base is 0.7 V. Moreover, the collector is 4.7 V (6.6 − 1.9) more positive than the base; that is, the reverse bias on the collector to base is 4.7 V. These facts are consistent with the biasing requirements of a CE amplifier.

The voltages shown in Fig. 34.2-2 are normal for that amplifier using a silicon transistor for the circuit parameters shown. But amplifier parameters and transistor types will differ, and the manufacturer of transistor equipment will usually provide the technician with a chart of voltages for the circuit measured under normal operating conditions, or the voltage data will be provided on circuit diagrams. In the absence of such information, these norms are suggested as a guide for a *small-signal linear amplifier*.

Voltages are usually measured with respect to the common return (ground). The emitter-base of a transistor amplifier should be forward-biased. The range of bias which may be expected depends on the signal level which an amplifier is designed to handle and on the transistor used. In a small-signal linear amplifier using silicon transistors, the bias voltage may vary from 0.65 to 0.75 V (approximately). In an NPN transistor the emitter should be negative relative to the base; in PNP the emitter should be positive with respect to the base.

The collector-base junction of a transistor should be reverse-biased. Thus in an NPN amplifier the collector is positive; in a PNP, it is negative relative to the base. The collector-base voltages will depend on circuit design and on the supply voltage, and will range from approximately half the supply voltage to almost the full battery voltage. For example, if a 9-V battery is used, the collector-base voltage range may be from 5 to 8.9 V.

Inferences from DC Voltage Measurements

DC voltage readings are used to draw inferences of proper or improper functioning in transistor circuits. To this end we shall assume certain abnormal voltages in the circuit of Fig. 34.2-2 and analyze the possible causes for these voltages. In all cases assume that the power supply voltage is 9 V.

$V_C =$ 9 V. The full supply voltage, measured at the collector, indicates that the collector circuit is not drawing any current. If the transistor is assumed to be good, the trouble could be (*a*) an open in the emitter circuit (R_3 or any of the connecting wires); (*b*) an open in the base circuit (open connection between the junction of R_1 and R_2 and the base); (*c*) base-emitter short circuit so that there is no forward bias on the emitter-base junction; (*d*) base short-circuited to ground so that there is no forward bias on the emitter-base junction.

$V_C = 0$ V. Possible troubles could include: (*a*) open collector circuit (including R_4 or the connective wiring), (*b*) collector short-circuit to ground.

$V_E = V_C = 2.9$ V. Collector-to-emitter short circuit.

$V_E = 0$ V. This would indicate that (*a*) there is no current flowing in the emitter or (*b*) that the emitter is short-circuited to ground. It would *not* indicate an open emitter circuit, for if there were an open in the emitter circuit, connecting the voltmeter from emitter to ground would complete the circuit, and the meter would record a voltage.

Caution on Voltage Measurements

Associated components or the transistor itself can be damaged during voltage measurements. Care must therefore be exercised to prevent the voltmeter probes from causing short circuits between transistor terminals, or between high-voltage sources and components carrying a low-voltage rating. Moreover, voltmeters with a high-input impedance should be used to prevent circuit loading. For transistor voltage measurements EVMs or 20,000 Ω/V VOMs are preferable to low-input-impedance meters.

SUMMARY

1. The order in which to make troubleshooting checks on a CE amplifier is: (*a*) Dynamic signal-injection and signal-tracing checks are made to determine if the amplifier is normal. (*b*) If the amplifier is not normal, the transistor is tested. (*c*) If the transistor is normal, dc voltage measurements are made with respect to common and compared with norms for that circuit. Abnormal readings may themselves suggest which component is defective. (*d*) Other suspected parts in the external circuit may be further isolated by resistance measurements.
2. In a normally operating small-signal silicon CE amplifier, voltages at the collector may vary from one-half of V_{CC} to a voltage only slightly lower than V_{CC}.
3. Voltages measured at transistor elements can be interpreted to indicate whether there is emitter current, whether there is collector current, whether the base-emitter junction is forward-biased, whether the collector-base junction is reverse-biased.
4. Use an EVM or a high ohms-per-volt meter in making voltage checks.

SELF-TEST

Check your understanding by answering these questions.

1. The voltages measured in Fig. 34.2-1 at the transistor elements with respect to common are as follows: $V_E = 0$ V, $V_C = 9$ V, $V_B = 1.96$ V. The most probable cause of trouble is a ________.
2. In Fig. 34.2-1, the voltage measured at the collector with respect to common is 7.0 V. This is within normal range. ________ (true/false)
3. In Fig. 34.2-1, if the emitter is shorted to the base, $V_C =$ ________ V.
4. In Fig. 34.2-1, V_C measures 9 V, V_B measures 1.96 V, and V_E measures 3 V. The most likely cause of trouble is a(an) ________ ________.

MATERIALS REQUIRED

- Power supply: Variable, regulated low-voltage dc source
- Equipment: Oscilloscope; EVM; AF sine-wave generator; transistor tester
- Resistors: ½-W 560-, 1000-, 8200-, 18,000-Ω
- Capacitors: Two 25-μF/50-V; 100-μF/50-V
- Semiconductors: 2N6004 or the equivalent
- Miscellaneous: SPST switch; defective parts for troubleshooting Fig. 34.2-4

PROCEDURE

DC Voltage Measurements

1. Connect the circuit of Fig. 34.2-4. Estimate and record in Table 34.2-1 the normal dc voltages at the emitter, base, and collector of the transistor. Assume 4 mA of emitter current.
2. Measure with respect to common (ground) the voltages at the emitter, base, and collector. Record your results in Table 34.2-1.
3. **Power off.** Remove R_3, the emitter resistor. Estimate the voltages at the transistor elements for this condition and record these estimated values in Table 34.2-1.

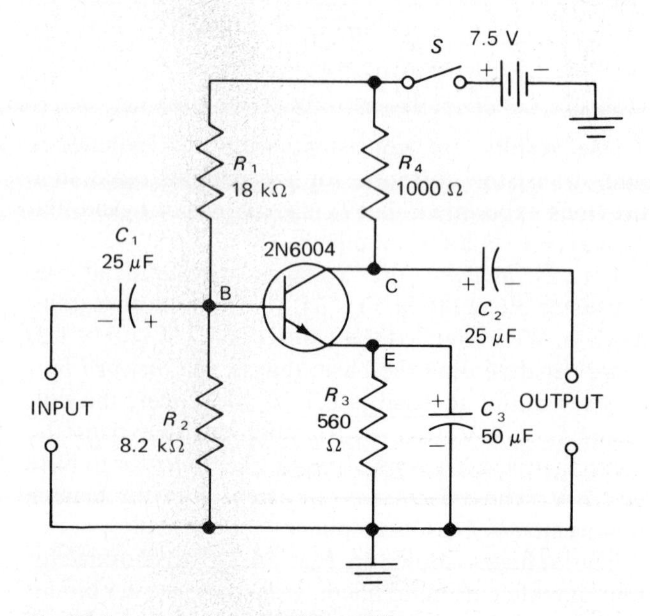

Fig. 34.2-4. Experimental CE amplifier.

TABLE 34.2-1. DC Voltages in CE Amplifier

	Voltage (Normal)		*Voltage (Emitter Open)*		*Voltage (Base Open)*		*Voltage (Collector Open)*	
Element	*Estimated*	*Measured*	*Estimated*	*Measured*	*Estimated*	*Measured*	*Estimated*	*Measured*
Emitter								
Base								
Collector								

4. **Power on.** Measure and record in Table 34.2-1 the voltages at the transistor elements.
5. **Power off.** Replace the emitter resistor. Disconnect the lead from the base to the voltage divider, leaving the base open. Estimate and record in Table 34.2-1 the voltages at the elements of the transistor for this condition.
6. **Power on.** Measure and record the voltages at the elements of the transistor.
7. **Power off.** Reconnect the base to the voltage divider. Remove R_4, the collector resistor, from the circuit. Estimate the voltages at the transistor elements for this condition and record in Table 34.2-1.
8. **Power on.** Measure and record the voltages at the transistor elements.
9. **Power off.** Replace the collector resistor.

Troubleshooting

10. Your instructor will insert a trouble into the experimental amplifier. Troubleshoot the circuit, keeping a step-by-step record of each check you make in the order performed. Record these in the standard troubleshooting report which will be used for all troubleshooting. When you have found the trouble, correct it and notify your instructor, who will put another trouble into the circuit. Service as many troubles as time will permit.

QUESTIONS

1. Refer to Table 34.2-1. Compare the estimated and measured voltage values for the amplifier operating normally. Comment on any unexpected results.
2. Explain the voltages measured at the transistor elements with the collector resistor open, Table 34.2-1.
3. Why is an understanding of the operation of a transistor amplifier helpful in troubleshooting it?

Answers to Self-Test

1. defective transistor
2. true
3. 9
4. open R_3

EXPERIMENT 34.3. Troubleshooting a CE Amplifier III

OBJECTIVE

To determine resistance norms at test points in an amplifier which is operating normally, and troubleshoot the amplifier

INTRODUCTORY INFORMATION

The dynamic test of an ac amplifier checks the ac operation of the amplifier and determines whether the amplifier is or is not operating normally. If the amplifier is defective, it is serviced by: (*a*) Checking the transistor, in or out of the circuit. (*b*) Voltage checks at the elements of the transistor. Inferences drawn from these measurements will suggest the part or parts which are defective. (*c*) Resistance measurements of the suspected component to verify that it is indeed defective.

Resistance Measurements

Resistance measurements in transistor circuits, always made with power OFF, are helpful in determining defective components. However, if a conventional ohmmeter is used, the readings obtained may be highly misleading because the energizing battery in the ohmmeter applies a voltage to the circuit under test which may cause current flow in the transistor. This parallel path through the transistor will affect readings in the external circuit.

Suppose, for example, that it is desired to measure the resistance of R_2, from base to ground, in Fig. 34.3-1. Assume that the ohmmeter M is connected in the circuit so that the positive terminal of the battery is applied to the base and the negative terminal to ground. Since this is the condition for forward bias in the emitter-base junction of this NPN transistor, current flows in this section. Therefore the resistance from emitter to base R_{EB} in series with R_3 is effectively in parallel with R_2 and will affect the resistance measured by the ohmmeter. The reading obtained does not give an accurate measure of R_2 in the external circuit.

To avoid such ambiguous measurements, the *low-power-ohms* function of an EVM or VOM should be used. This function is especially designed to produce such low voltages at the ohmmeter terminals that transistor or diode junctions cannot be forward-biased. Therefore a *low-power-ohms* meter will measure the actual external resistances at selected test points in a transistor circuit, and not parallel transistor or diode junction resistances.

Refer to the circuit of Fig. 34.3-1. What resistances may be expected if measurements are made at the transistor terminals with respect to common G, assuming that the low-power-ohms function of a multimeter is used and that the circuit is operating properly? The resistance measured at the emitter E to G should be 220 Ω, and at the base, point B to G, it should be 10,000 Ω. The resistance from collector C to G would be the sum of R_4, R_1, and R_2, in this case 50,100 Ω. These values, then, would be the standard or norm for the circuit of Fig. 34.3-1 (with switch S open). Measured values would be compared with these norms. Deviations from the norms would suggest defects.

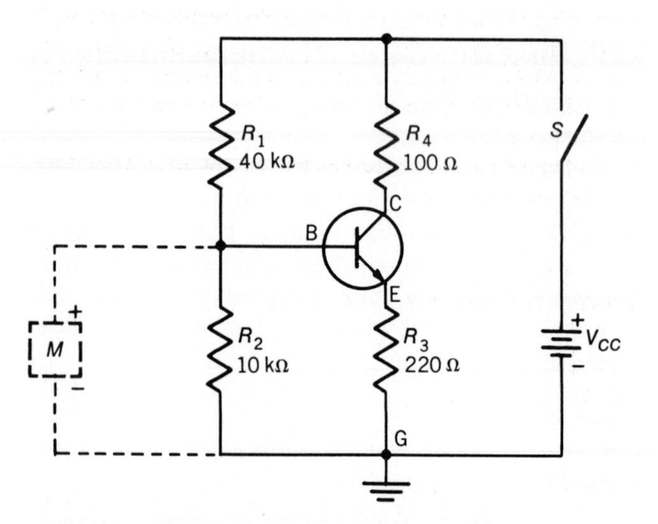

Fig. 34.3-1. Effect of conventional ohmmeter on resistance measurements in external circuit.

The most obvious resistor defects are opens, which can be spotted very easily. For example, if the resistance measured from E to G is infinite (∞), then either R_3 is open, or the connective wiring between the upper terminals of R_3 and the emitter is open, or the wiring between the lower terminal of R_3 and G is open. A resistance check across the terminals of R_3 will reveal whether the resistor is good. If it is, continuity checks of the wiring to E and to G will reveal which is open.

Similarly a resistance check from B and G will indicate whether the resistance in the base circuit is normal (10,000 Ω). If there is an open, further measurements will reveal if R_2 is open, or if the connective wiring from the terminals of R_2 to the base B or to common G is open.

If the resistance from point C to G is infinite and the resistance in the base circuit measures normal, then either R_4 or R_1 or the connective wiring is open. Resistance checks of R_4 and R_1 and continuity checks will reveal where the defect is.

SUMMARY

1. The order in which to make troubleshooting checks on a CE amplifier is: (*a*) *Dynamic* signal-injection and signal-tracing checks are made to determine if the amplifier is normal. (*b*) If the amplifier is not normal, the transistor is tested. (*c*) If the transistor is normal, dc voltage measurements are made at the transistor elements with respect to common and compared with the norms for that amplifier. If they are not normal, the measurements themselves may suggest the defective component. (*d*) A suspected component in the external circuit may then be identified by resistance measurements at the transistor elements, in the external circuit.
2. In making resistance checks, the *low-power-ohms* function of a multimeter should be used.

SELF-TEST

Check your understanding by answering these questions. Assume that all resistance checks are made with a low-power ohmmeter.

1. If R_4 in Fig. 34.3-1 is *open*, a resistance check from points C to G will measure ________ Ω.

2. If the emitter is shorted to the base in Fig. 34.3-1, an ohmmeter will measure ________ Ω from base to ground.
3. If the circuit in Fig. 34.3-1 is normal, the resistance from base to ground will measure ________ Ω.
4. If R_1 in Fig. 34.3-1 is open, a resistance check across R_1 will measure ________ Ω.
5. If the base is shorted to the collector in Fig. 34.3-1, the resistance from C to G will measure ________ Ω.

MATERIALS REQUIRED

- Power supply: Variable low-voltage dc source
- Equipment: Oscilloscope; EVM with *low-power-ohms* function; AF sine-wave generator; transistor tester
- Resistors: ½-W 560-, 1000-, 8200-, 18,000-Ω
- Semiconductors: 2N6004 or the equivalent
- Miscellaneous: SPST switch; defective parts for troubleshooting Fig. 34.3-2

PROCEDURE

Resistance Measurements

1. Estimate the resistance values at the elements of the experimental amplifier Fig. 34.3-2 and record your results in Table 34.3-1.

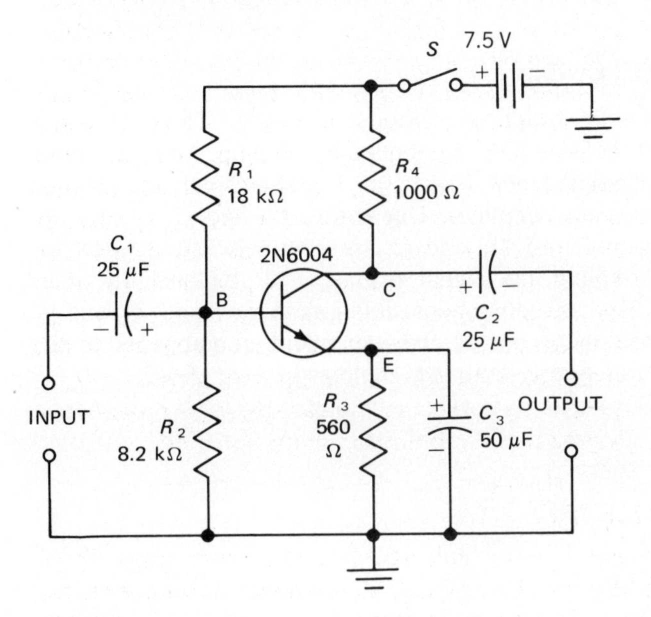

Fig. 34.3-2. Experimental CE amplifier.

TABLE 34.3-1. Resistance Measurements

Element	Resistance to ground (G)	
	Estimated	*Measured*
Emitter		
Base		
Collector		

2. **Power off.** Set your EVM to the low-power-ohms function. Measure and record in Table 34.3-1 the resistances from the base, emitter, and collector to common G.

Troubleshooting the CE Amplifier

3. Your instructor will insert trouble into the amplifier. Troubleshoot the circuit, keeping a step-by-step record of each check that you make in the order performed. Record these checks in the standard troubleshooting report. When you have found the trouble, correct it and notify your instructor, who will put another trouble into the circuit. Again service the circuit, using a separate troubleshooting report to record your results. Service as many troubles as time will permit.

QUESTIONS

1. Why is a *dynamic* test of an amplifier's operation more meaningful than static dc voltage and resistance checks?
2. Why is it necessary to use the low-power-ohms function of an EVM, rather than the conventional ohms function, in measuring resistance in the external circuit of a transistor stage?
3. What is the purpose of resistance tests in troubleshooting a transistor amplifier?

Answers to Self-Test

1. infinite
2. 220
3. 10,000
4. infinite
5. 10,000

CHAPTER 35 CASCADED TRANSISTOR AMPLIFIERS

EXPERIMENT 35.1. Cascaded Transistor Amplifier—Linear Operation

OBJECTIVE

To determine the range of linear operation of an RC-coupled two-stage amplifier

INTRODUCTORY INFORMATION

A single transistor amplifier, like the CE amplifier in the preceding experiments, does not produce as much gain as may be needed in an electronic device, as for example, in a TV receiver. To meet higher gain requirements, two or more transistor amplifiers are connected in cascade. A cascade connection is one in which the output signal of one stage serves as the input signal to the next. The block diagram in Fig. 35.1-1 illustrates three amplifiers connected in cascade. Notice how the output signal of each amplifier acts as the input for the next. Cascade amplifiers are used as audio amplifiers in sound-reproducing systems, as video (picture) amplifiers in television receivers, and in many other applications.

Coupling Methods

Transformer Coupling

Transformers are frequently used in coupling amplifier stages. Transformers make it possible to match the output impedance of the first stage to the input impedance of the next. Proper impedance matching ensures maximum transfer of power from one stage to the next.

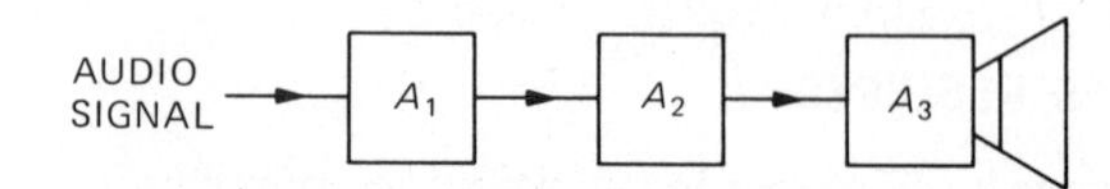

Fig. 35.1-1. Block diagram of a three-stage cascaded audio amplifier.

Figure 35.1-2 is the circuit diagram of a two-stage transistor amplifier using transformer coupling. The input signal is coupled by T_1 into the base of Q_1. C_1 is used to isolate the secondary of transformer T_1 from the dc base-bias circuit. Base bias and bias stabilization are provided by the voltage-divider arrangement of R_1 and R_2 across the battery V_{CC}, and resistor R_3 in the emitter. C_2 bypasses R_3 to prevent ac degeneration. T_2 couples the output signal of transistor Q_1 to the base circuit of Q_2. In the circuit of Q_2, the components C_3, C_4, R_4, R_5, and R_6 have the same functions, respectively, as C_1, C_2, R_1, R_2, and R_3 in Q_1. The amplified output of Q_1 is fed to Q_2, where additional amplification occurs.

RC Coupling

Other coupling methods are used. Thus there are direct-coupling, RC-coupling, and impedance-coupling arrangements. Figure 35.1-3 shows an RC-coupled cascade amplifier. Capacitors C_1 and C_3 couple the signal into Q_1 and Q_2, respectively. C_5 is used for coupling the signal from Q_2 to its load. In other respects, component functions in the stages Q_1 and Q_2 are the same as the equivalent components in the transformer-coupled amplifier in Fig. 35.1-2.

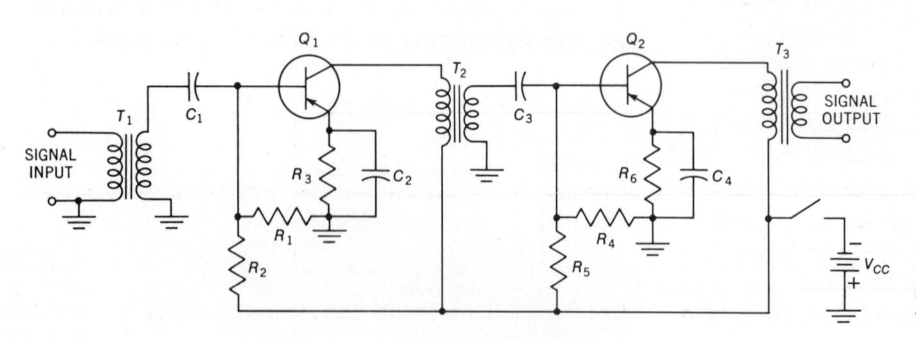

Fig. 35.1-2. Cascaded transistor amplifier with transformer coupling.

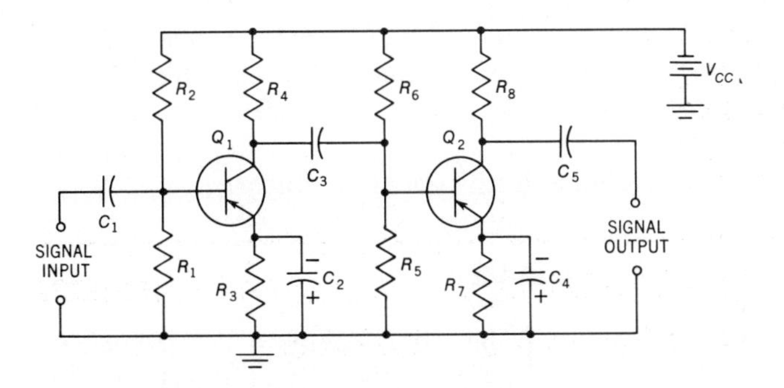

Fig. 35.1-3. *RC*-coupled transistor amplifier.

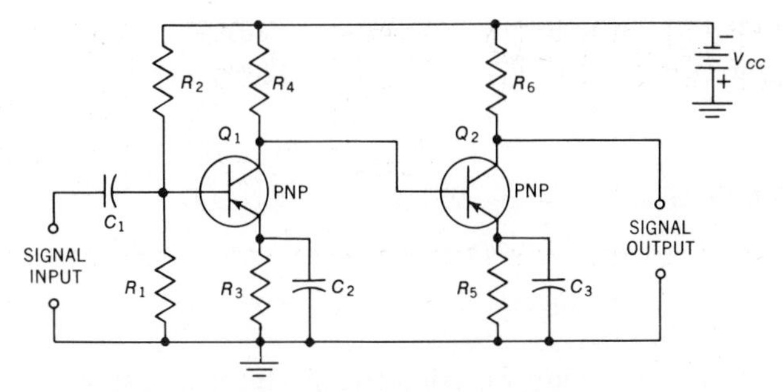

Fig. 35.1-4. Direct-coupled transistor amplifier.

If the operation of coupled amplifiers is considered, a complicating factor appears. The addition of a second stage may change the characteristics of the first stage and thus affect the level of signal fed to the second stage. For example, if C_3 in Fig. 35.1-3 were open, R_4 would act as both the ac (signal) load and the dc load for Q_1. When C_3 is connected as shown, R_4 is no longer the ac load for Q_1. Instead, the ac load now consists of R_4 in parallel with R_5, R_6, and the input resistance R_{in} of Q_2. The ac collector load R_L is therefore smaller in value than R_4. In this analysis it is assumed that the reactance of C_3 is very low at the frequency of the input signal and that the battery V_{CC} is a very low impedance path for the ac signal; that is, V_{CC} acts like an ac short circuit for the signal.

The signal voltage developed at the collector of Q_1 is the product of the ac collector current i_c and the ac load resistance R_L. That is, $v_{1,\text{out}} = i_c \times R_L$. An effect of the reduced value of ac load resistance is the reduction in signal voltage $v_{1,\text{out}}$ at the collector of Q_1. It is the signal voltage $v_{1,\text{out}}$ which determines the value of signal current flow in the input base circuit of Q_2. It is clear, then, that the signal current coupled to the base of Q_2 is changed by *RC* coupling of cascaded amplifiers.

Direct Coupling

Direct coupling is also used in cascaded transistor amplifiers. An advantage of direct coupling is the savings possible in components and the improvement in frequency response. Figure 35.1-4 is the circuit diagram of a direct-coupled two-stage amplifier. Bias and bias stabilization of Q_1 are conventional. Bias of Q_2 is determined by the collector voltage of Q_1 and the emitter voltage of Q_2. The values of R_4 (in the collector of Q_1), of R_5, and of the operating point of Q_1 must be so chosen that the collector of Q_1, and so the base of Q_2, are negative with respect to the emitter of Q_2. This establishes forward bias for Q_2. Bias stabilization for Q_2 depends on the direct connection between Q_2 and Q_1. Thus, if an increase in operating temperature occurs, the collector currents of Q_1 and Q_2 will increase. An increase in collector current of Q_1, however, will cause the voltage at the collector of Q_1 to become less negative, and hence more positive. This will make the base of Q_2 more positive and will reduce the forward bias of Q_2, reducing collector current in Q_2. Effectively, then, the direct-coupled circuit of Fig. 35.1-4 provides bias stabilization of Q_2.

Linear Operation

Two or any number of amplifiers operated in cascade may be considered as a single amplifier with a single input and a single output. Figure 35.1-4 shows such a two-stage amplifier. A cascaded amplifier of any number of stages is shown in block diagram form in Fig. 35.1-5.

When two or more amplifiers are operated in cascade, the characteristics of the total unit must meet the requirements of the application. For example, if two or more transistor amplifiers in cascade make up an audio amplifier, the amplifier must be operated over its linear characteristic for distortionless reproduction of sound.

An oscilloscope may be used to test linear operation. An audio sine-wave generator serves as the signal source. The output of the amplifier is observed with an oscilloscope. To determine the range of linear operation, the input-signal level is increased from zero to just below the point of distortion (clipping) in the output. The maximum generator signal which does not introduce distortion may then be measured.

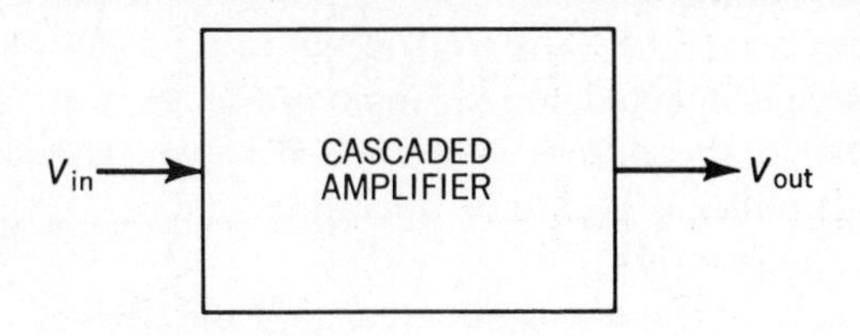

Fig. 35.1-5. Block diagram of a cascaded amplifier.

In this experiment you will measure the range of linear operation of an *RC*-coupled, two-stage audio amplifier.

SUMMARY

1. Two amplifiers are said to operate in cascade when the output signal of the first serves as the input signal to the second.
2. Amplifiers may be coupled in any of a variety of ways. There is *RC* coupling, direct coupling, transformer coupling, and impedance coupling.
3. Transformers are used when it is necessary to match the output impedance of the first stage to the input impedance of the second, ensuring a maximum transfer of power from the first to the second stage.
4. In *RC*-coupled amplifiers, capacitors couple the signal from one stage to the next.
5. In coupling two amplifiers, the effect of the input of the second stage on the ac load resistance of the first stage must be considered. The effect of the input impedance of the second stage is to lower the ac load impedance of the first stage, reducing the level of *ac* output signal of the first stage.
6. Direct-coupled stages are frequently found in transistor circuitry. One advantage is the elimination of coupling components (e.g., transformers, capacitors). Another advantage is the improvement in frequency response.
7. The range of linear operation of amplifiers connected in cascade may be determined by injecting a sine wave into the input of the first stage and observing the output of the last stage with an oscilloscope. The system is said to be linear over the range of inputs for which the output remains a true sine wave.

SELF-TEST

Check your understanding by answering these questions.

1. Coupling between Q_1 and Q_2 in Fig. 35.1-2 is accomplished by means of ________ ________.
2. In Fig. 35.1-3 ________ ________ couples the signal from the output of Q_1 to the input of Q_2.
3. In Fig. 35.1-3 the ac load in the output of Q_1 is the resistor R_4. ________ (true/false)
4. A sine wave is injected into the input of Fig. 35.1-4. The signal in the output is a clipped wave. The cascade amplifier is said to be operating ________ (linearly/nonlinearly).
5. By ________ (increasing/decreasing) the drive signal at the input of the amplifier in question 4, the output waveform can be made sinusoidal, in a properly designed cascade audio amplifier.
6. An oscilloscope is used to check the waveforms at the ________ and ________ of an amplifier to determine if it is operating linearly.

MATERIALS REQUIRED

- Power supply: Variable, regulated low-voltage dc source
- Equipment: Oscilloscope; EVM; AF generator
- Resistors: ½-W 100-, 470-, 560-, 1000-Ω; 8.2-, 10-, 18-, 33-kΩ
- Capacitors: Two 25-μF/50-V; two 100-μF/50-V
- Semiconductors: 2N6004, 2N2102 (or their equivalent)
- Miscellaneous: 2-W 500-Ω potentiometer; two SPST switches

PROCEDURE

Range of Linear Operation

1. Connect the circuit of Fig. 35.1-6. An AF signal generator set at 1000-Hz, 50-mV output is used as the signal source. Set level control R_1 at minimum.
2. **Power on.** Connect the oscilloscope at the collector of Q_2. Adjust R_1 until the sine wave at the collector of Q_2 distorts. Then reduce the signal level, *just below* the point of distortion.

   NOTE: If the circuit is unstable (oscillates), bypass the collector of Q_1 with a 0.1-μF capacitor. The capacitor will provide a low-impedance path for the oscillatory frequency and will eliminate it.

3. Measure and record in Table 35.1-1 the maximum undistorted output signal; also measure and record the input signal at TP_1, which permits the two-stage amplifier to operate without distortion.

   NOTE: You may not be able to measure, directly, the signal level at TP_1. If you cannot, proceed as follows:

   a. *Remove* the signal generator from the circuit. Do *not* vary the setting of the *level* control on the experimental amplifier. **Power off.**
   b. Disconnect the center arm of the *level* control from the circuit. Measure the resistance R_{CB} from the center arm (point C) to ground (point B). Measure also the total resistance R_{AB} of the level control. Compute v_{in}, the input signal, by

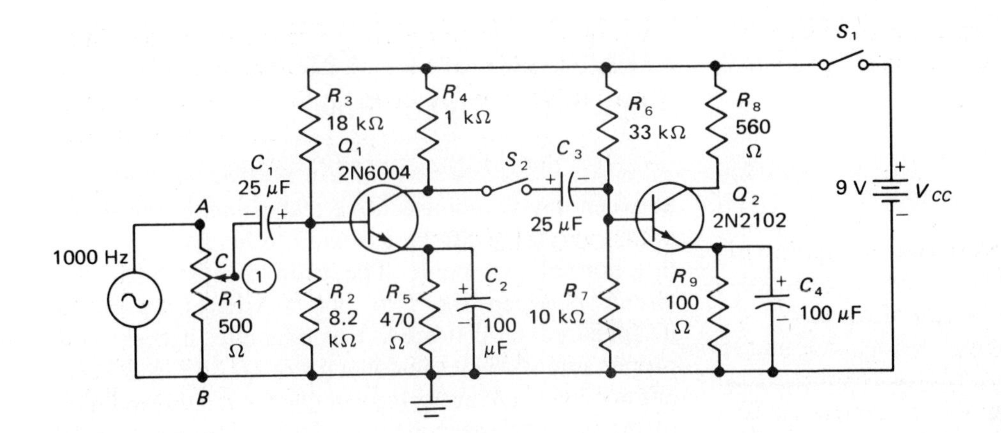

Fig. 35.1-6. Experimental *RC*-coupled audio amplifier.

TABLE 35.1-1. Range of Linear Operation

	Undistorted Signal Level, V p-p
	Maximum
v_{out}	
v_{in}	

substituting the measured values of resistance in this formula:

$$v_{in} = \frac{R_{CB}}{R_{AB}} \times 50 \text{ mV}$$

The 50 mV is the generator output. If a voltage other than 50 mV is used, substitute that value in place of 50 mV.

Loading Effect of Q_2 on Q_1

4. With the generator level set for the amplifier's maximum undistorted output, measure and record in Table 35.1-2 the signal voltage at the collector of Q_1 ($v_{1,\ out}$).
5. Open switch S_2. Again measure and record the signal voltage at the collector of Q_1.

TABLE 35.1-2. Loading Effect of Q_2

Condition	$v_{1,\ out}$, V p-p
S_2 closed	
S_2 open	

QUESTIONS

1. What is the purpose of cascading amplifiers?
2. Does the input circuit of Q_2 have any effect on the signal level at the collector of Q_1? Refer to your data.
3. What is the range of linear operation of the total amplifier?
4. What is the voltage gain of the total amplifier?

Answers to Self-Test

1. transformer T_2
2. capacitor C_3
3. false
4. nonlinearly
5. decreasing
6. input; output

EXPERIMENT 35.2. Cascaded Transistor Amplifiers—Signal Flow

OBJECTIVE

To observe the signal waveforms and their phase at the input and output of each stage in a two-stage amplifier

INTRODUCTORY INFORMATION

Signal Tracing

In troubleshooting electronic devices, it is desirable to isolate the defect first to a section of the device, and then to a *single stage* in the suspected section. If the device is, say, a television receiver, and if the suspected section is a multistage amplifier, the most effective method for finding the defective stage is signal tracing.

Assume the suspected section is the two-stage cascaded audio amplifier in Fig. 35.2-1. We can signal-trace the amplifier by injecting an AF sine-wave signal into the input, as shown. The progress of the signal through each stage is then followed with an oscilloscope. The dynamic signal test points in this circuit are 1, 2, 3, and 4. These are, respectively, the base and collector of Q_1 and the base and collector of Q_2. In a linear amplifier which is operating properly, the lowest amplitude signal appears at the base of Q_1, while a higher amplitude signal waveform is seen at the collector. At the base of Q_2 the signal level is the same, approximately, as at the collector of Q_1. Again, there is amplification by Q_2, and the highest-amplitude signal waveform is at the collector of Q_2. Since the cascade amplifier is operating linearly, the waveforms at each test point 1 through 4 are undistorted.

If the waveform disappears or becomes distorted between any two consecutive test points in the signal path, everything can be assumed to be okay up to the last normal test point. The trouble must be in the circuit between the two points where the signal disappeared or distorted. For example, if there is a proper sine wave at the collector of Q_1, in Fig. 35.2-1, but no sine wave at the base of Q_2, the trouble must be to the *right* of the collector of Q_1. (The most likely trouble here is an open coupling capacitor C_3.)

Tracing the signal through each stage of an amplifier will identify the stage or stages which are operating properly and will disclose the stage which is not.

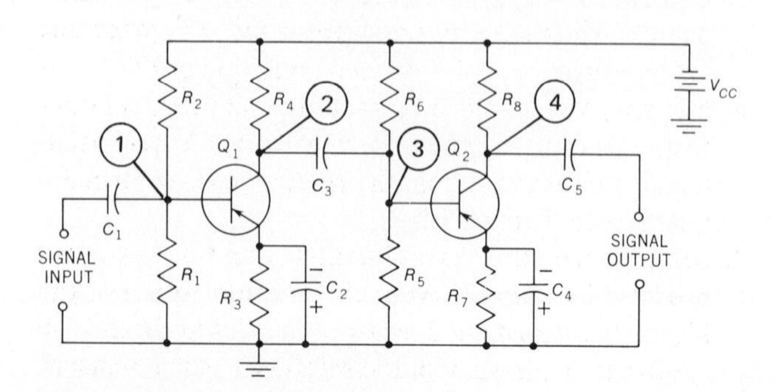

Fig. 35.2-1. Cascaded transistor amplifier.

Phase Relations

In signal-tracing an amplifier, it is also possible to determine the phase of the waveforms between the input and any test point in the signal path. In troubleshooting video (picture) amplifiers in television receivers, it is sometimes necessary to measure the phase difference between the waveforms at two signal points in the amplifier. One technique for following phase relations was used in earlier experiments. We shall briefly review it.

The oscilloscope is externally triggered/synchronized by the waveform from the *output* of one of the amplifiers under test. If the signal generator which acts as the signal source has an output labeled *Sync,* this output is used to externally trigger/synchronize the oscilloscope. It should be noted that the triggering/synchronizing signal must have sufficient amplitude to lock in the oscilloscope for a stable presentation. The oscilloscope is then adjusted for proper centering of the generator signal on the graticule, as in Fig. 35.2-2*a*. This is the reference signal. The phase of the

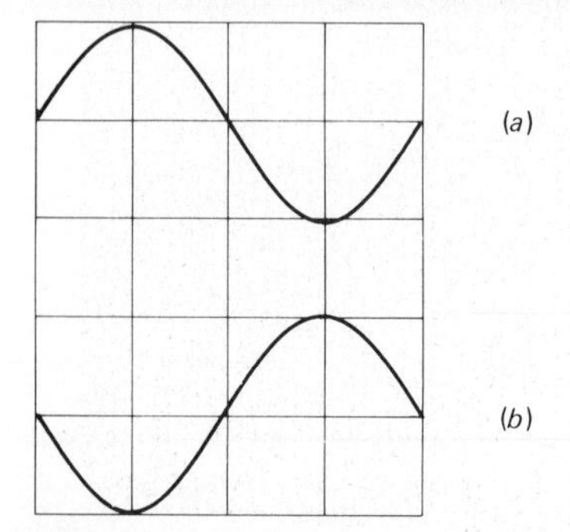

Fig. 35.2-2. Phase relations between (*a*) reference sine wave and (*b*) output waveform of Q_1.

signal waveform at each point in the signal path may then be compared with the reference. For example, the input and output signals in Q_1, Fig. 35.2-1, are 180° out of phase. So if the first alternation of the input sine wave is positive-going, the first alternation of the waveform at the collector will be negative-going, as in Fig. 35.2-2*b*.

A *dual-trace* oscilloscope may be used to display simultaneously the reference waveform and, say, the waveform at the collector of Q_1. This convenience feature is frequently found in triggered service-type oscilloscopes, like that in Fig. 35.2-3.

SUMMARY

1. Signal tracing is used in troubleshooting electronic devices, first to isolate the trouble to a specific section, then to isolate the trouble to a specific stage.
2. In signal-tracing an amplifier, the signal waveform is followed with an oscilloscope from the input to the output of each stage. The trouble in a defective amplifier lies between those two points in the signal path where a normal signal disappears or becomes distorted.
3. The phase of the waveforms in an amplifier may be observed by externally triggering/synchronizing the oscilloscope with the signal from the generator or from the output of one of the stages in the amplifier.

SELF-TEST

Check your understanding by answering these questions.

Fig. 35.2-3. Dual-trace oscilloscope. (*B. and K. Dynascan Corp.*)

1. In the amplifier of Fig. 35.2-1, the sine-wave signal at the collector of Q_1 is 50 mV, that at the collector of Q_2 is 5 V. The signal at the input to Q_1 is ½ mV. The waveforms at test points 1 through 4 are undistorted. It would appear that this amplifier ________ (is/is not) operating normally.
2. In the amplifier of Fig. 35.2-1, the signal at the collector of Q_1 is 50 mV, and that at the base of Q_2 is 10 mV. The signal voltage at the collector of Q_2 is 1 V. This amplifier ________ (is/is not) defective.
3. There is ________ degrees phase difference between the signal waveform at the collector of Q_1 and the collector of Q_2 in Fig. 35.2-1.
4. By signal tracing, trouble has been isolated to stage Q_2 in Fig. 35.2-1. The defective part may now be found by ________ and ________ checks.

MATERIALS REQUIRED

- Power supply: Variable, regulated low-voltage dc source
- Equipment: Oscilloscope; EVM; AF sine-wave generator
- Resistors: ½-W 100-, 470-, 560-, 1000-, 8200-, 10,000-, 18,000-, 33,000-Ω
- Capacitors: Two 25-μF/50-V; two 100-μF/50-V
- Semiconductors: 2N6004, 2N2102, or their equivalent
- Miscellaneous: 2-W 500-Ω potentiometer; SPST switch

PROCEDURE

1. Connect the circuit of Fig. 35.2-4. An AF generator set at 1000-Hz, 100-mV output is used as the signal source.
2. Set the level control in Fig. 35.2-4 for maximum undistorted output at the collector of Q_2.
3. Externally trigger/synchronize your oscilloscope using either the sync signal from the AF generator or the output signal from the collector of Q_2.
4. Let the AF generator signal, across the *level* control, serve as the reference waveform. Adjust oscilloscope controls so that the reference waveform appears as in Table 35.2-1.
5. Signal-trace the experimental amplifier at test points 1 through 4. At each test point observe the signal waveform and draw it in Table 35.2-1, in proper time phase with the reference. Also measure and record the peak-to-peak amplitude of the signal at each test point.

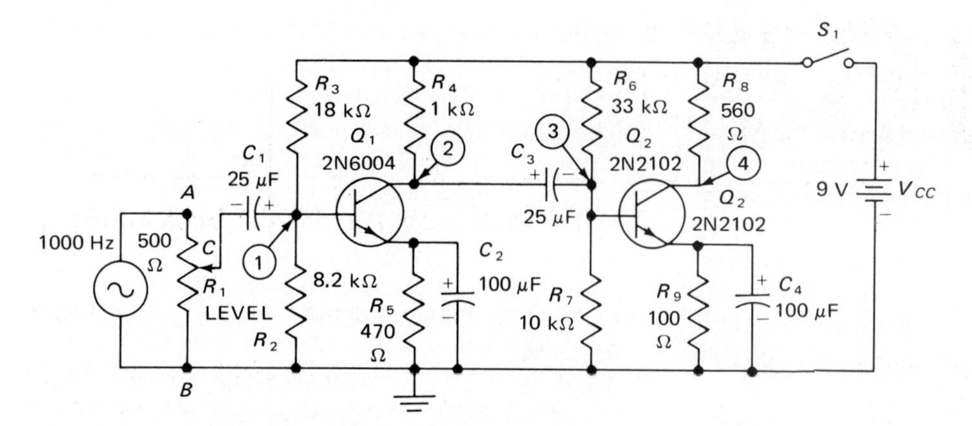

Fig. 35.2-4. Experimental amplifier.

TABLE 35.2-1. Waveforms in an Audio Amplifier

Test Point	*Waveform*	V p-p
Reference		
1		
2		
3		
4		

QUESTIONS

1. What is the purpose of signal-tracing a multistage amplifier?
2. What test instruments are required for signal-tracing an audio frequency amplifier?
3. In the experimental amplifier what is the gain of Q_1? of Q_2?
4. In testing the experimental amplifier, what conclusion can you draw if you find that stage 1 or stage 2 has no gain?
5. From your experimental data what can you say about
   (*a*) The phase of the waveforms at test points 1 and 2?
   (*b*) The phase of the waveforms at test points 1 and 4?

Answers to Self-Test

1. is
2. is not
3. 180
4. voltage; resistance

CHAPTER 36 CLASS A AUDIO POWER AMPLIFIERS

EXPERIMENT 36.1. The Loudspeaker

OBJECTIVE

To become familiar with the loudspeaker and its operation

INTRODUCTORY INFORMATION

The Dynamic Speaker

The loudspeaker, or simply speaker, changes electric audio-frequency energy into sound. The speaker is a *transducer,* one in a family of devices which convert energy from one form into another.

The speaker in greatest use today is the permanent-magnet (pm) dynamic speaker. The electrostatic speaker may be found as part of a speaker system in hi-fi installations. Electromagnetic dynamic speakers use electromagnets for the speaker field, but they are no longer used, though they were very popular early in the development of audio systems. Dynamic speakers are similar in operation. They differ in the method used to obtain the stationary magnetic field.

Figure 36.1-1 shows the mechanical construction of a pm speaker. A permanent magnet is used to concentrate a magnetic field at the pole pieces of a highly permeable housing. The pole pieces are very close together to obtain an intense magnetic field. A voice coil cemented to the speaker cone is freely suspended between the magnetic poles. A flexible membrane called the "spider" is attached to the voice-coil form and cemented to the speaker frame. The spider centers the voice-coil form between the speaker poles and keeps it from rubbing against them. The flared end of the cone is flexibly attached to the speaker frame.

The audio-frequency signal currents fed to the voice coil set up a moving magnetic field about the voice coil. This interacts with the fixed magnetic field of the permanent magnet and causes the voice coil and the speaker cone to move in and out of the magnetic gap. The rate of movement of the speaker cone is determined by the frequency of the audio current. The amplitude of movement, i.e., how far the cone moves, depends on the amplitude of audio current. The speaker cone moves the air mass surrounding it, producing sound.

Connections from the voice-coil ends are brought to insulated solder terminals on the speaker frame. When output transformers are used to match the impedance of the last audio amplifier to that of the speaker, the transformer may be mounted on the speaker frame. Leads from the secondary of the output transformer are then connected to the voice coil at the solder terminals on the frame.

Checking the Dynamic Speaker

A continuity check may be made on the voice coil of a dynamic speaker. The resistance is very low, usually 3 or 4 Ω, although 8- and 16-Ω voice coils are also common. The continuous motion of the voice

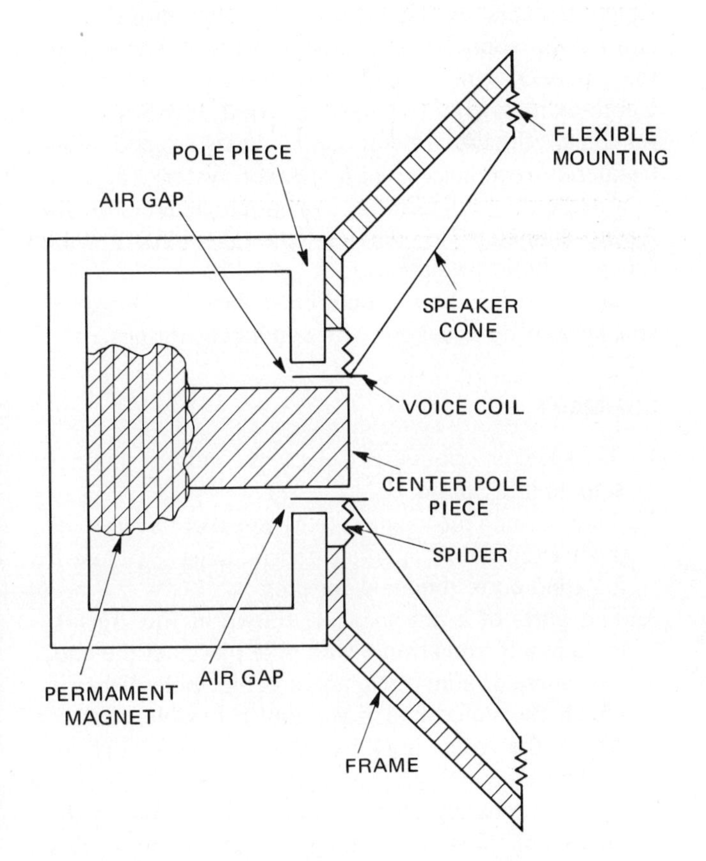

Fig. 36.1-1. Permanent-magnet loudspeaker construction.

coil may cause the coil wire to break, resulting in an "open" voice coil. This would check as an infinite resistance.

In checking the resistance of a voice coil, in a nondefective speaker, an interesting effect may be noted. As the ohmmeter leads are placed across the coil (ohmmeter should be set on its lowest range), the cone moves, and a click is heard. When the leads are removed, the cone moves back to its rest position, and again a click is heard. If the ohmmeter leads are reversed, the cone moves in the opposite direction.

A $1\frac{1}{2}$-V battery may be used instead of the ohmmeter for the click check, which is an indication that the speaker is operating.

Speaker Ratings

Permanent-magnet speakers are rated for their size, power-handling ability, frequency range, and voice-coil impedance. For hi-fi installations, the frequency response of the speaker is also specified. A 5-in speaker is one whose maximum cone diameter is 5 in. Large-speaker cones can move greater quantities of air, thus producing louder sound. More driving power is required, however. Hence large speakers have higher wattage ratings. Table-model receivers use small speakers. A 5-in speaker in a table-model radio can handle approximately 3 W of power. The size of the speaker cone affects the frequency response. Large-diameter cones have a good low-frequency response. Small-diameter cones have a good high-frequency response. A hi-fi speaker system enclosed in a baffle may contain three or more speakers of different diameters to provide coverage over a wide range of audio frequencies.

Defective speakers should be replaced by identical speakers or by speakers with equivalent ratings.

SUMMARY

1. A speaker converts electric audio-frequency signals into sound.
2. The permanent-magnet (pm) speaker is used almost exclusively in sound-reproducing systems. It is called a pm dynamic speaker.
3. The parts of a pm speaker, shown in Fig. 36.1-1, include a ferrous frame with pole pieces at the end, a permanent alnico magnet, a center pole piece on which the voice coil rides, and a flexible cone to which the voice coil assembly is physically attached.
4. Audio-frequency currents are made to flow in the voice coil. This causes the voice coil to move back and forth, driving the speaker cone back and forth. The motion of the speaker cone creates sound waves.
5. The resistance of the voice coil of a speaker is very low, about 3 to 4 Ω.
6. PM speakers are rated for their physical size, power-handling capability, voice-coil impedance, and frequency response.

SELF-TEST

Check your understanding by answering these questions.

1. The purpose of a loudspeaker is to ________ electric energy into ________.
2. Modern sound systems use ________ speakers.
3. The voice coil of a speaker carries the ________-frequency currents when the speaker is actuated.
4. The sound that issues from a loudspeaker is created by the movement of the ________, which is driven by the motion of the ________.
5. The resistance of the voice coil of a speaker is normally about ________ Ω.
6. A rough test to determine if a speaker is working is to place an ohmmeter across the ________ ________.
7. If the speaker is working, a ________ will be heard, in the test described in question 6.

MATERIALS REQUIRED

- Equipment: EVM or VOM; AF sine-wave generator
- Miscellaneous: PM loudspeaker, 3.2 Ω; AF output transformer Argonne type 119 or the equivalent

PROCEDURE

Speaker

CAUTION: *Carefully examine the pm speaker which you have received. The cone is fragile and can be easily damaged.* If possible, keep the speaker in a small enclosure to protect it.

1. Measure and record in Table 36.1-1 the resistance of the voice coil (listen for the click).
2. Observe the direction, forward or backward, in which the speaker cone moves when the ohmmeter (or $1\frac{1}{2}$-V battery) leads are placed across the voice coil. Repeat with ohmmeter leads reversed.
3. Connect AF generator output leads across the primary of an audio output transformer, whose secondary is connected to the voice coil of the

TABLE 36.1-1. Loudspeaker Tests

Step	Test Performed	
1	Voice Coil Resistance, Ω	
4	1000-Hz Signal	
	Loudness	Pitch
5	Increased Gen. Output, 1000 Hz	
	Loudness	Pitch
6	Lowest frequency, Hz	Highest frequency, Hz

speaker, as in Fig. 36.1-2. Set generator frequency at 1000 Hz, generator output at minimum.

4. Gradually increase generator output until sound is heard from the speaker. In Table 36.1-1, describe **(a)** the loudness of the sound, and **(b)** the pitch.
5. Increase generator output further. Indicate in Table 36.1-1 what happens to **(a)** loudness, and **(b)** pitch of sound.
6. Experimentally determine the lowest frequency coming from the speaker which you can hear; the highest frequency.

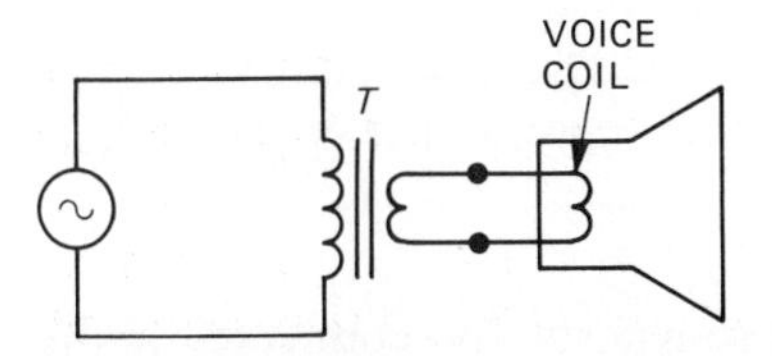

Fig. 36.1-2. Experimental circuit for loudspeaker tests.

QUESTIONS

1. How can you determine if there is an *open* voice coil in a speaker?
2. With the AF generator set at a specific frequency, say 1000 Hz, what is the effect of increasing the level of generator signal on (*a*) sound volume, (*b*) sound pitch?
3. Assume you have an AF generator which you know is working and no other equipment. How can you test a pm speaker to determine if the speaker is working?
4. In your experiment, which frequencies appeared to give a louder sound (for the same generator output), the low frequencies or the high?

Answers to Self-Test

1. convert, sound
2. pm
3. audio
4. cone; voice coil
5. 3 to 4
6. voice coil
7. click

EXPERIMENT 36.2. The Audio Output Transformer

OBJECTIVE

To become familiar with the operation of an output transformer and connect it to a speaker

INTRODUCTORY INFORMATION

The end product of an audio amplifier system is the sound that comes from the dynamic speaker. An audio system consists of a speaker which receives its driving power from audio amplifier stages. The audio amplifier receives the audio-frequency signals from some device such as a microphone, a tape head, the pickup arm of a phonograph, or other source. Frequently there is one other important component in an audio system, the *output transformer*. The output transformer delivers the audio power from the audio power output stage of the amplifier to the speaker. Figure 36.2-1 is a block diagram of an audio system.

Output Transformers

Dynamic speakers used in radio receivers are low-impedance devices. One way in which a low-impedance device is connected to the audio power amplifier from which it receives its driving power is through an output transformer. The high-impedance

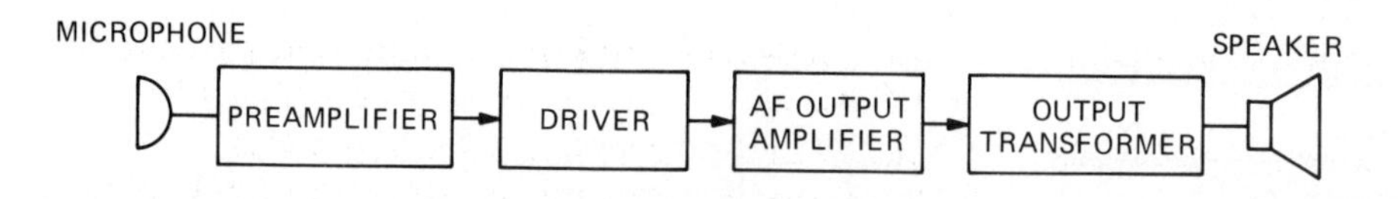

Fig. 36.2-1. Block diagram of an audio system.

primary of the output transformer is connected in the high-impedance collector circuit of the transistor; the low-impedance secondary is connected to the low-impedance voice coil of the speaker, as in Fig. 36.2-2. This provides a maximum transfer of audio power from the output transistor to the speaker. The audio output transformer, wound on an iron core, is a voltage-stepdown, current-stepup device.

The output impedance of a transistor varies with the transistor type and with the circuit arrangement. Therefore to match a particular output transistor stage to a given speaker requires an output transformer with the proper turns ratio. A variety of output transformers are manufactured to match the various combinations of output circuits and speakers. In replacing an output transformer in an audio system, an identical transformer or one with similar characteristics should be used.

The EIA color code for the leads of an audio output transformer is shown in Table 36.2-1. If there is any question about which is the primary winding and which the secondary, a resistance check of the windings will help identify them. The resistance of the secondary winding is much lower than that of the primary. There is a wide range of resistances of the windings of output transformers. For example, one manufacturer lists the range of primary resistance for a particular series of his output transformers from 600 to 2.3 Ω. For these, the range of secondary resistance is from 25 to 0.25 Ω. A good rule to remember is this: The lower the resistance of a winding for a given impedance, the higher is the power-handling capacity of that winding.

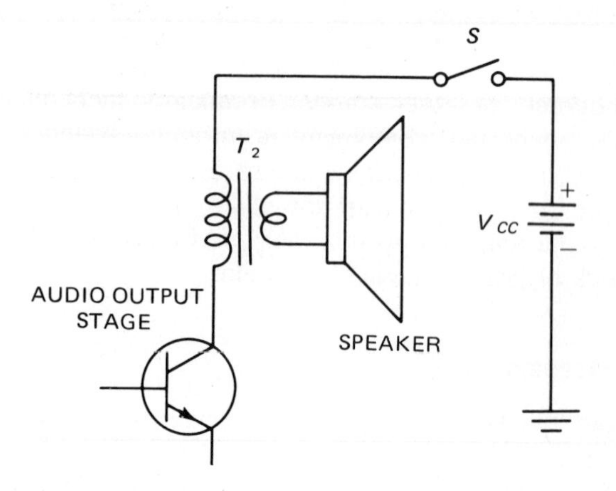

Fig. 36.2-2. Audio output transformer matches the high impedance of the collector circuit to the low impedance of the speaker.

TABLE 36.2-1. EIA Color Code for Audio Output Transformers

Transformer Winding Lead	*Color*
Primary:	
Collector	Blue
$V+$ or V_{CC}	Red
Collector (push-pull only)	Blue or brown or white
Secondary:	
High side	Green or yellow
Low side	Black

It should be noted that in present audio-system design the output transformer is frequently eliminated in favor of direct speaker drive. In this type of system, the speaker acts as the load in the output of a low-impedance transistor power amplifier called an *emitter follower*.

SUMMARY

1. An audio system consists of the following:
   (*a*) An audio-frequency source such as a tape head, microphone, or phonograph pickup.
   (*b*) Audio-frequency amplifiers.
   (*c*) Audio-frequency output transformer (except in direct-drive systems).
   (*d*) Dynamic speaker(s).
2. Output transformers are used to deliver maximum power from the output stage of an audio amplifier to the speaker.
3. An audio output transformer is a voltage-stepdown, current-stepup device. It is designed to match the output impedance of the power amplifier to the impedance of the speaker.
4. A defective output transformer must be replaced by an identical unit or one with identical characteristics.
5. The resistance of the primary of an output transformer is much greater than the resistance of the secondary.

SELF-TEST

Check your understanding by answering these questions.

1. The purpose of an audio output transformer is to

transfer ________ ________ from the audio output stage to the ________.

2. An audio output transformer is an ________-matching device.
3. The resistance of the primary of an output transformer is ________ (higher/lower) than the resistance of the secondary.
4. A rough test to determine if the windings of an output transformer are good is to make a(n) ________ test of these windings.

MATERIALS REQUIRED

- Equipment: Oscilloscope; EVM or VOM; AF sine-wave generator
- Miscellaneous: AF output transformer Argonne type AR-119 or equivalent; pm loudspeaker (3.2 Ω)

PROCEDURE

1. Examine the output transformer you have received. Identify the primary leads. Observe that the primary is center-tapped. *The tap will not be connected in this experiment.* Measure and record in Table 36.2-2 the total resistance of the primary winding, the resistance from the center tap to each primary connection, and the resistance of the secondary.
2. Connect an AF sine-wave generator set at 1000 Hz across the primary. Connect a 3.2-Ω speaker across the secondary. With an oscilloscope measure the AF signal voltage across the primary. Adjust the output of the generator for 4 V p-p (v_p).
3. With the oscilloscope measure the peak-to-peak voltage across the secondary (v_s) and record in Table 36.2-2.
4. Compute and record the turns ratio a of the transformer by substituting the measured values of v_p and v_s in the formula:

$$a = \frac{v_p}{v_s}$$

QUESTIONS

1. How can you determine if there is an open primary or secondary winding in an output transformer?
2. The turns ratio of a voltage output transformer is 50. There is 10 V across the primary. What is the voltage across the secondary?
3. What is meant by the *turns ratio* of a transformer?
4. In a voltage-stepdown, current-stepup output transformer, why is the diameter of the wire in the secondary winding larger than the diameter of the wire in the primary?

Answers to Self-Test

1. maximum power; speaker
2. impedance
3. higher
4. continuity

TABLE 36.2-2. Output Transformer Measurements

Resistance, Ω				*V p-p*		*Turns ratio a*
Primary top to CT	*Primary bottom to CT*	*Full primary*	*Secondary*	v_p	v_s	

EXPERIMENT 36.3. Class A Audio Power Amplifier

OBJECTIVE

To measure the voltage and signal waveforms in a class A audio power amplifier

INTRODUCTORY INFORMATION

Class A Power Amplifier

Preceding experiments have dealt with small-signal, high-gain voltage amplifiers. In an audio amplifier, the end product of the system is the sound which must be produced by the speaker. As we have seen, sound is caused by the vibration of the cone of a speaker. To cause this vibration, audio power is required. This power is the audio current and voltage delivered to the voice coil of a speaker by the final amplifier in the audio system. The last stage in an audio system is therefore called the *power stage* or *power amplifier*. It is usually a large-signal stage.

There are various classes of power amplifiers, including classes A, AB, B, and C. We shall be concerned here with the class A linear power amplifier. A class A amplifier is one whose emitter to base remains forward-biased during the entire input signal and in which collector current flows during all portions of the signal swing. If, in addition, the output signal is an exact reproduction of the input, the amplifier is said to be linear. A linear class A power amplifier is therefore one which does not distort or change the shape of the signal waveform and which delivers audio power to a load.

Figure 36.3-1 is the circuit diagram of a power amplifier Q which receives its input signal from a preamplifier or driver. An output transformer T delivers audio power to the speaker. R_1 and R_2 provide voltage-divider forward bias to Q. R_3 is required for bias stabilization, and C_1 is a bypass capacitor for R_3, to prevent degeneration of the audio signal. Proper operation of this stage as a class A amplifier requires that the bias be properly set so that an undistorted input signal delivered by the driver will be amplified without distortion by Q. As previously noted, for maximum transfer of audio power to the speaker, transformer T must match the output impedance of Q to the input impedance of the speaker.

Figure 36.3-2 is the circuit which will be used in this experiment to demonstrate an RC-coupled two-stage audio amplifier, including a 2N2102 power output amplifier. The design of small-signal amplifier Q_1 is very similar to that of the first stage in Fig. 35.2-4 in Exp. 35.2. Q_2, in Fig. 36.3-2, is a class A small-signal power amplifier similar in circuit arrangement to that in Fig. 36.3-1.

Test points in Fig. 36.3-2 include the bases, emitters, and collectors of Q_1 and Q_2 and the secondary of output transformer T. Assume that the audio amplifier is operating normally and a sine-wave test signal is coupled to the input. Signal-tracing this circuit with an oscilloscope will reveal a sine wave at test points 1, 2, 4, 5, and across 7–8, the secondary of T_2. Points 1 through 6 are also dc test points. The dc voltages measured at the bases (1,4) and emitters (3,6) should show that for these NPN transistors each base is positive relative to its emitter (for forward bias of the base-to-emitter circuit). The dc voltage at the collector of Q_1 is approximately one-half of V_{CC}. The voltage at the collector of Q_2 is approximately equal to V_{CC}, because of the low primary resistance of T.

When a sine wave or other audio signal is applied to the volume control, and when the control is set for the proper input level, sound should be heard from the speaker.

SUMMARY

1. A class A linear transistor amplifier is one which delivers an undistorted, amplified signal in its output.

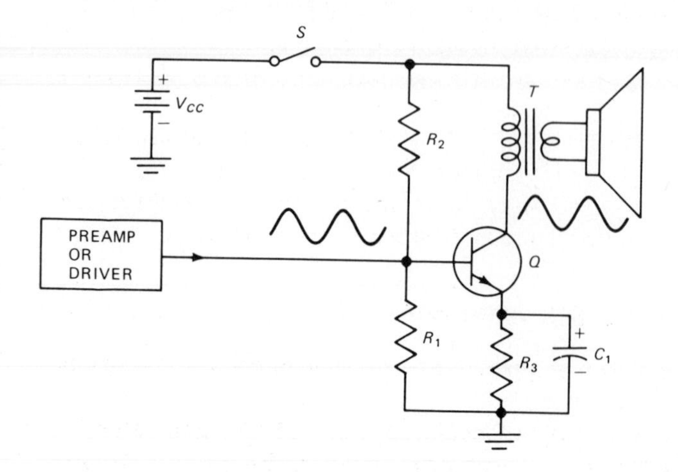

Fig. 36.3-1. Class A power output audio amplifier.

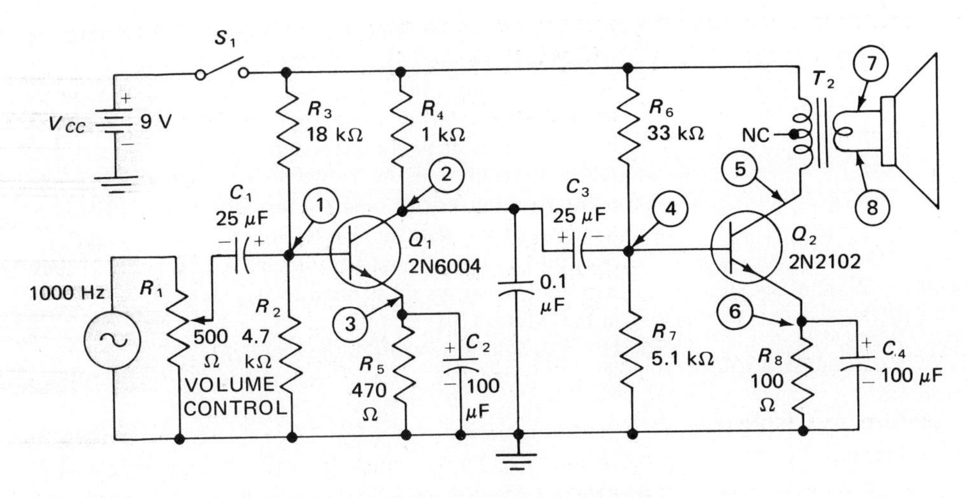

Fig. 36.3-2. Audio amplifier with output transformer and speaker.

2. A class A audio power amplifier delivers power to a load.
3. Frequently an output transformer is used to match the high impedance of the collector in a power amplifier to the low impedance of the speaker.
4. A multistage audio amplifier has both dc and ac test points. The ac test points are used in signal-tracing the amplifier. DC measurements are made in stages suspected of being defective.

SELF-TEST

Check your understanding by answering these questions.

1. In Fig. 36.3-2, the emitter of Q_2 must remain ________ (negative/positive) with respect to the base for class A operation.
2. The output stage Q_2 in Fig. 36.3-2 is called a power amplifier because it delivers ________ ________ to the speaker.
3. In signal-tracing the amplifier in Fig. 36.3-2, for a sine-wave input and linear operation, a sine wave will be observed at each test point 1 through 6. ________ (true/false)
4. In Fig. 36.3-2 the voltage of the waveform across the voice coil of the speaker (7–8) is ________ (higher/lower) than the voltage of the waveform at test point 5. (Assume a 3.2-Ω speaker.)

MATERIALS REQUIRED

- Power supply: Variable, regulated low-voltage dc source
- Equipment: Oscilloscope; EVM; AF generator
- Resistors: ½-W 100-, 470-, 1000-Ω; 4.7-, 5.1-, 18-, 33-kΩ
- Capacitors: Two 25-μF/50-V; two 100-μF/50-V; 0.1-μF
- Semiconductors: 2N6004, 2N2102 (or their equivalent)
- Miscellaneous: 2-W 500-Ω potentiometer; one SPST switch; AF output transformer Argonne type AR-119 or equivalent; pm loudspeaker (3.2 Ω); audio source such as a microphone or record player

PROCEDURE

1. Connect the circuit of Fig. 36.3-2. The AF generator, at 1000 Hz, is connected across the volume control. Set volume at zero. Switch S_1 is **off.**
2. Adjust the output of the power supply to 9 V. Close S_1 **(power on).** Do you hear any sound?
3. With an oscilloscope observe the signal voltage at the collector of Q_2 to ground. Adjust the volume control for the *maximum undistorted* sine wave. What do you hear?
4. With the oscilloscope measure and record the peak-to-peak signal voltage at every test point listed in Table 36.3-1.
5. Measure also and record the dc voltage with respect to ground at the test points shown in Table 36.3-1.
6. Measure and record the direct current in the collector of Q_2 with and without signal.
7. Remove the AF signal generator and replace it with a microphone or record player. Adjust volume control for comfortable sound level. With the oscilloscope signal-trace the amplifier and observe the waveforms at the base and collector of Q_1 and

TABLE 36.3-1. Audio Amplifier Measurements

Test Point	*1*	*2*	*3*	*4*	*5*	*6*	*7–8 (Voice Coil)*
Signal volts (peak-to-peak)							
DC volts							

DC in collector Q_2, mA	With signal	Without signal

Q_2. Observe whether the pattern is stable or varies with sound pitch and sound level.

QUESTIONS

1. Is output transformer T_2, in your experimental circuit, a current-stepup or current-stepdown device? Explain by referring to your test data.
2. From your measurements, what is the turns ratio of output transformer T_2?
3. Which of your measurements indicates if the direct current rating (4 mA) of the primary of the output transformer is or is not exceeded?
4. Refer to your data in Table 36.3-1. Are the base-to-emitter sections of Q_1 and Q_2 properly biased? Explain.
5. Refer to your data in Table 36.3-1. Are Q_1 and Q_2 operated as linear amplifiers in this experiment? If they were not, what would happen to the 1000-Hz sine wave in the circuit?
6. Is it possible to overdrive the amplifiers? How?

Answers to Self-Test

1. negative
2. audio power
3. false
4. lower

CHAPTER 37 CLASS B AUDIO POWER AMPLIFIERS

EXPERIMENT 37.1. Class B Push-Pull Audio Power Amplifier

OBJECTIVE

To connect and signal-trace a push-pull audio power amplifier

INTRODUCTORY INFORMATION

Class B Operation

In high-power audio systems it is necessary to develop more audio power than a single output stage can provide. One solution is to use two or more transistors connected in push-pull.

Push-pull audio amplifier circuits are operated as either class B or class AB. Let us see exactly what this means.

In the preceding experiment we studied the operation of a single-transistor class A audio power amplifier. You will recall that in class A operation the emitter-base section of the transistor is forward-biased throughout the entire period of the input signal. Current flows for 360°, and the output of a class A amplifier is undistorted, as in Fig. 37.1-1*a*.

In class B operation, the emitter-base section of a transistor is forward-biased by the signal during one-half of the input signal and reverse-biased during the other half. The collector-current waveform of a class B circuit appears as in Fig. 37.1-1*b*. You will note that current flows for approximately 180° and is cut off during the remainder of the cycle.

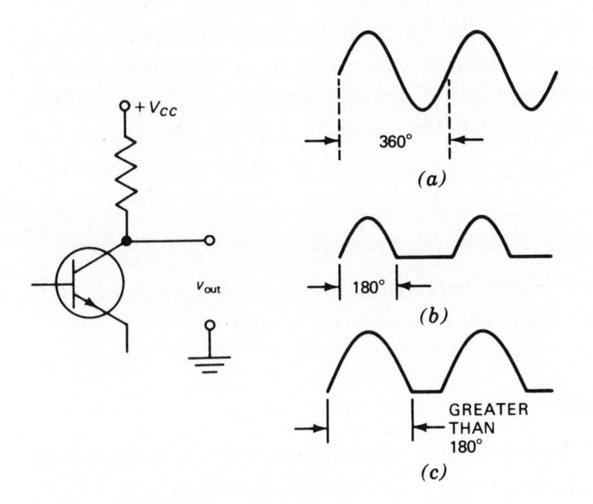

Fig. 37.1-1. Current waveforms in (*a*) class A, (*b*) class B, (*c*) class AB amplifier.

Class AB operation lies between class A and class B. Current flows for more than 180°, but less than 360°, as in Fig. 37.1-1*c*. It is clear from the current waveforms that if a single transistor operating into a resistive load were biased class B or AB, signal distortion would occur. A push-pull arrangement, properly designed, eliminates the distortion which would occur with a class B or class AB single-transistor power amplifier.

Push-Pull Amplifiers

When biased for class AB or class B operation, push-pull output amplifiers can handle a signal amplitude approximately twice as large as that which a conventional class A power amplifier can handle. For this reason class B or class AB output stages can deliver more power than a single-ended class A stage can produce.

The circuit of a push-pull output stage is shown in Fig. 37.1-2. Q_2 and Q_3 are two output transistors connected as common-emitter amplifiers in a balanced-circuit arrangement. T_2, an input transformer, couples the signal from Q_1 to stages Q_2 and Q_3. Q_1 is called the *driver*.

The bases of Q_2 and Q_3 are connected to opposite ends of the center-tapped secondary of T_2. Hence the bases receive two signals, equal in amplitude but 180° out of phase. Thus, for a sine-wave input, during the time that the base of Q_2 receives the positive alternation, the negative alternation is delivered to the base of Q_3. When the base of Q_2 swings negative, the base of Q_3 goes positive. As a result, when there is signal current flow in the collector of Q_2, there is no signal collector current in Q_3, and vice versa.

The collector currents of Q_2 and Q_3 flow in opposite directions through the primary of T_3, the output trans-

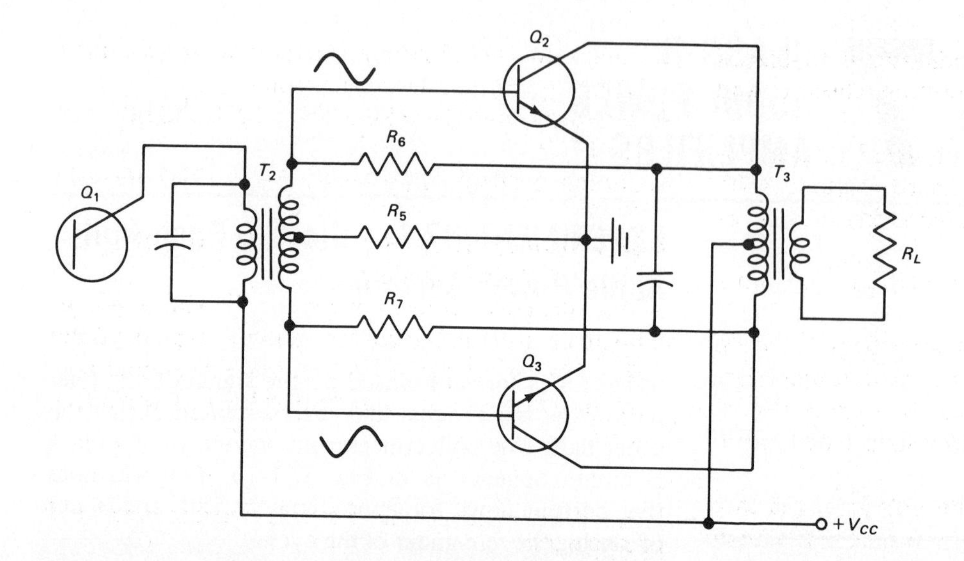

Fig. 37.1-2. Push-pull audio output amplifier.

former. Suppose the magnetic field that the collector current of Q_2 sets up about the primary is expanding when collector current in Q_2 is increasing. At this same time collector current in Q_3 is decreasing, and the resulting magnetic field is moving in the same direction as the field arising from Q_2. As a result, the two fields aid and induce a larger emf in the secondary than either could deliver alone.

Q_2 and Q_3 are two medium-power transistors, operated close to class *B*, with just enough forward bias supplied by R_5 and the combination of R_6 and R_7 to cause a small collector idling current to flow in Q_2 and Q_3, to prevent what is known as *crossover distortion.* The waveform in Fig. 37.1-3 shows the distortion which occurs when the transistors in a push-pull stage are dc-biased to cutoff. If there is cutoff (zero) bias on the base-emitter of a silicon transistor, no current will flow in the transistor until the input signal voltage has risen to about 0.7 V. Therefore there will be a period of time, when the signal is rising from 0 to 0.7 V, during which the transistor will not conduct. If the transistors of a push-pull stage are biased at cutoff, the current waveform will appear as in Fig. 37.1-3. The gaps in current, periods t_1 and t_2, represent the time when the signal polarity is changing, and when the signal is crossing over to activate one transistor while the other is turning off. No current flows during t_1 and t_2. The resulting current waveform is a distortion of the input waveform. Therefore to eliminate crossover distortion, the transistors in a push-pull stage are not biased class B (at cutoff), but are forward-biased slightly so that a small collector current flows even in the absence of any signal. However, they are biased close enough to cutoff for us to designate their mode of operation as class B.

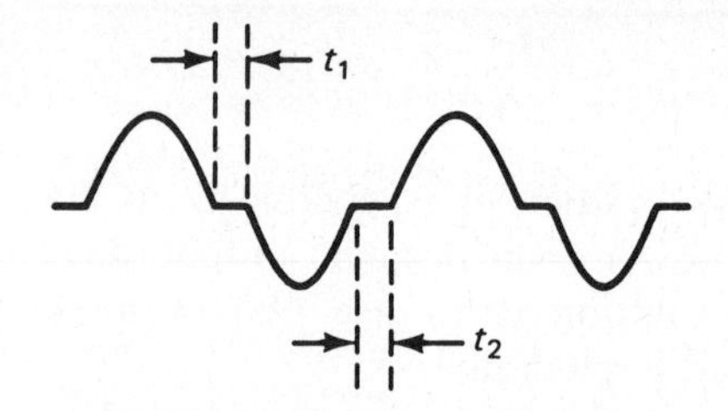

Fig. 37.1-3. Crossover distortion caused by biasing push-pull transistors at cutoff.

The emitters of Q_2 and Q_3 are returned directly to ground. The collectors are returned to $+V_{CC}$ through their respective load windings on the primary of T_3.

The low level of forward bias on Q_2 and Q_3 keeps the collector currents *very low* in the absence of signal. When a signal is applied, Q_2 and Q_3 conduct alternately on each half-cycle of the incoming signal. The average collector current is therefore *low without* signal and much higher in the presence of a signal.

The audio power delivered to a resistive load *R* may be computed, if the amplitude *V* of signal voltage appearing across the load is known, by using the formula

$$P = \frac{V^2}{R} \qquad (37.1\text{-}1)$$

where *V* is the rms voltage, given in volts, and *R* is the load resistance, measured in ohms.

SUMMARY

1. High-power audio systems normally use push-pull output stages dc-biased either class B or AB.
2. In class B operation the emitter-base section of a transistor is dc-biased at cutoff.

3. In class AB operation the transistor is dc-biased on, but the bias level is between class B and class A.
4. In class B operation the input signal causes collector current to flow for approximately 180°, during the interval when the *signal forward-biases* the base-to-emitter section.
5. In class AB operation collector current flows more than 180° but less than 360°.
6. The reason push-pull output amplifiers are used rather than single-ended class A output amplifiers in high-power audio systems is that push-pull stages can deliver more power to a load than a single-ended stage.
7. The power P delivered by the output stages to a resistive load R (Fig. 37.1-2) can be calculated using the formula

$$P = \frac{V^2}{R}$$

where V is the rms voltage measured across R, and R is the resistance in ohms of the load.

SELF-TEST

Check your understanding by answering these questions.

1. The output amplifiers in Fig. 37.1-2 are called ________ ________ amplifiers.
2. In a class A amplifier current flows during the ________ ________ of the input signal.
3. In a class B amplifier current flows for ________ degrees of the input signal.
4. Push-pull amplifiers are not dc-biased completely at cutoff, but close to cutoff, in order to prevent ________ ________.
5. The average direct current in the collector circuits of a class B push-pull amplifier is ________ ________ with signal than in the absence of signal.
6. In Fig. 37.1-2, T_2 is the ________ transformer, T_3 the ________ transformer.
7. The voltage divider R_5–R_6 provides some ________ ________ for the base-emitter section of ________.
8. The rms voltage measured across the secondary of T_3 (Fig. 37.1-2) is 1.5 V. The load resistance $R_L = 3.2\ \Omega$. The power delivered by the output stage to the load is ________ W.

MATERIALS REQUIRED

- Power supply: Variable, regulated low-voltage dc source
- Equipment: Oscilloscope; AF sine-wave generator; EVM; 0–100-mA dc milliammeter
- Resistors: ½-W 5-, 680-, 820-, 2200-, 8200-, two 18,000-Ω
- Capacitors: 0.02-μF; 0.1-μF; 25-μF/50-V; 100-μF/50-V
- Semiconductors: 2N6004, two 2N2102 with heat sinks (or their equivalents)
- Miscellaneous: SPST switch; 2-W 500-Ω potentiometer; push-pull input transformer (type Argonne #109 or equivalent); 3.2-Ω speaker push-pull output transformer (type Argonne #AR 119 or equivalent)

PROCEDURE

1. Connect the circuit of Fig. 37.1-4. M_1 is a 0–100-mA meter connected to measure the combined collector currents of Q_2 and Q_3. Check all circuit connections. S_1 is open. Output of power supply is set at zero.
2. Close S_1, applying power to the circuit. Gradually increase power supply output V_{CC} until milliammeter M_1 measures about 2 mA. (NOTE: V_{CC} will measure about 9 V; Q_2 and Q_3 will be biased close to cutoff.)
3. Connect an oscilloscope, set for proper viewing, across R_L, the output-load resistor. Connect a sine-wave generator set at 1000 Hz across R_1, the volume control. Generator output is at minimum level; R_1 is at maximum level.
4. You will observe a sine wave on the oscilloscope as the signal-generator attenuator is advanced. If, however, the output of the audio generator at its minimum setting overloads the amplifier, the signal on the oscilloscope will be distorted. In that case it will be necessary to reduce the input to the amplifier by readjusting R_1 until the distortion just disappears. When overload distortion has been eliminated, if there is crossover distortion, gradually increase the supply voltage V_{CC} until the crossover distortion disappears (V_{CC} should be about 9 V).
5. If initially the sine wave on the oscilloscope was undistorted, increase the output of the signal generator for *maximum* undistorted output.

   The audio generator and R_1 are now set to give maximum undistorted output at 1000 Hz. *Do not change their settings.*
6. Measure, and record in Table 37.1-1, the dc voltage with respect to ground at each of the test points (1 to 8) in Fig. 37.1-1.

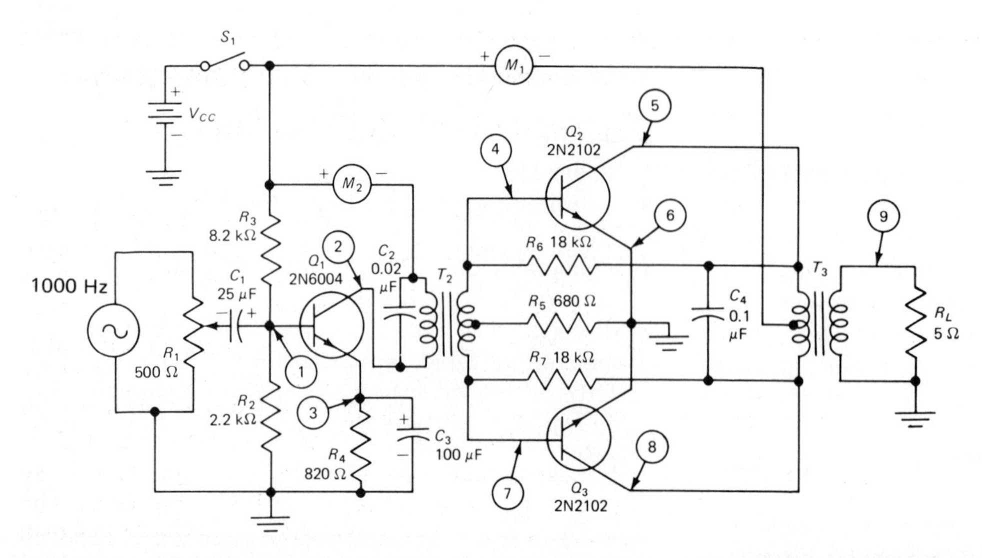

Fig. 37.1-4. Experimental push-pull amplifier.

7. With an oscilloscope, observe and measure the peak-to-peak amplitude of audio signal at test points 1 to 9. Record these in Table 37.1-1.
8. Observe and record the collector current I_C in stages Q_2 and Q_3 with signal applied as in step 4. Measure also the bias of Q_2 and Q_3. Record these data in Table 37.1-1.
9. Reduce R_1 to minimum (zero signal). Observe and record I_C in stages Q_2 and Q_3 with no signal. Measure also and record the bias of Q_2 and Q_3 without signal.
10. Replace R_L, the load resistor, with a 3.2-Ω speaker. Substitute an audio source, such as a record player or microphone, for the signal generator. Vary volume control from minimum to maximum undistorted level. Observe whether there is more audio power than with the single-ended class A power amplifier in Exp. 36.3.

TABLE 37.1-1. Transistor Push-Pull Amplifier Checks

Test Point	*Direct Current (to ground)*, V	*Signal*, V p-p
1		
2		
3		
4		
5		
6		
7		
8		
9		

	I_C, mA	*Bias* Q_2	*Bias* Q_3
With signal			
Without signal			

QUESTIONS

1. Explain the difference in operation between class A and class B push-pull amplifiers.
2. Refer to the data in Table 37.1-1. Are Q_2 and Q_3 in Fig. 37.1-4 operated class B? Explain why.
3. What are the operating advantages of a class B push-pull amplifier?
4. In Fig. 37.1-4 compute the power developed in R_L for maximum undistorted signal. Show your computations.
5. Interpret the meaning of the signal voltage readings, Table 37.1-1, at test points 3 and 6.
6. Explain the polarity of C_1, Fig. 37.1-4.
7. Explain the difference, if any, in collector current, Fig. 37.1-4, with and without signal. Refer to your data in Table 37.1-1.
8. Do the results of your experiment indicate that the class B push-pull amplifier can deliver more power than the class A output amplifier? Explain why.
9. Explain the differences between overload distortion and crossover distortion. Illustrate with waveforms showing each.
10. When does overload distortion occur? Crossover distortion?

Answers to Self-Test

1. push-pull
2. entire period
3. 180
4. crossover distortion
5. much greater
6. input; output
7. forward bias; Q_2
8. 0.703

EXPERIMENT 37.2. Complementary-Symmetry Push-Pull Amplifier

OBJECTIVE

To connect a class B push-pull complementary-symmetry audio power amplifier, measure the dc bias, and observe the waveforms in this amplifier

INTRODUCTORY INFORMATION

Complementary Symmetry (Two Power Supplies)

The class B push-pull audio power amplifier in Exp. 37.1 required the use of an input and output transformer. These transformers are not needed in a class B push-pull audio power amplifier employing complementary symmetry.

The complementary-symmetry circuit uses two transistors with identical characteristics. However, one of these is a PNP, the other an NPN type. Figure 37.2-1*a* shows an idealized complementary-symmetry push-pull amplifier. Q_2 is an NPN, Q_3 a PNP transistor, each connected as an emitter follower, with the emitters connected together. The load R_L in the emitter circuit is common to Q_2 and Q_3. The collector of Q_2 goes to $+V_{CC}$, a positive supply. The collector of Q_3 receives its *dc* voltage from $-V_{CC}$, a negative supply. The bases of Q_2 and Q_3, connected together, receive the input signal from some external circuit.

Assume that the dc bias keeps Q_2 and Q_3 just cut off. Now observe what happens when a sine wave is applied to the input of this amplifier. During the positive alternation (1) the base of Q_2 is driven positive relative to its emitter, turning on Q_2, the NPN transistor. Q_2 remains on during this positive alternation (1), and the current waveform in Q_2 is shown in Fig. 37.2-1*b*. During the positive alternation, Q_3, a PNP transistor, remains reverse-biased and is cut off. During the negative alternation (2), Q_3 is forward-biased by the signal and turned on while Q_2 is cut off. Figure 37.2-1*b* shows that current in Q_3 is opposite in direction to current in Q_2. This is so because Q_2 is an NPN, Q_3 a PNP transistor. The arrows in Fig. 37.2-1*a* show the direction of electron current in the external circuit of Q_2, Q_3, and the load R_L. The voltage developed across R_L is a sinc wave, like the input. Since the action of Q_2 and Q_3 complement each other and since the circuit is symmetrical, the arrangement is called *complementary symmetry*. It should be noted that complementary-symmetry circuits can be common-emitter type as well as common-collector type, and they may be operated class A, as well as class B.

Complementary Symmetry (One Power Supply)

The circuit in Fig. 37.2-1 utilized two power supplies, equal in voltage but opposite in polarity, with a common ground. The circuit of Fig. 37.2-2 uses a single supply for a push-pull complementary-symmetry amplifier. The symmetry of the circuit is maintained by two equal voltage dividers, R_1 and R_2. The top divider provides just enough forward bias for Q_2, the bottom divider for Q_3, to give each transistor a low idling current to eliminate crossover distortion.

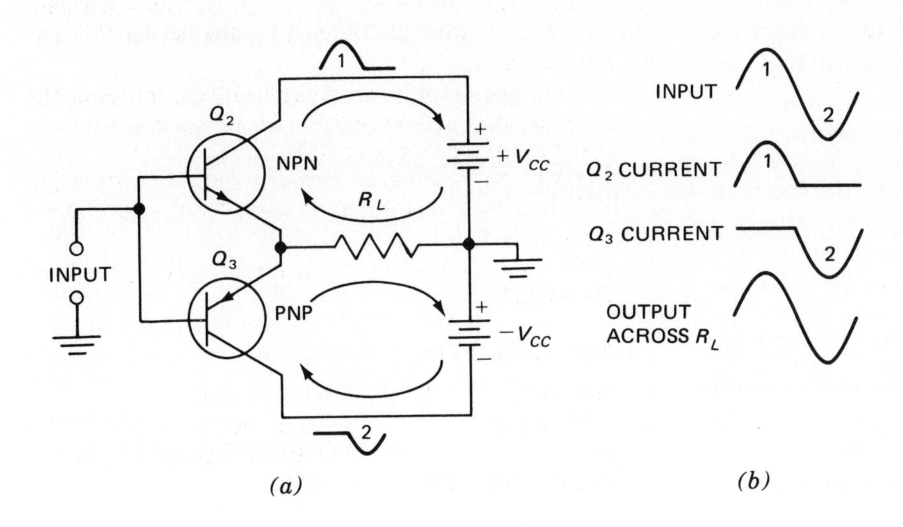

Fig. 37.2-1. (*a*) Idealized complementary-symmetry push-pull amplifier. (*b*) Current waveforms.

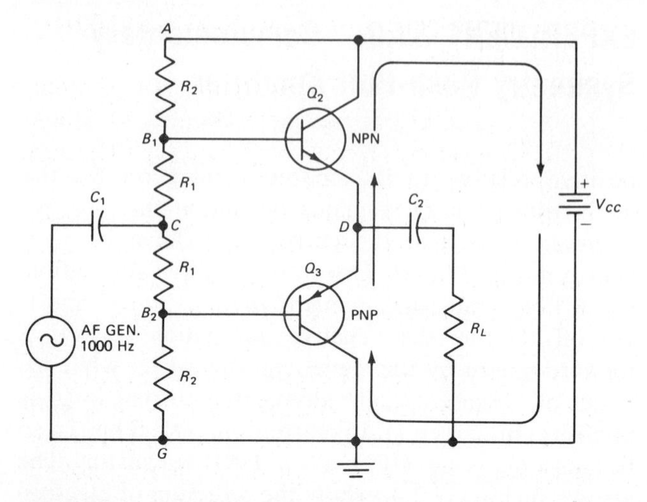

Fig. 37.2-2. Complementary-symmetry push-pull class B amplifier using one power supply.

The arrows in Fig. 37.2-2 show the path for this idling current in the external circuit of Q_2 and Q_3. Notice that the path for idling current includes Q_3, Q_2, and the power supply. Since Q_2 and Q_3 are assumed to have identical characteristics, point D is the *dc* voltage midpoint of the circuit, that is, $V_{AD} = V_{DG} = \frac{V_{CC}}{2}$. Similarly C is the dc voltage midpoint in the voltage divider between points A and G, assuming that both R_1s are equal in resistance, that both R_2s are equal in resistance, and that Q_2 and Q_3 have identical characteristics. Therefore a dc voltmeter connected between points C and D would measure 0 V because C and D are at the same dc potential with respect to ground G. By voltage divider action of top resistors R_1 and R_2, point B_1 is more positive than point C (and D). Hence the base of Q_2 (B_1) is positive relative to its emitter D, and this provides forward bias for Q_2 (NPN) to develop a low idling current. Similarly point B_2 is negative with respect to C (and D). Hence the base of Q_3 is negative relative to its emitter, providing forward bias for Q_3 (PNP).

The ratio of R_1 to R_2 is critical in setting the proper forward bias for a low idling current in Q_2 and Q_3.

The input signal is coupled by C_1 to the bases of Q_2 and Q_3 and is the same signal on each base. Q_2 is forward-biased by the positive alternation of the signal. The resulting current develops the positive alternation across R_L. Q_3 is forward-biased by the negative alternation of the signal and develops the negative alternation across R_L. The operation of Fig. 37.2-2 is, in all other respects, the same as that of the amplifier in Fig. 37.2-1 which uses two power supplies.

Complementary-symmetry push-pull class B amplifiers are frequently used in the output stage of high-power audio amplifiers. In such a system the speaker voice coil acts as the load and takes the place of R_L. The circuit may be so designed that capacitor C_2 may be eliminated, and the load is then directly connected between emitter and ground. This is possible because of the low-output impedance of the emitter-follower design.

SUMMARY

1. A complementary-symmetry amplifier uses two transistors, one a PNP, the other an NPN, with identical electrical characteristics.
2. A completely symmetrical circuit arrangement biases both transistors equally, so that each permits the same current to flow, in the absence of an input signal.
3. A complementary-symmetry push-pull amplifier may use one or two separate power supplies. It may be biased class B, or the push-pull amplifier may be operated class A.
4. The bias circuit of a class B push-pull complementary-symmetry circuit must provide a small idling current, in the absence of an input signal, to prevent crossover distortion.
5. The design of the complementary-symmetry audio output stage may permit direct connection of the speaker voice coil as the load in the output of the amplifier. This arrangement is possible where the amplifiers are designed as emitter-follower circuits.
6. Use of a complementary-symmetry class B audio output stage permits the advantages of class B operation, while eliminating costly input and output transformers.

SELF-TEST

Check your understanding by answering these questions.

1. In the class B circuit of Fig. 37.2-2, to prevent crossover distortion both Q_2 and Q_3 must be biased at cutoff in the absence of a signal. ________ (true/false)
2. A dc voltmeter connected across CD in Fig. 37.2-2 will measure ________ V, theoretically.
3. The dc voltage across AC, in Fig. 37.2-2, equals ________, if the circuit is completely balanced.
4. The ________ alternation of the input signal causes current to flow in Q_3.
5. In the absence of signal, in Fig. 37.2-2, the base of Q_2 is ________ (positive/negative) with respect to the emitter.

6. In Fig. 37.2-2 the dc voltage across Q_2, V_{AD}, is ________ (equal to/smaller than/greater than) the voltage across Q_3, V_{DG}.
7. In the circuit of Fig. 37.2-2, transistors Q_2 and Q_3 may be interchanged without affecting circuit performance. ________ (true/false)

MATERIALS REQUIRED

- Power supply: Two variable regulated low-voltage dc sources, or dual supply
- Equipment: Oscilloscope; EVM; AF sine-wave generator; 0–100-mA milliammeter
- Resistors: ½-W, 5-, two 100-, 470-, three 1000-, 1200-, 2200-, 4700-, 10,000-Ω
- Capacitors: 0.002-μF; 25-μF/50-V; four 100-μF/50-V
- Transistors: 2N2102 with heat sink, 2N4036 with heat sink, 2N6004 or their equivalents
- Miscellaneous: Two SPST switches; 3.2-Ω speaker; 500-Ω 2-W potentiometer; audio source (record player or microphone)

PROCEDURE

Single Power Supply Circuit

NOTE: If you have a single power supply, do steps 1 through 8. If a dual power supply is available, do steps 9 through 17.

1. Connect the circuit of Fig. 37.2-3. S_1 is **open.** Volume control is set for minimum output. Sine-wave signal generator is at 1000-Hz, 100-mV output. V_{CC} is at *zero* V.
2. Close S_1, **power on.** *Gradually* increase the output of V_{CC} until M_1 measures about 2 to 3 mA. (This is the idling current for Q_2 and Q_3.) Measure V_{CC}. It should be about 15 V.
3. Connect an oscilloscope across R_L, test point 8 to ground. Increase volume control output (and sine-wave generator level, if necessary) for the maximum undistorted signal across R_L.

NOTE: If the output signal shows evidence of crossover distortion, gradually increase V_{CC} until the distortion disappears. Recheck *idling current* in Q_2 and Q_3, *without signal.* It should not be greater than 5 mA. Distortion may also be due to improper biasing of Q_1. It may be necessary to change the 2.2-kΩ base resistor in Q_1 if this stage is not operating properly.

4. With maximum signal, measure and record in Table 37.2-1 V_{CC} and the dc voltages with respect to ground at test points 1 through 8. Measure also and record the current in Q_2 and Q_3.
5. With the oscilloscope, measure and record in Table 37.2-1 the peak-to-peak signal voltage at test points 1 through 8. Observe that Q_2 and Q_3 current is increasing. Why?
6. Remove the input signal. **Power off.** Permit transistors to cool about 3 min. Again measure, without signal, and record in Table 37.2-1 the dc voltages at test points 1 through 8. Measure also and record the idling current in Q_2 and Q_3. NOTE: If current is higher than original idling current, let current stabilize for about 3 to 5 min without

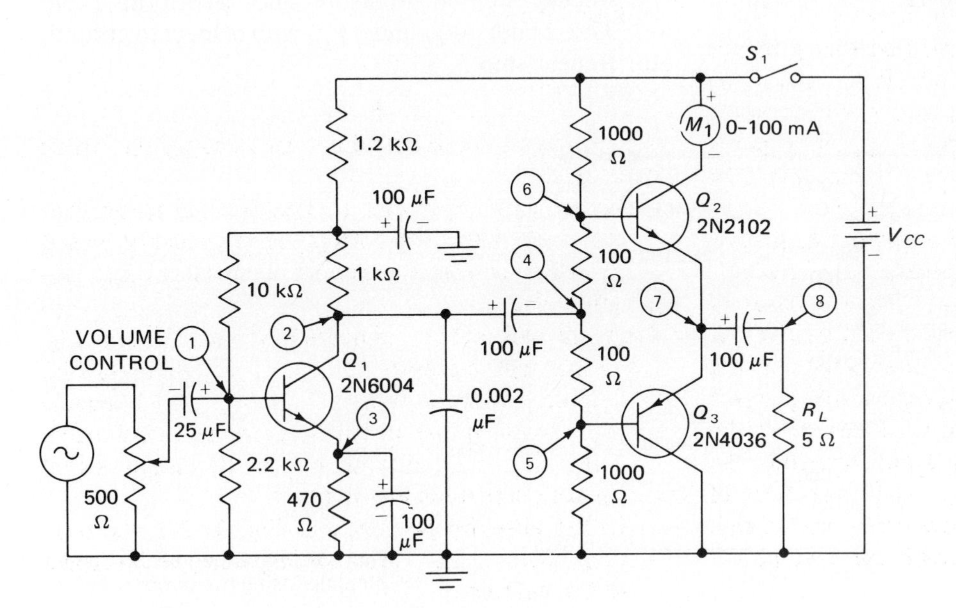

Fig. 37.2-3. Experimental audio amplifier using one power supply.

TABLE 37.2-1. Measurements in Audio Amplifier—One Power Supply

Test Point	*DC* V		*Signal* V p-p	*Current,* mA, *in* Q_2 *and* Q_3
	With Signal	*Without Signal*		
V_{CC}			X	With signal
1				
2				
3				
4				Without signal; idling
5				
6				
7				
8				

signal, or by shutting off power. Then measure idling current.

7. Replace R_L with a 3.2-Ω speaker. Replace the AF sine-wave generator with an audio source (record player or microphone). Adjust volume control for comfortable level. Check the quality of sound.
8. **Power off.**

Double Power Supply Circuit

9. Connect the circuit shown in Fig. 37.2-4. Two power supplies (or a dual supply) are used.
10. S_1 and S_2 are **open. Power off.** $+V_{CC}$ and $-V_{CC}$ are at zero V. Volume control is at minimum output. Sine-wave generator, at 1000 Hz, is at minimum output.
11. Close S_1 and S_2. **Power on.**

NOTE: In the power-supply adjustments which follow, it will be necessary to maintain the two supplies at the same voltage level. If a dual power supply with a *single* adjustable control is available, $+V_{CC}$ and $-V_{CC}$ will be automatically equal. However, if two separate supplies are used, the voltage of each must be monitored with a voltmeter to ensure that each delivers the same voltage.

12. Gradually increase the output of the two supplies, maintaining $+V_{CC}$ and $-V_{CC}$ equal in output, until the idling current, as measured by M_1, is about 2 to 3 mA. Measure $+V_{CC}$ and $-V_{CC}$. They should be about 7.5 V each.
13. Repeat step 3.
14. Repeat step 4. Measure and record in Table 37.2-2 both $+V_{CC}$ and $-V_{CC}$ with respect to ground.
15. Repeat step 5.

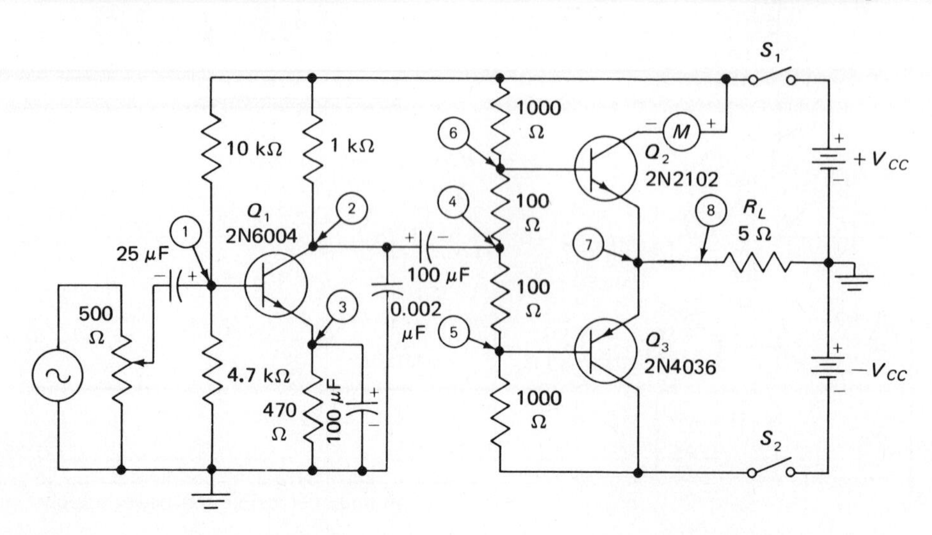

Fig. 37.2-4. Experimental audio amplifier using two power supplies.

TABLE 37.2-2. Measurements in Audio Amplifier—Two Power Supplies

Test Point	*DC* V		*Signal* V p-p	*Current,* mA, *in* Q_2 *and* Q_3
	With Signal	*Without Signal*		
$+V_{CC}$			X	With signal
$-V_{CC}$			X	
1				
2				
3				Without signal
4				
5				
6				
7				

16. Repeat step 6.
17. Repeat step 7.

QUESTIONS

Refer specifically to the data in Table 37.2-1 or 37.2-2, whenever possible, in answering these questions.

1. In the experimental amplifier, without signal, what was the base-emitter bias on Q_2? Was this forward or reverse bias? Explain.
2. In the experimental amplifier, without signal, what was the base-emitter bias on Q_3? Was this forward or reverse bias? Explain.
3. In Fig. 37.2-3, do your measurements confirm that the dc voltage, without signal at test points 4 and 7 with respect to ground, equals $V_{CC}/2$? If not, why not? (HINT: Was the circuit absolutely balanced?)
4. What was the voltage gain of Q_1? Show your computation.
5. In the experimental amplifier, compare the direct current in Q_2 and Q_3 with and without signal, and explain the difference, if any.
6. How does the experimental amplifier in Fig. 37.2-4 differ from that in Fig. 37.2-3?
7. Would driver Q_1 in Fig. 37.2-4 operate with S_1 ON and S_2 OFF? Explain.
8. Why do transistors Q_2 and Q_3 require heat sinks, when operated with signal, under full load?

Answers to Self-Test

1. false
2. 0
3. $V_{CC}/2$
4. negative
5. positive
6. equal to
7. false

CHAPTER 38 FREQUENCY RESPONSE OF AN AUDIO AMPLIFIER

EXPERIMENT 38.1. Frequency Response of an Audio Amplifier

OBJECTIVE

To measure the frequency response of an audio amplifier

INTRODUCTORY INFORMATION

Frequency Response

The range of sound frequencies that the average person can hear is 30 to 15,000 Hz (approximately). An audio amplifier which is designed to reproduce audio signal voltages faithfully should be able to amplify *equally* the signal voltages in this range of frequencies. Frequently, however, because of price and other practical considerations, audio amplifiers do not permit the entire range of audio frequencies to pass equally. In particular, the low frequencies (up to 500 Hz) and the high frequencies (beyond 5,000 Hz) may be attenuated.

An important characteristic of an audio amplifier, then, is the output voltage that it provides at each of the frequencies within the audio range for a given level of input signal. This characteristic can be determined by using an audio sine-wave generator as the signal source and an oscilloscope for observing and measuring the input and output signals. The gain is then computed at each frequency.

One method for checking the gain of an amplifier at a specific frequency is to introduce a measured signal voltage at that frequency into the amplifier and measure the output signal voltage. The gain of the stage is

$$\frac{v_{out}}{v_{in}} = \text{gain}$$

The frequency response of an amplifier is a plot of gain or output versus frequency at many frequency points. When these points are connected by a smooth curve, we have the response curve of the amplifier.

Figure 38.1-1 is the connection diagram for measuring the frequency response of an amplifier. The signal source is v_{in}, and the output signal voltage is v_{out}, whose peak-to-peak output is measured by an oscilloscope or EVM. If the output of the audio generator were constant over its entire frequency range, it would be possible to set v_{in} at some value, say, 1 V p-p, and simply monitor the output v_{out} with an oscilloscope or EVM. However, the output of the audio oscillator may vary under load with frequency. Therefore it is necessary to measure the output of the generator under load (v_{in}) for every frequency setting. This is the value of v_{in} that is used in substituting in the gain formula v_{out}/v_{in}.

Another method that can be used is to maintain the output of the audio generator at a fixed value of v_{in}, say, 1 V p-p. This again requires the use of the oscilloscope for monitoring v_{in} for every frequency setting of the generator and adjusting the output control of the generator to maintain v_{in} at the desired level.

If a 1-V p-p signal is used for v_{in}, the value of v_{out} is also the gain of the amplifier because

$$\text{Gain} = \frac{v_{out}}{v_{in}} = \frac{v_{out}}{1} = v_{out}$$

In plotting the response curve of an audio amplifier on graph paper, values of v_{out} are plotted vertically as the ordinates of the graph. The corresponding frequency settings are the abscissas.

SUMMARY

1. The frequency response of an audio amplifier describes how faithfully the amplifier reproduces

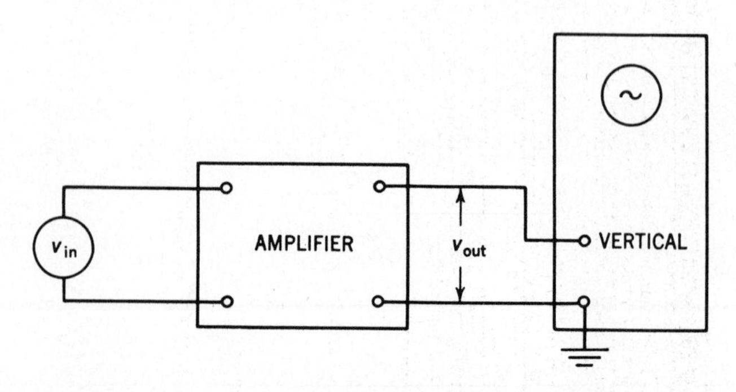

Fig. 38.1-1. Test setup for measuring frequency response of an amplifier.

the range of frequencies that people can hear, approximately 30 to 15,000 Hz.

2. This is the characteristic which indicates what output voltage the amplifier develops for the frequencies within the audio range, for a fixed level of input signal.
3. An audio sine-wave generator is used as a signal source, and an oscilloscope or EVM acts to measure the input and output voltages.
4. The frequency response is a graph of amplifier gain versus frequency, where gain is the ratio v_{out}/v_{in}.
5. *If the input signal to the amplifier is kept constant,* then the frequency response can also be a graph of v_{out} versus frequency.
6. In plotting the response curve, values of v_{out} or gain are plotted vertically as the ordinates of the graph; values of frequency are plotted horizontally as the abscissas.

SELF-TEST

Check your understanding by answering these questions.

1. The block diagram in Fig. 38.1-1 illustrates a method of connecting test equipment to check the __________ __________ of an amplifier.
2. The test instruments used in checking the frequency response of an amplifier are a (an) __________ and __________ or __________.
3. In graphing the frequency response of an amplifier, the horizontal axis is used for __________, the vertical axis for __________ or __________.
4. The range of audio frequencies which people can hear is approximately __________ to __________ Hz.
5. Is it possible to use a sine-wave generator as the signal source in making a frequency response check of an amplifier, even though the output of the generator is not constant over the range of test frequencies? __________ (yes/no)

MATERIALS REQUIRED

- Power supply: Variable, regulated low-voltage dc source
- Equipment: Oscilloscope; AF sine-wave generator; EVM
- Resistors: ½-W, 100-, 150-, 330-, 560-, 1000-, 8200-, 10,000-, 18,000-, 33,000-Ω
- Capacitors: Two 25-μF/50-V; two 100-μF/50-V; 0.005-μF
- Transistors: 2N2102, 2N6004, or their equivalents
- Miscellaneous: 2-W/500-Ω potentiometer; SPST switch

PROCEDURE

1. Connect the circuit of Fig. 38.1-2. S_1 is open. Set the output of an AF sine-wave generator at zero and the volume control for minimum; that is, set arm C at point B. Connect an oscilloscope from the collector of Q_2 to ground.
2. Close S_1. **Power on.** Close S_2. Set the frequency of the generator to 1000 Hz and increase the output until a 100-mV sine wave v_{AB} is applied across AB. Now gradually increase the volume control setting until a 4-V p-p output sine-wave v_{out} is measured at the collector of Q_2. *Do not change the setting of the volume control during the remainder of this experiment.*

NOTE: If a 100-mV signal is maintained across AB and if the setting of the volume control is not changed, the signal measured between point C and ground will

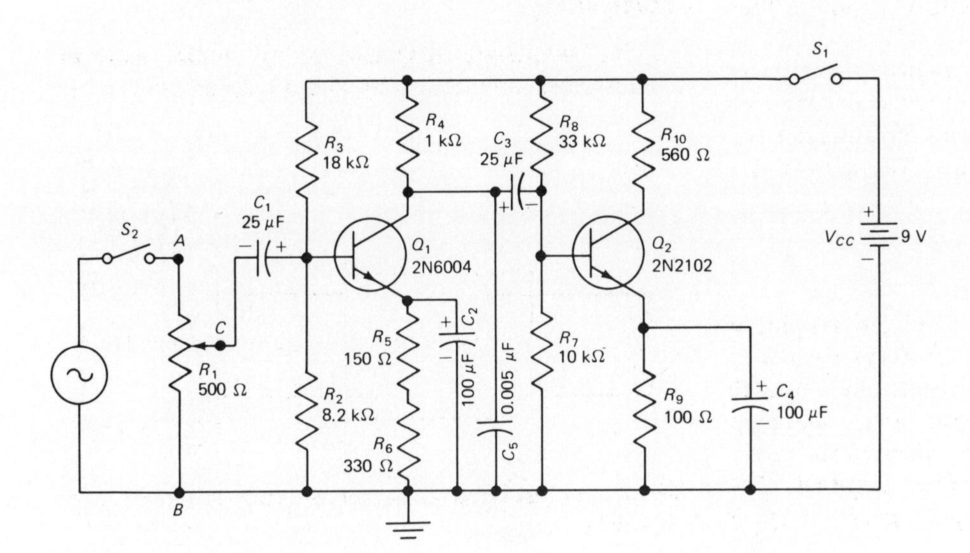

Fig. 38.1-2. Circuit used in determining frequency response of two-stage audio amplifier.

TABLE 38.1-1. **Amplifier Frequency Response**

v_{AB}: 100 mV	R_{CB}, Ω:	R_{AB}, Ω:
Frequency, Hz	v_{out}, p-p	
30		
40		
60		
100		
200		
400		
600		
1000	4.0	
2000		
3000		
4000		
5000		
6000		
8000		
10,000		
12,000		
15,000		

remain constant within the range of audio frequencies delivered by the generator.

3. Set the frequency of the generator at 30 Hz. Measure the signal across *AB*, and readjust generator output, if necessary, until the signal level reaches 100 mV as in step 2.
4. Measure and record the output signal voltage v_{out} from the collector of Q_2 to ground.
5. Repeat steps 3 and 4 for every frequency setting of the sine-wave generator as shown in Table 38.1-1.
6. Plot the response curve of the amplifier on graph paper and label it 1.

QUESTIONS

1. What is meant by the "frequency response" of an amplifier?
2. Why is it necessary to maintain a constant-level input signal across the volume control when measuring frequency response?
3. Compute the total gain of the amplifier at 1000 Hz. Show your computations. Why is it necessary to know the ratio R_{CB}/R_{AB} in computing the gain?
4. If the amplifier response is considered "flat" for those frequencies whose output is 70 to 100 percent of maximum, what is the range of these frequencies? Refer to your graph.

Answers to Self-Test

1. frequency response
2. sine-wave generator; oscilloscope; EVM
3. frequency; signal output; gain
4. 30; 15,000
5. yes

CHAPTER 39 TROUBLESHOOTING AN AUDIO AMPLIFIER

EXPERIMENT 39.1. Troubleshooting an Audio Amplifier

OBJECTIVES

1. To isolate a defective stage in an audio amplifier by signal tracing
2. To isolate a defective component by voltage and resistance checks

INTRODUCTORY INFORMATION

Troubleshooting Procedure

Locating a trouble in a defective electronic device, such as a multistage audio amplifier, requires a logical, systematic procedure. An effective first step in such a procedure is to assume that there is just *one* defect causing the trouble. The technician tries to locate that defect in the simplest and fastest way. When the trouble has been found and corrected, amplifier operation should be restored to normal. If it is not, then the technician must seek a second possible source of trouble, and if necessary, a third, etc. The steps in a logical procedure are:

1. Listen to the amplifier's performance. For this, an audio source, such as an AF sine-wave generator or a record player, will be needed. The symptoms, coupled with past knowledge and experience, may lead to a rapid diagnosis.
2. Inspect the audio amplifier for obvious defects, such as open connections, improper connections, burned resistors, broken parts, overheated parts. Burned-out resistors and transformers have a characteristic odor. *Determine the reason for the failure of a defective part before replacing it.*
3. Measure the supply voltage to determine if that is the source of trouble.
4. *Isolate the defective stage* by signal-tracing the amplifier. When you find the defective stage, the trouble must be in one or more of the components in that stage.
5. *Isolate the defective part.* Make dc voltage checks in the defective stage and compare them with the rated values for this stage. If no ratings are available, the technician should be able to determine approximate values from a knowledge of the circuit. Refer to Exp. 34.2 for a discussion of voltages which may be expected in an amplifier. Analyze any differences between rated and measured values for inferences as to which is the defective part. Then test the suspected part, or substitute a known good part to verify that the trouble has been located. Supplement dc voltage checks with resistance checks in the suspected circuit. See Exp. 34.3 for a discussion of resistance norms.
6. *Correct the defect.* The replacement part must meet the electrical and physical specifications of the original. In servicing a commercial amplifier, use the manufacturer's recommended replacement parts.

Isolating a Defective Stage by Signal Tracing

Figure 39.1-1 is a block diagram of a three-stage audio amplifier which includes a predriver, driver, and complementary-symmetry output amplifier. The dynamic test points for signal-tracing this amplifier are numbered 1 through 7. The detailed block diagram does not show the circuitry but does identify the transistor elements where the test points are located. Assume that the amplifier is dead (no sound is heard in the speaker) and we wish to isolate the defective stage by signal tracing. The procedure which follows describes the process.

1. Connect an AF generator to the input of the amplifier. A 1000-Hz signal set at a moderate level acts as the signal source.
2. With an oscilloscope trace the progress of the signal from the input (base of Q_1) to the output (voice coil of the speaker).
3. If the amplifier were operating normally, we would expect to find the 1000-Hz sine wave at each test point 1 through 7. The level (amplitude) of the sine wave at each test point (TP) is approximately:

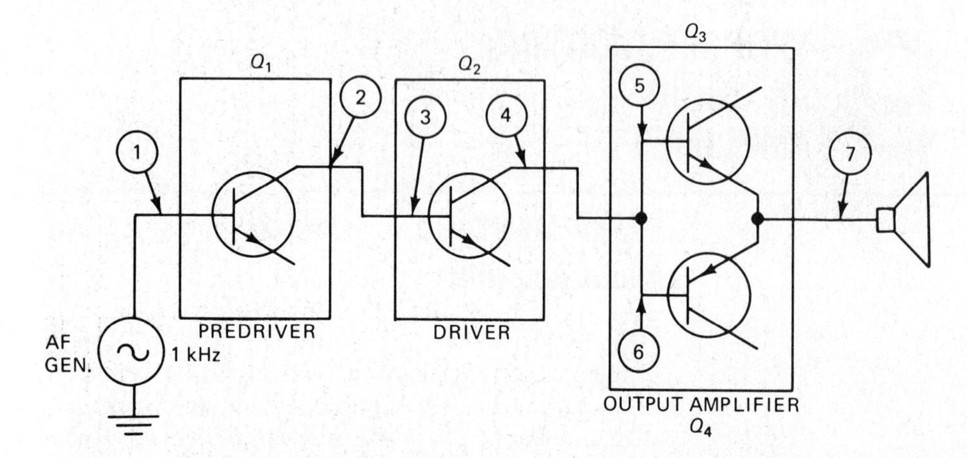

Fig. 39.1-1. Detailed block diagram of three-stage audio amplifier showing dynamic signal-tracing test points.

TP 1 Low level, determined by generator output.
TP 2 Higher peak-to-peak signal than at TP 1 (because of gain of Q_1).
TP 3 The signal at TP 2 and at TP 3 should be approximately the same level.
TP 4 Higher peak-to-peak signal than at TP 3.
TP 5 Same level signal as at TP 4.
TP 6 Same level signal as at TP 4.
TP 7 We have assumed a complementary-symmetry, emitter-follower output stage. In this type of circuit the output sine wave at the emitter will be somewhat lower in level than the signal at test points 5 and 6.

Circuit variations from the amplifier in the block diagram must be treated individually, depending on the actual circuit used. For this reason, the technician should have at hand the diagram of the circuit being signal-traced in order to identify the dynamic test points and know what to expect at each.

4. If the amplifier were defective (completely dead), it is clear that the signal would disappear at some point in the circuit. Suppose, for example, that the signal is normal up to TP 3, but disappears at TP 4. A logical inference is that either the driver stage (Q_2) is defective or a defective output stage (Q_3–Q_4) is loading Q_2 and preventing it from functioning properly. If possible, unload Q_3–Q_4 and check the signal again at TP 4. If it is *now* normal, the trouble is in Q_3–Q_4. But if there is still no signal at TP 4, the trouble is in Q_2. In this *signal-tracing* process, the trouble has been isolated to the defective stage. Now it is possible to examine the suspected stage to determine the defect.

Isolating the Defective Part by Measurement

Assume that trouble has been isolated to the driver Q_1 in Fig. 39.1-2. A 60-mV sine wave is measured at the base, TP 1, but there is no signal at the collector, TP 2. A dc voltage check at the base, emitter, and collector of Q_1 may suggest the trouble.

Normal dc voltages at the test points are: collector, +6.6 V; base, +1.8 V; emitter, +1.2 V. Assume that the measured voltages in the defective amplifier are: collector, +9 V; base, +1.9 V; emitter, 0 V. The tran-

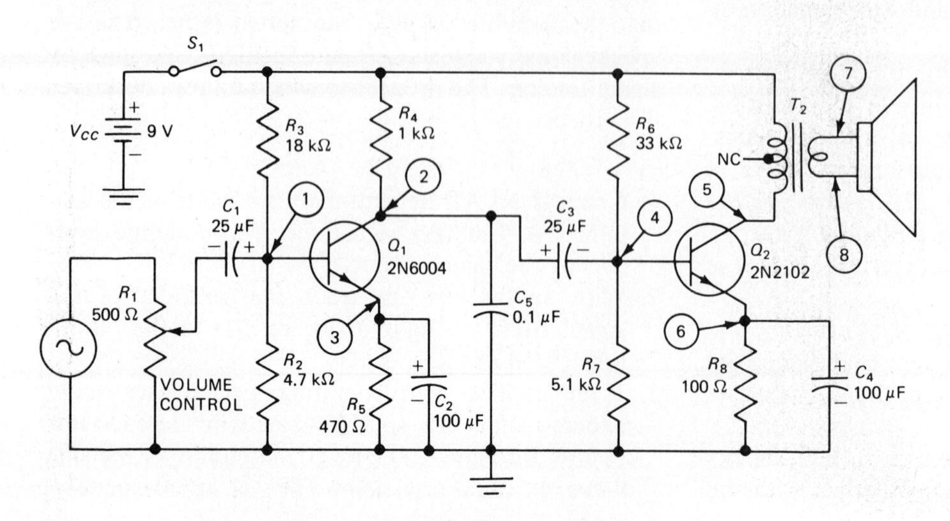

Fig. 39.1-2. Isolating the defective stage by signal tracing.

sistor is obviously not conducting, since there is no voltage drop across the emitter or collector resistors. The measurements suggest a defective transistor. A replacement of Q_1 with a known good 2N6004 transistor should clear up the trouble.

A further check of the transistor is possible before replacement, with an in-circuit transistor checker or curve tracer.

In the trouble just discussed it was very evident that the transistor was defective. More subtle troubles frequently require both voltage and resistance checks to find the bad part.

NOTE: Solid-state audio amplifiers are generally more complex than the circuit in Fig. 39.1-2, using some form of dc bias and ac stabilization. These feedback circuits make troubleshooting more challenging. Chain reaction failures may occur, and feedback paths must be opened to isolate the trouble.

SUMMARY

1. Troubleshooting an electronic device, such as an audio amplifier, requires a logical, systematic procedure.
2. A troubleshooting assumption which helps simplify the process is that there is just *one* defect. When that defect is found, if the circuit is still not operating properly, the technician looks for another defect, continuing until all the troubles have been found and corrected.
3. The steps in a logical troubleshooting procedure for an audio amplifier are:

   (*a*) Listen to the amplifier for a clue to the trouble.
   (*b*) Inspect the amplifier for obvious defects such as broken, burned, or open parts.
   (*c*) Measure the dc supply voltage to be sure that it is normal.
   (*d*) Isolate the defective stage in a multistage amplifier by signal tracing, that is, by following with an oscilloscope the progress of the signal from the input to the output.
   (*e*) When a discontinuity in the signal occurs, we assume that everything is working properly through the last test point where a normal signal was found. The trouble must be in the stage where the signal disappeared, or in the loading of that circuit by the next stage.
   (*f*) Voltage and resistance checks of the suspected stage should lead to the defective part.

4. An in-circuit transistor tester or an in-circuit curve tracer is helpful in locating defective transistors.

SELF-TEST

Check your understanding by answering these questions.

1. The amplifier shown in Fig. 39.1-2 is dead. An inspection of the circuit does not reveal any obvious defects. The first measurement to make is the ________ ________.
2. The amplifier, Fig. 39.1-2, is dead. In signal-tracing the circuit a normal waveform appears at TP 2, but there is no waveform at TP 4. A possible cause of trouble is an ________ ________.
3. Trouble has been isolated to Q_2, Fig. 39.1-2. A dc voltage check of Q_2 shows the following: collector, 0 V; base, 1.2 V; emitter, 0 V. The most likely cause of trouble is: (*a*) defective 2N2102, (*b*) open R_8, (*c*) open primary in T_2, (*d*) primary to secondary short in T_2.
4. Volume in the amplifier, Fig. 39.1-2, is very low. Signal tracing reveals that a normal signal exists up to TP 4. There is a sine wave at TP 5, but its amplitude is very low. There is a sine wave at TP 6, almost equal in amplitude to the signal at TP 4. DC voltage measurements are all normal. The most likely cause of trouble is: (*a*) defective 2N2102, (*b*) open secondary in T_2, (*c*) open R_8, (*d*) open C_4.
5. The amplifier in Fig. 39.1-2 is dead. A check shows that there is +9 V at the output of the power supply, but 0 V at the junction of R_3, R_4, R_6, and the primary of T_2. A likely cause of trouble is: (*a*) primary of T_2 is shorted to the frame, (*b*) defective S_1, (*c*) defective Q_1, (*d*) shorted C_5.
6. Trouble in Fig. 39.1-2 has been isolated to Q_1. A resistance check to ground at the elements of Q_1 shows the following: collector, 0 Ω; base, 4700 Ω; emitter, 470 Ω. Most likely cause of trouble is: (*a*) shorted C_5, (*b*) open C_5, (*c*) open R_4, (*d*) shorted R_4.

MATERIALS REQUIRED

- Power supply: Variable, regulated low-voltage dc source
- Equipment: Oscilloscope; EVM; AF sine-wave generator; transistor tester
- Miscellaneous: A commercial low-power audio amplifier and speaker with accompanying circuit diagram which is defective or into which a defect has been inserted

NOTE: If a commercial audio amplifier is not available, the student may connect any of the audio amplifiers (including that shown in Fig. 39.1-2) used in previous experiments. The instructor will insert defects in the amplifier for troubleshooting purposes.

PROCEDURE

1. Your instructor will give you a defective amplifier or will introduce trouble into an audio amplifier which you have connected, and you will service it, following the techniques discussed in this experiment. Keep a step-by-step record of the checks made, as you make them, recording these steps in the following troubleshooting form. When you have found the trouble, correct it and notify your instructor, who will give you another amplifier to service. Service it and fill out a separate troubleshooting report. Service as many troubles as time will permit.

Answers to Self-Test

1. supply voltage
2. open C_3
3. (*c*)
4. (*d*)
5. (*b*)
6. (*a*)

TROUBLESHOOTING REPORT

Date ________

Name ____________________ Amplifier Make ____________________

Teammates ____________________ Model # (or Fig. #) ____________________

1. Describe symptoms. ____

____

2. Preliminary inspection. ____

____

3. Supply-voltage(s) measurement: __________ V

4. Describe procedure (list steps in order performed).

Test point	*Voltages, ac or dc*	*Other*

5. Describe trouble found and list parts used for repair. ____________________

____

____

6. Was amplifier operation normal after repair? ____________________

If not, describe trouble: ____

____

7. Time required to make repair: ____________________

8. Approved: ____________________

9. Instructor's signature: ____________________

CHAPTER 40 FIELD-EFFECT TRANSISTORS AND AMPLIFIERS

EXPERIMENT 40.1. Field-Effect Transistor Characteristics

OBJECTIVE

To observe the drain characteristic curves of a metal-oxide semiconductor field-effect transistor (MOSFET) using a curve tracer and to graph these curves, using a point-by-point method

INTRODUCTORY INFORMATION

Junction Field-Effect Transistor (JFET)

The transistors you have studied to this point are called *bipolar,* because their operation depends on the action of *two* types of charge carriers, holes and electrons. Field-effect transistors are unipolar devices because their action depends on only *one* type of charge carrier. FETs include the junction type (JFET) and the metal-oxide semiconductor type (MOSFET). JFETs were developed before MOSFETs.

FETs are used in electronic voltmeters, in television tuners, in FM receivers, and in other electronic applications.

Figure 40.1-1 illustrates an N-channel JFET. The three elements of the transistor are called the *source, gate,* and *drain.* The body or channel of the transistor is an N-type semiconductor. Terminal leads for the drain and source make ohmic contacts at the top and bottom of the channel. They are *not* semiconductor junctions. The P-type material, called gate, embedded on both sides of the channel does form a semiconductor junction, hence the name *junction* FET.

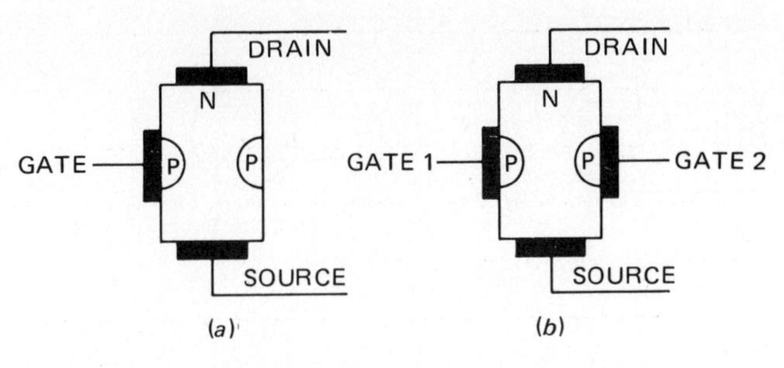

Fig. 40.1-1. N-channel JFET: (*a*) single gate; (*b*) dual gate.

Ohmic contacts to the P-type material serve as terminal leads for the gates. When gates 1 and 2 are internally connected in the manufacturing process, the device is a single-gate FET, Fig. 40.1-1*a*. When separate leads are brought out at each junction, a dual-gate FET results, Fig. 40.1-1*b*.

In the FET the drain corresponds to the collector of a bipolar transistor, the source to the emitter, and the gate to the base. However, the operation of a unipolar is completely different from a bipolar transistor. The chief operational difference is that drain current I_D in the JFET is controlled by gate-to-source *voltage* (V_{GS}), whereas collector current in the bipolar transistor is controlled by base *current.*

To understand the operation of a JFET, consider the N-channel semiconductor with ohmic contacts at the top (drain) and bottom (source) of the channel as in Fig. 40.1-2. If a battery V_{DD} is connected across the channel, with the polarity shown, the negative charge carriers (electrons) in the N channel will move toward the positive terminal of the battery, and electrons from the negative terminal of the battery will move through the source into the N channel to replace those that left at the drain. Current in this circuit will continue as long as the circuit is complete. A *limited control* of current is possible by varying V_{DD}.

A much more effective way of controlling drain current is by adding a gate and *reverse-biasing* the

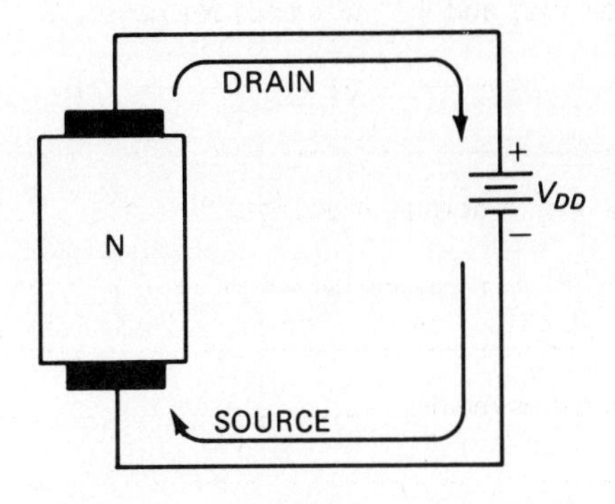

Fig. 40.1-2. Voltage source across an N-channel semiconductor will cause current flow in the circuit.

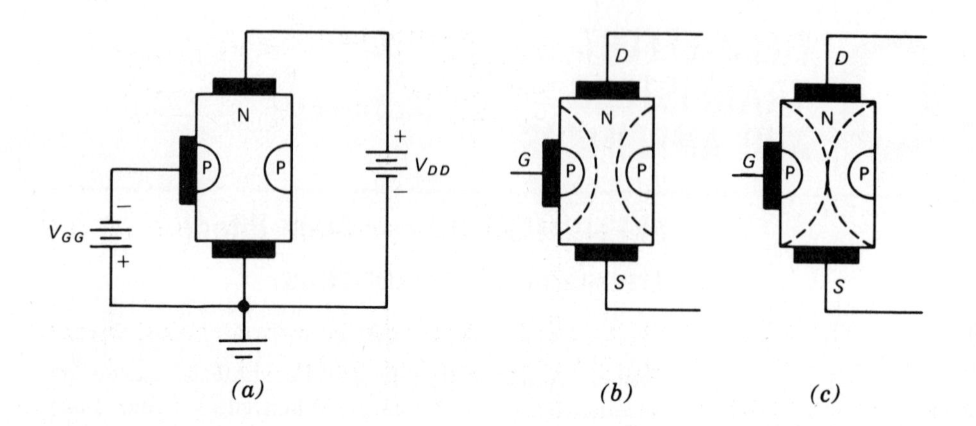

Fig. 40.1-3. (*a*) Negative-biasing the gate to source has the effect of (*b*) restricting the channel. (*c*) If the negative gate bias is made high enough, drain current is cut off.

gate with respect to the source, as in Fig. 40.1-3*a*. Observe that the first effect of adding the gate, even when unbiased, is to *narrow* the channel somewhat physically. This immediately restricts the current in the channel. By reverse-biasing the gate, the electric field at the junction has the effect of further reducing the channel width, as in Fig. 40.1-3*b*. If the negative bias is increased sufficiently, the gate becomes so wide, Fig. 40.1-3*c*, that the channel is blocked and *no* drain current flows.

Note that there is no gate current because the gate junction is reverse-biased. So gate control of drain current I_D results from varying the *negative voltage* on the gate, *not* from gate current.

The FET we have been discussing is an N-channel, P-gate device whose symbol is shown in Fig. 40.1-4. Observe that the gate arrow points toward the N channel (vertical line). The gate arrow may be centered on the channel, Fig. 40.1-4*a*, or the gate arrow may be drawn close to the source, Fig. 40.1-4*b*.

It is also possible to manufacture a P-channel, N-gate JFET whose symbol is shown in Fig. 40.1-5*a* and *b*. For a P-channel device the arrow points away from the channel. For a P-channel JFET all battery polarities V_{GG} and V_{DD} must be reversed in connecting the FET in a circuit.

MOSFET

Like the JFET, the MOSFET is a field-effect transistor whose *drain current* I_D *is controlled by the voltage on the gate*. There are physical differences between the MOSFET and the JFET and differences in operation. The manner in which the MOSFET is constructed determines whether it is a depletion type or an enhancement type. The MOSFET has a much higher input impedance than the JFET.

Enhancement-type MOSFET

In this type there is no channel between the drain and source, but separating the N-type drain and source is a P-type substrate, as in Fig. 40.1-6. Deposited over the substrate is a *very thin layer* of silicon dioxide (SiO_2), an insulator. Deposited on the SiO_2 is a metallic film, which acts as the gate. Ohmic contacts are brought out for the gate, drain, source, and substrate. When the substrate is internally connected to the source, there is no substrate lead.

From Fig. 40.1-6 it is clear that the gate is insulated from the body of the FET, and for this reason the MOSFET is also called an insulated-gate FET (IGFET). Though there is no physical channel, the MOSFET in Fig. 40.1-6 is called an N-channel en-

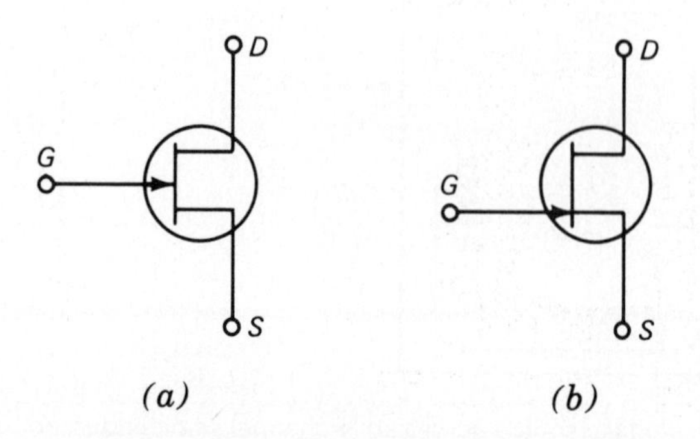

Fig. 40.1-4. Symbols for N-channel JFET.

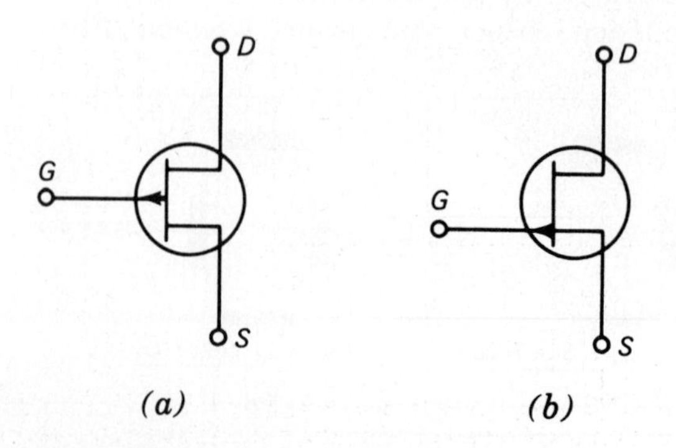

Fig. 40.1-5. Symbols for P-channel JFET.

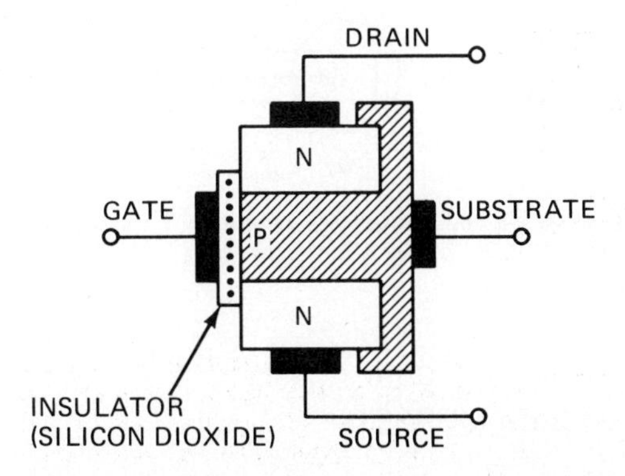

Fig. 40.1-6. Construction of an enhancement-type MOSFET.

hancement type, for reasons which will become clear.

The gate and the substrate act as the plates of a capacitor separated by the SiO_2 insulator. When the gate is made positive relative to the source to which the substrate is connected, as in Fig. 40.1-7, the capacitor charges. Since the gate is positive, negative charges appear in the substrate between drain and source, Fig. 40.1-7*b*, in effect creating an N channel which will permit current in the source-to-drain circuit.

The N-channel enhancement-type MOSFET will conduct only when the gate is positive relative to the source. It is cut off when there is zero or negative gate-to-source bias. For this reason it is designated a normally OFF MOSFET. Because the gate is insulated from the substrate, there will be no dc gate current even though the gate is positive relative to the substrate. Therefore the MOSFET is a high-impedance transistor.

It should be noted that a P-channel normally OFF MOSFET may be made by using an N substrate and P-type drain and source. The symbols for N- and P-channel enhancement types are shown in Fig.

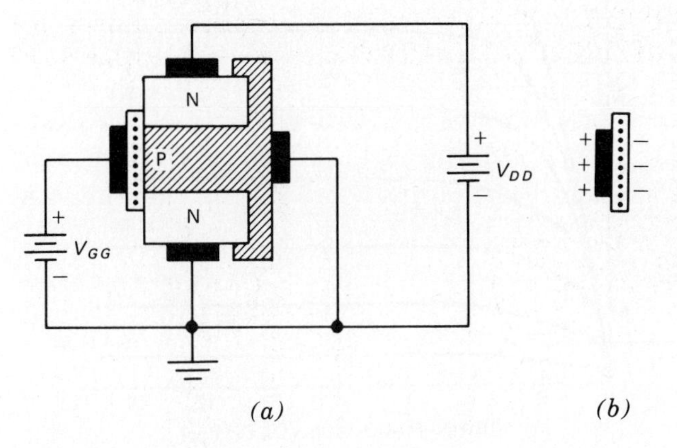

Fig. 40.1-7. (*a*) Biasing an enhancement-type MOSFET. (*b*) By capacitor action an N channel is induced between source and drain.

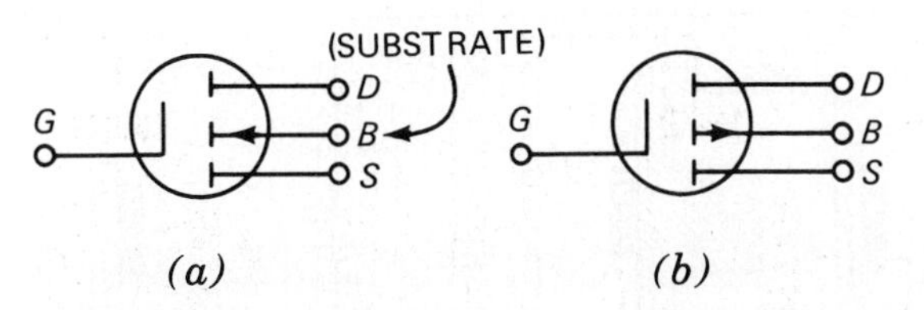

Fig. 40.1-8. Symbols for enhancement-type MOSFET: (*a*) N channel, (*b*) P channel.

40.1-8*a* and *b*. Note that the broken vertical channel line indicates a normally OFF MOSFET.

Depletion-type MOSFET

The MOSFET in Fig. 40.1-9 is constructed like the one in Fig. 40.1-6 with one exception. The one in Fig. 40.1-9 *has* an N-type channel, while the one in Fig. 40.1-6 does not. The FET in Fig. 40.1-9 can be operated with either a positive or a negative gate. When the gate is positive, the FET operates in the enhancement mode; when negative, in the depletion mode.

Consider the effect of a positive gate on the charge carriers in the N channel, shown in Fig. 40.1-10*a*. The gate-to-channel capacitor charges, inducing negative charge carriers in the N channel, Fig. 40.1-10*b*, thus increasing the conductivity of the channel and increasing the drain current. The more positive the gate is made, the more drain current flows. This is the enhancement mode.

When the gate is shorted to the source, that is, when there is zero gate voltage, drain current will flow in the channel, from source to drain, if a voltage source V_{DD} is connected as in Fig. 40.1-10*a*. The drain current with zero gate voltage is less than the drain current with positive gate voltage. Since there is drain current, even with 0 V gate bias, this is a normally ON MOSFET.

What happens when the gate is made negative relative to the source, as in Fig. 40.1-11*a*? The

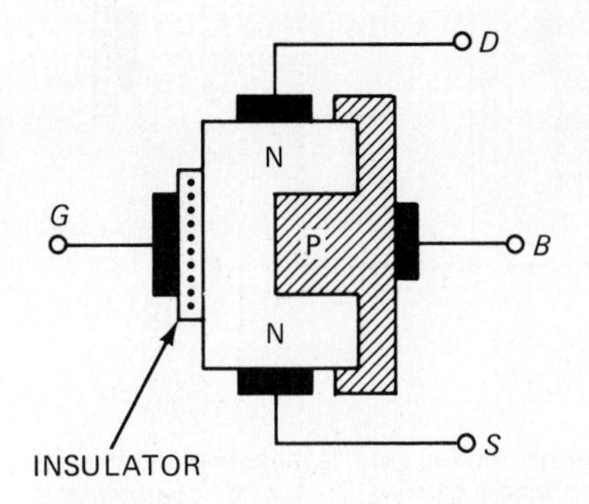

Fig. 40.1-9. Depletion enhancement-type MOSFET.

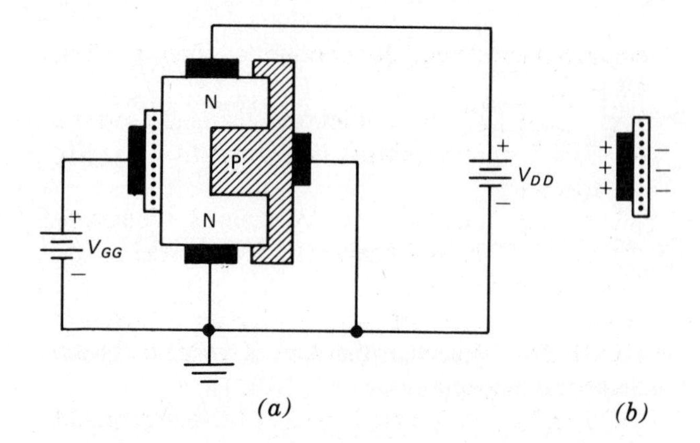

Fig. 40.1-10. (a) Enhancement mode; gate is positive relative to source. (b) Negative charge carriers are induced in the N channel.

capacitor charge distribution on the gate and N-type channel is shown in Fig. 40.1-11*b*. Electrons on the gate repel the negative-charge carriers on the N channel, in effect depleting the channel of negative carriers, reducing drain current. The more negative the gate, the less the drain current. When V_{GS} is made sufficiently negative, drain current is cut off.

The depletion MOSFET may be constructed with a P channel and N substrate. The symbols for N- and P-channel depletion MOSFETs are shown in Fig. 40.1-12. Note that the vertical channel line is not broken here because the device is normally ON.

Figure 40.1-13 shows a family of drain curves for an N-channel depletion MOSFET. Notice that gate voltage practically determines drain current, and that drain voltage has little effect on drain current. Notice also that for this N-channel depletion-type MOSFET, the more positive the gate voltage, the greater is the drain current.

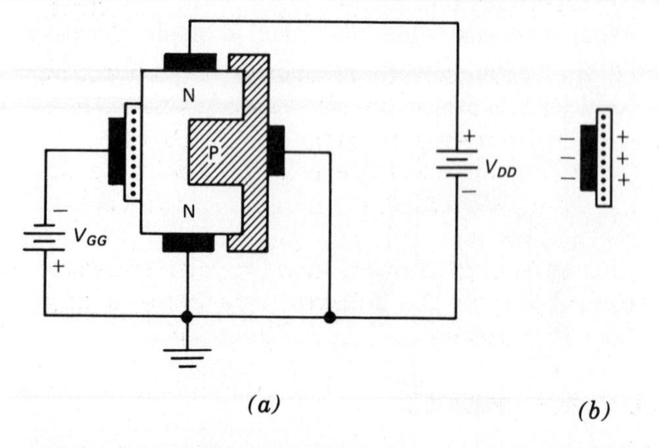

Fig. 40.1-11. (a) Depletion mode; gate is negative relative to source. (b) Negative charge carriers in the N channel are depleted.

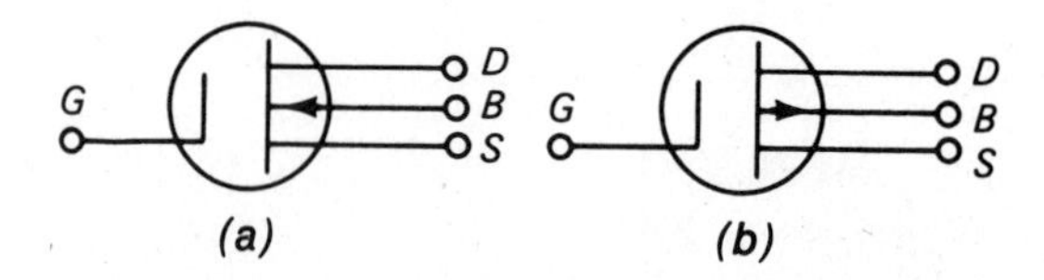

Fig. 40.1-12. Symbols for depletion-type MOSFET: (a) N channel; (b) P channel.

Dual-Insulated-Gate MOSFET

For special applications MOSFETs may be constructed with two gates, G_1 and G_2, as in Fig. 40.1-14. This is the diagram of a four-terminal 3N187, a device which may be used as an amplifier for frequencies up to 300 MHz. Gate 1 is at terminal 3, gate 2 at terminal 2. The drain is terminal 1, the source is terminal 4. Note that the substrate is internally connected to the source. Two sets of back-to-back diodes are built into this device. One set of diodes is internally connected between gate 1 and the source, the other set between gate 2 and the source. The manufacturer states that the diodes bypass any voltage transients which exceed approximately ± 10 V, protecting the gates against damage in normal handling and use.

You will recall that the silicon dioxide insulator in a MOSFET is a thin glass-like deposit on the substrate. It may be easily fractured by static voltages to which the MOSFET is subject in ordinary handling. The back-to-back diodes connected between gates and source protect the MOSFET against static charges and other voltage transients.

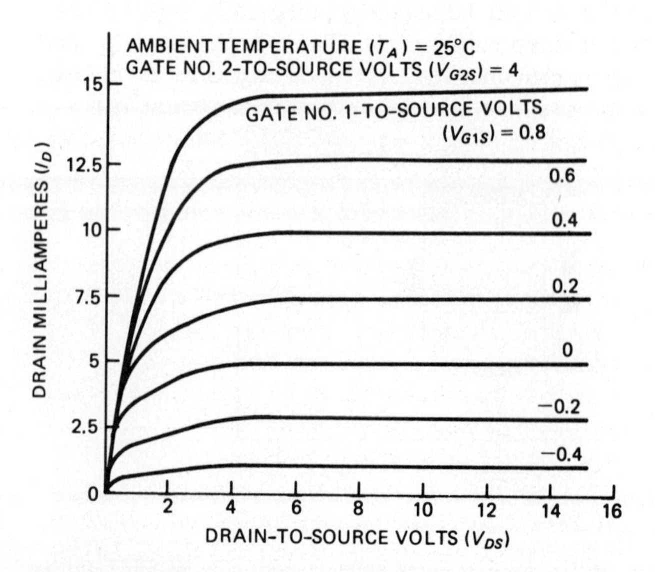

Fig. 40.1-13. Drain curves for 3N200, N-channel depletion MOSFET. (*RCA*)

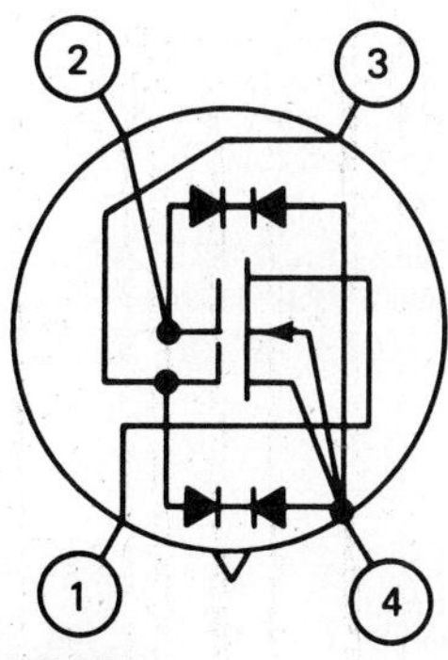

LEAD 1-DRAIN
LEAD 2-GATE NO. 2
LEAD 3-GATE NO. 1
LEAD 4-SOURCE, SUBSTRATE, AND CASE

Fig. 40.1-14. Terminal diagram of a 3N187, dual-insulated-gate MOSFET. (*RCA*)

SUMMARY

1. FETs are called unipolar transistors because they use only one charge carrier, that in the channel, for their operation.
2. The two types of FETs are the *junction* (JFET) and the *metal-oxide semiconductor* (MOSFET).
3. The elements of a FET are the drain, source, and gate. These correspond, respectively, to the collector, emitter, and base of a bipolar transistor.
4. A FET is a voltage-controlled device, whereas a bipolar transistor is a current-controlled semiconductor.
5. In the JFET the gate is biased negatively, relative to the source.
6. The two basic types of MOSFET are (*a*) depletion and (*b*) enhancement types.
7. The gate of a MOSFET is insulated from the substrate, and the device is therefore also called an insulated-gate field-effect transistor (IGFET).
8. The enhancement-type MOSFET has no physical channel between drain and source. The channel, consisting of charge carriers, is electrically induced in the substrate by biasing the gate.
9. Since there is no physical channel in an enhancement MOSFET, its symbol contains a broken (vertical) channel line (see Fig. 40.1-8). This figure shows that an enhancement-type MOSFET may have an N channel or a P channel, depending on whether the semiconductor substrate of the FET is P or N type, respectively.
10. The gate of an N-channel *enhancement*-type MOSFET must be biased *positively*. The gate of a P-channel *enhancement*-type MOSFET must be biased *negatively*.
11. Depletion-type MOSFETs, also called depletion-enhancement type, do contain a channel, Fig. 40.1-10.
12. The gate of a depletion-enhancement-type MOSFET may be biased either positively, zero, or negatively.
13. When the gate of an N-channel depletion MOSFET is biased positively, it operates in the enhancement mode.
14. When the gate of an N-channel depletion MOSFET is biased negatively, it operates in the depletion mode.
15. MOSFETs constructed with two independent gates are called dual-gate.
16. MOSFETs are very fragile, since the glass-like insulator between gate and substrate may readily rupture when a transient voltage appears across it, in handling or in use. To protect the MOSFET against destruction from transient voltages, back-to-back diodes may be built into the device as in Fig. 40.1-14.

SELF-TEST

Check your understanding by answering these questions.

1. Drain current in a JFET can be cut off by increasing the ________ (forward/reverse) bias on the gate to source.
2. The semiconductor material of the gate in a P-channel JFET is ________ (P/N).
3. In an enhancement-type MOSFET there ________ (is/is not) a physical channel between drain and source.
4. The gate of an enhancement-type MOSFET must be biased ________ (positive/negative) relative to the source, in order to permit drain current flow.
5. In an N-channel depletion-type MOSFET the substrate acts as the channel. ________ (true/false)
6. If the gate of an N-channel depletion-type MOSFET is biased positive relative to the source, the transistor operates in the ________ mode.
7. The vertical channel line in the symbol for a depletion-type MOSFET ________ (is/is not) a broken line.
8. The drain characteristic of a unipolar transistor corresponds to the collector characteristic of a bipolar transistor. ________ (true/false)

MATERIALS REQUIRED

- Power supply: Dual regulated, variable low-voltage dc source, or two independent regulated variable dc sources

- Equipment: EVM; 0–10-mA dc milliammeter or equivalent range on VOM
- Transistor: 3N187 or equivalent
- Resistors: ½-W 1-, 10-kΩ
- Miscellaneous: Two SPST switches

PROCEDURE

Curve Tracer

Display the family of drain characteristics of the 3N187 for 0.2 V (V_{GS}) steps. Limit V_{DS} to 10 V. Identify the curves, the axes, and units of measurement.

Point-by-Point Method

CAUTION: Do *not* exceed the voltages listed in Table 40.1-1.

1. Figure 40.1-15 is a bottom view of the terminals of the 3N187. Connect the circuit of Fig. 40.1-16.

TABLE 40.1-1. Drain Characteristics (Gate 1 Control)

V_{DS}, V	I_D, mA						
	0	*1*	*3*	*5*	*7*	*9*	*10*
V_{G1S}, V							
−0.6							
−0.4							
−0.2							
0							
+0.2							
+0.4							
+0.6							

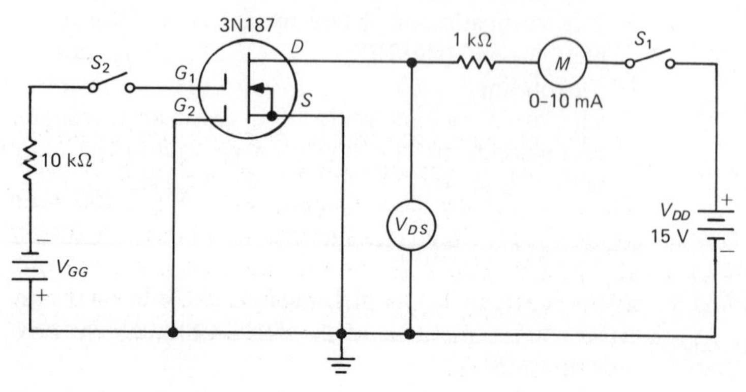

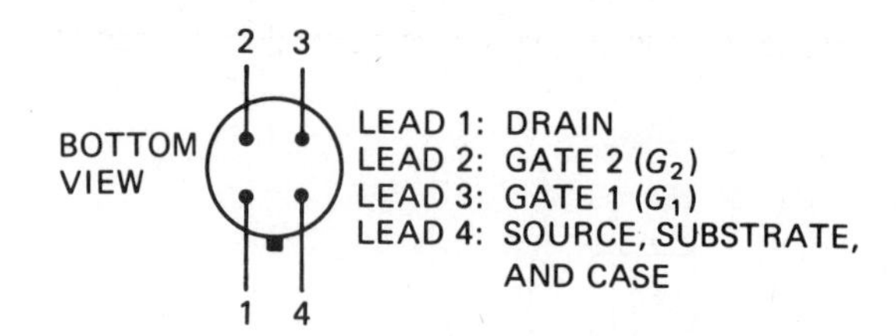

Fig. 40.1-15. Bottom view of 3N187.

S_1 and S_2 are **off.** Set V_{DD} at 0 V, V_{GG} at −0.6 V.

2. Close S_1 and S_2. Measure I_D for $V_{G1S} = -0.6$ V, $V_{DS} = 0$ V and record your measurement in Table 40.1-1.
3. Maintain V_{G1S} at −0.6 V. Increase V_{DS} to +1 V, +3 V, +5 V, etc., as in Table 40.1-1, and record the measured values of I_D.
4. Repeat steps 2 and 3 for each negative value of V_{G1S} in Table 40.1-1 and record I_D.
5. S_1 and S_2 **off.** Reverse the polarity of V_{GG}, the gate bias supply, and repeat steps 2 and 3 for each value of V_{G1S} listed in Table 40.1-1. Record I_D.

QUESTIONS

1. From your data in Table 40.1-1, what is the value of V_{G1S} at which drain current is cut off?
2. If I_D was not cut off in this experiment, how can you determine what value of V_{G1S} will cut off I_D?
3. What is meant by a depletion-type MOSFET?
4. Do the data in Table 40.1-1 confirm that the 3N187 is a depletion-type transistor? Explain.
5. In which mode of operation of the 3N187 is the transistor likely to be cut off? Why?
6. What is the effect on I_D of biasing the 3N187 positively? Negatively? Refer to your data.

Answers to Self-Test

1. reverse
2. N
3. is not
4. positive
5. false
6. enhancement
7. is not
8. true

Fig. 40.1-16. Experimental circuit to determine gate 1 control characteristics of 3N187 MOSFET.

EXPERIMENT 40.2. MOSFET Common-Source Amplifier

OBJECTIVE

To construct a MOSFET common-source amplifier and measure its voltage gain

INTRODUCTORY INFORMATION

MOSFETs are used as amplifiers in radio and television tuners. Their high-impedance input is especially advantageous in electronic voltmeters and in many other electronic applications.

JFET Biasing

Biasing the gate of a JFET is different from biasing the base of a bipolar transistor, in that the JFET gate must be reverse-biased, while the bipolar base is forward-biased. Three bias arrangements are in popular use, voltage-divider bias, source bias, and self-bias.

Voltage-Divider Bias

Figure 40.2-1*a* shows an N-channel JFET gate-bias arrangement employing voltage-divider bias. The voltage at the gate to ground V_G is given by the formula

$$V_G = \frac{R_1}{R_1 + R_2} \times V_{DD} \qquad (40.2\text{-}1)$$

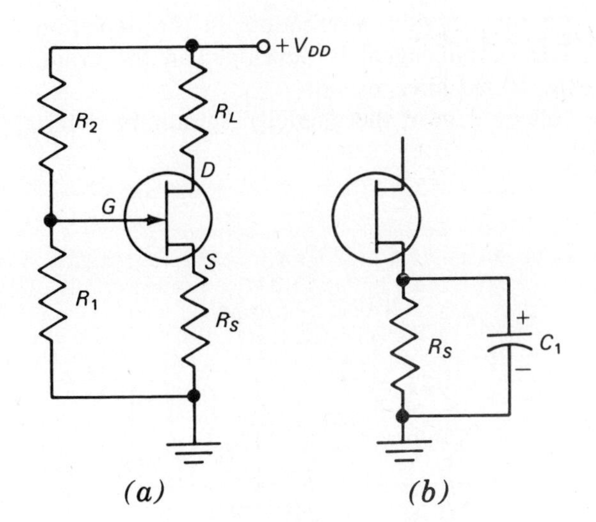

Fig. 40.2-1. (*a*) Voltage-divider bias for a JFET. (*b*) Bypassing the source resistor prevents ac degeneration.

The voltage on the source to ground V_S is the IR drop across the source resistor R_S and depends on the value of drain current I_D (since drain and source current are the same) and on R_S.

$$V_S = I_D \times R_S \qquad (40.2\text{-}2)$$

The gate bias V_{GS} is the difference between V_G and V_S. That is,

$$V_{GS} = V_G - V_S \qquad (40.2\text{-}3)$$

To ensure reverse bias of the gate, V_S must be larger than V_G, so that $V_G - V_S$ is a negative quantity. The component values are chosen to ensure this and to provide the proper operating point for the circuit.

By injecting an ac signal v_g between gate and ground, the circuit in Fig. 40.2-1*a* acts as an ac amplifier. However, the gain of the circuit is very low, because an ac voltage v_g is developed across R_S and the gate-to-source signal v_{gs} which the amplifier "sees" is the difference between v_g and v_s. That is,

$$v_{gs} = v_{\text{in}} = v_g - v_s \qquad (40.2\text{-}4)$$

The degenerative signal voltage developed across R_S can be eliminated by providing a bypass capacitor C_1 across R_S, as in Fig. 40.2-1*b*. To approximate the value of C_1, we use the relationship

$$X_{C1} = \frac{R_S}{10} \qquad (40.2\text{-}5)$$

for the *lowest* frequency which the amplifier must handle. You will recall that this is the same arrangement used to bypass the emitter resistor in a bipolar transistor amplifier.

In Fig. 40.2-1, the output signal is developed across R_L and is taken from the drain.

Source Bias

If two independent power supplies are available, source bias, as illustrated in Fig. 40.2-2, may be used. Here $-V_{SS}$ is the source supply, and $+V_{DD}$ is the drain supply. The circuit parameters are chosen to supply the proper gate bias for the required drain current. If the voltage $-V_{SS}$ is high enough, drain current may be made independent of the variation in characteristics of the FET.

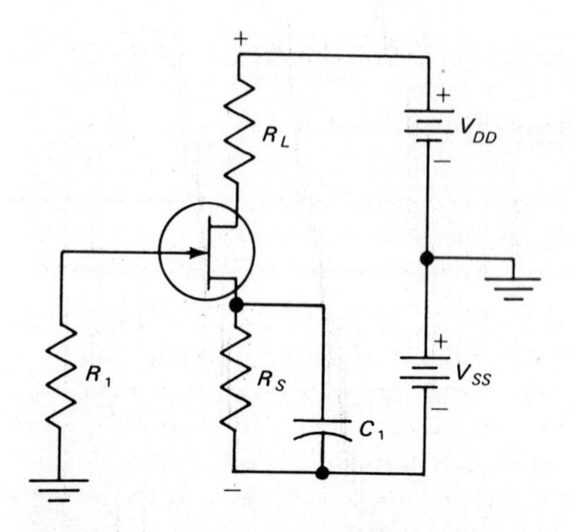

Fig. 40.2-2. Source bias requiring two independent power supplies.

Self-Bias

The circuit of Fig. 40.2-3 illustrates a self-bias arrangement for an N-channel JFET. It is completely different from bipolar transistor bias circuits. Since the gate is returned to ground through R_1, and there is no gate current, the voltage on the gate V_G is 0 V. Current I_D in the FET and in the external circuit (R_S and R_L) will develop a voltage drop V_S across R_S.

$$V_S = I_D \times R_S \qquad (40.2\text{-}6)$$

Since $V_G = 0$, the voltage difference between V_G and V_S is the gate bias V_{GS}. That is,

$$V_{GS} = 0 - V_S = -I_D \times R_S \qquad (40.2\text{-}7)$$

For example, if $I_D = 1$ mA and $R_S = 2.2$ kΩ, the gate bias is

$$V_{GS} = -(1)10^{-3} \times 2.2 \times 10^3 = -2.2 \text{ V}$$

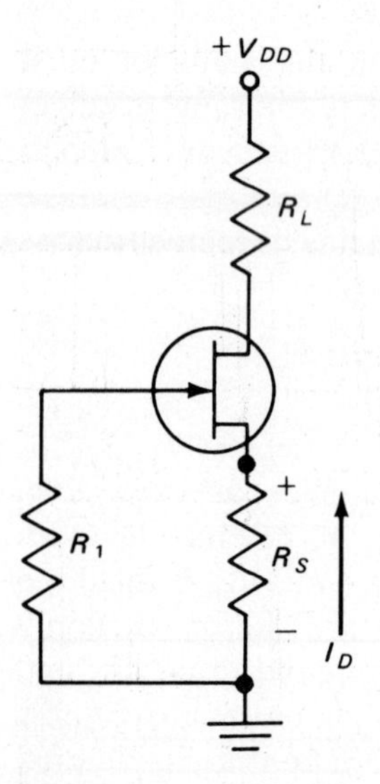

Fig. 40.2-3. JFET self-bias arrangement.

Normally, the source resistor R_S would be bypassed by a capacitor C to prevent ac signal degeneration.

The reason self-bias can be used with FETs but not with bipolar transistors is that FETs will permit drain current with zero gate-to-source voltage, but bipolar transistors are cut off when the base-to-emitter voltage is zero.

MOSFET Biasing

Voltage-divider gate bias, self-bias, and source bias may be used in biasing MOSFETs. Circuit arrangements are similar to those shown in Figs. 40.2-1 to 40.2-3. The polarity of the bias required will depend on the nature of the channel, whether it is N or P, and on the type of MOSFET, whether it is an enhancement or a depletion type.

Depletion MOSFETs may be operated at zero bias, that is, $V_{GS} = 0$ V. An ac signal fed to the gate of an N-channel depletion MOSFET will cause the FET to operate in the enhancement mode on the positive alternation and in the depletion mode on the negative alternation. The opposite is true of P-channel depletion MOSFETs.

MOSFET Common-Source Amplifier

The circuit shown in Fig. 40.2-4 is an experimental common-source amplifier, using an N-channel depletion type dual-gate MOSFET. Gate 1 receives the 1-kHz sine-wave test signal. Gate 2 is connected to gate 1. Gate 2 is used in special applications requiring a transistor with two inputs, for example, a mixer in a radio and TV receiver. The amplifier is operated at zero bias. On the positive alternation of the input sine wave, the amplifier operates in the enhancement mode; on the negative alternation, in the depletion mode. The output signal is developed in the drain, across the 10-kΩ load resistor.

The voltage gain of the amplifier A_v may be deter-

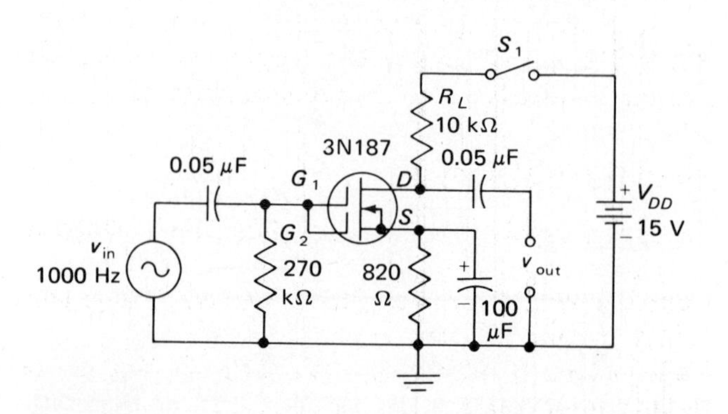

Fig. 40.2-4. Experimental MOSFET common-source amplifier.

mined experimentally by measuring the output and input signals, and substituting the measured values in the formula

$$A_v = \frac{v_{out}}{v_{in}} \qquad (40.2\text{-}8)$$

The voltage gain of the MOSFET amplifier is relatively low. The phase relationship between the input and output signals may also be determined with an oscilloscope, by the methods used in previous experiments.

SUMMARY

1. FETs may use voltage-divider bias, source bias, and self-bias.
2. FET voltage-divider and source bias are like bipolar transistor voltage-divider and emitter bias, but self-bias is found only in FET circuits.
3. FET amplifiers have a high input impedance and a low voltage gain.
4. The voltage gain A_v of a FET amplifier may be determined experimentally by measuring the input (v_{in}) and output (v_{out}) signals and substituting these values in the gain formula

$$A_v = \frac{v_{out}}{v_{in}}$$

SELF-TEST

Check your understanding by answering these questions.

1. The bias arrangement used in FET amplifiers, but not used in bipolar transistor amplifiers, is ________ ________.
2. In Fig. 40.2-1, the gate-bias voltage is the difference between the voltage on the ________ and the voltage developed across ________.
3. The purpose of C_1 in Fig. 40.2-1*b* is to prevent ________ ________.
4. The FET in Fig. 40.2-4 is cut off because the gate-to-source voltage is zero. ________ (true/false)
5. In Fig. 40.2-4, control of the output signal depends on G_1 signal ________ (voltage/current).

MATERIALS REQUIRED

- Power supply: Variable, regulated low-voltage dc source
- Equipment: Oscilloscope; AF sine-wave generator; EVM; curve tracer
- Resistors: ½-W 820-Ω; 10-, 270-kΩ, and as required for step 5
- Capacitors: Two 0.05-μF; 100-μF/50-V
- Semiconductors: 3N187 (MOSFET) or the equivalent
- Miscellaneous: SPST switch

PROCEDURE

1. Connect the circuit of Fig. 40.2-4. Set the sine-wave generator at 1000 Hz, *minimum* output.
2. **Power on.** With an oscilloscope observe v_{out}. Gradually increase the input signal for the *maximum undistorted output*, v_{out}.
3. Measure, and record in Table 40.2-1, the peak-to-peak voltage of v_{out} and v_{in}.
4. Measure also and record the dc voltages V_{G1S} and V_{DS}. Compute and record the gain of the amplifier.

NOTE: Before removing any component from the circuit, turn power **off**.

5. Experimentally determine and record in Table 40.2-2 the effect of decreasing and increasing the drain load resistor R_L on input signal level which the amplifier can handle without distortion, on amplifier output, and on gain. Record the values of R_L used, v_{in}, and v_{out}. Compute and record gain.

QUESTIONS

1. How is the experimental circuit of Fig. 40.2-4 different from the usual bipolar transistor amplifier?

TABLE 40.2-1. MOSFET Amplifier Gain

v_{in}, V p-p	v_{out}, V p-p	V_{G1S}, V dc	V_{DS}, V dc	*Gain*

TABLE 40.2-2. Gain versus Load Resistor

R_L	v_{in}, V p-p	v_{out}, V p-p	*Gain*

2. How does the signal gain of the experimental amplifier compare with that of the bipolar amplifiers you have connected in previous experiments?
3. What is the effect of drain load resistor R_L on amplifier output? Gain?

Answers to Self-Test

1. self-bias
2. gate; R_S
3. ac degeneration
4. false
5. voltage

CHAPTER 41 TRIODE VACUUM TUBES

EXPERIMENT 41.1. Triode-Vacuum-Tube Plate Characteristics

OBJECTIVE

To determine and plot a graph of the effects on plate current of a plate-voltage change if the grid voltage is kept constant ($V_p - I_p$ plate characteristic)

INTRODUCTORY INFORMATION

The rapid growth of transistor technology was made possible by the vacuum tube, which was developed before the transistor. The body of engineering knowledge based on this tube stimulated the growth and development of transistors and transistor circuits. The transistor, in turn, stimulated the development of solid-state microelectronics, the integrated circuits. And so the chain of discovery and growth will continue.

The vacuum tube is seldom used in the design of new electronic equipment. But there are many older electronic devices, such as radio and television receivers, still in use which contain vacuum tubes. Therefore the technician must understand how tubes operate and how they are used.

The bipolar transistor, a current-driven device, is capable of *current* amplification. The unipolar (FET) transistor, a voltage-driven device, is capable of *voltage* amplification. The vacuum tube is like the FET. It is voltage-driven and can be used as a voltage amplifier.

Amplification in a tube is accomplished by introducing a control grid into a diode, creating a triode, as shown in Fig. 41.1-1*a*.

The triode has three active elements: the cathode, control grid, and plate, or anode. The cathode, when heated by a filament, emits electrons. When the plate voltage is made positive relative to the cathode, these electrons move toward and strike the plate. This electron flow is cathode-to-plate current, or simply I_p plate current. Up to this point the action of the cathode and plate in a triode is similar to the action of these elements in a vacuum-tube diode. However, the *grid* dramatically changes the *control of plate current.*

The grid is a wire winding placed between cathode and plate, physically closer to the cathode. Because of its closeness to the cathode the grid controls plate current more effectively than does the plate. A similarity between the bipolar vacuum-tube triode and a transistor triode, Fig. 41.1-1*b*, is immediately apparent. The cathode corresponds to the emitter, the control grid to the base, and the plate to the collector. Compared with the FET, Fig. 41.1-1*c*, the control grid corresponds to the gate, the cathode to the source, and the plate to the drain.

Like the bipolar transistor, the triode may be considered as two diodes; the cathode and grid make up one diode, the cathode and plate the other. Proper biasing of these two diodes is required for normal operation. Figure 41.1-2 shows the triode connected in a common-cathode configuration. The grid-to-cathode diode (input) is reverse-biased; that is, the grid is made negative with respect to the cathode. The plate-to-cathode diode is forward-biased; that is, the plate is made positive with respect to the cathode. This bias arrangement is the exact opposite of that in a bipolar transistor, but it is normal for operation of a vacuum-tube triode.

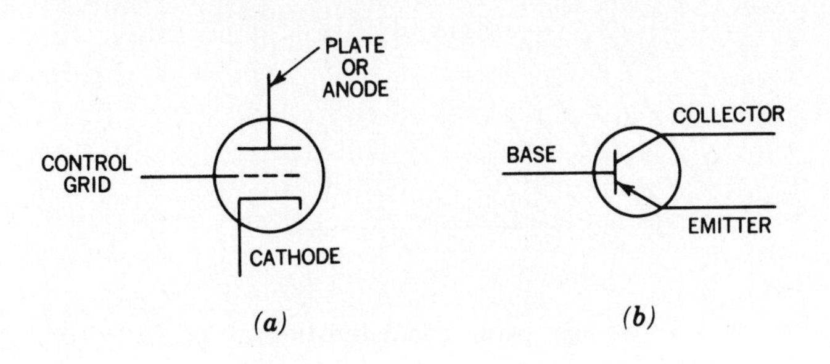

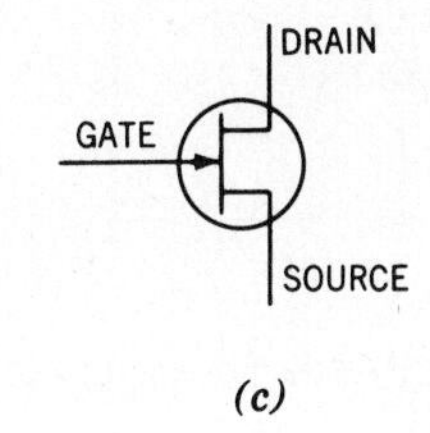

Fig. 41.1-1. (*a*) Vacuum-tube symbol and vacuum-tube elements; (*b*) equivalent elements in a bipolar transistor; (*c*) equivalent elements in an FET.

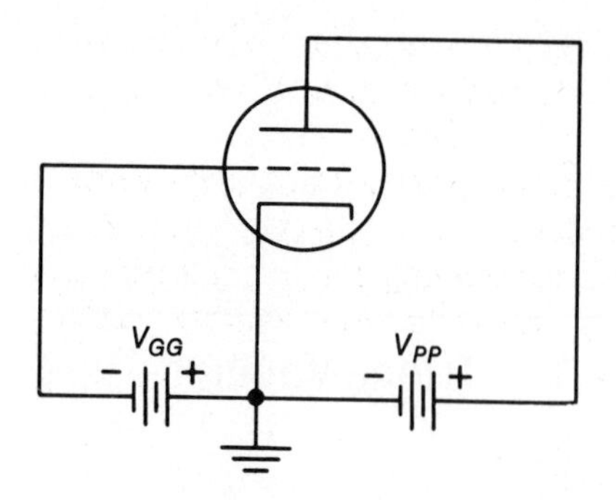

Fig. 41.1-2. Biasing a vacuum-tube triode.

Static Characteristics

Consider the circuit of Fig. 41.1-3. V_{PP} and V_{GG} are sources of variable dc voltages which can be manually set at a desired level. The polarity of V_{PP} is such that the plate is maintained at a positive (+) potential with respect to the cathode. V_{GG}, however, is connected so that the control grid may be kept at a negative (−) potential with respect to the cathode. V_1 is a dc voltmeter which measures V_P, the plate voltage. V_2 is a dc voltmeter which measures V_G. V_G is referred to as "grid bias," or simply "bias." M is a milliammeter which measures I_P.

As in the diode, an increase or decrease in V_P with the grid bias of the triode held constant will result in an increase or decrease, respectively, in I_P. Moreover, I_P may be increased or decreased, if V_P is held constant, by decreasing grid bias, i.e., making grid voltage less negative with respect to the cathode, or increasing it, respectively.

The effects on I_P of increasing or decreasing V_P may be determined experimentally and plotted in the form of a graph. The graph of Fig. 41.1-4 is the "plate characteristic" of a particular triode. It is a static characteristic obtained with different dc voltages applied between plate and cathode while bias is kept constant.

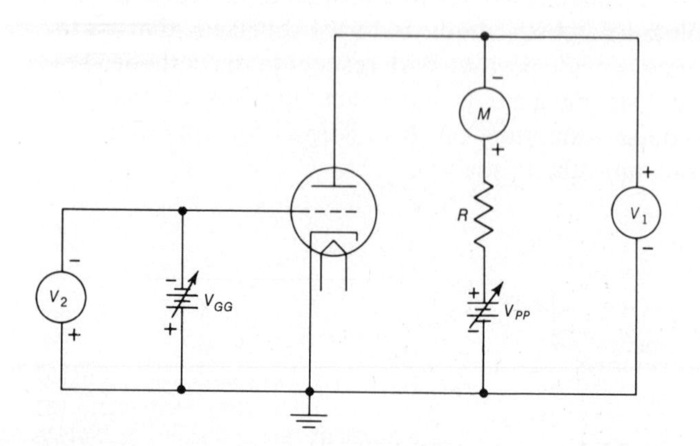

Fig. 41.1-3. Circuit to determine the family characteristics of a triode.

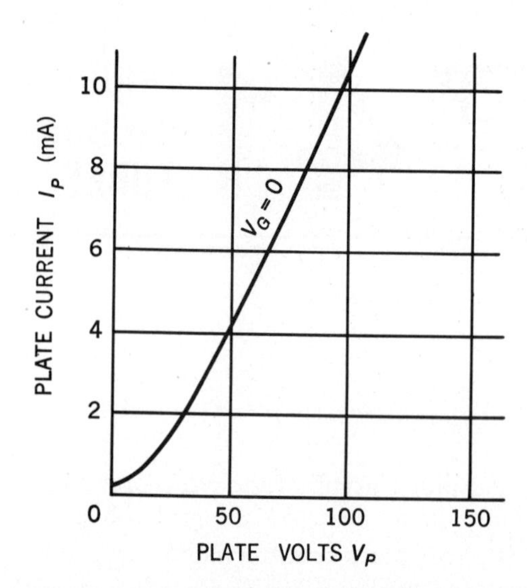

Fig. 41.1-4. Static plate characteristics.

Additional plate-characteristic curves may be obtained by setting the bias at −2 V, −4 V, etc., and noting effects on I_P as the dc plate voltage is varied at each bias setting. If these curves are plotted on the same graph, a family of plate characteristics will result as in Fig. 41.1-5. This is the family of plate characteristics for each triode in the 6CG7, which is a dual-triode tube.

The effects on I_P of increasing or decreasing grid bias may also be determined experimentally and graphed. These curves are known as "mutual" or "transfer" characteristics. They are static characteristics (see Fig. 41.1-6).

The graphs of Figs. 41.1-5 and 41.1-6 give the same information about a tube's static characteristics but in different form.

SUMMARY

1. The triode, a three-element vacuum tube, is a voltage-driven device which is capable of *voltage* amplification.
2. The *active elements* of the triode are the control

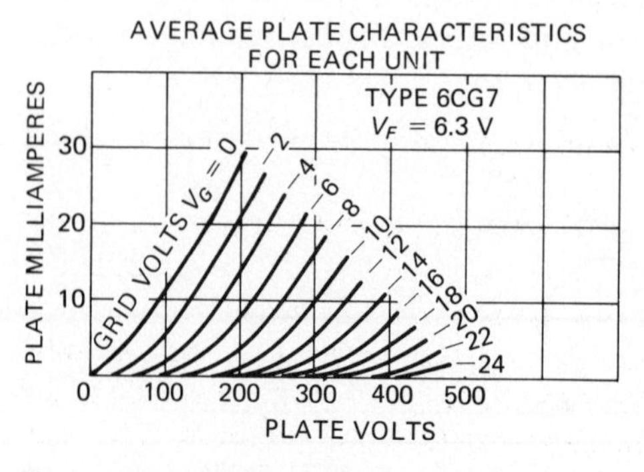

Fig. 41.1-5. Average plate characteristics of 6CG7. *(RCA)*

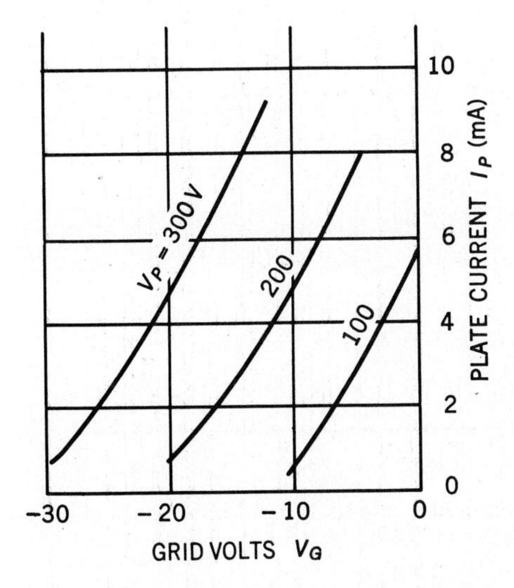

Fig. 41.1-6. Family of mutual characteristics.

grid, cathode, and plate (or anode). If a triode is compared with a bipolar transistor, the control grid corresponds to the base, the cathode to the emitter, and the plate to the collector. Compared with an FET, the control grid corresponds to the gate, the cathode to the source, and the plate to the drain.

3. Current flow in a triode, called plate current I_P, is brought on by filaments which heat the cathode. Electrons are emitted by the hot cathode. If the plate is made positive with respect to the cathode, the electrons are attracted to the plate. Plate current is the flow of electrons from cathode to plate.
4. The number of electrons which reach the plate is determined by the control grid, which is reverse-biased with respect to the cathode. The more negative the grid, the smaller the plate current.
5. Plate characteristic curves, similar to collector characteristics in a transistor, show how plate current varies with a variation in plate voltage, for some constant value of grid bias.
6. Mutual or transfer characteristic curves show the effect of grid bias on plate current for a fixed value of plate voltage.

SELF-TEST

Check your understanding by answering these questions.

1. The element of a triode which corresponds to the gate of an FET is the ________ ________.
2. In a triode the control grid is biased ________, the plate ________ with respect to the cathode.
3. The triode resembles the FET more than it does the bipolar transistor because the vacuum tube is a ________-driven device, like the FET.
4. Thermionic emission, or the emission of electrons due to heat, results from the heating of the ________ by the ________ in a triode.
5. Plate current in a triode results from the movement of electrons from the ________ to the ________.
6. One condition for maintaining current in a triode is that the ________ must be kept positive relative to the ________.

MATERIALS REQUIRED

- Power supply: Variable direct current with a positive output from 0 to 100 V and a negative output for bias voltage from 0 to −3 V; 6.3-V rms source
- Equipment: Two voltmeters; 0- to 20-mA milliammeter
- Resistor: ½-W 1000-Ω
- Tube: 12AU7A and appropriate socket

PROCEDURE

NOTE: The 12AU7A used in this experiment contains two identical but separate triode tubes. Only one will be used here. Figure 41.1-7 is the circuit symbol for the 12AU7A. The triode we shall use is identified as follows: plate, pin 1; control grid, pin 2; and cathode, pin 3.

The filament arrangement of the 12AU7A, Fig. 41.1-8, makes it possible to use either a 12.6-V rms or 6.3-V rms source. For a 12.6-V source, pin 9 is not connected, but pins 4 and 5 are. For a 6.3-V source, the filament circuit of Fig. 41.1-9 is used.

Plate Characteristics

1. Connect the circuit of Fig. 41.1-10. Set $V_G = 0$ V. Do *not* change this grid bias until instructed to do so.
2. Measure and record in Table 41.1-1 the plate current I_P for each value of plate voltage V_P listed in the table.

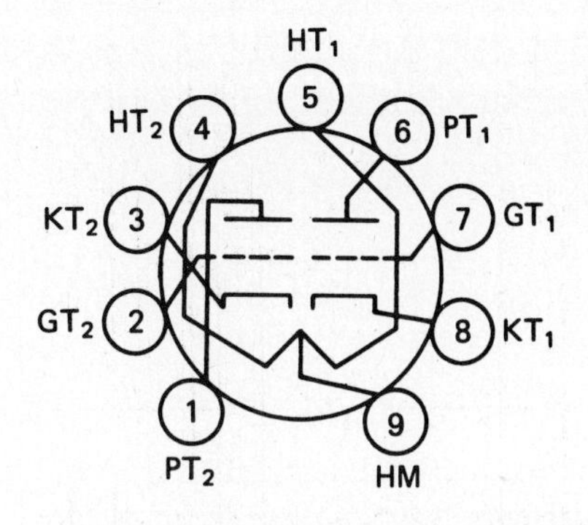

Fig. 41.1-7. Circuit symbol for 12AU7A. (*RCA*).

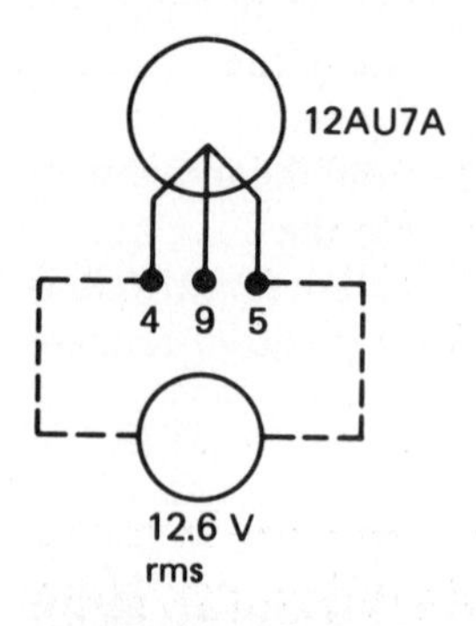

Fig. 41.1-8. Filaments of 12AU7A connected for 12.6 V rms.

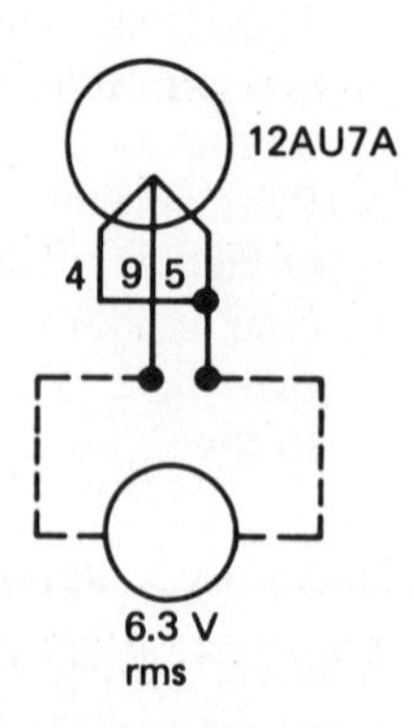

Fig. 41.1-9. Filaments of 12AU7A connected for 6.3 V rms.

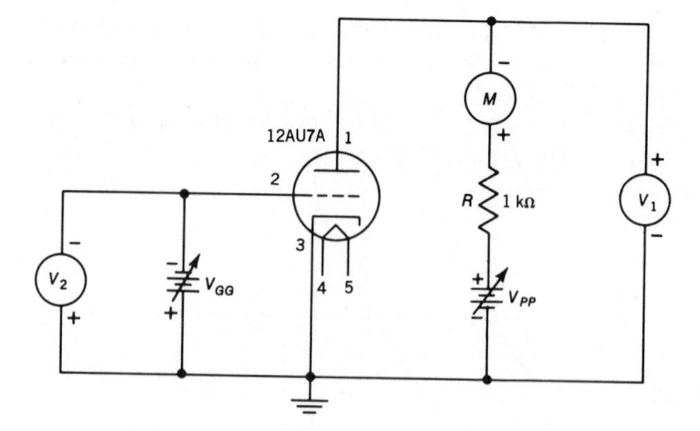

Fig. 41.1-10. Experimental circuit to determine plate characteristics of a 12AU7A.

3. Repeat step 2 for $V_G = -0.5$ V, -1 V, etc.
4. Graph the plate characteristics from the data in Table 41.1-1.

NOTE 1: If you have a calibrated curve tracer, use it, instead of steps 1 to 4, to obtain the family of plate characteristics, for the range of values in Table 41.1-1. Photograph these curves, or draw them from the curve tracer. Indicate on the X and Y axes of the graphs the values of plate voltage and current, respectively.

NOTE 2: Keep the data and graphs for use in the next experiment.

QUESTIONS

1. From Table 41.1-1, when $V_G = -1$, what is the change in I_P for a 70-V change in plate voltage (between $V_P = 30$ and 100 V)?
2. From Table 41.1-1, at $V_P = 100$, what is the change in I_P between $V_G = -1$ V and $V_G = -1.5$ V (a change of 0.5 V)?

TABLE 41.1-1. Plate Characteristics

	I_P, mA						
V_P	$V_G = 0$	$V_G = -0.5$	$V_G = -1$	$V_G = -1.5$	$V_G = -2$	$V_G = -2.5$	$V_G = -3$
0							
10							
20							
30							
40							
50							
60							
70							
80							
90							
100							

3. From your answers to questions 1 and 2, which appears to be more effective in controlling I_P, plate voltage or grid voltage? Explain?
4. What other family of curves may be derived from the family of plate characteristics.

Answers to Self-Test

1. control grid
2. negative; positive
3. voltage
4. cathode; filaments
5. cathode; plate
6. plate; cathode

EXPERIMENT 41.2. Triode-Vacuum-Tube Parameters

OBJECTIVE

To determine the parameters of a triode: μ, r_p, and g_m

INTRODUCTORY INFORMATION

The plate characteristics of a triode were determined in the last experiment. From this family of static characteristics may be obtained important information about the tube's parameters or dynamic characteristics. These are the μ (mu) or amplification factor, the r_p or dynamic plate resistance, and the g_m or transconductance. A knowledge of these parameters helps the technician predict the circuit operation of a triode because they are concerned with the effect of an input signal on the output.

DYNAMIC CHARACTERISTICS

Amplification Factor μ

The μ of a tube is a number which tells how much more effective the grid is in controlling plate current than the plate is. It may be compared with the *beta* of a bipolar transistor, which describes how much more effective the base is than the collector in controlling collector current. An example will clarify this idea.

Example: To determine the effect on I_P of changes in V_p, refer to the family of plate characteristics in Fig. 41.2-1.

For a particular combination of V_p (say, 100 V) and V_G (say, −2 V), there is 5.5 mA of plate current in the tube. Now if V_p is increased to 150 V while V_{GG} is kept constant at −2 V, I_p increases to 12.5 mA.

Thus, for a 50-V change in V_p, there is a corresponding increase in I_P of 7 mA. If we assume a linear relationship between plate voltage and plate current, for a fixed value of grid bias (−2 V in this case), we can see that a 7-V change in plate voltage (approximately) causes a 1-mA change in plate current.

To determine the effect of grid bias on I_P, V_P is brought back to 100 V. If V_G is reduced to 0 V, while V_p is kept constant at 100 V, I_P increases from 5.5 to 11.5 mA. It is apparent therefore that for a 2-V change in grid bias, there is a 6-mA change in plate current for this tube under the operating conditions specified. Again, assuming a linear relationship between grid bias and plate current, for a fixed value of plate voltage, we note that in this same tube, a $\frac{1}{3}$-V change in grid bias causes a 1-mA change in plate current.

The control grid is therefore 21 times more effective than the plate in controlling plate current ($7 \div \frac{1}{3} = 21$). This property of the control grid to control I_P more effectively than the plate is called its *amplification factor*. The formula for μ is

$$\mu = \frac{\Delta V_P}{\Delta V_G} \quad (I_P \text{ constant}) \qquad (41.2\text{-}1)$$

where ΔV_G represents a *change* in bias voltage and ΔV_P represents the *change* in plate voltage required to offset ΔV_G and maintain I_P constant.

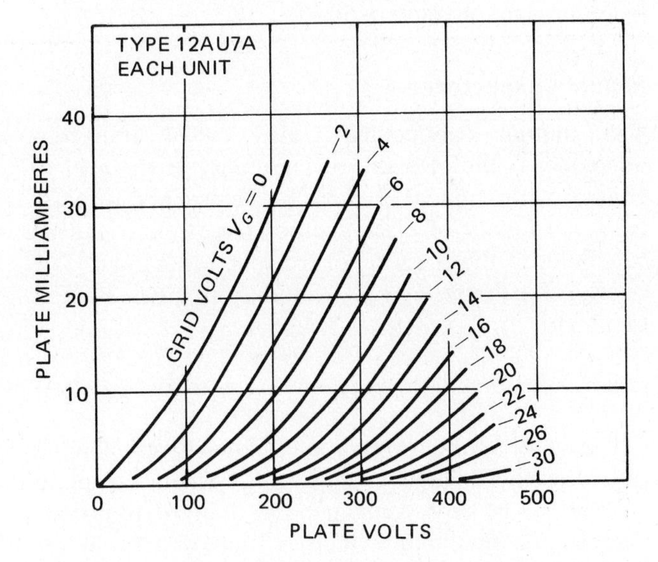

Fig. 41.2-1. Average plate characteristics of 12AU7A. (*RCA*)

It is possible to apply Eq. (41.2-1) directly to the family of plate characteristics to determine μ of one of the triodes in the 12AU7A. Thus, note in Fig. 41.2-1 that when $V_P = 300$ V and $V_G = -12$ V, $I_P = 10$ mA. When $V_G = -14$ V and V_P is 340 V (approximately), $I_P = 10$ mA. For these operating conditions (with $I_P = 10$ mA),

$$\mu = \frac{\Delta V_P}{\Delta V_G} = \frac{340 - 300}{14 - 12} = \frac{40}{2} = 20$$

The two values of μ are slightly different, which illustrates that the μ of a tube is not constant over its entire operating range. However, it is relatively constant over that portion of its operating characteristic called the linear portion.

AC Plate Resistance r_p

Another dynamic characteristic of a tube is its r_p. AC plate resistance r_p is the plate-to-cathode resistance that a tube offers to the flow of alternating current. It is expressed in ohms. The formula for r_p is

$$r_p = \frac{\Delta V_P}{\Delta I_P} \quad (V_G \text{ constant}) \qquad (41.2\text{-}2)$$

where ΔV_P represents a change in plate voltage which gives rise to a change in plate current ΔI_P, while grid bias V_G is kept constant. For example, in Fig. 41.2-1, for $V_G = -6$, a change in plate voltage of 35 V (200 to 235) gives rise to a current change of 5 mA (10 to 15). The ac plate resistance of this tube is then

$$r_p = \frac{35}{0.005}\ \Omega = 7000\ \Omega$$

It is clear, then, that the r_p of a tube may be found from its plate characteristics.

Mutual Conductance g_m

The mutual conductance, also called transconductance, is the measure of the ability of the grid to control plate current. It may be determined directly from the family of plate characteristics, or it can be calculated if the μ and r_p of the tube are known. The relationship between g_m, μ, and r_p is given by Eq. (41.2-3):

$$g_m = \frac{\mu}{r_p} \qquad (41.2\text{-}3)$$

The g_m of a tube is the ratio of plate-current change (ΔI_P) to grid-voltage change (ΔV_G) when the plate voltage V_P is kept constant. The unit of transconductance is the *mho* (which is ohm written in reverse). The formula is

$$g_m = \frac{\Delta I_P}{\Delta V_G} \quad (V_P \text{ constant}) \qquad (41.2\text{-}4)$$

Tube Specifications

The 6AL5 used in an early experiment and the 12AU7A are examples of the wide variety of vacuum tubes in use. Each type of tube has unique characteristics which are listed by the manufacturer in a tube manual.

A listing of triode-tube characteristics usually includes:

1. A brief description, including applications
2. A circuit symbol, designating the pin number and function of each element
3. Heater (filament) voltage, current, and warm-up time
4. Peak heater-to-cathode voltage (positive and negative)
5. Direct interelectrode capacitance between plate and cathode, plate and grid, and grid and cathode
6. Maximum ratings of: (*a*) Plate voltage; (*b*) grid voltage, positive-bias value; (*c*) plate dissipation (wattage) for each plate; and (*d*) cathode current.
7. Characteristics as a class A amplifier, including μ, r_p, g_m, plate voltage, and grid bias
8. Characteristics as an oscillator (if applicable)
9. Other unique application characteristics, when pertinent

SUMMARY

1. A triode has three characteristics which describe its operation. These are: (*a*) Amplification factor μ; (*b*) AC plate resistance r_p (measured in ohms); and (*c*) transconductance g_m (measured in mhos).
2. $\mu = \Delta V_P / \Delta V_G$ (I_P constant)
   Here ΔV_P is the change in plate voltage required to offset the change in grid voltage ΔV_G. Each different triode will have a different μ, with the lowest value about 10. This characteristic shows how much more effective the grid is in controlling plate current than the plate.
3. $r_p = \Delta V_P / \Delta I_P$ (V_G constant)
   where ΔV_P is the change in plate voltage which gives rise to a change in plate current ΔI_P.
4. $g_m = \mu / r_p$
   This characteristic is a measure of the ability of the grid to control plate current.

SELF-TEST

Check your understanding by answering these questions.

1. A change from 150 to 200 V is required to increase plate current from 2.0 to 3.0 mA. To bring plate current back to 2 mA, the grid bias must be raised from −1.5 to −2.5 V. The μ of the tube is therefore ________.
2. For a particular tube, when $V_G = -6$ V, and $V_P =$ 200 V, $I_P = 7.5$ mA. When $V_G = -6$ V and $V_P =$ 250 V, $I_P = 15$ mA. The r_p of the tube is ________ Ω.
3. The μ of a tube = 20, the $r_p = 6000$ Ω. The g_m of this tube is ________ μmhos.

MATERIALS REQUIRED

- Miscellaneous: Table 41.1-1 and graph of family of plate characteristics drawn in Exp. 41.1; tube manual

PROCEDURE

1. From the family of plate characteristics of the 12AU7A and Table 41.1-1 (Exp. 41.1) determine the μ of the 12AU7A. Show all graphical work. Record in Table 41.2-1.

TABLE 41.2-1. Dynamic Characteristics of 12AU7A

μ	r_p, Ω	g_m, μmhos	g_m, μmhos

2. As in step 1, determine the r_p of the 12AU7A. Show all graphical work. Record in Table 41.2-1.
3. Determine the g_m of the 12AU7A by two methods. Show all work. Record both results in Table 41.2-1.

QUESTIONS

1. What is meant by the μ of a triode tube?
2. How do the values of μ, r_p, and g_m obtained in this experiment compare with the ratings for a 12AU7A listed in a tube manual?
3. From your tube manual for the 12AU7A, what are the maximum values of (*a*) plate voltage? (*b*) cathode current? (*c*) plate dissipation?

Answers to Self-Test

1. 50
2. 6667
3. 3333

CHAPTER 42 TRIODE CLASS A VOLTAGE AMPLIFIERS

EXPERIMENT 42.1. Triode-Tube Class A Voltage Amplifier Gain

OBJECTIVE

To measure the effect on voltage gain of plate load resistors of different values, *with and without ac load*

INTRODUCTORY INFORMATION

In bipolar-transistor class A amplifiers, current gain is the important consideration. In vacuum-tube class A amplifiers, voltage gain is the first concern. The size of the plate load resistor is one factor which determines voltage gain. This experiment will demonstrate the relationship between voltage gain and the size of the plate load.

Class A Amplifier

Figure 42.1-1 is the circuit diagram of a voltage amplifier. Here a triode tube is used. The load is a resistor R_L connected from the plate to $V+$. R_g is the grid resistor which is in series with the bias voltage V_{GG}. C is a capacitor which couples the signal voltage v_{in} to the grid of the tube. This arrangement is similar to the grounded-emitter bipolar transistor amplifier.

The class of operation of the amplifier depends on the bias V_G. A class A vacuum-tube amplifier is one in which the grid bias is set so that plate current in that amplifier flows at all times for the range of signal voltages that it is capable of handling.

The grid bias sets the *operating point* of the amplifier. Consider the V_G/I_P characteristic curve (Fig. 42.1-2) of this amplifier. This curve shows that the bias has been set at −10 V and that the grid signal varies around this voltage. The "operating point" is −10 V. The linear portion of the amplifier's characteristic curve lies between points *A* and *B*. At point *A* the grid-to-cathode voltage is −19 V and the plate current is 1 mA. At point *B* the grid-to-cathode voltage is −1 V and the plate current is 21 mA. The operating point for this amplifier was chosen so that it is in the middle of the linear portion of the characteristic. The point *P* on the curve is the operating point, and at *P* there is 11 mA of plate current (I_P).

Without any signal voltage, the grid-to-cathode voltage is −10 V and I_P is 11 mA. If an 18-V peak-to-peak sine-wave signal is coupled to the grid of the amplifier, the signal voltage will vary the grid-to-cathode voltage from −10 to −1 V on the positive

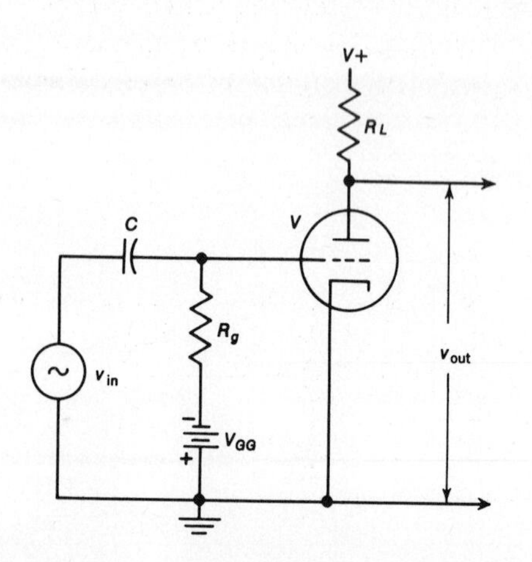

Fig. 42.1-1. Grounded-cathode amplifier.

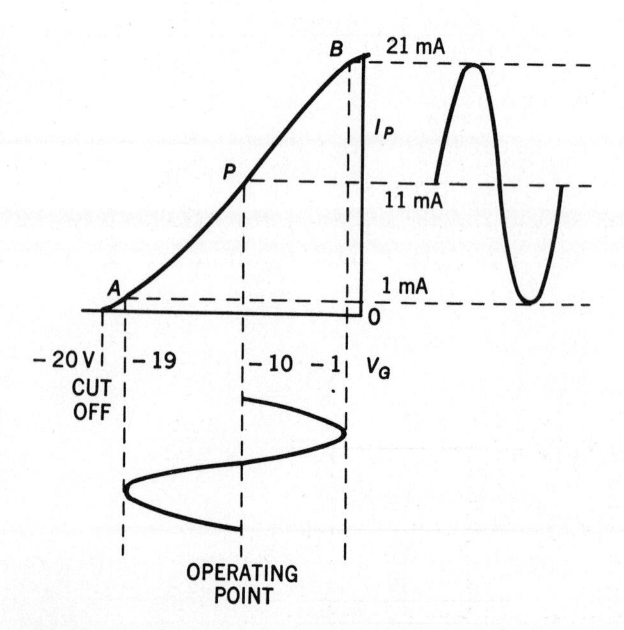

Fig. 42.1-2. Amplifier V_G/I_p characteristic.

alternation, and from −10 to −19 V on the negative alternation. The current in the plate circuit will vary sinusoidally between 1 and 21 mA.

Any voltage greater than 20 V p-p will cause the tube to cut off during that part of the negative alternation which drives the grid-to-cathode voltage beyond −20 V. Therefore, in class A amplification the maximum signal that the amplifier can handle is 20 V p-p. However, for distortionless operation the signal voltage must be limited within the points *A* and *B*, that is, 18 V p-p.

So, for class A amplification the operating point is set in the middle of the linear portion of the amplifier's dynamic V_G/I_P characteristic. Moreover, the signal swing is limited to the end points *A* and *B* of the linear portion.

Gain and Size of R_L

The equivalent circuit of a triode amplifier (Fig. 42.1-3) shows a generator applying a signal voltage μv_g to the plate resistance r_p of the tube in series with the plate load resistor R_L. Of course, v_g is the signal voltage applied to the grid, and μ is the amplification factor of the tube. The output voltage v_{out} is taken across R_L, and by voltage-divider action it is equal to

$$v_{out} = \frac{R_L}{r_p + R_L} \mu v_g \qquad (42.1\text{-}1)$$

The gain of the tube, which is the ratio v_{out}/v_{in}, may be found by substituting the formula value of v_{out} in the equation

$$\text{Gain} = \frac{v_{out}}{v_{in}} \qquad (42.1\text{-}2)$$

This then becomes

$$\text{Gain} = \frac{R_L/(r_p + R_L)}{v_g} \mu v_g = \frac{\mu R_L}{r_p + R_L} \qquad (42.1\text{-}3)$$

If i_p is the ac signal current in the plate of the amplifier, then the signal voltage v_{out}, developed across the plate load resistor, is

$$v_{out} = i_p \times R_L \qquad (42.1\text{-}4)$$

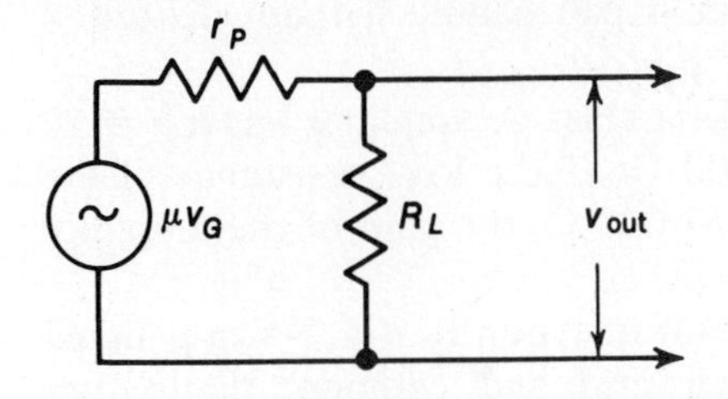

Fig. 42.1-3. Equivalent circuit of a triode amplifier.

It would appear from Eqs. (42.1-3) and (42.1-4) that for efficient operation R_L must be large compared with r_p. However, the size of R_L is limited by the fact that an increased R_L reduces i_p and causes the plate-to-cathode voltage of the tube to decrease. The tube then cannot handle so large a signal voltage without distortion. The value of R_L will depend on the size of the input signal which the amplifier must handle and on the gain required.

Equation (42.1-3) also shows that the higher the μ of the tube, the greater is the stage gain possible with proper circuit arrangement.

AC Plate Load

Consider the circuit of Fig. 42.1-4. V_1 is a class A amplifier. The input signal is coupled to the grid of V_1 by C_1 and R_1. The grid signal v_g is amplified by the tube and appears in the plate circuit, that is, across the plate load R_L, as v_p. This is then coupled by C_2 and R_2 to the grid of the succeeding amplifier V_2.

The ac equivalent circuit of V_1 must now include R_L in parallel with the series combination of C_2 and R_2 (Fig. 42.1-5). R_L is the dc plate load. The ac plate load Z_L consists of the combination of R_L in parallel with the impedance Z_2 of the branch C_2R_2.

If Z_2 is 10 (or more) times greater than R_L, the ac and dc plate loads are effectively the same, namely, R_L. So only R_L need be considered in determining the gain of the tube. However, if Z_2 is less than $10 \times R_L$, the ac plate load Z_L must be considered in computing the gain of the tube (see Fig. 42.1-6). In this case Z_L is less than R_L, and the gain of the tube will be less than when R_L alone was considered.

Vacuum-Tube Audio-Voltage Amplifiers

As we found in previous experiments, sound waves are transformed into electrical voltages by a microphone. These are called *audio-signal voltages.* The

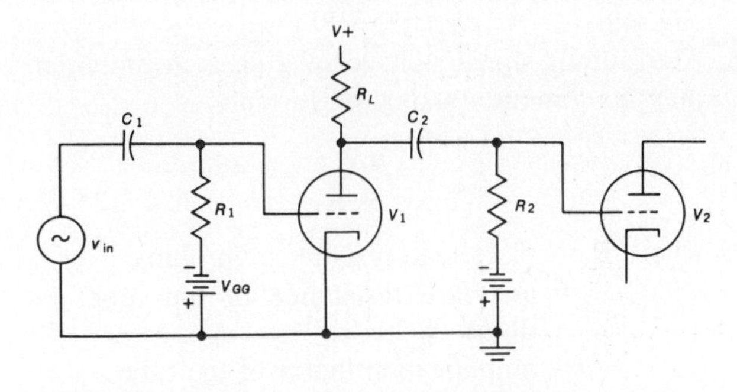

Fig. 42.1-4. Output of V_1 is developed across R_L and the parallel impedance of C_2 in series with R_2.

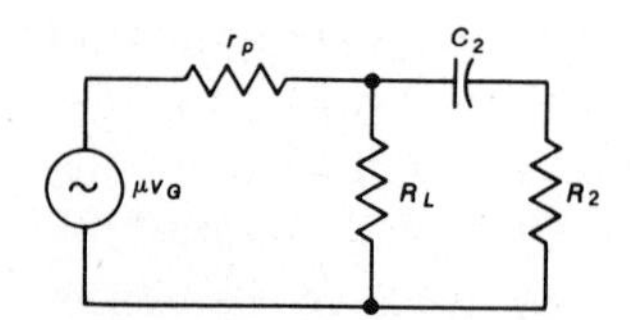

Fig. 42.1-5. Equivalent circuit of V_2 ac plate load.

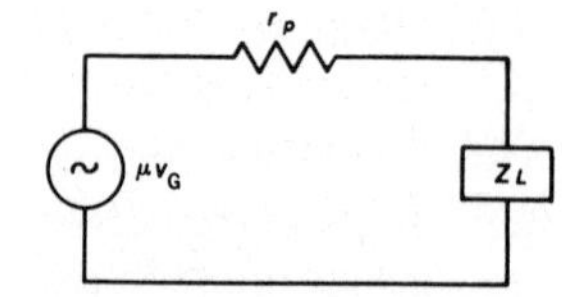

Fig. 42.1-6. AC plate load.

audible frequency range of these audio signals is approximately 20 to 16,000 Hz, depending on the source of the sound. Audio-signal voltages may consist of a single frequency or unique combinations of frequencies within the range of audio frequencies.

The audio-voltage-tube amplifier is therefore a special application of the ac voltage amplifier in Fig. 42.1-1. Audio-voltage amplifiers are usually operated class A.

SUMMARY

1. A vacuum-tube triode may be connected in a circuit to act as a voltage amplifier. The usual arrangement is that of the grounded or common-cathode amplifier. This is comparable with the grounded or common-emitter transistor amplifier.
2. In the circuit of Fig. 42.1-1, V_{GG} supplies the grid bias, $V+$ is the plate voltage source, and R_L acts as the plate load resistor, across which the output signal is developed. The input signal appears between grid and cathode. The output signal is taken from the plate-to-cathode circuit.
3. A vacuum-tube class A voltage amplifier is one in which there is plate current at all times for the range of signal voltages that the amplifier is designed to handle.
4. The grid bias sets the operating point of the tube. The input signal carries the grid above and below the level of the operating point.
5. In a class A vacuum-tube voltage amplifier, the operating point is set in the middle of the linear portion of the tube's V_G/I_P characteristic, as in Fig. 42.1-2.
6. The output signal voltage of a class A amplifier may be computed using the formula

$$v_{\text{out}} = \frac{\mu R_L}{r_p + R_L} \times v_g$$

where R_L = plate load resistance, in ohms
r_p = ac plate resistance of the tube, in ohms
μ = amplification factor of the tube
v_g = signal voltage at the input grid

7. The gain of an amplifier may be determined experimentally by measuring the output signal voltage v_{out}, the input signal voltage v_{in} (or v_g), and substituting the measured values in the formula

$$\text{Gain} = \frac{v_{\text{out}}}{v_{\text{in}}}$$

8. The gain may also be computed theoretically using the formula

$$\text{Gain} = \frac{\mu R_L}{R_L + r_p}$$

9. The output signal voltage v_p appears across the plate load resistor R_L and is equal to

$$v_p = i_p \times R_L$$

where i_p is the signal current in the plate of the tube. Within the design limits of the circuit v_p will increase, with an increase in R_L.

10. If there is an ac load across R_L, the ac load impedance Z_L is the parallel combination of R_L and the impedance of the *ac* load. In that case, the output signal voltage v_p is

$$v_p = i_p \times Z_L$$

SELF-TEST

Check your understanding by answering these questions.

1. The device which "amplifies" the signal in a common-cathode amplifier is the ________.
2. Bias voltage in the circuit of Fig. 42.1-1 is determined by the value of ________.
3. The amplifier in Fig. 42.1-1 could operate without any plate voltage, that is, with $V+ = 0$. ________ (true/false)
4. If the linear portion of a class A amplifier's V_G/I_p characteristic lies between the limits (see Fig. 42.1-2) −7 V (point A) and −1 V (point B), the amplifier should be biased at ________ V.
5. For the amplifier in question 4, the maximum input signal that it can handle without distortion is ________ V p-p.
6. A triode used as a class A amplifier has a $\mu = 20$ and an $r_p = 7000\ \Omega$. If the load resistance of the amplifier $R_L = 33{,}000\ \Omega$, the gain of the amplifier is ________.
7. In the amplifier of question 6, if a 2-V p-p signal appears between grid and cathode, the output signal at the plate will measure ________ V p-p.

MATERIALS REQUIRED

- Power supply: Regulated 250-V+ source; variable, negative-bias supply; 6.3-V rms source
- Equipment: Oscilloscope; EVM; audio-frequency sine-wave signal generator
- Resistors: ½-W 10,000-Ω; 100-, 270-kΩ; 4.7-MΩ
- Capacitors: Two 0.01-μF
- Tubes: 6AV6 and tube socket
- Miscellaneous: 2-W 500-kΩ potentiometer

PROCEDURE

Plate Load and Gain

1. Connect the circuit of Fig. 42.1-7*a*. The triode section of a 6AV6 is used. Initial value of R_L is 10,000 Ω. Filament pins 3 and 4 are connected to a 6.3-V rms source. Set variable bias supply V_{GG} so that grid bias is −2 V. Maintain this bias throughout the experiment.

   Connect an audio-signal generator, set at 1000 Hz, to the input terminals of the amplifier.

   Set $V+$ at 250 V and maintain it at this value throughout the experiment.
2. Observe with an oscilloscope, and adjust the signal-generator output control so that v_g measured from grid to cathode is ½ V p-p.
3. With an oscilloscope observe and measure the output signal voltage v_p at the plate of the amplifier. Record the data in Table 42.1-1.
4. Measure the dc voltage, plate (pin 7) to cathode (pin 2). Record the result in Table 42.1-1.
5. Repeat steps 1 through 4 for each value of R_L in the table. Record the results.
6. From the relationship

$$\text{Gain} = \frac{v_{\text{out}}}{v_{\text{in}}} = \frac{v_p}{v_g}$$

   compute amplifier gain for each of the plate loads. Record them in Table 42.1-1. Show a sample computation.
7. Using the formula

$$\text{Gain} = \frac{\mu R_L}{R_L + r_p}$$

   compute the amplifier gain for each of the plate loads. Record the data. Show a sample computation.

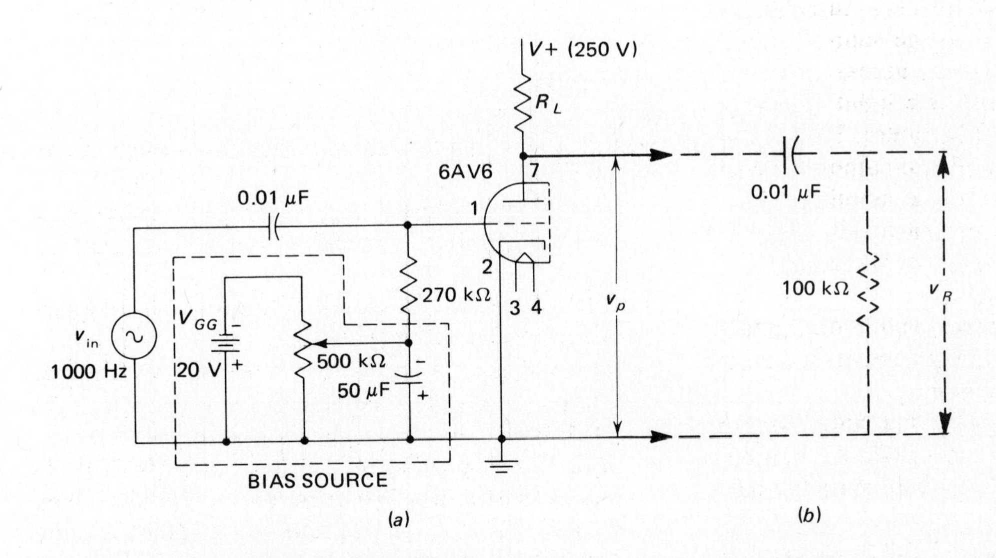

Fig. 42.1-7. (*a*) Experimental class A amplifier; (*b*) ac load.

TABLE 42.1-1. Amplifier Gain and Plate Load (R_L)

R_L, Ω	v_g, V p-p	v_p, V p-p	*Plate voltage, dc*	$Gain = \frac{v_p}{v_g}$	$Gain = \frac{\mu R_L}{R_L + r_p}$
10,000					
270,000					
4,700,000					

TABLE 42.1-2. Effect of AC Load on Output and Gain

R_L, Ω	v_g, V p-p	v_p, V p-p	v_r, V p-p	*Gain*, v_p/v_g
10,000	½			
270,000	½			
4,700,000	½			

AC Plate Load

8. To the circuit of Fig. 42.1-7*a*, add the *ac* load shown in Fig. 42.1-7*b*. Initial value of R_L is 10,000 Ω.
9. Set grid bias at −2 V and $v_g = ½$ V p-p.
10. With an oscilloscope observe, measure, and record in Table 42.1-2 v_p and v_r.
11. Repeat steps 9 and 10 for each value of R_L in the table.
12. Compute and record the gain (v_p/v_g) of the amplifier for each value of R_L.

QUESTIONS

1. What effect does the plate load resistor R_L in the experimental amplifier have on the level of output signal?
2. What effect does the plate load resistor R_L in the experimental amplifier have on the gain of the amplifier?
3. What effect does the ac load in the experimental amplifier have on the level of output signal? Discuss effect for the three values of R_L.
4. What effect does the ac load in the experimental amplifier have on the gain of the amplifier? Discuss effect for the three values of R_L.

Answers to Self-Test

1. vacuum tube
2. V_{GG}
3. false
4. −4
5. 6
6. 16.5
7. 33

CHAPTER 43 CATHODE BIAS

EXPERIMENT 43.1. Cathode Bias

OBJECTIVE

To determine experimentally the value of cathode resistor required to cathode- (self-) bias a grounded-cathode amplifier

INTRODUCTORY INFORMATION

Cathode Bias

In the early days of radio, C batteries were used to provide bias for amplifier circuits. These C batteries were soon eliminated by a circuit arrangement which provided bias without any external bias-voltage source. The circuit arrangement utilizes normal electron flow in the tube to provide self-bias.

One fact which makes cathode or self-bias possible in a triode is that current will flow in a triode when the grid-to-cathode voltage (bias) is zero. In the circuit of Fig. 43.1-1 there is current in the tube, though the control grid and cathode are both returned to common ground and there is no external bias source. This action is completely different from that in a bipolar transistor, and no current would flow if the transistor were to replace the tube, as in Fig. 43.1-2.

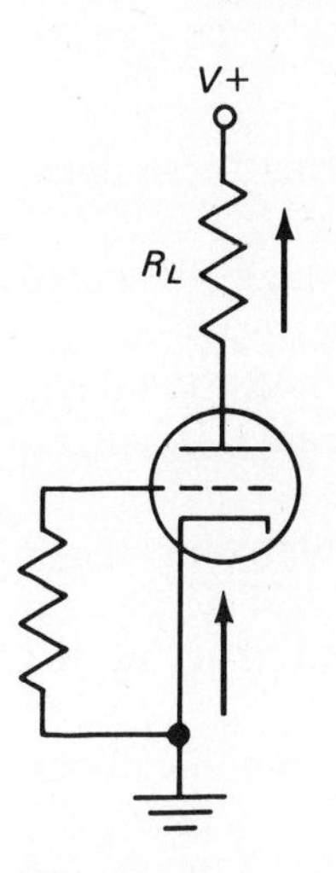

Fig. 43.1-1. With zero bias there *is current* flow in the tube.

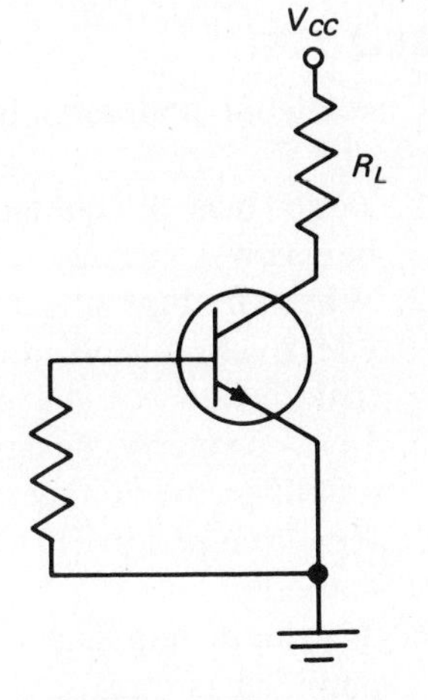

Fig. 43.1-2. With zero base-to-emitter bias there is *no* current flow in the bipolar transistor.

Mathematical Method for Determining R_K

Consider the circuit of Fig. 43.1-3. When power is applied, there is electron flow through the tube in the direction indicated by the arrows. There is a voltage drop V_K across R_K, with polarity shown, as a result of this electron flow. V_K may be measured, from K to ground, with a dc voltmeter. In the grid circuit of this tube, there is no direct current. DC voltage measured from grid to ground (V_G) is therefore zero, or G is at ground potential. The cathode K is positive with respect to ground and therefore positive with respect to the grid. If in Fig. 43.1-3 $V_K = +2$ V, K is 2 V positive with respect to the grid. Therefore the grid is 2 V negative with respect to the cathode.

Tube bias is defined as grid-to-cathode voltage. Here then is another means of biasing the tube, namely, by cathode bias, or self-bias.

The value of voltage V_K at the cathode with respect to ground depends on the size of R_K and on the value of the current in the cathode circuit. In the triode circuit of Fig. 43.1-3, there is no grid current. Therefore cathode current and plate current are the same. Assume that there is 1 mA of plate current. The size of R_K which will give 2 V bias (that is, $V_K = +2$ V) can now easily be found. From Ohm's law,

$$R_K = \frac{V_K}{I} \tag{43.1-1}$$

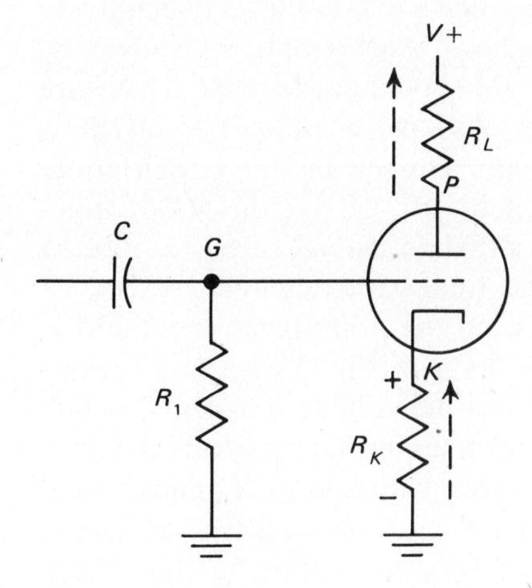

Fig. 43.1-3. Cathode and plate current are the same.

Substituting $I = 0.001$ A and $V_K = +2$ V in Eq. (43.1-1), we have

$$R_K = 2000\ \Omega$$

Nonmathematical Method for Determining R_K

A simple nonmathematical method for determining the value of R_K is available.

In the circuit of Fig. 43.1-3, if the value of R_L and of $V+$ are known, it is possible to connect a rheostat in place of R_K. An EVM is connected across the rheostat to measure the dc voltage (hence the bias) in the cathode. The rheostat is adjusted for the value of V_K required. Without its setting being changed, the rheostat is removed from the circuit, and its resistance is measured (R_K). A fixed resistor, whose value is approximately equal to the measured value of R_K, is then connected in the cathode circuit.

Measuring Grid Bias (Meter-Loading Effects)

The procedure which follows requires that low dc voltage readings be made in cathode circuits and in high-impedance grid circuits. The input impedance of an EVM is high enough to eliminate circuit-loading effects. The ground lead of the EVM, however, may disturb a circuit if it is placed at a high-impedance test point. So, when you measure a voltage between grid and cathode, it is best to place the EVM ground lead to chassis ground, read the grid voltage, read the cathode voltage, and compute the total cathode-to-grid voltage. If the grid circuit happened to be open, however, unreliable indications might be obtained. Therefore, a voltage measurement between grid and cathode would be advisable with the EVM ground lead connected to the cathode, which is the lower-impedance point.

A VOM with low input impedance connected from grid to cathode may disturb a circuit sufficiently to cause erroneous readings. An example will show why a low-impedance VOM should not be used to measure grid-to-cathode bias. Assume a 1000-Ω/V VOM is set on the 10-V range. This means the circuit under test becomes loaded by 10,000 Ω, the VOM input impedance. If the VOM is placed across a 5000-Ω cathode resistor, the total cathode-circuit resistance becomes 3300 Ω. If, on the other hand, the VOM is placed from cathode to grid (Fig. 43.1-4*a*), it creates a voltage divider across the voltage source V_K, which gives an incorrect bias reading. The equivalent circuit (Fig. 43.1-4*b*) shows that with a large R_G and a small R_M the voltage across R_M, grid-to-cathode voltage, is small compared with V_K. In addition to this effect, when a meter with a low ohms-per-volt characteristic

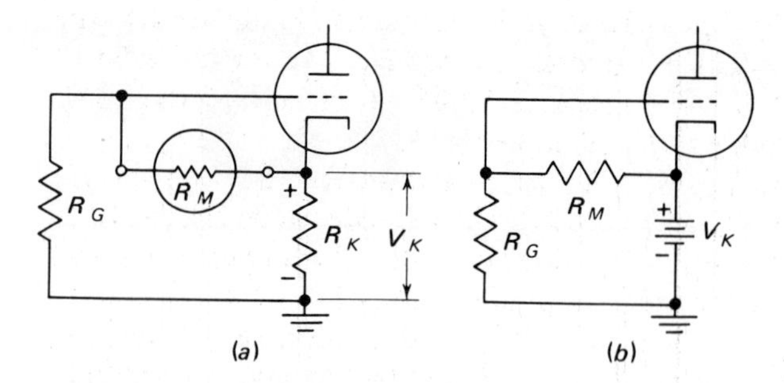

Fig. 43.1-4. VOM circuit loading.

is used to read grid bias in a high-impedance circuit, the voltage V_K will also change because of the change in grid bias. Therefore, a low-input-impedance VOM should *not* be used to measure *grid-to-cathode* voltage.

SUMMARY

1. Triode tubes may be cathode biased or self-biased by connecting a resistor R_K in the cathode of the tube, as in Fig. 43.1-3.
2. It is possible to set the bias of a tube experimentally by connecting a variable resistor R_K in the cathode and adjusting the size of the resistor for the desired value of bias voltage.
3. It is also possible to determine the value of the cathode bias resistor mathematically if the tube bias and tube current are known. Thus if the bias is 3 V and the tube current is 0.001 A,

$$R_K = \frac{V_K}{I} = \frac{3}{0.001} = 3000\ \Omega$$

4. To measure the bias (grid-to-cathode voltage) of a triode amplifier, a high-impedance voltmeter, such as an EVM, must be used to prevent loading the high-impedance input circuit of the triode.

SELF-TEST

Check your understanding by answering these questions.

1. Tube bias is defined as the voltage measured between ________ and ________.
2. When the bias of a triode tube is 0 V, the tube is effectively cut off, drawing no current. ________ (true/false)
3. In a properly designed triode common-cathode amplifier, the voltage on the control grid is ________ (negative/positive) with respect to that on the cathode.
4. The input impedance of a triode common-cathode amplifier is ________ (high/low).
5. In the circuit of Fig. 43.1-3, if $R_K = 2200\ \Omega$ and tube current is 0.004 A, the bias on the tube is ________ V.

6. In Fig. 43.1-3 it is necessary to determine the value of cathode resistor R_K, to bias the tube at 6 V, when there is 0.0015 A of tube current. The value of R_K must be ________ Ω.
7. In Fig. 43.1-4*b*, R_M is a voltmeter used to measure bias on the tube. If the cathode voltage $V_K = 10$ V and $R_m = R_g = 100$ kΩ, the value of bias measured by the meter will be ________ V.
8. The input impedance of the voltmeter in question 7 is too ________ (high/low) to give an accurate reading of bias voltage.

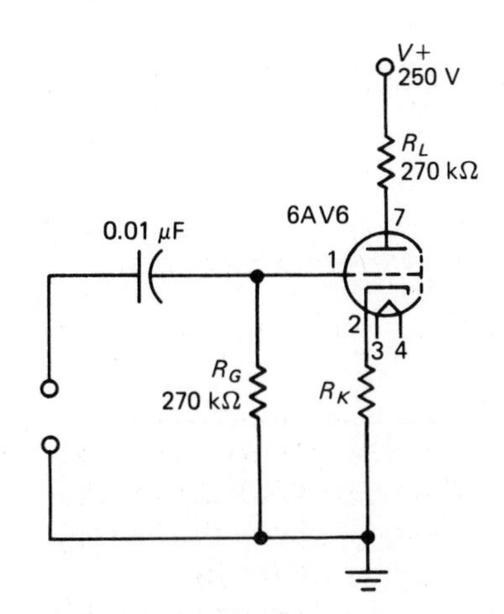

Fig. 43.1-5. Experimental circuit.

MATERIALS REQUIRED

- Power supply: Regulated 250-V dc source; 6.3-V rms source
- Equipment: EVM; VOM
- Resistors: ½-W two 270-kΩ; resistors required in steps 2, 5, and 9
- Capacitor: 0.01 μF
- Tube: 6AV6 and socket
- Miscellaneous: 2-W 10-kΩ potentiometer

PROCEDURE

1. Compute and record in Table 43.1-1 the value of R_K which will be needed to give a 1-V bias to the circuit of Fig. 43.1-5, if $R_L = 270{,}000$ Ω, $V+$ is 250 V, and $I_P = 0.55$ mA. Show your computations.

NOTE: Plate and cathode current are the same.

2. Connect the circuit of Fig. 43.1-5 using the value of R_K found in step 1.
3. **Power on.** With an EVM measure and record in Table 43.1-1 the *dc* voltage from cathode to ground V_K.
4. If V_K, measured in step 3, does not equal 1 V, determine experimentally (using a 10-kΩ potentiometer as a rheostat) the value of R_K required to develop $V_K = 1$ V. Remove the rheostat from the circuit and measure the resistance R_K. Record the experimentally determined value of R_K in Table 43.1-1.
5. Connect a resistor equal (or as close as possible) to the experimental value found in step 4 in the cathode of the circuit.
6. With an EVM, measure and record in Table 43.1-1 V_K, V_G (grid to ground), and the bias voltage from grid to cathode V_{GK}.
7. Repeat step 6 with a VOM set on the 2.5-V, or closest, range.
8. Experimentally determine and record in Table 43.1-2 the value of R_K required to bias the circuit of Fig. 43.1-5 at 2 V if $R_L = 270{,}000$ Ω.
9. Connect a resistor equal (or as close as possible) to the value of R_K found in step 8 in the circuit and measure and record V_K.

TABLE 43.1-1. Self-Bias Measurements

R_K, Ω *Computed Value*	V_K, V	R_K, Ω *Experimental Value*	V_K, V

	V_K, V	V_G, V	V_{GK}, V
EVM			
VOM			

QUESTIONS

1. Did the computed value of R_K (step 1) produce a 1-V bias? If not, explain why not.
2. What information is needed to *compute* the value of R_K in a tube circuit, for a required value of cathode bias?
3. What equipment and components are required to find the value of R_K experimentally to produce a required cathode bias?
4. How effective is the experimental method in determining R_K?

TABLE 43.1-2. R_K, Experimental Value

R_K, Ω	V_K, V

5. Which of the bias measurements V_{GK} in Table 43.1-1 is closer to the actual tube bias? How do you know?
6. What problem, if any, is there in measuring grid bias V_{GK} with a low-impedance voltmeter?

Answers to Self-Test

1. grid; cathode	5. 8.8
2. false	6. 4000
3. negative	7. 5
4. high	8. low

EXPERIMENT 43.2. Cathode Bypass Capacitor

OBJECTIVE

To verify that the cathode resistor must be properly bypassed to avoid ac degeneration

INTRODUCTORY INFORMATION

Negative Feedback due to Unbypassed Cathode Resistor

In earlier experiments it was established that an unbypassed emitter resistor in a common-emitter transistor amplifier caused negative feedback, resulting in signal degeneration and reduction in amplifier gain. In this experiment it will be observed that the same effect occurs with an unbypassed cathode resistor. Bypassing the cathode resistor with the proper size capacitor will eliminate this problem.

How does the cathode resistor R_K in Fig. 43.2-1 act to reduce the gain of the amplifier? By providing degeneration, or negative feedback. Consider the effect of a sine-wave signal voltage applied between grid and ground. On the positive alternations, plate current (and hence cathode current) increases. On the negative alternations, plate current decreases. This causes the cathode voltage to vary in phase with grid voltage, and a sine-wave signal voltage now appears in the cathode, which "follows" the grid voltage. The signal voltage which the tube "sees" is the effective grid-to-cathode voltage. Therefore, to determine the effective signal-voltage input to the tube, it is necessary to subtract the cathode signal voltage from the grid signal voltage. Thus the tube "sees" a smaller amplitude signal v_g than was actually injected. Therefore, the output signal at the plate, v_p, is smaller than it would be if v_k did not vary with the input signal.

It is therefore necessary to eliminate the feedback voltage at the cathode, if cathode bias is to be used.

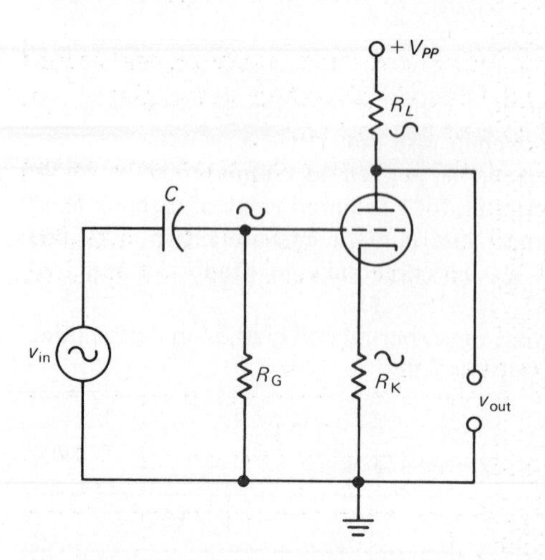

Fig. 43.2-1. Unbypassed cathode resistor causes ac degeneration.

Cathode Bypass Capacitor

"Bypassing" R_K with a suitable capacitor C_K as in Fig. 43.2-2 provides another path for alternating current in the cathode circuit. For cathode-current variations can flow through R_K *and* C_K, and the distribution of current between R_K and C_K will depend on

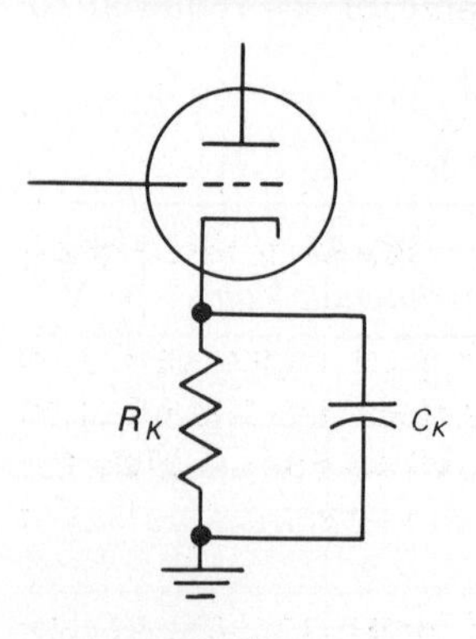

Fig. 43.2-2. Cathode bypass capacitor.

the X_C of C_K as compared with R_K. For example, if at a certain frequency $X_{CK} = 200\ \Omega$ and $R_K = 2000\ \Omega$, $^{10}/_{11}$ of the alternating current will flow through C_K and only $^{1}/_{11}$ through R_K. Bypassing R_K with C_K, in this instance, reduces the effects of degeneration. However, it does not affect the *direct* current through R_K. Therefore, V_K, the cathode-bias voltage, is maintained at its required value.

If an amplifier is required to handle more than one frequency, then the value of C_K must be chosen so that it provides adequate bypassing for the lowest of all the frequencies. It will then be a good bypass for all the higher frequencies also, because the impedance of C_k for alternating current is lower at higher frequencies.

To determine the size of C_K, it is therefore necessary to know the lowest frequency F_o it is required to pass and the size of the cathode resistor R_K in parallel with it. C_K is considered a good bypass if at F_o

$$X_{CK} = \frac{R_K}{10} \qquad (43.2\text{-}1)$$

It will be recalled that the same formula applies in determining the size of the emitter resistor bypass capacitor.

SUMMARY

1. An unbypassed cathode resistor causes negative or degenerative feedback in a triode amplifier, reducing the effective input signal of the amplifier.
2. To prevent ac feedback, the cathode resistor R_K must be bypassed by a capacitor C_K; that is, C_K must be connected in parallel with R_K.
3. The size of C is determined by the lowest frequency which the tube must amplify and by the value of the cathode resistor. The reactance X_C of the bypass capacitor must be such that at the lowest frequency,

$$X_C = \frac{R_K}{10} \quad \text{(approx)}$$

SELF-TEST

Check your understanding by answering these questions.

1. The value of the cathode bypass capacitor C_K, in Fig. 43.2-2, required to prevent ac degeneration depends on the ________ frequency which the circuit must amplify and on the ________ of ________.
2. The circuit of Fig. 43.2-2 must amplify all frequencies in the range 100 to 20,000 Hz. If $R_K = 2200\ \Omega$, the reactance of the bypass capacitor at 100 Hz must be equal to or less than ________ Ω.

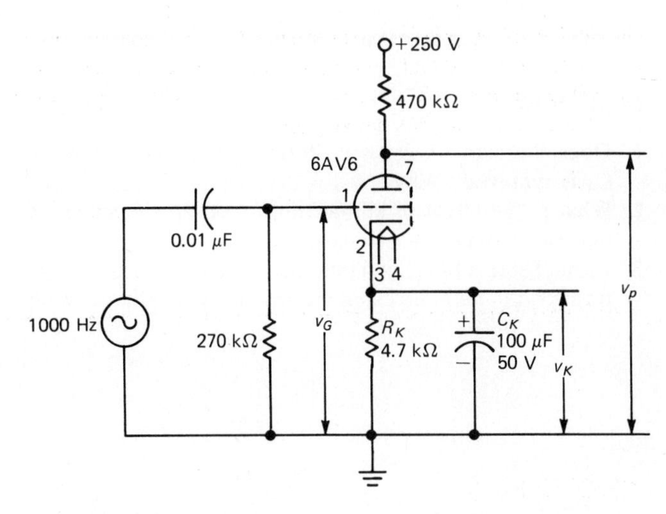

Fig. 43.2-3. Experimental circuit to determine effects of cathode bypass capacitor.

MATERIALS REQUIRED

- Power supply: Regulated 250-V dc source; 6.3-V rms source
- Equipment: Oscilloscope; EVM
- Resistors: ½-W 4700-, 270,000-, and 470,000-Ω
- Capacitors: 0.01-μF; 100-μF/50-V
- Tube: 6AV6 with socket

PROCEDURE

1. Connect the circuit of Fig. 43.2-3. **Power on.**
2. Inject a 1000-Hz sine-wave signal, v_{in} 0.5 V p-p, into the input.
3. With an oscilloscope observe, measure, and record in Table 43.2-1 the signal voltage at the plate v_p, at the grid v_g, and at the cathode v_k.
4. Remove the cathode bypass capacitor and repeat steps 2 and 3.
5. Compute and record the gain of the amplifier (v_p/v_g) with the cathode bypassed and unbypassed.

TABLE 43.2-1. Amplifier Gain

Condition	V p-p			Gain
	v_p	v_g	v_k	v_p/v_g
R_K bypassed				
R_K not bypassed				

QUESTIONS

1. Does the signal voltage v_k in the cathode change when C_K is removed? Why?
2. What is the effect of eliminating C_K on v_p, the signal at the plate? Why?
3. From Table 43.2-1, compute the effective signal voltage from grid to cathode using the unbypassed cathode. With this effective signal input, compute the gain of the amplifier. How does this compare with the gain when the cathode is bypassed with a 100-μF capacitor?
4. What is the phase of v_g and v_k? Explain.

Answers to Self-Test

1. lowest; resistance; R_K
2. 220

CHAPTER 44 TROUBLESHOOTING A VACUUM-TUBE AUDIO AMPLIFIER

EXPERIMENT 44.1. Resistance Analysis of a Vacuum-Tube Audio Amplifier

OBJECTIVE

To evaluate and measure the resistances at the various tube-socket terminals of a vacuum-tube audio amplifier

INTRODUCTORY INFORMATION

The techniques in servicing many-stage vacuum-tube amplifiers are very similar to those used in troubleshooting transistor amplifiers. By a signal-tracing process the trouble is isolated to a specific stage or stages. Then voltage and resistance checks are made at the tube-socket terminals. If the rated values of voltage and resistance are supplied, the technicians have a reference source for comparing the measured with the rated values. They can then spot abnormal readings, and by further measurement or parts substitution they can determine the defective component.

Frequently, however, rated values are not available. Technicians must then know what values to expect in order to recognize an incorrect reading.

You now have enough circuit knowledge to determine the resistance and voltage values you may expect at every test point in the amplifier.

Two-Stage Vacuum-Tube Audio Amplifier

The circuit of Fig. 44.1-1 is that of a two-stage audio amplifier. V_1 is a triode voltage amplifier, V_2 is a pentode power amplifier. Some explanation is necessary about V_2. So far you have studied two-element tubes, called *diodes,* and three-element tubes, called *triodes.* There are other tube types: tetrodes, or four-element tubes; pentodes, or five-element tubes; and other specialized tubes. The pentode in Fig. 44.1-1 consists of five active elements (and filaments not shown). The active elements are the cathode, pin 8; control grid, pin 5; screen grid, pin 4; suppressor grid, shown internally tied to the cathode; and plate, pin 3. The screen grid and suppressor greatly change the characteristics of the tube, as compared with a triode. However, it still is used as a device to amplify signals. The pentode in Fig. 44.1-1 serves as the final, or power output, stage in the audio amplifier and delivers audio currents by way of the transformer to the speaker. If an audio signal is fed to the input of V_1, sound will be heard in the speaker.

This circuit, of course, is very similar to a two-

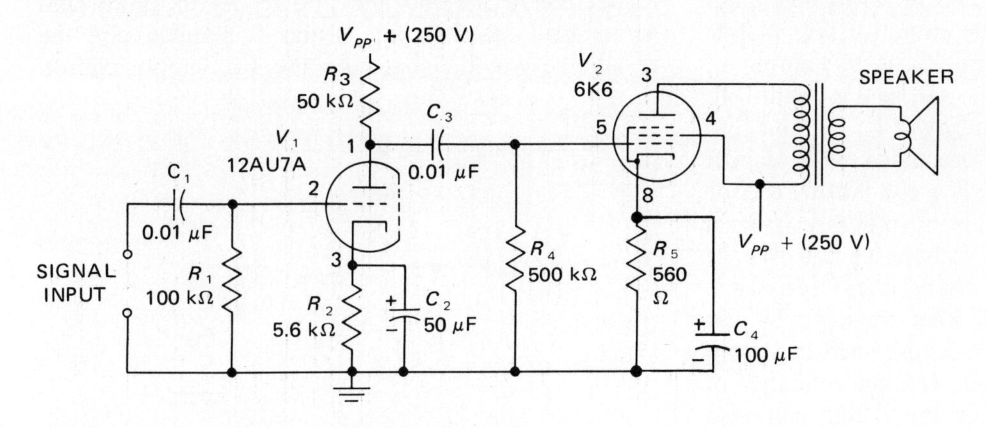

Fig. 44.1-1. Audio amplifier, including output stage.

stage transistor audio amplifier. Of interest, however, are the differences in component values and voltages used.

The first differences to observe are the values of C_1 and C_3, the coupling capacitors to the control grids of V_1 and V_2. The 0.01-μF value is very small compared with the 50- and 100-μF coupling capacitors found in transistor audio stages. The reason is that the input impedance of a vacuum-tube audio amplifier is very high, while that of a transistor amplifier is low. C_1 and R_1 in V_1, Fig. 44.1-1, make up a voltage divider. If a 1-V p-p 1000-Hz signal is applied to the input of V_1, about 0.99 V will appear across R_1. The signal voltage across R_1 is even higher for frequencies above 1000 Hz, becoming 1 V (as far as measurement is concerned) at 10,000 Hz. It is clear then that small-valued capacitors do couple the signal to the control grid of a vacuum-tube audio amplifier. If a 0.01-μF capacitor were used in the input of a bipolar transistor audio stage, most of the signal would be lost across the capacitor, and very little would reach the base.

Another difference between the vacuum tube and the usual transistor audio amplifier is the value of dc voltage which is used to power the amplifier. Note the 250 V used in Fig. 44.1-1. In an equivalent bipolar transistor amplifier, a 10- or 15-V source would be found. Most tubes require higher dc voltages than transistors. (Twelve-volt tubes were used in automobile radios before they were replaced by transistors.) As a third difference, we see that high-valued control-grid and plate resistors are used in vacuum-tube audio amplifier circuits, while smaller-valued resistors are employed in bipolar transistor circuits.

Resistance Analysis

As in the case of transistor circuits, resistance computations in vacuum-tube circuits are based on simple knowledge of series, parallel, and series-parallel combinations of resistances. In the circuit of Fig. 44.1-1, assume that the power source V_{PP} is the output of an ac rectifier and filter connected in the circuit but not shown here. Assume that power is OFF and that all resistance measurements are made with respect to ground. At pin 2 of V_1, R_1 is a 100,000-Ω resistor connected to ground. There is no other resistor in parallel with R_1. The grid-to-cathode circuit of V_1 is an open circuit. Therefore, there is no resistance from pin 2 of V_1 to ground other than R_1. So the resistance that an ohmmeter would measure from pin 2 of V_1 to ground is 100,000 Ω ± the tolerance of the resistor. The accuracy of the meter will also determine the actual value read. However, practically

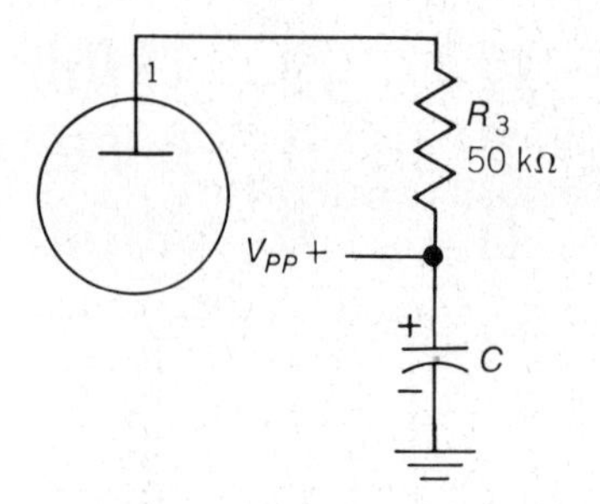

Fig. 44.1-2. Plate circuit.

it may be assumed that the resistance from pin 2 of V_1 to ground is approximately 100,000 Ω.

The resistance from pin 3 of V_1 to ground is 5600 Ω. The leakage resistance of C_2 is so high compared with the value of R_2 that it will not affect the reading.

The resistance from pin 1 of V_1 to ground presents a slightly different problem. It depends on the V_{PP} supply circuit. Figure 44.1-2 is another view of the plate circuit (pin 1) of V_1 and the V_{PP} supply, which is filtered by electrolytic capacitor C. When the ohmmeter leads are placed across pin 1 and ground, C starts to charge to the battery voltage in the ohmmeter. As a result, the resistance read depends on the polarity of the meter leads and how long the meter leads are kept in the circuit. The meter will register an increasing resistance with time. When the meter lead polarity is the same as the polarity of the capacitor, the final resistance reading will be 50,000 Ω plus the electrolytic leakage resistance.

If, however, the supply contains a bleeder resistor R across C, as in Fig. 44.1-3, and if $R = 10{,}000$ Ω, the resistance measured from pin 1 to ground would be 60,000 Ω. R_4 will not affect resistance readings at pin 1 of V_1 because as far as direct current is concerned, C_3 acts as an open circuit.

The previous considerations apply also to V_2 in Fig. 44.1-1. Resistance at pins 5 and 8 of V_2 would be, respectively, 500,000 and 560 Ω. At the plate and screen-grid circuits (pins 3 and 4, respectively) the resistance would depend on the V_{PP} supply circuit.

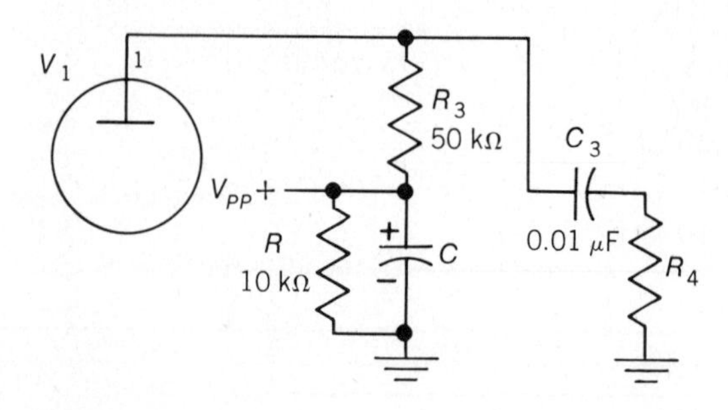

Fig. 44.1-3. Plate circuit with power supply bleeder.

The resistance of the transformer primary is low compared with the V_{PP} to ground resistance, so that the readings at pins 3 and 4 will be approximately the same.

Resistance checks in the filament circuits will depend on how the filaments are connected. If the filaments are supplied by a 6.3-V winding on the power transformer and if one lead of this winding is grounded, as in Fig. 44.1-4, the resistance from pins 2 of V_2, or 4–5 of V_1, to ground is the resistance of the filament winding, which is very low, about 0.1 Ω. Filaments connected differently must be individually analyzed.

Parallel connections must be considered in computing resistance. For example, in Fig. 44.1-5, the resistance at pin 2 or at pin 7 is determined by the parallel combination of the two branches shown. The R_T of this combination is approximately 4000 Ω. In measuring resistance in a vacuum-tube circuit the technician is not restricted to the use of a low-power ohmmeter. An ordinary ohmmeter may be used. Power is always turned OFF before making resistance checks. When resistance is measured across an electrolytic capacitor, meter lead polarity should be the same as that of the capacitor.

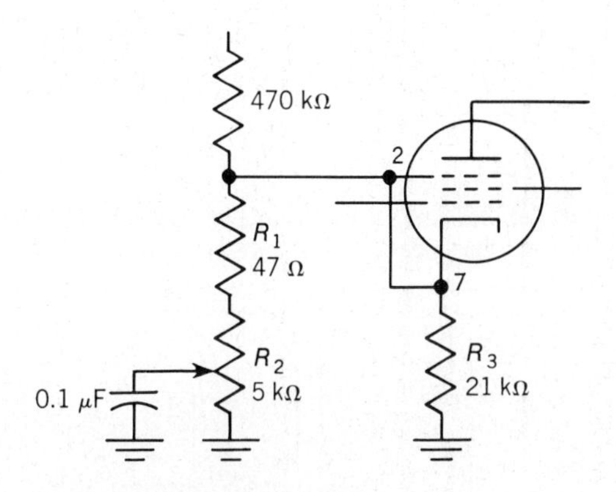

Fig. 44.1-5. Parallel branches.

SUMMARY

1. In servicing vacuum-tube audio amplifiers, a signal-tracing procedure is used, as in transistor circuits, to isolate the defective stage or stages.
2. After the defective stage is found, dc voltage and resistance checks (if necessary) are made in that stage. Abnormal readings help identify the defective component, which can then be further checked by measurement or by substituting a known good component for the suspected part.
3. Resistance checks, of course, are made with power OFF.
4. To determine if resistance measurements at the tube-socket terminals of an amplifier are normal, the technician must refer to the diagram of that circuit.
5. The resistance, measured to ground, should be the value shown going to ground in the circuit diagram.
6. If there are parallel resistors to ground at any tube-socket terminal, the measured value is the parallel combination of these resistors.
7. The technician should look for series and series-parallel combinations in estimating the normal resistance in the circuit.

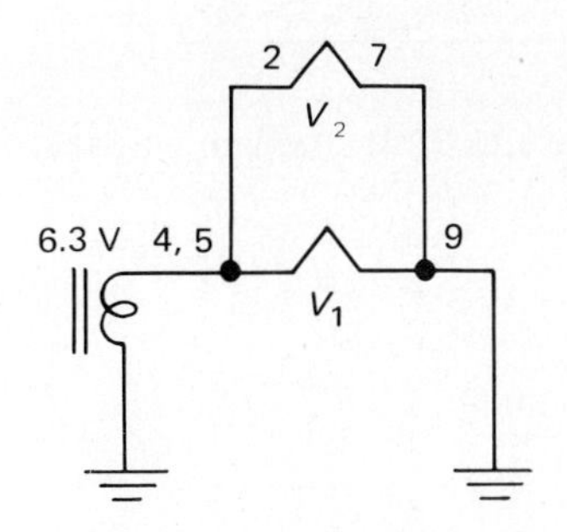

Fig. 44.1-4. Filament circuit.

SELF-TEST

Check your understanding by answering these questions.

1. In V_1 of Fig. 44.1-1 assume $R_1 = 560{,}000$ Ω, $R_2 = 3300$ Ω, $R_3 = 27{,}000$ Ω. The estimated resistance to ground measured from (*a*) the control grid is ________ Ω; (*b*) the cathode is ________ Ω; (*c*) the plate of V_1 is (assume a power supply bleeder resistor of 5000 Ω) ________ Ω.
2. In Fig. 44.1-5, the resistance to ground from the suppressor grid (pin 2) is ________ Ω.
3. In Fig. 44.1-4, if the resistance of the secondary winding of the 6.3-V transformer is 0.1 Ω and the cold resistance of the filaments of $V_1 = 14$ Ω and of $V_2 = 32$ Ω, the resistance measured from the top of the secondary winding to ground is ________ Ω.

MATERIALS REQUIRED

- Power supply: 250-V regulated dc source; 6.3-V rms filament supply
- Equipment: EVM or VOM
- Resistors: ½-W 3300-, 6800-, 56,000-, 120,000-, two 270,000-Ω

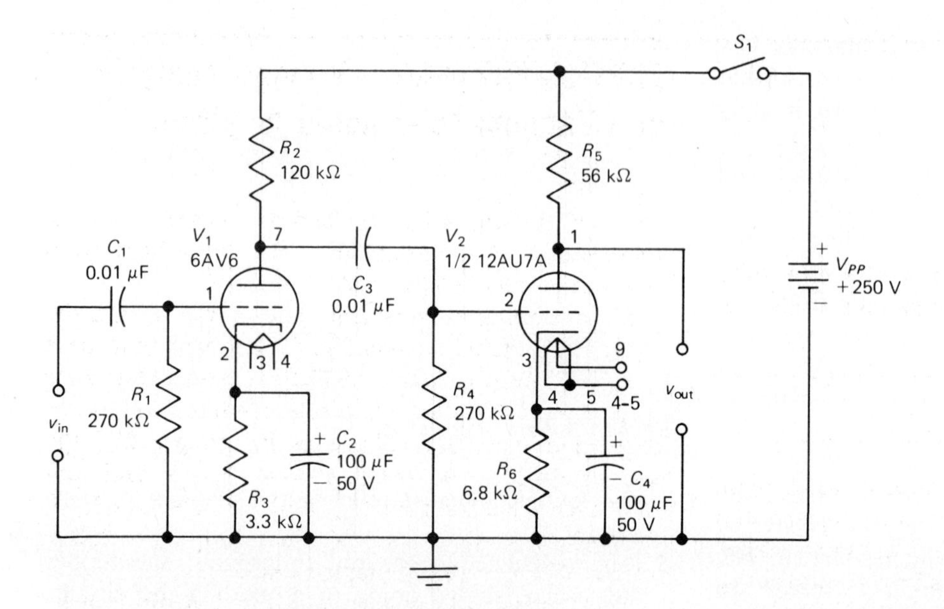

Fig. 44.1-6. Experimental audio amplifier.

- Capacitors: Two 0.01-μF; two 100-μF/50-V
- Tubes: 6AV6, 12AU7A, and sockets
- Miscellaneous: SPST switch

PROCEDURE

NOTE: All resistances are measured with respect to ground.

1. Connect the circuit of Fig. 44.1-6. Ground one end of the filament supply. **Power off.** S_1 open.
2. **a.** Estimate and record the values of resistance with respect to ground at every test point listed in Table 44.1-1.
   b. Fill in the blanks under "element" corresponding to each pin number in the table.
3. Measure and record the resistance at every test point in Table 44.1-1.

TABLE 44.1-1. Resistance Analysis

Tube	Element	Pin Number	Resistance, Ω	
			Estimated	*Measured*
V_1 6AV6		1		
		2		
		3		
		4		
		5		
		6		
		7		
V_2 12AU7A		1		
		2		
		3		
		4		
		5		
		9		

QUESTIONS

1. In the circuit of Fig. 44.1-1, what would be the meaning of an infinite resistance between pins 3 and 4 of V_2?
2. The resistance to ground at the plate of V_1, Fig. 44.1-1, measures 50,000 Ω when the ohmmeter is first connected but gradually increases if the meter is left connected for a period of time. What does this change mean?
3. In Fig. 44.1-4 the resistance from pin 2 to pin 7 measures less than 1 Ω when V_2 is in the circuit. When V_2 is removed from the circuit, the resistance measured across pins 2 to 7 of the tube measures 32 Ω. Explain why the two measurements differ.

Answers to Self-Test

1. (*a*) 560,000; (*b*) 3300; (*c*) 32,000
2. 4070 (approx)
3. 0.1

EXPERIMENT 44.2. Voltage Analysis of a Vacuum-Tube Audio Amplifier

OBJECTIVE

To evaluate and measure the voltages at the tube-socket terminals of a vacuum-tube audio amplifier

INTRODUCTORY INFORMATION

As in transistor amplifiers, voltage measurements in vacuum-tube circuits are useful in troubleshooting a defective circuit. If an amplifier has many stages, the defective stage is first localized by signal tracing. Voltage measurements in the suspected stage then pinpoint the defective component. When necessary, resistance measurements (with voltage OFF) are made to identify the bad part. The suspected part is then replaced with a known good part to determine whether the circuit has been restored to proper operation.

DC Voltage Analysis

Approximations, wherever possible, simplify the job of estimating dc voltages at tube-socket terminals.

In audio amplifiers it may be assumed that the amplifier is operating class A. Where push-pull amplifiers are used, the class of operation should be known.

In the circuit of Fig. 44.2-1, V_1 is one-half of a 12AU7A tube whose plate supply is a 250-V source. It is now necessary to refer to the normal operating conditions of this tube. A tube manual shows that if the 12AU7A is operated class A with a plate supply of 250 V, the bias is −8 V (approximately). Since V_1 in Fig. 44.2-1 is cathode-biased, there is, therefore, 8 V at the cathode. So, the voltage from pin 3 to ground in V_1 is 8 V.

Since cathode current and plate current I_P are the same in this tube, we can find I_P, knowing that there is an 8-V drop across a 5600-Ω resistor. Thus $I_P = 8/5600 = 0.00143$ A (to slide rule accuracy). Now, to determine the plate voltage of V_1, simply find the voltage drop across the plate-load resistor and subtract this value from the V_{PP} supply. Thus $I_P \times R_L = 0.00143 \times 50{,}000 = 71.5$ V. Therefore, $V_P = 250 - 71.5 = 178.5$ or 180 V (approximately).

The voltage with respect to ground at pin 2 of V_1 is zero, since the grid, pin 2, is returned to ground and there is no direct current in the grid circuit.

Similarly in V_2, grid voltage is zero (pin 5), and a tube manual shows that the cathode (pin 8) voltage is 18 V approximately. The plate current I_P is therefore about 32 mA. The resistance of the primary of the output transformer is several hundred ohms. There is therefore a small voltage drop (about 10 V) across this primary resistance. The plate voltage (pin 3) is about 240 V, and the screen voltage (pin 4) is 250 V.

From actual experience the technician learns what percentage of the power-supply voltage will be found at the plate of a voltage amplifier, under normal operating conditions. The technician also knows that the plate and screen voltages of a power amplifier are almost the same and are close to the V_{PP} supply source, unless a voltage-dropping resistor is used in the plate or screen. As far as the cathode voltage is concerned, where cathode bias is used, the technician

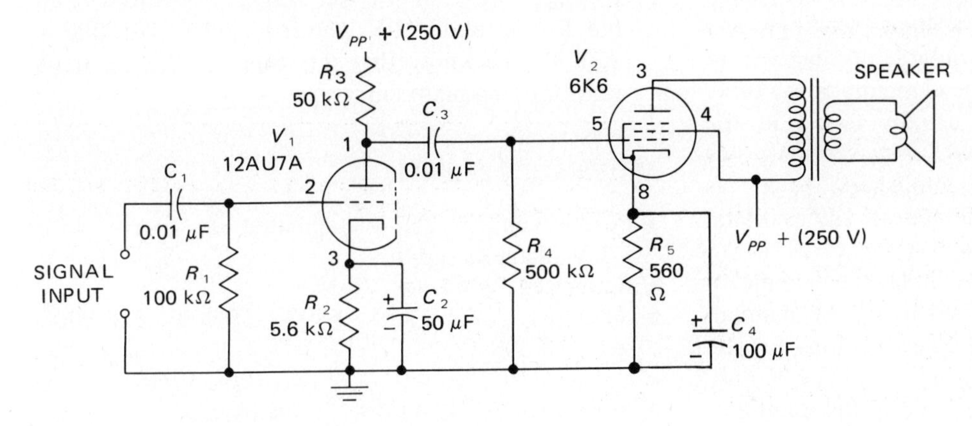

Fig. 44.2-1. Audio amplifier with output stage.

knows that it is low in voltage amplifiers and higher in power amplifiers.

Filament circuits are checked for filament voltage. The technician must know how the filaments are connected and fed. In most cases alternating current is used for heating the filaments. In some early applications, such as old portable radios, direct current was used as the filament source. Filament voltage is checked *across* the filaments. The tube-type designation indicates the amount of filament voltage which may be expected.

Certain facts now become apparent. These are:

1. In an amplifier which is cathode-biased (has a resistor in the cathode circuit), current through the tube will result in a voltage drop across the cathode. Absence of voltage at the cathode would mean that there is no current in the tube or that there is a short circuit across the cathode resistor.
2. Similarly, current in a plate and screen-grid circuit results in a voltage drop across the plate and screen-grid resistor. The voltage at the plate and screen is lower than the supply source by an amount equal to the drop across each resistor. If the plate or screen voltage is the same as the supply voltage, it is because there is no IR drop across the resistors in these circuits. This may be caused either by no current flow or by a short circuit across the resistors in the plate or screen-grid circuits.
3. If the plate or screen-grid resistor is open, there will be no voltage at the plate or screen, respectively.

SUMMARY

1. In troubleshooting a defective vacuum-tube amplifier, dc voltage and resistance checks are made in the suspected stage to find the defective part.
2. In order to evaluate dc voltage readings, the technician should have available a diagram of the circuit and voltages at the elements of the tube, for reference. These are usually found in the manufacturer's service notes, or in service notes supplied by commercial publishers, such as Howard W. Sams, an affiliate of International Telephone and Telegraph.
3. If service notes are not available, the technician has still other resources, such as a tube manual. The manual would give the specifications for the tube used, say, as a class A audio amplifier. Knowing the plate voltage source used in the amplifier, the technician checks the recommended bias for the tube, serving in this instance as a class A amplifier.
4. If the tube is self-biased, as in Fig. 44.2-1, the technician can calculate the cathode current by dividing the recommended bias voltage by the resistance of the cathode resistor. For example, in Fig. 44.2-1, for V_1, the recommended bias is 8 V. Cathode current is then:

$$I_K = \frac{8}{R_k} = \frac{8}{5600} = 0.00143 \text{ A}$$

5. Since cathode and plate currents are the same in a triode tube, the technician can now calculate the voltage to expect at the plate by multiplying I_K by R_L and subtracting this voltage from V_{PP}, the plate voltage source. Again, in Fig. 44.2-1, $0.00143 \times 50{,}000 = 71.5$ V. The voltage measured from plate of V_1 to ground should therefore be 250 V − 71.5 V = 178.5 V (approximately).
6. Filament voltages should be the rated values shown in a tube manual. Tube filaments are heated by alternating current in most cases.

SELF-TEST

Check your understanding by answering these questions.

1. In an audio amplifier stage, where is the highest dc voltage in the stage: (*a*) V_{PP}, the plate voltage source, (*b*) V_P, the plate voltage, (*c*) V_g, the grid voltage, or (*d*) V_k, the cathode voltage?
2. If in the circuit of Fig. 44.2-2 the bias voltage for V_2, a class A audio amplifier, is 12.5 V, then the voltages (measured to ground or as indicated) are as follows:
   (*a*) Control grid: ________ V
   (*b*) Cathode: ________ V
   (*c*) Grid to cathode: ________ V
   (*d*) Plate: ________ V
3. In Fig. 44.2-1, in V_1 the voltage measured from pin 3 to ground = 0 V, and from pin 1 to ground = 250 V. We know that this tube ________ (is/is not) drawing plate current.

MATERIALS REQUIRED

- Power supply: 250-V regulated dc source with 6.3-V filament supply
- Equipment: EVM or VOM
- Resistors: ½-W 3300-, 6800-, 56,000-, 120,000-, two 270,000-Ω
- Capacitors: Two 0.01-μF; two 100-μF/50-V
- Tubes: 6AV6, 12AU7A, and sockets
- Miscellaneous: SPST switch

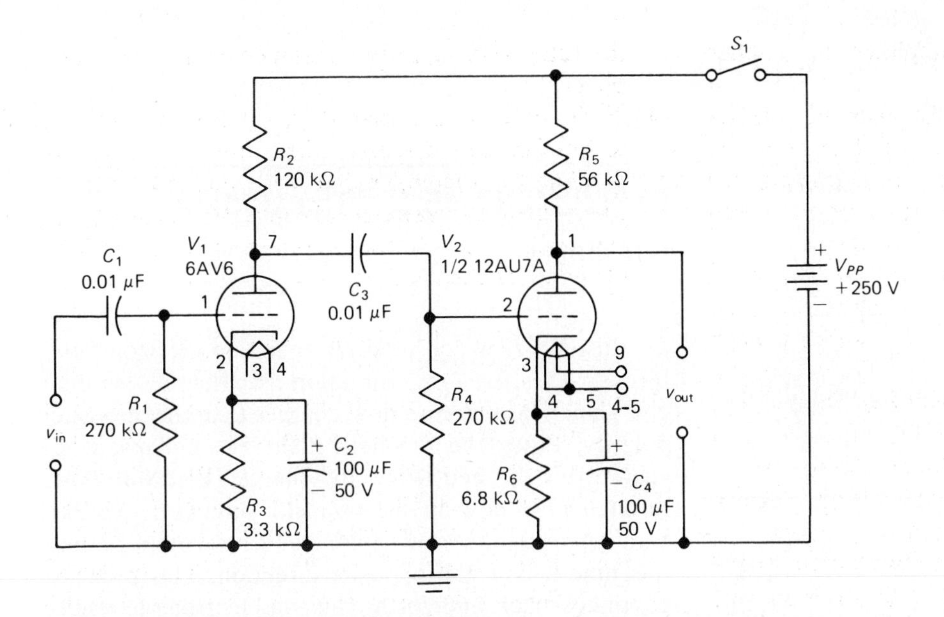

Fig. 44.2-2. Experimental audio amplifier.

PROCEDURE

NOTE: All voltage measurements are made with respect to ground. Avoid finger contact with the measuring probe ends, or your readings may be affected.

1. Connect the circuit of Fig. 44.2-2. Ground one end of the 6.3-V filament supply.
2. **a.** Estimate the values of voltage at every test point shown in Table 44.2-1. Record these values in the appropriate column in the table. Assume that the tube is operating normally. If there is an ac voltage at a specific test point, indicate it. Otherwise, the current is presumed to be direct.
   b. Fill in the tube element corresponding to each number in the appropriate column.
3. **Power on.** Measure and record in Table 44.2-1 the voltage at every test point shown.

TABLE 44.2-1. Voltage Analysis

			Voltage, V	
Tube	*Element*	*Pin Number*	*Estimated*	*Measured*
		1		
		2		
		3		
V_1 6AV6		4		
		5		
		6		
		7		
		1		
		2		
		3		
V_2 12AU7A		4		
		5		
		9		

QUESTIONS

1. Zero volts is measured at the cathode, pin 2 of V_1 in Fig. 44.2-2. Is this normal? If not, what are the defects which could cause it?
2. The voltage at the plate of V_1 in Fig. 44.2-2 is 250 V. Is this normal? If not, what are the troubles which could cause it?
3. The voltage at the control grid of V_1 in Fig. 44.2-2 is 0 V. Is this normal? If not, what are the troubles which could cause it?
4. You measure 4 V instead of 8 V at pin 3 of V_1, Fig. 44.2-1. When C_2 is disconnected, the voltage at pin 3 is 8 V. What is wrong, if anything?

Answers to Self-Test

1. (*a*) V_{pp}
2. (*a*) 0 (*b*) 12.5 (*c*) −12.5 (*d*) 147
3. is not

CHAPTER 45 THE HARTLEY OSCILLATOR

EXPERIMENT 45.1. The Hartley Oscillator

OBJECTIVE

To connect a Hartley oscillator and observe and compare the collector and base waveforms

INTRODUCTORY INFORMATION

Oscillatory "Tank" Circuit

An oscillator is an electronic device for generating (creating) an ac signal voltage. The frequency of the generated signal depends on the circuit constants.

Oscillators are used in radio and television receivers, in radar, in all transmitting equipment, and in military and industrial electronics.

Oscillators may generate sinusoidal or nonsinusoidal waveforms, from very low frequencies up to very high frequencies. The local oscillator in present-day broadcast-band radio receivers will "cover" a range of frequencies from 1000 through 2100 kHz (approximately).

An oscillation is a back-and-forth motion. In mechanics a pendulum or swing illustrates the principle of oscillation. Once a pendulum is started, it would continue swinging indefinitely if it were not for the energy lost in overcoming friction. It is necessary to add energy to replace this loss and keep the pendulum moving.

In a parallel LC circuit, electrons will oscillate when the circuit is excited. In the circuit of Fig. 45.1-1, when S_1 is closed, capacitor C will charge to the battery voltage V. If S_1 is then opened and S_2 closed, C will discharge through L, creating an expanding magnetic field about L. After C has discharged, the magnetic field collapses and induces a voltage in L which tends to maintain electron flow through L in the same direction as when C was discharging. This electron flow charges C in the opposite polarity. After the magnetic field has collapsed, C again tries to neutralize its charge. Electron flow through L is now in the opposite direction. An expanding magnetic field again appears around L, but this time it is in the opposite direction. This process continues back and forth, causing electrons to oscillate in the tuned circuit, also called "tank circuit." However, owing to the resistance in the circuit and the resulting heat losses (I^2R), the amplitude of oscillation is damped, as in Fig. 45.1-1, although the period of every cycle is the same.

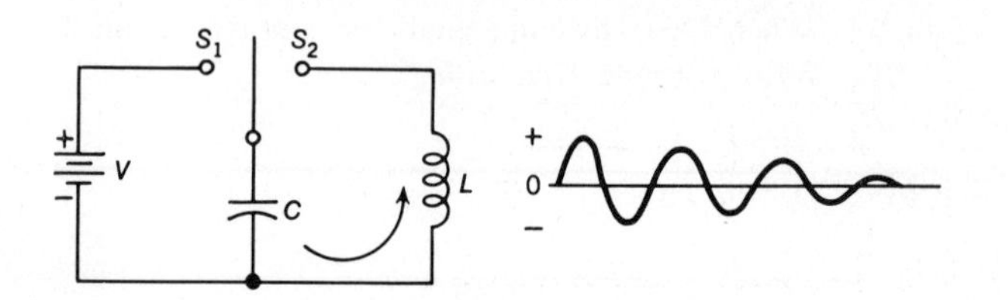

Fig. 45.1-1. Exciting a parallel LC circuit into oscillation.

Overcoming Losses in an Oscillatory Tank Circuit

When the energy fed into the circuit has been used up, it is necessary to supply more energy by recharging capacitor C from the power supply and again permitting it to discharge through L. By switching S_1 and S_2 at the proper time, we can maintain oscillation. Moreover, a sine wave of constant amplitude may be generated. In this process dc energy is used by the circuit to overcome losses.

Another method of keeping the LC tank circuit oscillating is to connect the tank circuit in the output of an amplifier, as in Fig. 45.1-2. The transistor amplifier is cut off by V_{BB}, which reverse-biases the base-emitter circuit. A sine wave is injected into the base circuit with such amplitude that collector current flows at the peak of the negative alternation. This shock-excites the LC circuit in the collector of Q, and the tank circuit oscillates. If the input sine

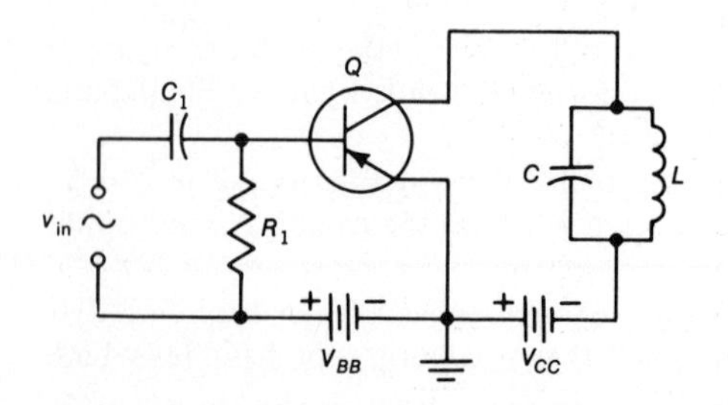

Fig. 45.1-2. Amplifier with resonant LC circuit.

wave has the same frequency as the frequency of oscillation of the tank circuit, the oscillation in the *LC* circuit is maintained.

Tickler-Coil Oscillator

A transistor or vacuum-tube amplifier may be connected in a circuit in such a manner that it will cause oscillations without requiring any external signal source. A self-excited electronic oscillator is illustrated in Fig. 45.1-3. This diagram shows the circuit hookup of a self-excited oscillator. L_1 is inductively coupled to L. When power is first applied, current flow starts in the transistor. As the current flows through L, it induces a voltage in L_1, which is coupled to the base of the PNP transistor and amplified. If the feedback voltage is in proper phase, there is an increase in collector current. This action rapidly builds up a large current pulse, which shock-excites the *LC* tank into oscillation. The signal fed back by L_1 to the base is a sine wave of the same frequency as that in the *LC* circuit and of the proper phase to sustain oscillation. The signal induced in the base thus eliminates the need for an input signal, and the *LC* tank will oscillate as long as the circuit is not upset.

The regenerative feedback, that is, feedback in proper phase to sustain oscillation, provided by L_1, causes capacitor C_1 to charge with the polarity shown and keeps the base positive with respect to the emitter. This biases the base-emitter circuit to cut off, and current in this circuit flows only during the peaks of the negative alternations fed back to the base. These current pulses in the collector circuit provide the energy to overcome the losses in the tank circuit, which therefore keeps on oscillating. L and C determine the frequency of oscillation. R_1 is a bias-limiting resistor.

The requirements for oscillation are:

1. Power-supply source
2. Amplifier
3. Feedback in proper phase from output to input
4. Frequency-determining constants

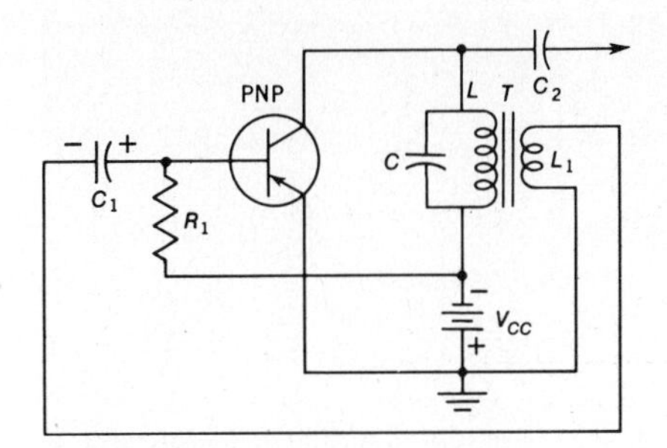

Fig. 45.1-3. Tickler-coil oscillator.

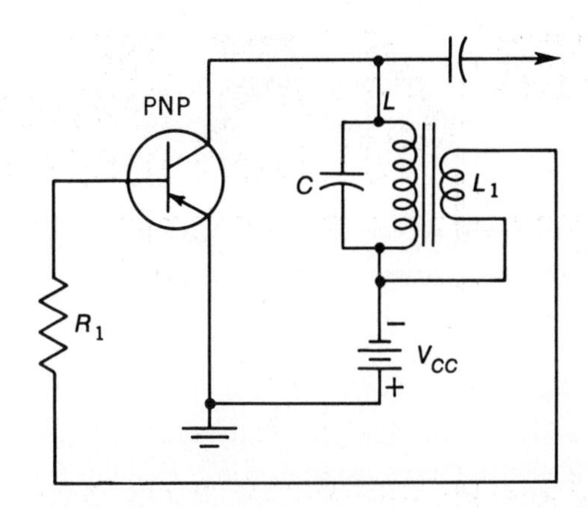

Fig. 45.1-4. Variation of tickler-coil oscillator.

A variation of the tickler-coil oscillator is shown in Fig. 45.1-4. The difference between this and the previous circuit is that C_1 has been removed. This changes the class of transistor operation. Another difference is the return of the base circuit to the negative terminal of V_{CC}, rather than to the positive.

Series-Fed Hartley Oscillator

Figure 45.1-5 is a series-fed Hartley oscillator. This circuit is similar to the tickler-coil oscillator, but the tickler coil L_1 is physically connected to, and is part of, L, which now becomes an autotransformer.

The NPN transistor used in Fig. 45.1-5 is biased as a conventional amplifier, with forward bias on the emitter-base circuit and reverse bias on the emitter-collector circuit. Current in the collector circuit flows through L_1 and produces a regenerative current in L which is fed to the base. By design, the tap on autotransformer L is at the proper point to sustain oscillation in the tank circuit. L–L_1 and C determine the resonant frequency. R_1 sets the base-emitter bias. C_1 charges because of current flow in the emitter-base circuit. When an NPN transistor is used, as in Fig. 45.1-5, the polarity of charge on C_1 is as shown. The base is maintained at a negative potential with respect to the emitter, biasing the transistor to cut off, except during the positive peaks of the oscillations. If a PNP

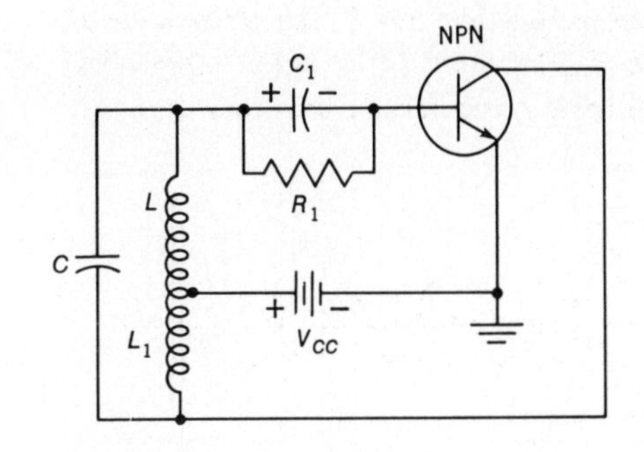

Fig. 45.1-5. Series-fed Hartley oscillator.

transistor were used in a similar circuit arrangement, the charge on C_1 would be opposite in polarity, and the transistor base would be maintained positive with respect to the emitter.

This oscillator is called *series-fed* because the radio frequency (rf) and dc paths are the same, just as they would be in a series circuit.

Parallel or Shunt-Fed Hartley Oscillator

Figure 45.1-6*a* is the circuit of a Hartley oscillator frequently used as the local oscillator in a broadcast-band superheterodyne radio receiver. Capacitor C_2 and L_1 make up the path of rf current in the collector-to-emitter circuit. This is really an amplifier circuit with provision for feedback in phase to keep the oscillator going. Feedback is again provided by L_1, the tap on L, the autotransformer. The position of the tap on the coil (junction of L_1 and L) determines how much signal current is fed back to the base.

The autotransformer and C are the main frequency-determining components. C_1 and R_1 have the same use as in the circuit of Fig. 45.1-5. The use of the rf choke in the collector circuit and C_2 ensures separate paths for direct current and for the generated rf signal current.

The circuit of Fig. 45.1-6*a* may be dc-stabilized by the use of a resistor R_2 connected in the emitter leg as in Fig. 45.1-6*b*. R_2 is properly bypassed by C_3 to prevent ac (signal) degeneration.

The Hartley oscillator coil has three connections. These are usually coded on the coil. If they are not, it is generally possible to identify them by a resistance check. The resistance between the taps T and P is small compared with the resistance between T and G (Fig. 45.1-6*b*). If the coil connections are not made properly, the oscillator will not operate.

SUMMARY

1. Electronic oscillators generate sinusoidal and nonsinusoidal ac voltages.
2. They are used in every branch of electronics, for example, radio, television, radar, transmitters.
3. An *LC* circuit may be shock-excited into oscillation, but unless the losses of the circuit are replaced, the oscillations are damped and die out.
4. Special arrangements of vacuum tubes or transistors, together with coils, may be used to supply the energy required to overcome tank circuit losses and sustain oscillation.
5. One type of oscillator circuit employs tickler coils to feedback energy in proper phase, from the output to the input of the oscillator. Examples are shown in Figs. 45.1-3 and 45.1-4. At the start, the transistor must be forward-biased to begin operation.
6. A Hartley oscillator uses a three-terminal autotransformer-type coil to sustain oscillation.
7. A series-fed Hartley oscillator is shown in Fig. 45.1-5. Here the dc and ac paths are the same.
8. In parallel- or shunt-fed Hartley oscillators the generated ac signal current path is in parallel with the dc current.

SELF-TEST

Check your understanding by answering these questions.

1. In Fig. 45.1-1, when S_2 is closed and C discharges through L, a damped oscillation is generated. The waveform is damped because of the ________ in the circuit.
2. In Fig. 45.1-2, Q conducts during the most ________ (negative/positive) part of the input cycle.
3. In Fig. 45.1-3, when power is first applied and the transistor starts to conduct, the top of winding L_1 must be driven ________ (negative/positive) to start the oscillatory cycle.
4. The Hartley oscillator in Fig. 45.1-5 uses a ________-terminal coil to sustain oscillation.
5. In Fig. 45.1-5, the bias developed by C_1R_1 after oscillation begins keeps the transistor OFF except during the top of the ________ (positive/negative) alternation of the sine wave.

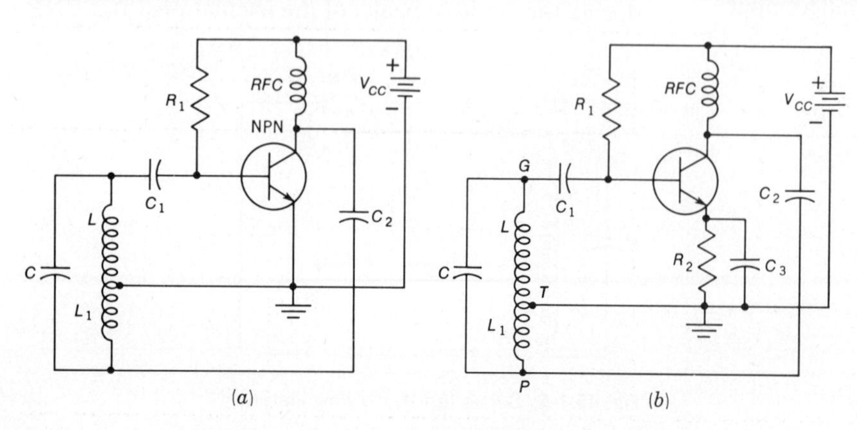

Fig. 45.1-6. Parallel-fed Hartley oscillator.

6. In Fig. 45.1-5 the voltage measured from base to emitter, when the oscillator is ON, will be ________ (positive/negative).
7. In measuring the voltage from base to emitter in Fig. 45.1-5, a ________ (high/low) impedance voltmeter should be used.
8. In Fig. 45.1-5, an oscilloscope connected from collector to emitter will show a ________.
9. The purpose of the *RFC* in Fig. 45.1-6 is to keep the ________ out of the ________ ________.

MATERIALS REQUIRED

- Power supply: Variable, regulated low-voltage dc source
- Equipment: Oscilloscope; EVM; 0–10-mA milliammeter
- Resistors: ½-W 390-, 270,000-Ω
- Capacitors: 47-pF; 250-pF; 0.001-μF; 0.01-μF
- Semiconductor: 2N6004 transistor
- Miscellaneous: 30-mH rf choke; Hartley oscillator coil (Miller #2065 or the equivalent–broadcast band); SPST switch

PROCEDURE

1. Connect the circuit of Fig. 45.1-7. *L* is a broadcast-band tapped oscillator coil. *M* is a dc milliammeter or a VOM set on the 10-mA current range. Check circuit connections before applying power.
2. **Power on.** With an oscilloscope whose Time/cm switch is set at 1 μs/cm (or a nontriggered scope set on the highest sweep frequency range), observe and measure the waveform at test points (TP) 1 and 2. Record the results in Table 45.1-1. NOTE: On narrowband oscilloscopes, peak-to-peak measurements will be relative because of high-frequency signal attentuation by the amplifiers in the oscilloscope.

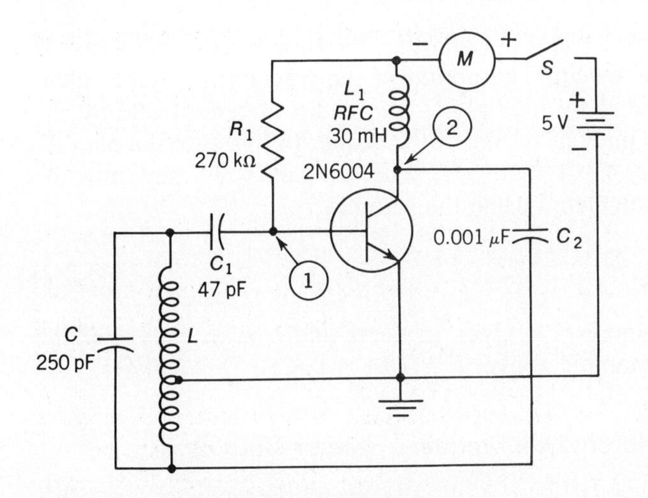

Fig. 45.1-7. Experimental Hartley oscillator.

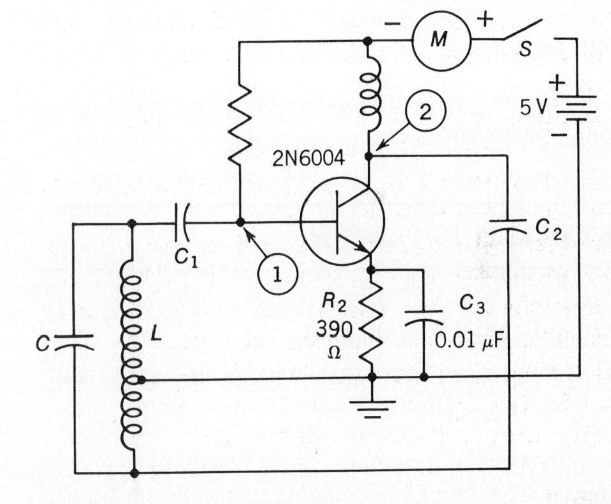

Fig. 45.1-8. Adding bias stabilization to experimental oscillator.

TABLE 45.1-1. Experimental Oscillator Checks

Step	*TP 1*			*TP 2*			I_T
	Waveform	V p-p	V_{BE}	*Waveform*	V p-p	V_{BE}	
2,3,4							
6							
7							

3. With an EVM measure base-to-emitter dc voltage V_{BE}. Record in Table 45.1-1.
4. Measure and record the circuit current.
5. **Power off.** Add the bias stabilization network, R_2 and C_3, as in Fig. 45.1-8.
6. **Power on.** Repeat steps 2, 3, and 4.
7. Open C_3. Repeat steps 2, 3, and 4.

QUESTIONS

1. What are the requirements for oscillation?
2. How are these requirements met in the experimental oscillator?
3. What are the effects of removing C_3 in the experimental oscillator? Why?
4. What is the purpose of C_2 in the experimental oscillator?

Answers to Self-Test

1. losses
2. negative
3. negative
4. three
5. positive
6. negative
7. high
8. sine wave
9. rf; dc supply

EXPERIMENT 45.2. Hartley Oscillator Frequency

OBJECTIVE

To determine experimentally the frequency of a Hartley oscillator

INTRODUCTORY INFORMATION

Checking Oscillator Frequency

Oscilloscope with Calibrated Time Base as Frequency Standard

The approximate frequency of an oscillator may be computed from the LC constants using the formula

$$F = \frac{1}{2\pi\sqrt{LC}} \qquad (45.2\text{-}1)$$

F is given in hertz, when L is in henries, and C in farads. The frequency of an oscillator may also be measured in other ways. Several methods will be discussed here and applied in the experimental procedure which follows.

One method uses an oscilloscope with a calibrated time base to measure the period of the waveform. From the period, the frequency is calculated. The periodic waveform whose frequency is to be measured is observed on the oscilloscope. The Time/cm controls are set at "calibrated," and the width of the waveform is measured horizontally along the time base. Suppose, for example, that the width of the sine wave in Fig. 45.2-1 is 4 cm and that the time base is calibrated at 1 ms/cm. The period of the waveform may be calculated by multiplying the width in centimeters by the Time/cm setting of the scope. In this case the period t is $4 \times 1 \times 10^{-3}$ s. The frequency F may now be calculated using the formula

$$F = \frac{1}{t} \qquad (45.2\text{-}2)$$

where F is given in hertz and t in seconds. For the example above, $F = 1/(4 \times 1 \times 10^{-3}) = 250$ Hz.

Heterodyne Frequency-Meter Method

Speaker as Null Indicator. Frequency can also be checked experimentally by the use of a heterodyne frequency meter, called a "beat-frequency" meter.

The operation of the frequency meter involves the

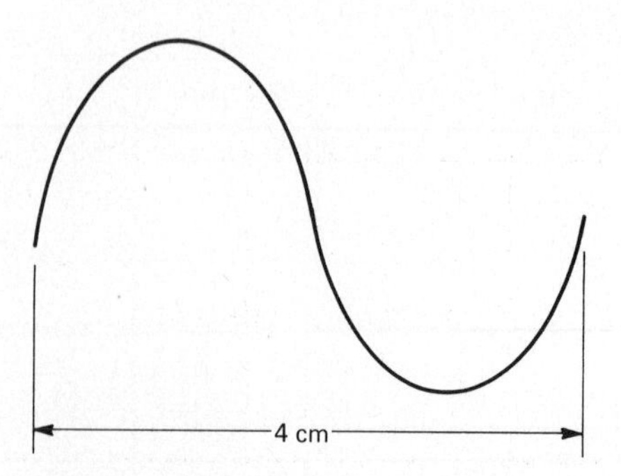

Fig. 45.2-1. Width of sine wave is 4 cm.

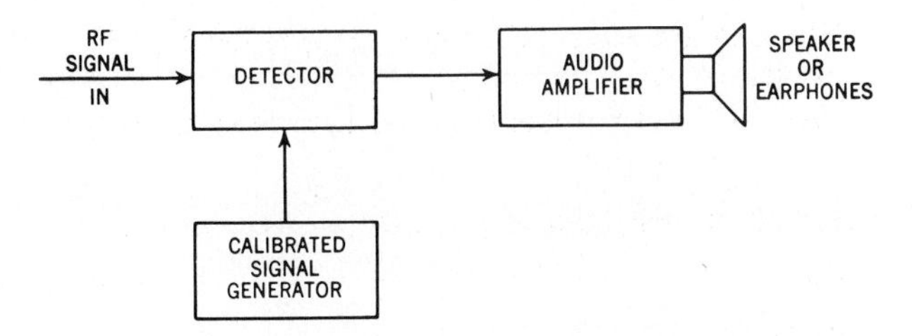

Fig. 45.2-2. Heterodyne frequency-meter block diagram.

heterodyning, or beating together, of two signal frequencies. When these signals are mixed in a non-linear device, such as a rectifier or detector, beat notes are created. The newly created frequencies consist of the difference and sum of the two frequencies, and harmonic combinations. When the difference in frequency between the two signals is in the *audio range,* that is, 30 to 15,000 Hz, the beat note can be heard by using a pair of earphones or a loudspeaker. As the two signals approach the same frequency, the frequency difference approaches zero. The two signals are assumed to be equal in frequency at zero beat. Zero beat is not heard but is located between the two points on the frequency-meter dial which produce the lowest audible frequency.

Figure 45.2-2 is the block diagram of a beat-frequency meter. Included in this device are a calibrated signal generator, a detector, an AF amplifier, and speaker. The unknown RF signal is fed into the meter, and the frequency-calibrated dial is adjusted for zero beat. The unknown RF signal frequency is read from the meter dial.

Oscilloscope as Null Indicator. Heterodyne frequency measurements require a null, or zero, indicator. In the method just described a speaker was used as the null indicator. It is possible to employ an oscilloscope as the indicator instead of a speaker. In this case the output of the audio amplifier, in Fig. 45.2-2, would be fed to an oscilloscope rather than to a speaker. This procedure assumes the use of an oscilloscope with a heterodyne frequency meter. As a matter of fact, it is possible to eliminate the heterodyne frequency meter when an oscilloscope is used, because the oscilloscope has built into it two of the elements in a beat-frequency meter, namely, the amplifier and null indicator. What is required, therefore, is an external detector and a calibrated RF generator.

An external *demodulator probe* acts as the detector. Figure 45.2-3 shows the circuit diagram of a demodulator probe. A 1N34 or equivalent diode is connected as an RF rectifier. The R and C values have been chosen to give RF filtering. The demodulator probe replaces the vertical input probe of the oscilloscope. The probe tip is connected to the unknown frequency signal source. Also coupled to the probe tip is the output of an accurately calibrated signal generator, set at the amplitude of the unknown signal. The generator is varied until zero beat is indicated on the oscilloscope.

The beat patterns on the oscilloscope are shown in Fig. 45.2-4. Zero beat is shown at (b). (This is an idealized drawing. Some signal will be observed at zero beat.) At (a) and (c) are audio sine waveforms, which are observed on either side of zero beat. As the generator frequency is moved further from zero beat, the frequency of the beat note increases.

SUMMARY

1. The frequency of an LC oscillator depends on the inductance and capacitance of the resonant or tank circuit and may be computed from the equation

$$F = \frac{1}{2\pi\sqrt{LC}}$$

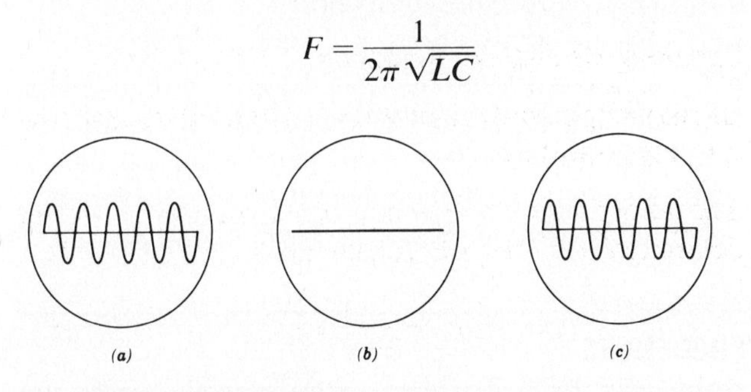

Fig. 45.2-4. Beat-frequency patterns on the oscilloscope.

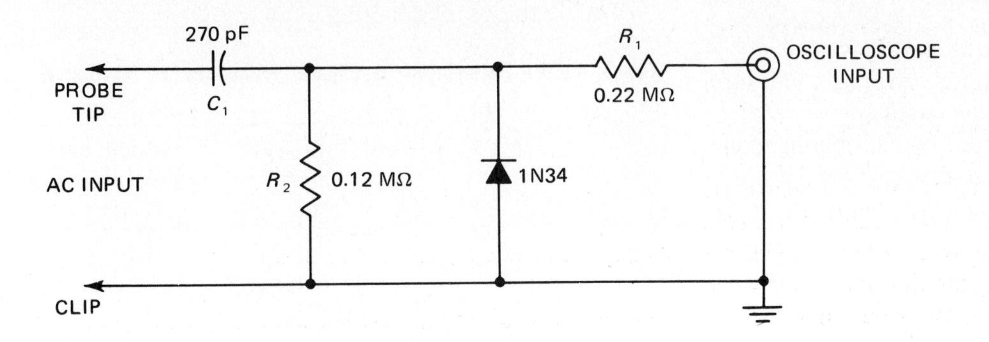

Fig. 45.2-3. Demodulator probe.

where F is in hertz, L in henries, and C in farads.

2. The frequency F of an oscillator may be measured by an oscilloscope with a calibrated time base.
3. F may also be measured by use of heterodyne frequency meters (zero-beat method).
4. F may be measured using the zero-beat method without a heterodyne frequency meter. What is required is an accurately calibrated signal generator, a demodulator probe, and an oscilloscope.

SELF-TEST

Check your understanding by answering these questions.

1. An oscilloscope with a calibrated time base is used to measure the frequency of an unknown rf signal. If the Time/cm control of the scope is set at 1 μs/cm, and if the width of one cycle of the signal is 2 cm, the frequency of the signal is ________ Hz.
2. If the inductance L of an LC oscillator is 200 μH, and the capacitance C is 250 pF, the frequency of the oscillator is ________ Hz.

MATERIALS REQUIRED

- Power supply: Variable, regulated low-voltage dc source
- Equipment: Oscilloscope (calibrated time base, if possible) and demodulator probe; EVM; accurately calibrated rf signal generator or heterodyne frequency meter
- Resistors: ½-W 390-, 270,000-Ω
- Capacitors: 47-pF; 250-pF; 0.001-μF; 0.01-μF
- Semiconductor: 2N6004 transistor or the equivalent
- Miscellaneous: 30-mH rf choke; Hartley oscillator coil (Miller #2065 or equivalent-broadcast band); SPST switch; if a demodulator probe is not available, the components in Fig. 45.2-3 will be needed

PROCEDURE

Measuring Oscillator Frequency (Oscilloscope with Calibrated Time Base)

1. Connect the circuit of Fig. 45.2-5. **Power on.**
2. Using a scope with a calibrated time base, measure the width of one cycle of the oscillator waveform. Convert this width into time t by multiplying cycle width (in centimeters) by the setting of the Time/cm switch. Determine the oscillator frequency F by using the formula

$$F = \frac{1}{t}$$

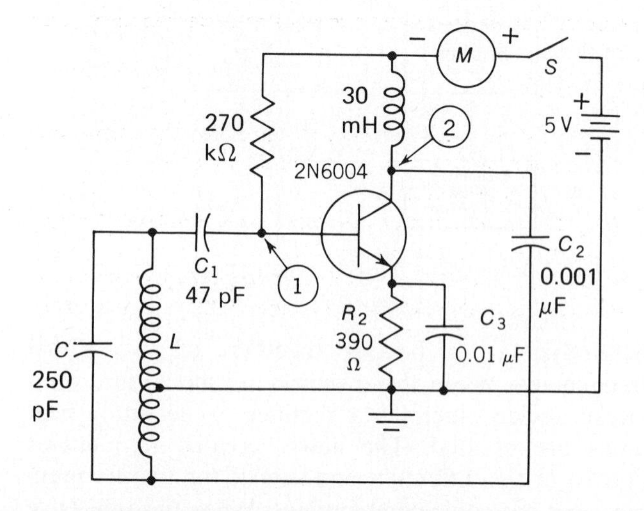

Fig. 45.2-5. Experimental Hartley oscillator.

where F is in hertz and t is in seconds. F = ________ Hz. (Show all your computations.)

Measuring Oscillator Frequency (Zero-Beat Method)

3. With an oscilloscope observe the rf waveform at TP 1, Fig. 45.2-5. Record waveform in Table 45.2-1.
4. Replace the oscilloscope probe with an rf demodulator probe. Bypass the oscilloscope vertical input with a 0.01-μF capacitor. Set oscilloscope Time/cm control on 10 ms/cm. Connect the probe tip to TP 1 in Fig. 45.2-5. If a demodulator probe is not available, use the circuit of Fig. 45.2-3.
5. Connect a 15-pF capacitor in series with the "hot" lead of a calibrated rf signal generator, and inject the output of the generator into TP 1. Set the generator at 600 kHz, and vary the generator frequency control on either side of 600 kHz until zero beat is observed on the oscilloscope. *Generator output* control should be set at the *lowest level* which will permit a null indication to be seen on the oscilloscope. Oscilloscope sensitivity is set at maximum. In Table 45.2-1 record the waveforms at zero beat (*b*), and on either side of zero beat, (*a*) and (*c*). Record also the generator frequency at zero beat. This is the oscillator frequency.

TABLE 45.2-1. Measuring Oscillator Frequency by Zero-Beat Method

RF Waveform	*Beat Waveforms*			*Oscillator Frequency,* kHz
	(*a*)	(*b*)	(*c*)	

QUESTIONS

1. (*a*) When is a *null* observed in using the zero-beat method of determining oscillator frequency?
   (*b*) What does *null* mean?
2. Why did you encounter many zero-beat indications when measuring oscillator frequency?
3. How did the measured oscillator frequency compare in the two methods you used experimentally: (*a*) Measuring the period with an oscilloscope with calibrated time base? (*b*) Zero-beat method? Explain any difference in the two frequencies.
4. How can you determine whether an oscillator circuit is oscillating?

Answers to Self-Test

1. 500,000
2. 712,000

CHAPTER 46 ICs: TROUBLESHOOTING THE LINEAR AMPLIFIER

EXPERIMENT 46.1. Integrated Circuits: The Linear Amplifier

OBJECTIVES

1. To become familiar with IC construction techniques
2. To connect an IC medium-power audio-frequency amplifier and check its operation

INTRODUCTORY INFORMATION

Integrated circuits (ICs) have greatly changed the world of electronics. The effect of these tiny "giants" was first felt in the computer field. The first-generation electronic computers used vacuum tubes. Vacuum tubes were replaced by transistors in second-generation computers. Third-generation computers used digital ICs, greatly reducing computer size and increasing computer speed and reliability.

As the state of the art advanced, specially designed ICs were used in *linear* circuits in communications, military, and industrial applications. ICs and large-scale ICs (LSICs) have changed the design of electronic devices from the use of just discrete components to hybrid solid-state devices which mix discrete components with ICs. *Modular electronics* owes its growth to ICs.

The speed at which electronics is moving and changing is breathtaking. The vacuum tube, which introduced electronics, was discovered early in the twentieth century. Knowledge and applications expanded gradually, compared with today's pace. The Second World War speeded up the process. The transistor, discovered in 1948, foreshadowed the obsolescence of vacuum tubes. It was also the forerunner of IC technology. The transistor made possible reductions in product weight, size, power requirements, cost, and it improved reliability. Integrated circuits have reduced further product dimensions and cost, while ensuring even greater reliability.

A Look at an Integrated Circuit

What is an integrated circuit? It is not just a more efficient transistor or several transistors inside a small case. It is a complete "circuit," consisting of active and passive elements connected in a unique circuit arrangement, but no larger than a small transistor. The "active" elements are transistors and diodes. The passive components are resistors and capacitors. The elements in an IC are *not* discrete components wired together in a miniature circuit. Rather, the IC is a complete circuit formed by an intricate chemical process, on a silicon chip no larger than the head of a pin. The IC "roadmap" of circuits can be seen only under high magnification. Think of a $\frac{1}{2}$-W audio amplifier, complete except for a few external, discrete components, mounted on a pinhead! The relative size of an IC can be appreciated by noting Fig. 46.1-1, which shows an IC (RCA's 3502) next to a dime. It has been estimated that a thimbleful of ICs contains enough circuitry to build thousands of radios.

How ICs Are Made

IC manufacturing processes are based on the production technology of silicon-diffused, or "planar," transistors. The student is referred to standard transistor texts which describe in detail the process for making planar transistors.

Briefly, by way of summary, a silicon IC is made on a silicon disk, approximately 1 in in diameter, about the thickness of this paper. The disk is processed in a series of individual steps. The top of the disk is first oxidized, then covered with a light-sensitive lacquer, called a "resist." Circuit patterns are etched into the oxides by a photographic process. After heating, minute quantities of "impurities" or dopants, such as boron or arsenic, are diffused into the silicon to form

Fig. 46.1-1. IC 3502 and dime for comparison. (*Radio Corp. of America*)

the P and N islands on the disk. The process is repeated many times until all the circuit elements, transistors, diodes, resistors, and capacitors, are made on the disk. The elements are connected by depositing vaporized aluminum in the desired pattern to form the circuits (see Fig. 46.1-2). The completed disk may have as many as 500 ICs on it. These are then diced. The chips are separated, tested, and mounted on ceramic or metal bases. Then aluminum wires, about one-third the thickness of a human hair, are connected between the IC contacts and the header leads. Then the package is sealed. The familiar transistor TO-5 case, flatpack, and dual-in-line packages, shown in Fig. 46.1-3, are used.

IC Circuit Arrangements

The reason IC circuitry is different from conventional circuits is closely related to the methods and economics of the manufacturing process and to the size of the chip. In conventional circuit design the "active" elements (transistors and diodes) are more costly than resistors and capacitors. In IC technology transistors and diodes are less costly than resistors and capacitors, respectively. The reason is that transistors and diodes take up very little space on the chip, while resistors and capacitors require a relatively large amount of room.

For this reason IC designers make extensive use of *active* elements and sparing use of resistors and capacitors. Direct coupling is favored over capacitive coupling. Integrated circuits use *RC* "tuning." Coupling transformers are avoided wherever possible.

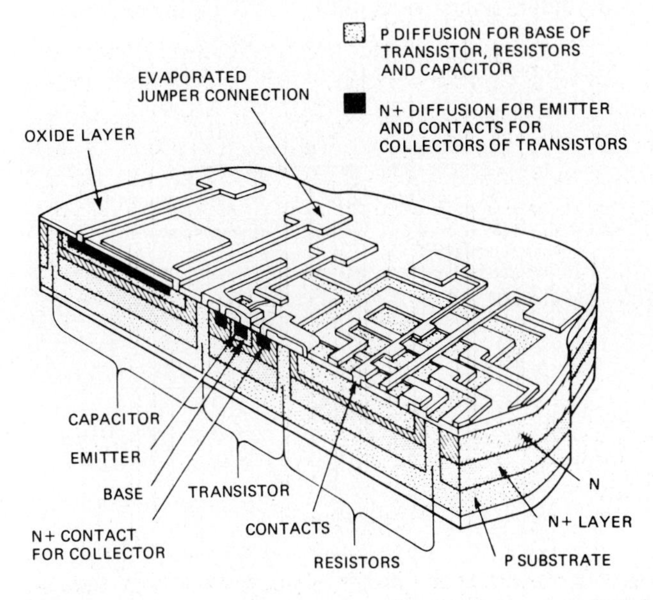

Fig. 46.1-2. IC chip cross section. (*Radio Corp. of America*)

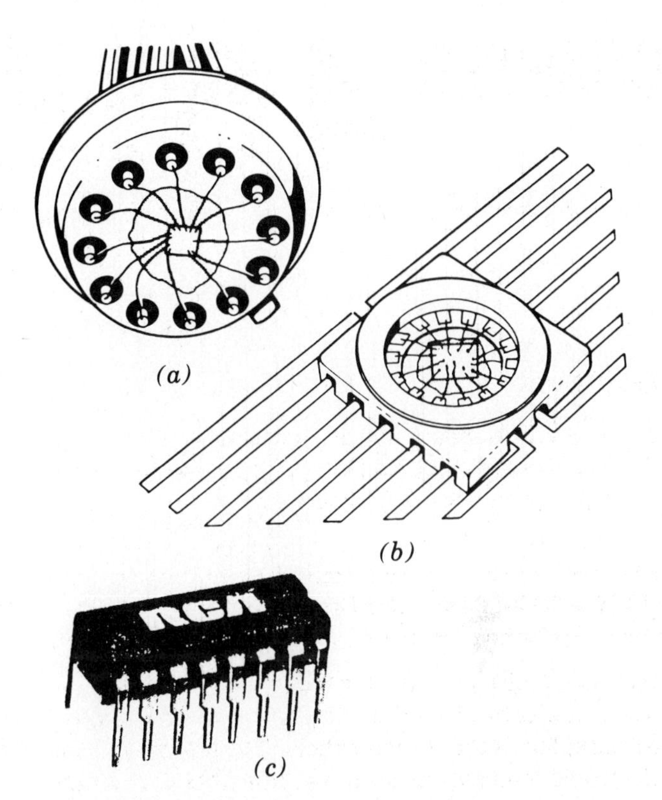

Fig. 46.1-3. ICs are housed in (*a*) transistor type TO-5 case, (*b*) flatpack, (*c*) dual-in-line package. (*Radio Corp. of America*)

Where it is not possible to avoid using coupling transformers, *external* transformers are employed. In the future even greater changes in circuitry are anticipated.

An important consideration in IC economics is the need for flexibility. Chips are designed to be used in more than one circuit arrangement. This is made possible by arrangements involving discrete external components. For example, the same chip in different external circuit arrangements may be used as a high-frequency wideband amplifier, as a low-frequency amplifier, as an oscillator and amplifier combined, or even as a high-frequency amplifier-detector audio amplifier. This flexibility is brought about by the circuit leads on the body of the IC which make it possible to "break in" on the circuit at critical points within the IC.

Linear IC Audio Power Amplifier

The CA 3020 is a linear IC audio power amplifier designed for use in portable and fixed audio communication equipment and servo control systems. The schematic diagram of the CA 3020 is shown in Fig. 46.1-4, and the block diagram of this circuit appears in Fig. 46.1-5. This is a direct-coupled amplifier which acts as a preamplifier, phase-inverter, driver, and

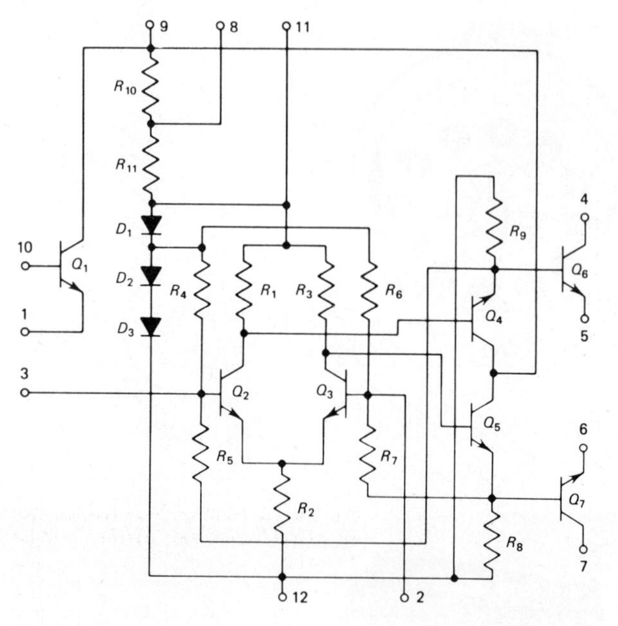

Fig. 46.1-4. Schematic of a CA 3020 audio amplifier. (*Radio Corp. of America*)

power-output stage without transformers. The circuit can operate from a +3- to +9-V power source. The power output is determined by the supply voltage. The range of direct output is from 65 mW at +3 V to 550 mW at +9 V. A temperature-compensating voltage regulator permits operation over the temperature range −55 to +125°C.

Series-connected diodes D_1, D_2, and D_3 together with resistors R_{11} and R_{10} make up the voltage regulator. The external power-supply voltage is connected between terminals 9 and 12 and supplies relatively constant voltages of 1.4 V (junction of D_1 and D_2) for base supply and 2.1 V (junction of D_1 and R_{11}) for the collectors.

Transistors Q_2 and Q_3, together with collector resistors R_1 and R_3, emitter resistor R_2, and base-biasing resistors R_4, R_5, R_6, and R_7, are a differential amplifier and phase inverter. The input ac signal may be capacitor-coupled into terminal 3 or terminal 10. If applied to terminal 10, Q_1 is operated as a buffer amplifier, an emitter follower whose output is then coupled to terminal 3, the input of Q_2 and Q_3. Coupling from Q_2 and Q_3 is by way of the common unbypassed emitter resistor R_2. Two equal-amplitude signals, 180° out of phase, are developed at the collectors of Q_2 and Q_3. These signals are direct-coupled to the bases of emitter-follower amplifiers Q_4 and Q_5. Negative feedback to Q_2 is provided by R_5 and R_7 for dc and ac stability of Q_2 and Q_3. The signal is direct-coupled from the emitters of Q_4 and Q_5 to the bases of Q_6 and Q_7, the power amplifiers. Q_6 and Q_7 operate in a class B push-pull mode to deliver power to the load, which may be a high-impedance center-tapped speaker fed directly as in Fig. 46.1-6, or a center-tapped output transformer feeding a low-impedance speaker as in Fig. 46.1-7.

In this experiment you will have the opportunity to try out this IC amplifier with a variety of audio sources.

Summary

1. An integrated circuit (IC) is a complete circuit photographically and chemically etched into a silicon "chip."
2. An IC consists of active elements (transistors and diodes) and passive elements (resistors and capacitors) connected in a desired circuit arrangement and housed in a small case. The case may be that of the familiar transistor TO-5 package, or the flatpack or dual-in-line packages.
3. ICs are economical because they are manufactured in large quantities, not singly. For example, in one manufacturing cycle 500 ICs may be formed on one disk. This disk is then diced, the ICs are tested, leads are added, and the ICs are mounted in cases.

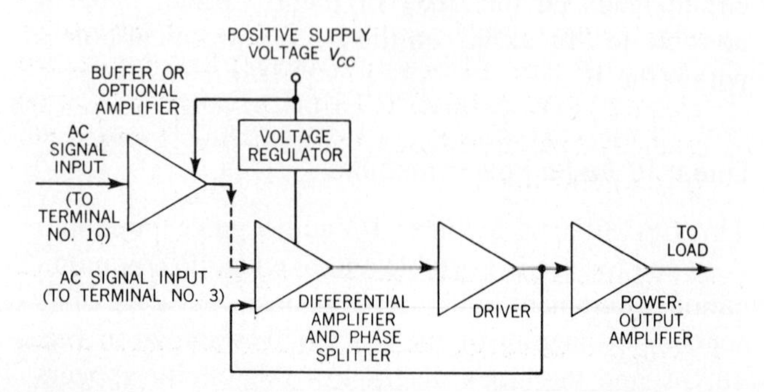

Fig. 46.1-5. Functional block diagram of CA 3020. (*Radio Corp. of America*)

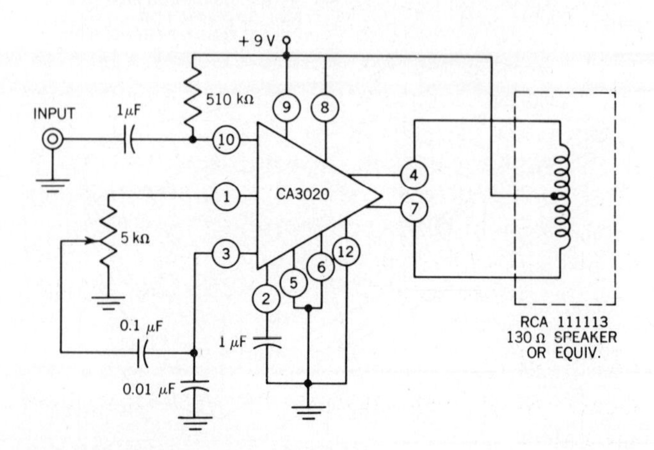

Fig. 46.1-6. CA 3020 audio amplifier without transformer. (*Radio Corp. of America*)

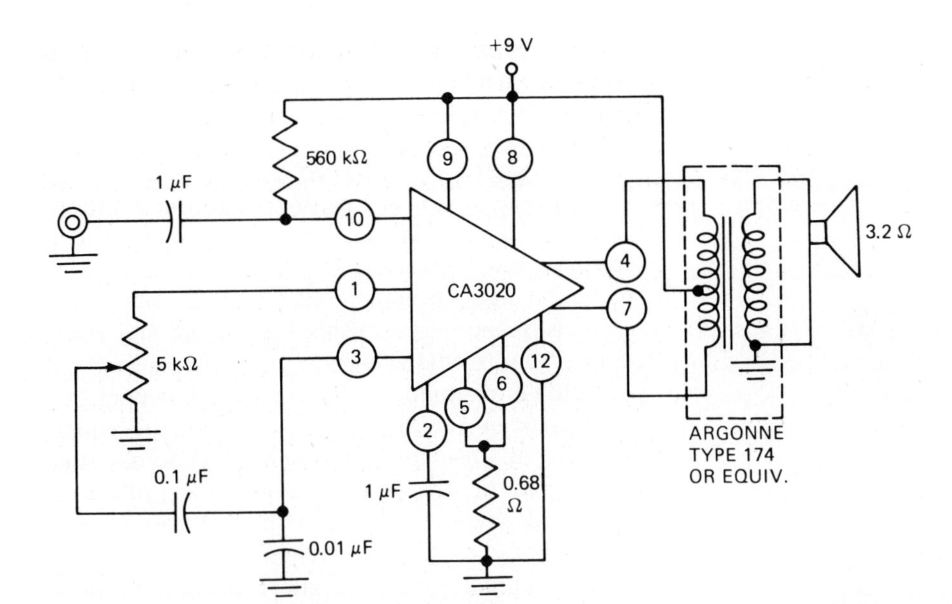

Fig. 46.1-7. IC amplifier driving a low-impedance speaker. *(Radio Corp. of America)*

4. There are many IC circuit arrangements. However, these may be broadly classified as linear ICs and digital ICs. Linear ICs are used mainly in signal-processing circuits (amplifiers, oscillators, etc.) in communications electronics, for example in TV, radio receivers and transmitters, and hi-fi equipment. Digital ICs are found in computers, calculators, and other electronic "counting" devices.
5. Because of economic considerations and the nature of the IC manufacturing process, IC circuit arrangements differ markedly from circuits using discrete components. Transistors, diodes, resistors, and capacitors are the components which can be etched into silicon chips. Transistors take the least "chip" area, diodes are next, then come resistors, and finally capacitors. The design of ICs therefore makes greatest use of transistors and diodes since these are least expensive. Resistors are used more sparingly, and capacitors rarely.
6. ICs make wide use of direct coupling.
7. IC chips are either highly specialized, as for example the chips used in a calculator, or they are designed with the greatest flexibility in mind. Where flexibility is necessary, the addition of external discrete components can change the circuit of an IC for a variety of purposes.
8. The CA 3020 is an example of a linear IC which can be used as a medium-power audio amplifier.

SELF-TEST

Check your understanding by answering these questions.

1. An IC is a complete circuit consisting of discrete components. ________ (true/false)
2. ICs usually come in one of three different cases. These are: ________, ________, and ________.
3. The most economical component to fabricate in an IC is a ________.
4. The active components in an IC are ________ and ________.
5. ICs may be classified as ________ or ________ devices.
6. An ________ ________ is an example of a linear IC.
7. In Fig. 46.1-4, the inputs to the IC amplifier are at either terminal ________ or terminal ________.

MATERIALS REQUIRED

- Power supply: Variable, regulated low-voltage dc source; isolation transformer
- Equipment: Oscilloscope; EVM
- Resistors: ½-W 0.68- and 560,000-Ω
- Capacitors: 0.01-μF; 0.1-μF; two 1.0-μF/12-V
- Semiconductors: RCA CA 3020 integrated circuit with mounting, or equivalent
- Miscellaneous: Push-pull output transformer (Argonne type 174 or equivalent); low-impedance speaker; 2-W 5000-Ω potentiometer; audio source such as high-impedance microphone, phonograph, or AM/FM tuner

PROCEDURE

NOTE: Figure 46.1-8*a* is a side and bottom view of the CA 3020. The leads are identified by reading

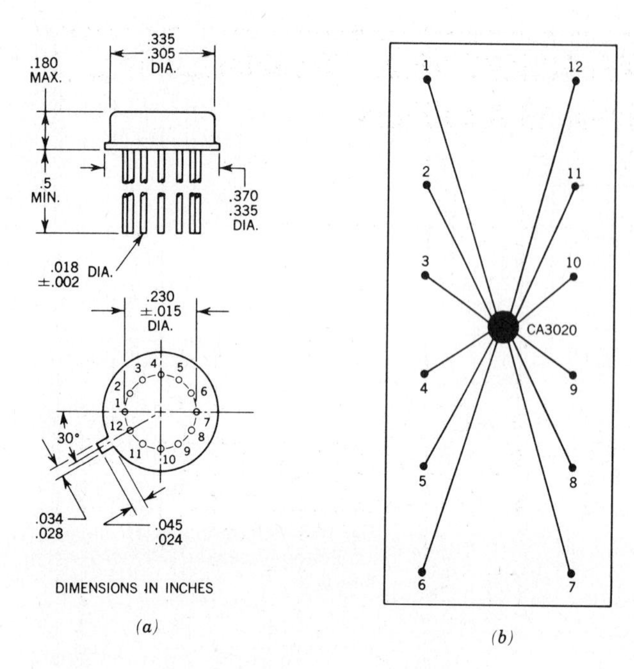

Fig. 46.1-8. (*a*) Identifying the leads of CA 3020 integrated circuit. (*b*) Suggested mount for CA 3020.

clockwise from the tab. The tab is adjacent to contact 12 on the IC.

The CA 3020 is a high-gain audio amplifier. In wiring the external circuit, extreme care must be used to prevent oscillation. There is no problem in its commercial use, where lead lengths can be kept short and external components can be arranged to minimize undesired regenerative feedback. Breadboarding the circuit by ordinary means used in the preceding experiments leads to instability.

The method described below was successful in eliminating instability and is therefore recommended. A special mount was prepared for the CA 3020 from an etched board 3 in × 1½ in. See Fig. 46.1-8*b*. The terminal pins on this board, numbered 1 to 12 in a counterclockwise direction, correspond to the terminal leads of the CA 3020. *Miniature* capacitors and resistors were placed on the board in accordance with the circuit required in this experiment. The capacitor and resistor pigtails were wound around the terminal pins, making good mechanical and electrical connections. Placement of parts was arranged to keep the shortest lead length. Where connection to external components was required, such as to the transformer, lead length was also kept short. Connection between terminals of the CA 3020, where required, was made by wrapping bare flexible solid wire jumpers around the terminals.

CAUTION: In the procedure which follows it will be necessary to plug any audio source which is not line-isolated into an isolation transformer.

1. Connect the circuit of Fig. 46.1-7. In this arrangement the CA 3020 serves as a ½-W amplifier driving a low-impedance speaker.

2. a. To the amplifier input connect a high-impedance microphone. Try out this PA system and comment on its effectiveness.
Volume of sound: ________ (good or poor).
Quality of sound: ________ (good or poor).

b. With an oscilloscope connected across the primary of the output transformer (observe ground connections), note the patterns (waveform) of each of the vowel sounds.

3. To the amplifier input connect the output of a phonograph pickup. Comment on the
a. Volume of sound ________
b. Quality of sound ________

4. To the amplifier input connect the output of an AM or FM tuner. Comment on the
a. Volume of sound ________
b. Quality of sound ________

QUESTIONS

1. How does an integrated circuit differ from a conventional transistorized circuit?
2. What advantages does an IC offer over conventional transistorized circuitry?
3. What is the relationship, if any, between amplifier power output and V_{CC}?
4. Why would you expect that a CA 3020 amplifier would be wideband?
5. (*a*) What advantages does the IC audio amplifier used in this experiment have over discrete component AF amplifiers?
   (*b*) What disadvantages does it have?

Answers to Self-Test

1. false
2. TO-5; flatpack; dual-in-line
3. transistor
4. transistors; diodes
5. linear; digital
6. audio amplifier
7. 10; 3

EXPERIMENT 46.2. Troubleshooting Linear IC Amplifiers

OBJECTIVE

To troubleshoot a linear IC amplifier

INTRODUCTORY INFORMATION

Troubleshooting Procedure

In troubleshooting electronic devices containing linear ICs, the same procedures and techniques are used as in troubleshooting any electronic device. These include *signal tracing* the circuit to isolate the trouble to a specific stage, then *voltage* and *resistance* checks in the suspected stage.

In signal tracing, the IC is treated like a discrete component, as for example a transistor, with known input and output signals. If the input signal to an IC is found to be normal, but there is no output or the output is *not* normal, then the IC, and the *external components* connected to the IC, must be checked.

Voltage Checks

The technician should know the dc voltages at the pins of an IC in order to compare actual measurements with the voltages under normal operation. Voltage norms are usually printed on the diagram of the device which is being serviced. Where norms are not available, a circuit diagram of the IC and of the circuit in which it is connected is helpful in deciding what to expect.

Consider the circuit diagram of the CA 3020 IC, Fig. 46.2-1, and the amplifier in which it is connected, Fig. 46.2-2. Q_1 is an emitter follower whose collector

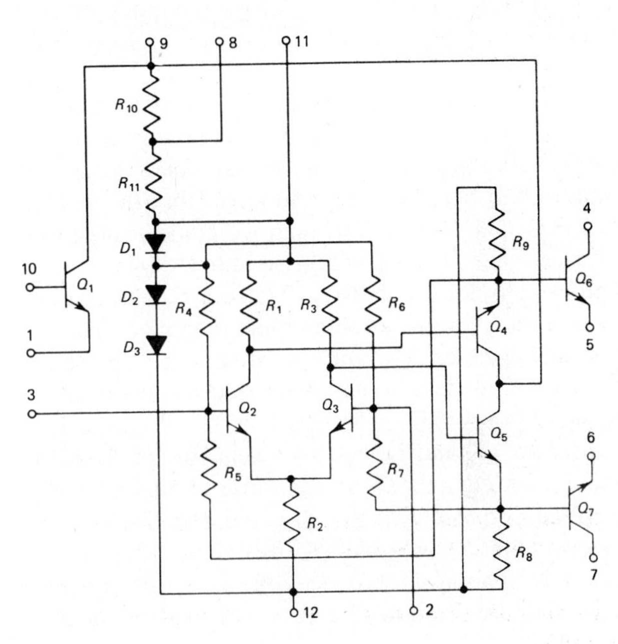

Fig. 46.2-1. Schematic of CA 3020 audio amplifier. (*RCA*)

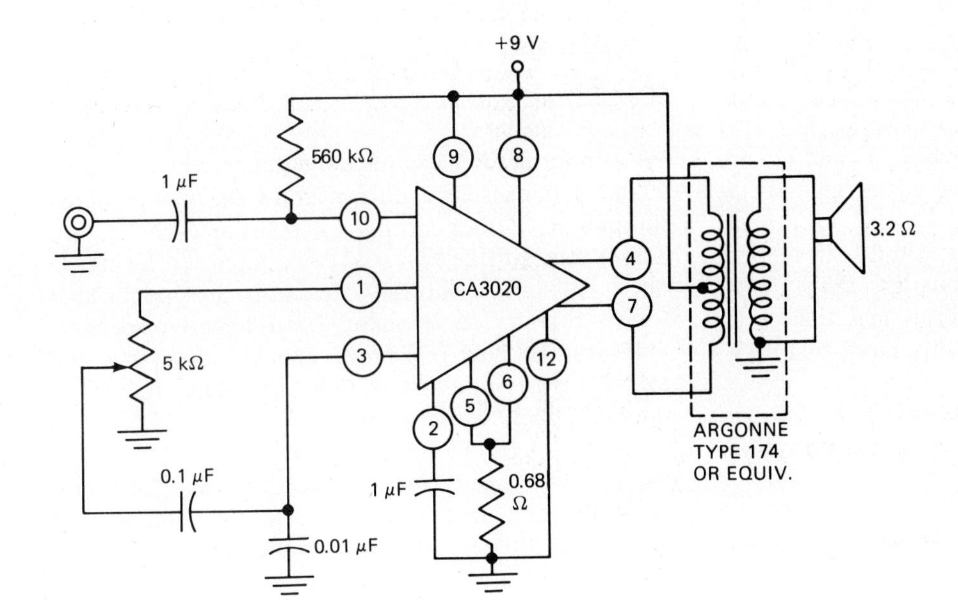

Fig. 46.2-2. Experimental IC amplifier. (*RCA*)

is connected to pin 9 of the IC and receives +9 V from the dc supply. The base of Q_1 is pin 10, and it is connected through a 560-kΩ biasing resistor to +9 V. The emitter of Q_1 is pin 1, and it is connected by a 5-kΩ potentiometer to ground. Under normal operation we should expect to measure a positive dc voltage at pins 10 and 1, and since it is an NPN transistor the base should be about 0.6 V more positive than the emitter. We do not know the current characteristics of Q_1, so we cannot say what the actual voltages will be. If we were to guess that the normal voltage at pin 10 is +4 V, then we could expect that the dc voltage at pin 1 would be about +3.4 V.

Now consider Q_2 and Q_3. These are balanced circuits connected as a differential amplifier. The dc *base* voltages on Q_2 and Q_3 should be the same. Therefore the dc voltages measured at pins 2 and 3 should be approximately equal. Now, the base bias circuits for Q_2 and Q_3 receive their voltage from the junction of D_1 and D_2 in the series voltage divider consisting of R_{10}, R_{11}, D_1, D_2, and D_3. Each of the silicon diodes D_1, D_2, and D_3 has about 0.6 to 0.7 V across it. Therefore the *dc* voltage at the junction of D_1 and D_2 is about 1.2 to 1.4 V with respect to ground (pin 12). So the *dc* voltages at the bases of Q_2 and Q_3 will be less than this value, say, about 1.0 V.

Q_6 and Q_7 are the output transistors which deliver audio power to the speaker. Their emitters (pins 5 and 6) are connected through a 0.68-Ω resistor to ground; their collectors (pins 4 and 7) are connected to the ends of the center-tapped primary of the output transformer. Since the resistance of the transformer primary is low, we can expect that there will be only a small voltage drop across it, when Q_6 and Q_7 are conducting. So we can look for *equal* voltages at the collectors of Q_6 and Q_7, pins 4 and 7. Moreover, the voltages at pins 4 and 7, measured with respect to ground, should be close to but less than +9 V. Finally, since the emitter resistor is less than 1 Ω, the voltage across it will be slightly more than 0 V, possibly 0.1 V.

To summarize our conclusions about the IC amplifier in Fig. 46.2-2, we have, under normal operation, the following conditions:

1. The dc voltage at pin 10 is about 0.6 V more positive than that at pin 1.
2. The dc voltage at pins 2 and 3 should be equal and less than 1.2 V.
3. The dc voltages at pins 4 and 7 should be equal and close to +9 V.
4. The dc voltage at pins 5 and 6 should be close to, but slightly higher than, 0 V.
5. The dc voltage at pins 8 and 9 should be +9 V.

If there is any appreciable variation from these norms, something is wrong in the circuit.

Resistance Checks

The voltage measurements should give some clue as to the nature of the defect. For example, if the voltage at pin 11 with respect to ground is zero and there is +9 V at pins 8 and 9, then R_{11} or the connection from the junction of R_{11} and D_1 to pin 11 must be open. If that is the defect, the IC must be replaced. Of course, the defect will not always be so easily determined.

Abnormal voltages in the IC amplifier of Fig. 46.2-2 can result from a defective IC or defective external component. If the defect is not obvious, it will be less time-consuming to resistance-check the external components than to unsolder the IC. As in other solid-state circuits, a *low-power ohmmeter* should be used to prevent forward-biasing PN junctions.

In the circuit of Fig. 46.2-2 the resistances measured to ground at the pin numbers of the IC should be as follows (assume that the lead from the +9 V supply has been disconnected from the circuit):

Pin Number	*1*	*2*	*3*	*4*	*5*	*6*	*7*	*8*	*9*	*10*	*11*	*12*
Resistance, Ω	5000	?	?	∞	0.68	0.68	∞	∞	∞	∞	∞	0

We cannot predict what the resistance will be at pins 2 and 3 because we do not know the values of R_4 through R_9. Other meaningful resistance checks are across each of the *external* components to locate an open resistor, a shorted capacitor, an open transformer primary or secondary, or a defective speaker.

If voltage and resistance checks show that the external parts and connections are okay, the IC is the most likely cause of trouble.

NOTE: The signal-tracing procedure may help pinpoint the defective part. For example, suppose the input is a 1000-Hz sine wave which appears at pins 10, 1, 3, 4, and 7, but no sound is heard from the speaker. We can conclude that either the output transformer is defective (say, an open secondary winding) or the speaker is defective. So we would immediately check the suspected *two parts* instead of making voltage or resistance checks in other parts of the circuit.

SUMMARY

1. In troubleshooting an electronic device containing ICs, signal tracing is used to isolate the trouble to a stage.
2. If the suspected stage uses a plug-in IC, the IC is replaced with a known good IC. If circuit operation is restored to normal, the repair is completed.
3. If the replacement IC does not restore normal operation, or if the IC is soldered into the circuit, voltage and resistance checks follow.
4. The signal-tracing process may indicate that a particular part is defective. In that case the suspected part is checked first.
5. Voltages, measured at the pins of an IC, are compared with the normal readings for that circuit. These checks may point to a defective external part. In that case the suspected part is checked next.
6. After voltage checks, resistance measurements are made at the pins of the IC and across components in the external circuit of the suspected stage. Of course, power must be OFF, and a low-power ohmmeter should be used.

SELF-TEST

Check your understanding by answering these questions. All questions refer to Figs. 46.2-1 and 46.2-2.

1. In the IC the voltage measured at pin 11 should be ________ V, approximately.
2. The amplifier does not work (no sound is heard). Each of the voltages at pins 1 and 10 measures +8.8 V. The most likely cause of trouble is ________.
3. In signal-tracing the amplifier, the signal voltage at pins 10 and 1 measures about 0.15 V p-p. There is no signal at pin 3. The dc voltage measured from pin 3 to ground is 1.0 V. A possible cause of trouble is ________.
4. The resistance measured from pin 2 to ground is 0 Ω. The most likely cause of trouble is ________ ________.

MATERIALS REQUIRED

- Same as for Exp. 46.1. In addition, defective parts for troubleshooting the amplifier.

PROCEDURE

1. Connect the circuit of Fig. 46.2-2. **Power on.**

Signal Tracing

2. Signal-trace the amplifier, using a 1000-Hz signal. The 5-kΩ potentiometer is set for maximum gain and generator signal level for maximum undistorted output.

   Measure and record the peak-to-peak signal voltages at the IC pins shown in Table 46.2-1.

Voltage Norms

3. Measure and record (without signal) the dc voltages with respect to ground at the IC pins shown in Table 46.2-2.

Resistance Measurements

4. **Power off.** *Disconnect the power source from the amplifier.* Measure and record in Table 46.2-3 the resistances with respect to ground at each of the IC pins. Measure also and record the resistance of the transformer primary: ________ Ω and transformer secondary: ________ Ω.

TABLE 46.2-1. Signal Levels

Pin Number	*10*	*1*	*2*	*3*	*4*	*5–6*	*7*	*Voice Coil*
V p-p								

TABLE 46.2-2. Voltage Norms

Pin Number	*1*	*2*	*3*	*4*	*5–6*	*7*	*8–9*	*10*	*11*	*12*
V dc										

TABLE 46.2-3. Resistance Norms

Pin Number	*1*	*2*	*3*	*4*	*5–6*	*7*	*8–9*	*10*	*11*	*12*
Resistance, Ω										

Troubleshooting the IC Amplifier

5. Your instructor will insert trouble into the amplifier. Troubleshoot the circuit, keeping a record of your checks, in the order that you make them. Use the Standard Troubleshooting Report.

   When you have found and corrected the trouble, notify your instructor. Troubleshoot as many defects as time permits.

QUESTIONS

1. Compare the techniques used in servicing an amplifier containing ICs with those used in servicing an amplifier with discrete parts.
2. What would be the effect, if any, on amplifier operation in the circuit of Fig. 46.2-2 of: (*a*) An open 1-μF capacitor on pin 10? (*b*) An open 0.1-μF capacitor on pin 3? (*c*) A shorted 0.01-μF capacitor on pin 3? (*d*) An open voice coil? (*e*) A shorted voice coil? (*f*) A shorted transformer primary?

Answers to Self-Test

1. 2.1
2. an open 5-kΩ emitter resistor
3. an open 0.1-μF coupling capacitor to pin 3
4. a shorted 1-μF capacitor on pin 2

CHAPTER 47 DIGITAL ICs: AND, OR, AND NOT GATES

EXPERIMENT 47.1. Digital ICs: AND Gate

OBJECTIVE

To become familiar with the characteristics of an AND gate and experimentally determine its truth table

INTRODUCTORY INFORMATION

Digital ICs—Logic Circuits

In the last two experiments you were introduced to linear ICs and their operation. In this and future experiments, you will work with digital ICs. Digital ICs were first used in and are still the building blocks of electronic computers. However, digital ICs have also found their way into television receivers, electronic instruments (digital voltmeters, counters, frequency meters, etc.), and other communications devices. It is therefore necessary for the technician to have an understanding of their operation.

Modern computers consist of complex arrangements of simple "logic circuits" contained in digital ICs. A computer's ability to solve a problem depends on its ability to make decisions as it works through the steps of a problem. The circuits in a computer that make comparisons and decisions are the logic circuits.

Computer decisions are of the YES, NO variety, that is, of the two-state type. Logic circuits can be in one of two positions, such as ON or OFF, HIGH or LOW. A toggle switch is a simple example of a two-state device.

The two-position nature of computer circuits, therefore, makes it possible to represent information and logical conditions by a dc level on a signal line. As an example, the presence of information may be indicated by a positive voltage level; the absence of information, by a negative level. These levels are also referred to as UP (positive) and DOWN (negative). A signal line is *active* when it has an UP level and *inactive* when it has a DOWN level.

Logic circuits analyze a combination of line levels at their input and produce a desired output when the input combination is correct for that particular circuit.

The three basic logic circuits are the AND gate, the OR gate, and the inverter. In this experiment you will study the AND gate.

AND Gate

A gate is a circuit which has *two or more inputs* and *one output*. Whether the output is UP or DOWN depends on the type of gate and on the input combinations. *An* AND *gate is one whose output is* UP *only when all its inputs are* UP. If *any* of the inputs to an AND gate is *down*, the output is down.

The following logical conclusions may therefore be drawn by measuring the output of an AND circuit:

1. If the output is UP, then all inputs must be UP.
2. If the output is DOWN, then any one or all inputs must be DOWN.

AND Gate Truth Table

The logic of digital circuits is usually given in the form of truth tables. We shall say that a digital UP state is represented by the binary number 1, a DOWN state by the binary number 0. Then all the conditions possible in a two-input AND gate are given in Table 47.1-1.

Transistor AND Gate

AND gates may be made by using diodes or transistors and resistors. An example of a two-input grounded-

TABLE 47.1-1. AND Gate Truth Table

Inputs		*Output*
0	0	0
1	0	0
0	1	0
1	1	1

collector AND gate is shown in Fig. 47.1-1. This circuit is designed to accept $+V$ or $-V$ inputs, where $+V$ represents the binary 1 and $-V$ is the binary 0. If the input to Q_1 or Q_2 (or to both) is $-V$ volts, the base-to-emitter junction of that transistor is forward-biased and that transistor is saturated. There is enough emitter-to-collector current in the forward-biased transistor to bring the output level at the common emitter to $-V$ volts, and the output represents a binary 0. On the other hand, if both inputs A and B are $+V$ volts, then both Q_1 and Q_2 are cut *off*, and the output level at the common emitter is $+V$ volts. It is evident therefore that the logic of Fig. 47.1-1 is that of an AND gate.

Because AND gates may be much more complex than the circuit in Fig. 47.1-1, "logic" circuit diagrams use standard *logic symbols,* rather than discrete component symbols. The symbol for a two-input AND gate is shown in Fig. 47.1-2*a*. Figure 47.1-2*b* is that of a three-input AND gate.

IC AND Gates

Discrete components such as those in Fig. 47.1-1 are no longer used in computers to form digital gates. Instead ICs and Large Scale Integrated Circuits (LSICs) are employed.

IC gate circuits are usually more complex than the preceding circuit. Moreover, the IC arrangements may follow one of several different types of logic, as for example, resistor transistor logic (RTL), diode transistor logic (DTL), transistor transistor logic (TTL).

An AND gate IC may contain one or more gates, as for example, the 7411 used in this experiment. This IC consists of three identical three-input AND gates. Figure 47.1-3 is the block symbol for this dual-in-line IC, showing the inputs to and outputs from each of the gates. Note that pin 14 is the common $+V_{CC}$ terminal for each of the gates, while terminal 7 is the common ground return. The inputs to the first gate are terminals 1, 2, and 13; the output is terminal 12.

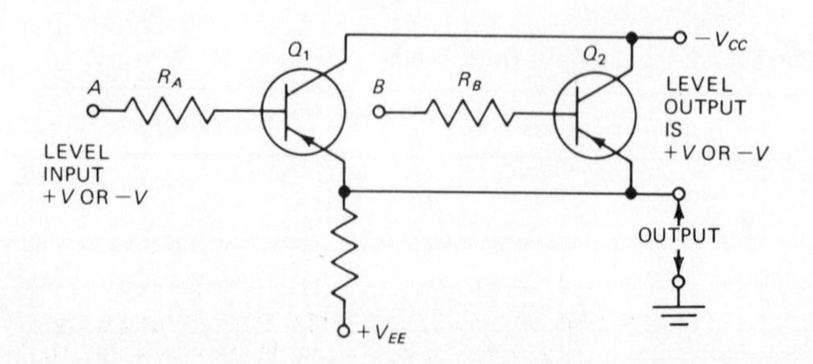

Fig. 47.1-1. Two-input grounded-collector AND gate.

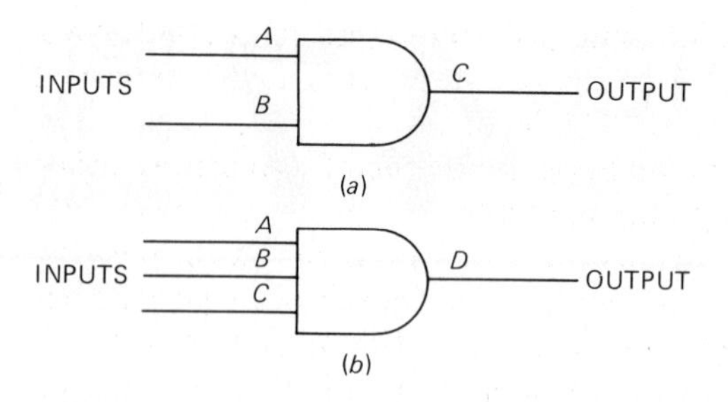

Fig. 47.1-2. AND gate symbols: (*a*) two-input, (*b*) three-input.

This IC requires +5.0 V between terminals 14 and 7. The logic levels are 0 V for binary 0 (DOWN) and +5 V for binary 1 (UP). These are not *absolute* values. There is some variation which will still permit the AND gate to operate satisfactorily. This IC is an example of *positive* logic where the digital 1 is represented by a positive voltage, a digital 0 by a lower positive voltage, a zero, or a negative voltage.

A shorthand statement (equation) for each of the three-input AND gates is

$$A \cdot B \cdot C = D \qquad (47.1\text{-}1)$$

The letters A, B, and C represent the inputs. The dot (·) between letters is read AND. The statement simply means "This is a three-input (A, B, C) AND gate with output D." It is read *"A and B and C equals output D."*

SUMMARY

1. Digital circuits use elementary logic building blocks to make up highly complex computing and counting circuits.
2. Digital circuits are called logic circuits because

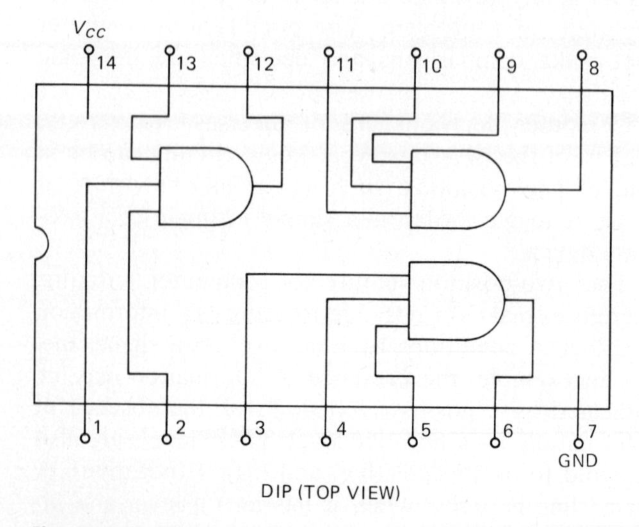

Fig. 47.1-3. IC circuit diagram for the triple, three-input AND gates in the 7411.

the type of input will determine the output. That is, the *logic* of the circuit ensures a predetermined output for specific inputs.

3. Digital circuits operate in one of two states, either ON or OFF, UP or DOWN.
4. In *positive* logic an UP level, the most positive voltage level which can appear on a signal line, represents the binary number 1. A DOWN level, which can be a lower voltage, 0 V, or a negative voltage, represents the binary number 0. The manufacturer of a digital device will specify the voltage which corresponds to binary 1 and that corresponding to binary 0.
5. The most elementary of the logic blocks are the AND gate, the OR gate, and the inverter.
6. A gate is a circuit with *two or more inputs* and *one* output.
7. The logic characteristics of an AND gate are:
   (*a*) When *all* the inputs are UP, the output is UP.
   (*b*) The output is DOWN when any of or all the inputs are DOWN.
8. An AND circuit is identified mathematically by the expression (equation)

$$A \cdot B \cdot C = D$$

9. The logic of digital circuits is given in truth tables.

SELF-TEST

Check your understanding by answering these questions.

1. Digital circuits are linear circuits. ________ (true/false)
2. In a four-input AND circuit ________ inputs must be ________ for the output to be UP.
3. The symbol for a four-input AND circuit is ________.
4. The 7411 IC used in this experiment has ________ three-input ________ gates.
5. In the 7411 the inputs to the gate whose output is terminal 6 are ________.
6. The operating voltage between pins 14 and 7 in the 7411 is ________ V.

MATERIALS REQUIRED

- Power supply: Variable regulated low-voltage dc source
- Equipment: EVM or VOM
- Integrated circuit: 7411 (AND gates) or the equivalent
- Miscellaneous: Four SPDT, one SPST switch

PROCEDURE

CAUTION: Do not use a voltage higher than the +5.0 V required for the IC. Observe the polarity of the $+V_{CC}$ source.

Three-Input AND Gate

1. Connect the circuit of Fig. 47.1-4. S_4 is open. Set the power supply for +5.0 V. Close S_4. S_1, S_2, and S_3 are SPDT switches which will deliver either 0 V (binary 0) or +5.0 V (binary 1) to inputs 3, 4, and 5 of the AND gate.
2. Perform the logic and complete the truth table for this circuit.

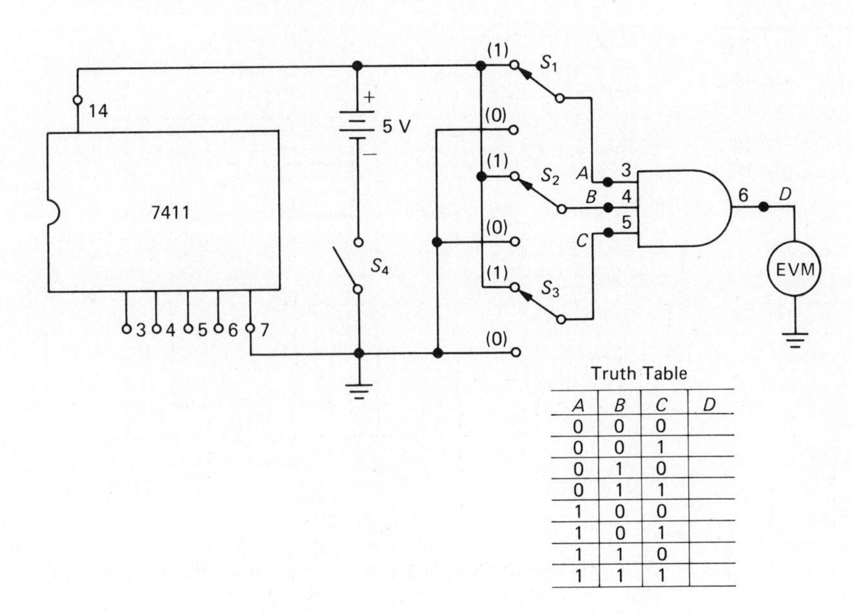

A	B	C	D
0	0	0	
0	0	1	
0	1	0	
0	1	1	
1	0	0	
1	0	1	
1	1	0	
1	1	1	

Fig. 47.1-4. Experimental three-input AND gate and truth table. Terminal numbers on the AND gate refer to terminals on the 7411.

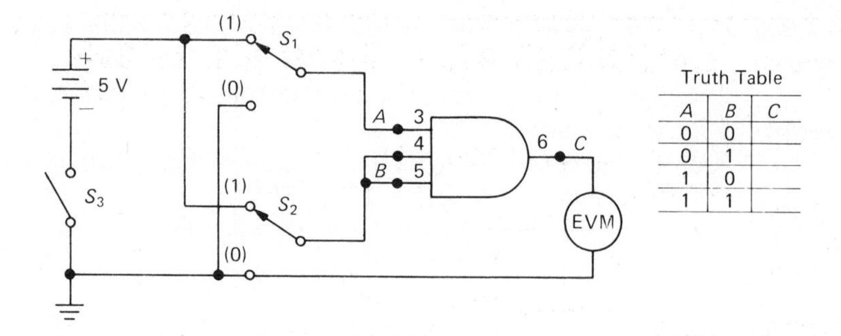

Truth Table

A	B	C
0	0	
0	1	
1	0	
1	1	

Fig. 47.1-5. Experimental two-input AND gate and truth table.

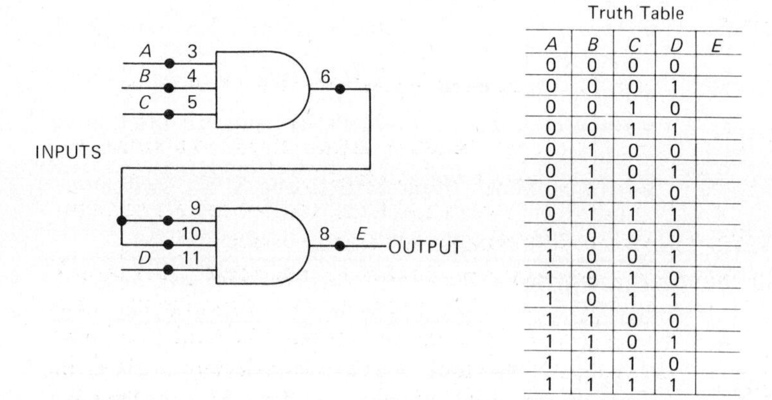

Truth Table

A	B	C	D	E
0	0	0	0	
0	0	0	1	
0	0	1	0	
0	0	1	1	
0	1	0	0	
0	1	0	1	
0	1	1	0	
0	1	1	1	
1	0	0	0	
1	0	0	1	
1	0	1	0	
1	0	1	1	
1	1	0	0	
1	1	0	1	
1	1	1	0	
1	1	1	1	

Fig. 47.1-6. Experimental four-input AND gate and truth table.

Two-Input AND Gate

3. Connect the circuit of Fig. 47.1-5. Perform the logic and complete the truth table for this circuit.

Four-Input AND Gate

4. The circuit of Fig. 47.1-6 is a four-input AND gate. Connect the circuit, including the dc input, as in Figs. 47.1-4 and 47.1-5. Perform the logic and complete the truth table for this circuit.

QUESTIONS

1. What is a logic circuit?
2. What were the two states of the logic circuits used in this experiment?
3. What is the largest number of inputs which a single AND gate can have constructed from the AND gates in the 7411?
4. What is the maximum number of inputs a five-input AND gate can have? A six-input AND gate?

Answers to Self-Test

1. false
2. all; UP
3. 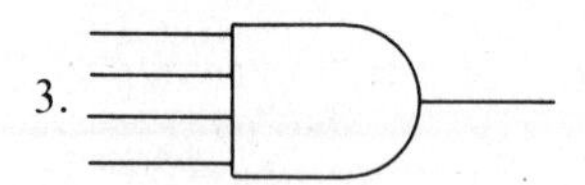
4. three; AND
5. 3, 4, 5
6. +5

EXPERIMENT 47.2. Digital ICs: OR Gate

OBJECTIVE

To become familiar with the characteristics of an OR gate, and experimentally determine its truth table

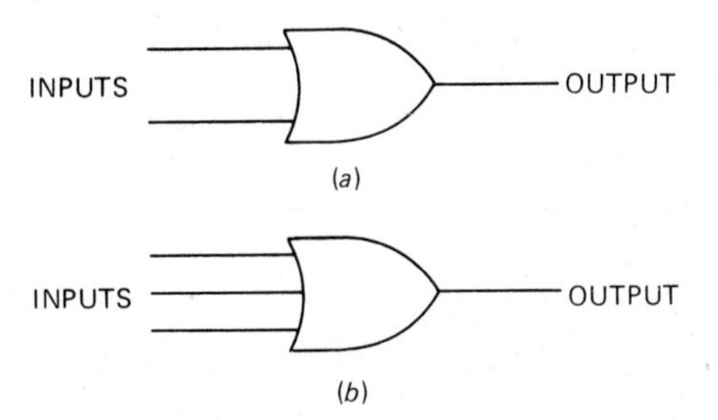

Fig. 47.2-1. OR gate symbols: (*a*) two-input, (*b*) three-input.

INTRODUCTORY INFORMATION

OR Gate

The OR gate is another logic circuit with unique characteristics. These are:

1. If any or all of the inputs to an OR gate are UP, the output is UP.
2. When, and only when, *all* the inputs to an OR gate are DOWN, the output is DOWN.

The following logical conclusions may therefore be drawn by measuring the output of an OR gate:

1. If the output is DOWN, then *all* inputs must be down.
2. If the output is UP, then *any one* or *all* inputs must be UP.

The OR gate may be described as a logic circuit whose *output* is equal to its *highest input*. Compare this with an AND gate, whose output is equal to its *lowest* input.

The truth table for a two-input OR gate is shown here.

TABLE 47.2-1. OR Gate Truth Table

Inputs		*Output*
A	*B*	*C*
0	0	0
0	1	1
1	0	1
1	1	1

IC OR Gates

OR gates may be constructed of diodes, transistors, and resistors, connected as discrete components, but these are no longer used. ICs and LSICs are employed. IC OR gates are fairly complex circuits, but all we need be concerned with is the logic symbol for an OR gate, because digital circuit diagrams use logic symbols for logic circuits.

Figure 47.2-1*a* is the logic symbol for a two-input OR gate; Fig. 47.2-1*b* for a three-input OR gate.

OR gate ICs may contain one or more gates, as for example, the 7432 used in this experiment. This positive-logic IC consists of four identical two-input OR gates. The block symbol identifying each gate, inputs and outputs, is shown in Fig. 47.2-2. This IC requires +5.0 V between terminals 14 (V_{CC}) and 7 (ground). The logic levels are 0 V for binary 0 and +5.0 V for binary 1. This IC is identified as a *quad* (four) two-input OR gate.

The equation which represents the logic of each of the two input OR gates in this IC is

$$A + B = C \qquad (47.2\text{-}1)$$

The + between letters is read OR. It does not mean the same as *plus* (+) in ordinary arithmetic or algebra. The letters *A* and *B* represent the inputs. The output is *C*.

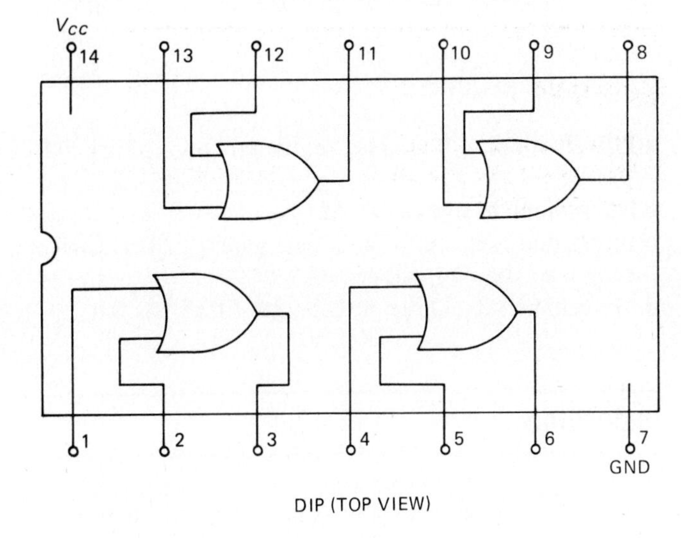

Fig. 47.2-2. IC circuit diagram for the quad two-input OR gates in the 7432.

The equation simply means, "This is a two-input (A,B) OR gate, with output C." It is read "A *or* B equals output C."

SUMMARY

1. The characteristics of an OR gate are:
   (*a*) If *any* input to an OR gate is UP, the output is UP.
   (*b*) For the output to be DOWN, *all* inputs to the OR gate must be DOWN.
2. The logic equation of a four-input OR gate is:

$$A + B + C + D = E$$

   The letters A, B, C, and D represent the inputs, E the output. The symbol + between input letters is read OR. It is *not* to be confused with the *plus* sign in arithmetic.
3. OR gates may be constructed of discrete components, diodes, transistors, and resistors. However, this type of logic circuit is obsolete. OR gates are contained in ICs, just as other logic circuits are.

SELF-TEST

Check your understanding by answering these questions.

1. In a three-input OR gate at least ________ input(s) must be ________ for the output to be UP.
2. The symbol ________, when used in logic equations, stands for OR.
3. A triple three-input OR gate contains ________ independent gates.
4. In the 7432 IC used in this experiment there are ________ OR gates and each has ________ inputs.

MATERIALS REQUIRED

- Power supply: Variable, regulated low-voltage dc source
- Equipment: EVM or VOM
- Integrated circuits: 7432 (OR gates), 7411 (AND gates), or the equivalent
- Miscellaneous: Three SPDT, one SPST switch

PROCEDURE

CAUTION: Do not use a voltage higher than the +5.0 V required for the IC. Observe the polarity of the $+V_{CC}$ source.

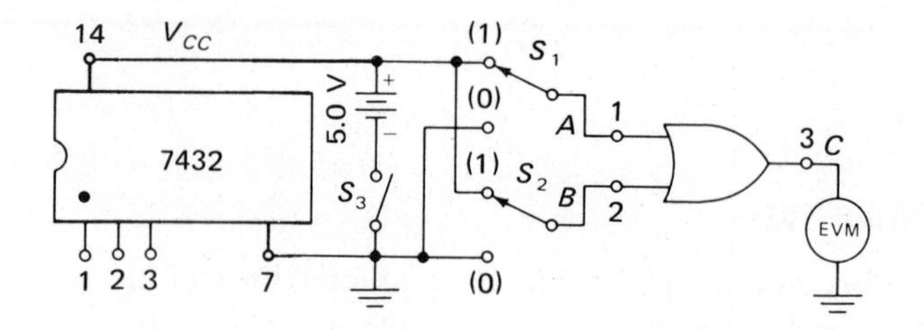

Truth Table

A	B	C
0	0	
0	1	
1	0	
1	1	

Fig. 47.2-3. Experimental two-input OR gate and truth table.

Two-Input OR Gate

1. Connect the circuit of Fig. 47.2-3. S_3 is open. Set the power supply for +5.0 V. Close S_3.
2. Perform the logic and complete the truth table for this circuit.

Three-Input OR Gate

3. Figure 47.2-4 shows how 2 two-input OR gates are connected to form a three-input OR gate. Connect the circuit of Fig. 47.2-4. In the 7432, connect pin 14 to +5 V and pin 7 to ground. Use an input similar to Fig. 47.2-3. Perform the logic and complete the truth table for this circuit.

Combined AND-OR Gates

4. Connect the circuit of Fig. 47.2-5, a combined AND-OR gate. In both the 7411 and the 7432, connect pin 14 to +5 V and pin 7 to ground. Perform the logic and complete the truth table for this circuit.

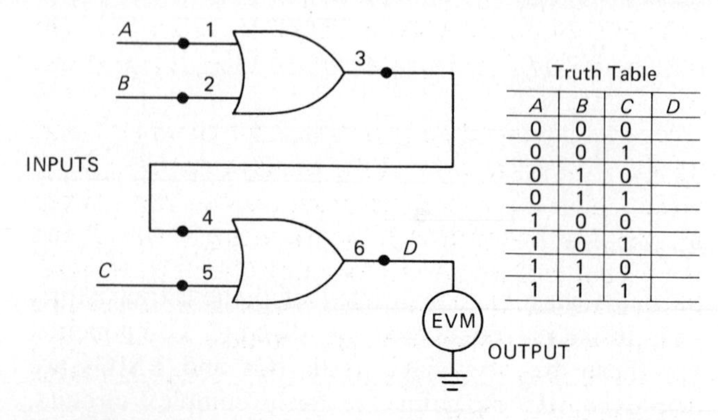

Truth Table

A	B	C	D
0	0	0	
0	0	1	
0	1	0	
0	1	1	
1	0	0	
1	0	1	
1	1	0	
1	1	1	

Fig. 47.2-4. Experimental three-input OR gate and truth table.

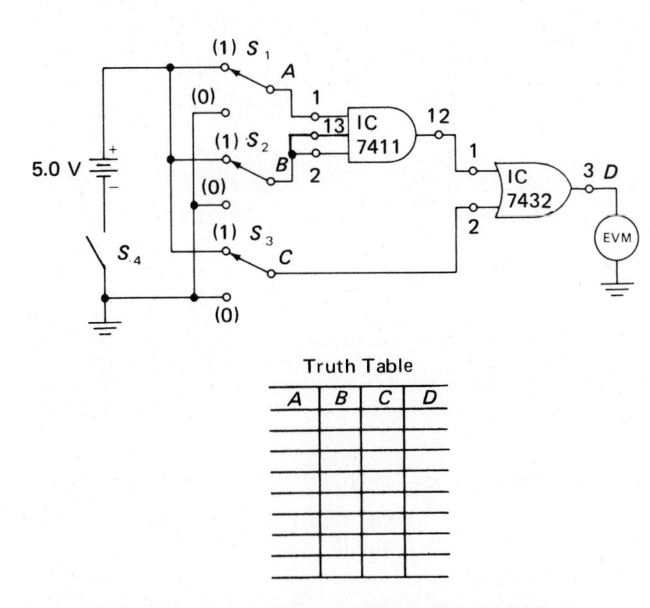

Fig. 47.2-5. Experimental AND-OR gate and truth table.

QUESTIONS

1. From the results in this experiment, what conclusions can you draw about the logic characteristics of an OR circuit?
2. What are the logic characteristics of the experimental AND-OR gate?

Answers to Self-Test

1. one; UP
2. +
3. three
4. four; two

EXPERIMENT 47.3. Digital ICs: NOT Logic

OBJECTIVE

To become familiar with the characteristics of an inverter (NOT circuit) and its logic

INTRODUCTORY INFORMATION

NOT Logic

In addition to the logic operations AND and OR, the third basic logic operation is called NOT. A NOT circuit is simply an *inverter*, as in Fig. 47.3-1*a*, an amplifier biased to cutoff whose output is 180° out of phase with its input.

In the grounded-emitter transistor amplifier, Fig. 47.3-1*a*, the input to the base can be a dc voltage or a pulse. The output is taken from the collector. When the input is DOWN (0 V), the transistor is cut off and the output is at $+V_{CC}$, UP. When the input is UP (positive voltage to the base), the transistor saturates and its output goes DOWN (0 V, approximately). A positive *pulse* will also cause the output to go DOWN during the time of the pulse.

The characteristics of a NOT circuit then are:

1. When the input is DOWN, the output is UP.
2. When the input is UP, the output is DOWN.

Mathematically, these characteristics are given by logic Eq. (47.3-1).

$$\text{If } V_{in} = A,\ V_{out} = \overline{A} \qquad (47.3\text{-}1)$$

The bar over the A represents NOT (logic inversion). For example, if the letter A represents an UP level (1), $\overline{A}$ represents DOWN (0), and if $A = 0$, $\overline{A} = 1$. We may also say that $\overline{0} = 1$ and that $\overline{1} = 0$.

Figure 47.3-1*b* is the symbol for a NOT circuit. A NOT circuit has a *single input* and a *single output*. The truth table for a NOT circuit is shown here.

TABLE 47.3-1. NOT **Circuit Logic**

A	$\overline{A}$
1	0
0	1

The output of a NOT circuit is said to be the *complement* of the input.

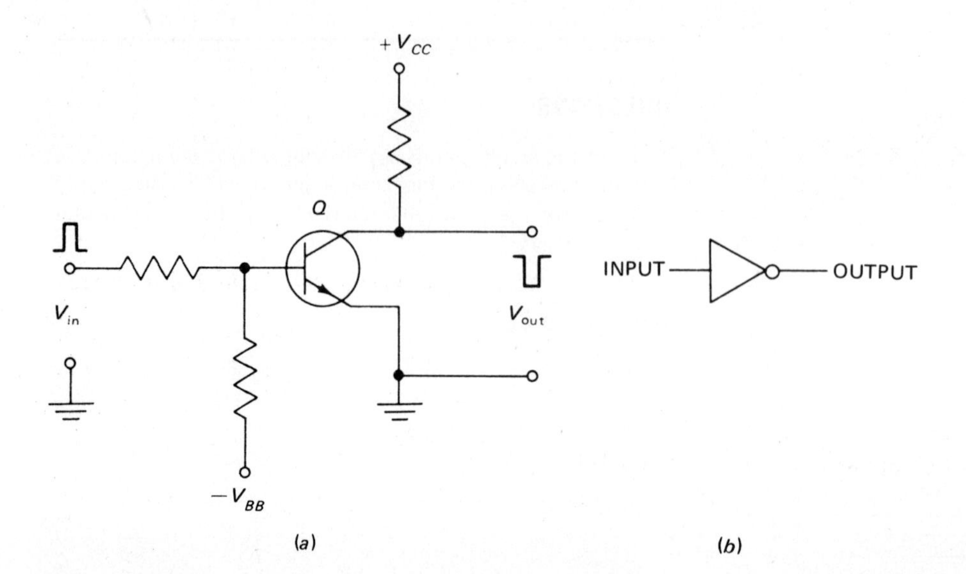

Fig. 47.3-1. (a) NOT or inverter circuit, and (b) logic symbol.

IC Inverters

The inverter in Fig. 47.3-1*a* is made of discrete components, but as in the case of other logic circuits, IC NOT circuits are used rather than discrete components. The diagram of the 7404 in Fig. 47.3-2 is that of an IC containing *six* inverters. It is therefore called a *hex* (six) inverter. The 7404 which will be used in this experiment uses positive logic, in which a binary 0 is represented by 0 V, a binary 1 by +5 V (approximately).

SUMMARY

1. A NOT circuit inverts logic levels. If the input to a NOT circuit is UP, the output is DOWN; if the input is DOWN, the output is UP.
2. A NOT circuit has a *single input* and a *single output*.
3. The NOT operation is mathematically represented in logic equations by a bar (–) over a letter or an expression, for example, $\overline{A}$ is read "NOT of *A*."
4. $\overline{0} = 1$, and $\overline{1} = 0$.

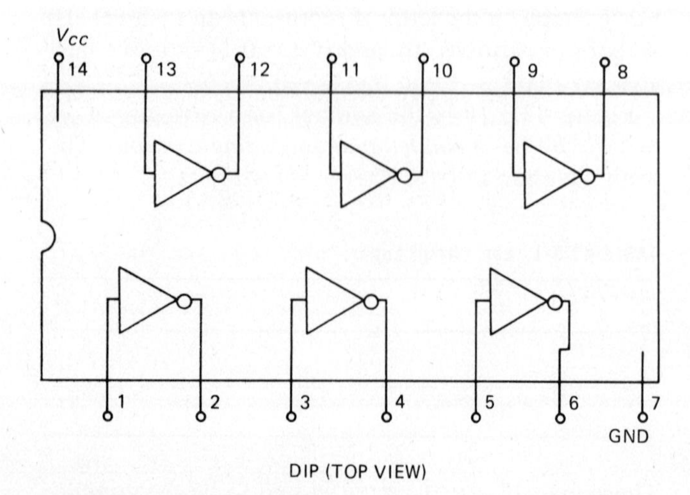

Fig. 47.3-2. IC 7404 hex inverter.

SELF-TEST

Check your understanding by answering these questions.

1. In the circuit of Fig. 47.3-1*a*, *Q* is normally ________ (ON/OFF) in the absence of a positive input.
2. The circuit in Fig. 47.3-1*a* ________ the logic of the input signal.
3. A two-input AND circuit whose output is connected to an inverter is represented mathematically as follows:
   $\overline{A \cdot B} = C$. If $A = 1$, $B = 0$, $C =$ ________.
4. The diagram for the circuit in question 3 is: ________.

MATERIALS REQUIRED

- Power supply: Variable, regulated low-voltage dc source
- Equipment: EVM or VOM
- Integrated circuits: 7404 and 7432 or the equivalents
- Miscellaneous: Two SPDT, SPST switches

PROCEDURE

NOT Logic

1. Connect the circuit of Fig. 47.3-3. Pin 14 of the 7404 is connected to +5.0 V and pin 7 to ground. Complete its truth table.

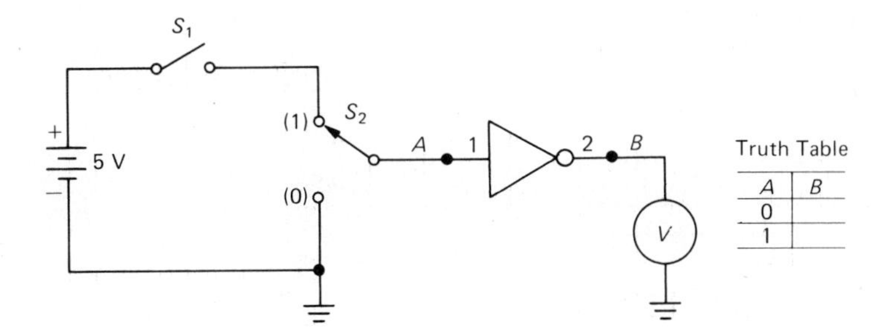

Truth Table

A	B
0	
1	

Fig. 47.3-3. Experimental NOT circuit and truth table.

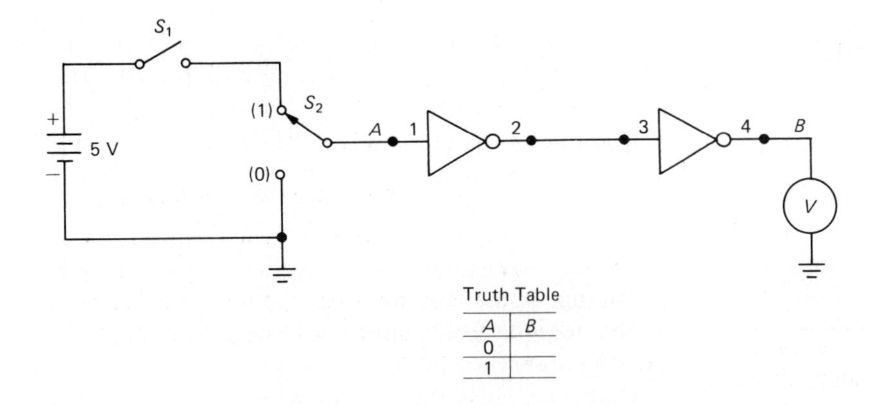

Truth Table

A	B
0	
1	

Fig. 47.3-4. NOT-NOT logic circuit and truth table.

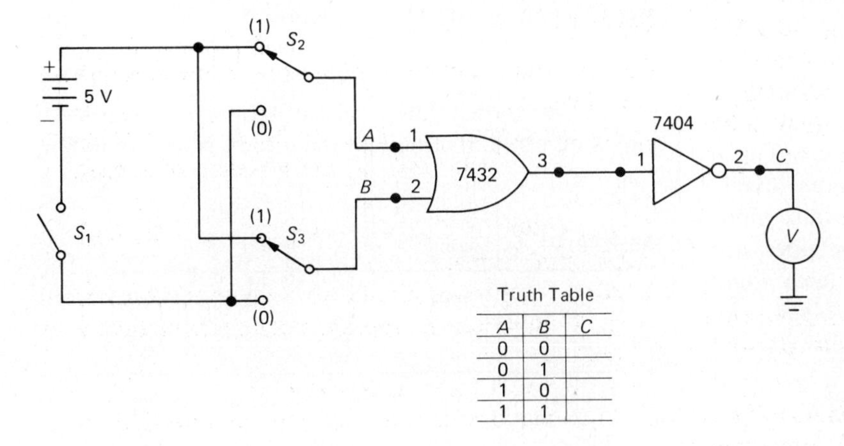

Truth Table

A	B	C
0	0	
0	1	
1	0	
1	1	

Fig. 47.3-5. NOT - OR gate and truth table.

NOT-NOT Logic

2. Connect the circuit of Fig. 47.3-4 and complete its truth table.

NOT-OR Logic

3. Connect the circuit of Fig. 47.3-5 and complete its truth table.

QUESTIONS

1. Can a NOT circuit be classified as a gate? Why?
2. Complete the logic equation for the circuit in Fig. 47.3-4, $\overline{\overline{A}}$ = ________. NOTE: The double bar (=) over the A means NOT NOT. Figure 47.3-4 is such a circuit.
3. How does the logic of the NOT-OR gate, Fig. 47.3-5, differ from that of an OR gate?

Answers to Self-Test

1. OFF
2. inverts
3. 1
4. INPUTS A, B — C OUTPUT

CHAPTER 48 DIGITAL ICs: NOR AND NAND LOGIC

EXPERIMENT 48.1. Digital Logic: The NOR Gate

OBJECTIVES

1. To determine experimentally the truth table for a NOR gate
2. To use NOR logic to construct a logic inverter and an OR gate

INTRODUCTORY INFORMATION

Logic Building Blocks

Experiments 47.1 to 47.3 introduced three basic logic building blocks, the AND gate, the OR gate, and the NOT circuit (inverter). The *gated* circuits were found to have *two or more inputs* and a single output. The output depended on specific input conditions. These circuits have the ability to evaluate input-signal levels, and they respond predictably when certain input conditions are met. It is as though these gates are making logical decisions, therefore the term *logic* circuits. The third basic logic building block, the single-input, single-output NOT circuit, inverts the logic level on its input.

NOR Gate

The three basic logic circuits can be combined to form other logic blocks with unique characteristics. One such arrangement is the NOR gate, a combination of a NOT circuit and an OR gate. The NOT circuit operates on the output of the OR gate, as in Fig. 48.1-1*a*. The logic symbol for the NOR gate is shown in Fig. 48.1-1*b*. Note that the triangle which is part of the symbol of a NOT circuit is left out in Fig. 48.1-1*b*. The circle acts as the NOT symbol, which here converts OR to NOR logic. Truth tables 48.1-1 and 48.1-2 give a comparison between two-input OR and NOR logic. It is clear that for every combination of inputs, the NOR gate inverts the output of the OR gate.

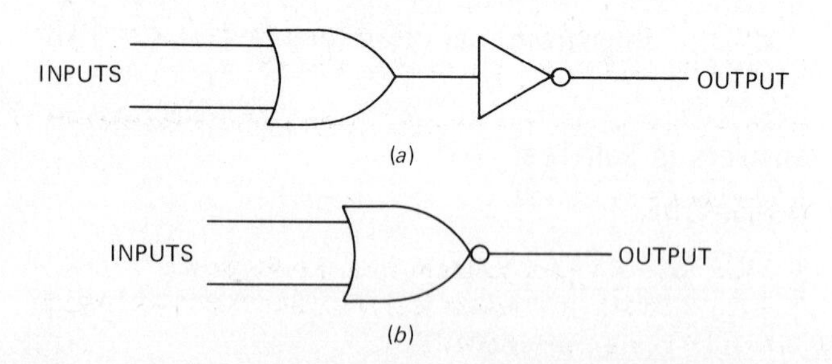

Fig. 48.1-1. (*a*) OR gate and inverter. (*b*) Logic symbol for NOR gate.

The logic symbol shown in Fig. 48.1-1*b* will be used for the NOR gate in this book.

Mathematically, the operation of a two-input NOR gate is given by the logic equation

$$\overline{A + B} = C \qquad (48.1\text{-}1)$$

It can be shown that if each of the inputs to an AND gate is inverted, the logic of the circuit is also that of a

TABLE 48.1-1. OR Logic

Input		Output
A	*B*	*C*
0	0	0
0	1	1
1	0	1
1	1	1

TABLE 48.1-2. NOR Logic

Input		Output
A	*B*	*C*
0	0	1
0	1	0
1	0	0
1	1	0

NOR gate. The equation of a NOR gate can therefore also be written as

$$\bar{A} \cdot \bar{B} = C \qquad (48.1\text{-}2)$$

IC NOR Gate

Because NOR gates are so widely used in computer and other digital circuits, there are ICs available which contain one or more NOR gates. The positive-logic IC 7427, used in this experiment, holds 3 three-input NOR gates, and is therefore called a triple three-input NOR gate. Figure 48.1-2 is the block symbol for this IC showing the inputs to and output from each gate. This IC requires that +5 V direct current be connected between terminals 14 ($+V_{CC}$) and 7 (ground). A binary 1 is equal to +5 V (approximately) on a signal line. A binary 0 is equal to 0 V. See Binary Addition, Exp. 49.1.

It is possible to change a NOR gate into a NOT circuit. Also, NOR gate combinations, and combinations with other logic blocks, can be used to create other *logic* circuits with unique characteristics.

SUMMARY

1. A NOR gate is an OR circuit whose output is inverted. It is a NOT OR gate, or simply a NOR gate.
2. The truth table of a NOR gate is that of an OR gate but with the output inverted (see Table 48.1-2).
3. The logic symbol for a two-input NOR gate is shown in Fig. 48.1-1*b*.
4. NOR gates may have two, three, or more inputs. The IC used in this experiment, the 7427, is a triple three-input NOR gate.
5. Mathematically, the logic equation for a NOR gate is

$$\overline{A + B} = C$$

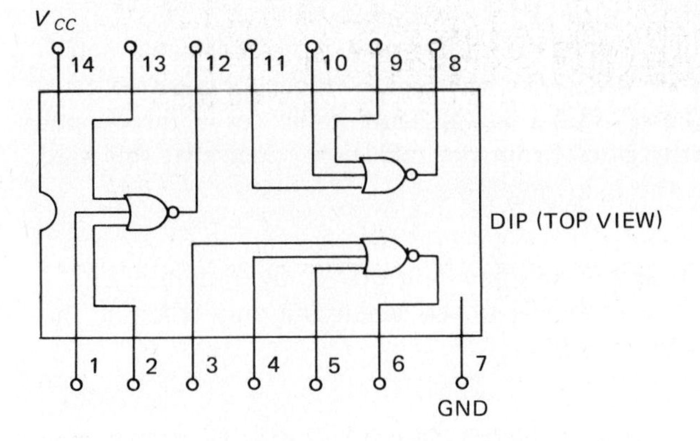

Fig. 48.1-2. Block diagram of IC 7427, triple three-input NOR gate.

SELF-TEST

Check your understanding by answering these questions.

1. If one or more of the inputs of a three-input NOR gate is UP, the output is ________.
2. The output of a NOR gate is UP when and only when all the inputs are ________.
3. The logic expression $\overline{1 + 0}$ represents a(an) ________ gate, one of whose inputs is ________, the other ________.
4. A circuit whose logic is the inverse of OR logic is called a(an) ________ gate.

MATERIALS REQUIRED

- Power supply: Variable, regulated low-voltage dc source
- Equipment: EVM or VOM
- Integrated circuit: 7427 (NOR) or equivalent
- Miscellaneous: Four SPDT, one SPST switch

PROCEDURE

NOTE: Do *not* exceed the +5 V required.

NOR Gate (Three-Input)

1. Connect the circuit of Fig. 48.1-3. Pin 14 receives +5 V, pin 7 is connected to the ground return.

2. Perform the logic and complete the truth table.

NOT Circuit

3. Connect the circuit of Fig. 48.1-4. Perform the logic and complete the truth table.

NOR Gate to OR Gate

4. The circuit in Fig. 48.1-5 converts two NOR gates into an OR gate. Connect the circuit. Perform the logic and complete the truth table. Use an input circuit similar to that in Fig. 48.1-3.

QUESTIONS

1. How is a NOR gate different from an OR gate?
2. Is it possible to convert a three-input NOR gate into a two-input NOR gate? How?
3. Draw the circuit which will satisfy this logic equation:

$$\bar{A} \cdot \bar{B} = C$$

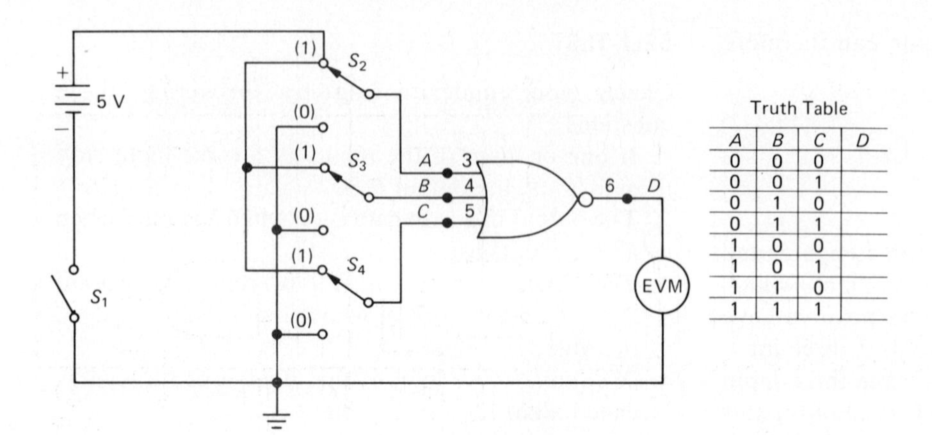

Truth Table

A	B	C	D
0	0	0	
0	0	1	
0	1	0	
0	1	1	
1	0	0	
1	0	1	
1	1	0	
1	1	1	

Fig. 48.1-3. Experimental NOR gate and truth table.

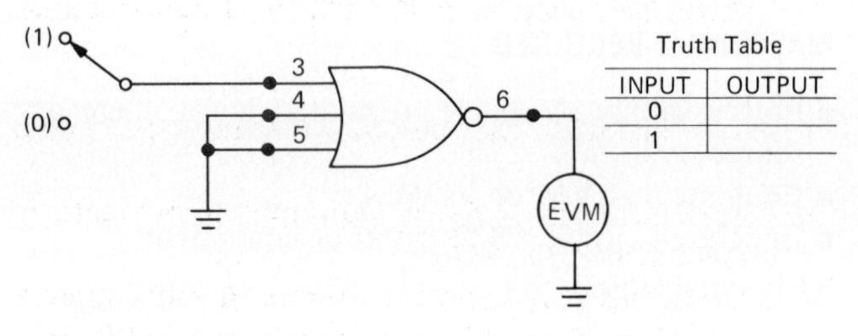

Truth Table

INPUT	OUTPUT
0	
1	

Fig. 48.1-4. NOT circuit from NOR gate, and truth table.

4. Complete the truth table for the logic block in question 3. What kind of gate is it?
5. Is it possible to convert a triple three-input NOR gate into a four-input NOR gate? How?

Answers to Self-Test

1. DOWN
2. DOWN
3. NOR; UP; DOWN
4. NOR

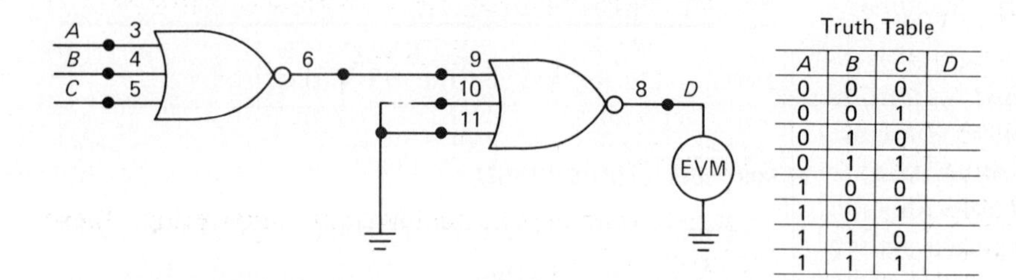

Truth Table

A	B	C	D
0	0	0	
0	0	1	
0	1	0	
0	1	1	
1	0	0	
1	0	1	
1	1	0	
1	1	1	

Fig. 48.1-5. NOR gates connected to form OR gate, and truth table.

EXPERIMENT 48.2. Digital Logic: NAND Gate

OBJECTIVES

1. To determine experimentally the truth table for a NAND gate
2. To use NAND logic to construct other logic blocks

INTRODUCTORY INFORMATION

NAND Gate

If the output of an AND gate is inverted, the result is a NOT AND or simply a NAND gate. Figure 48.2-1*a* shows in block form how a NAND gate is constructed. Figure 48.2-1*b* is the logic symbol for a NAND gate.

Table 48.2-1 is the truth table for a three-input NAND gate. From the table it is clear that this gate has the following logic:

1. If any or all inputs are DOWN, the output is UP.
2. The output is DOWN when and only when all the inputs are UP.

The logic equation for a NAND gate is

$$\overline{A \cdot B} = C \qquad (48.2\text{-}1)$$

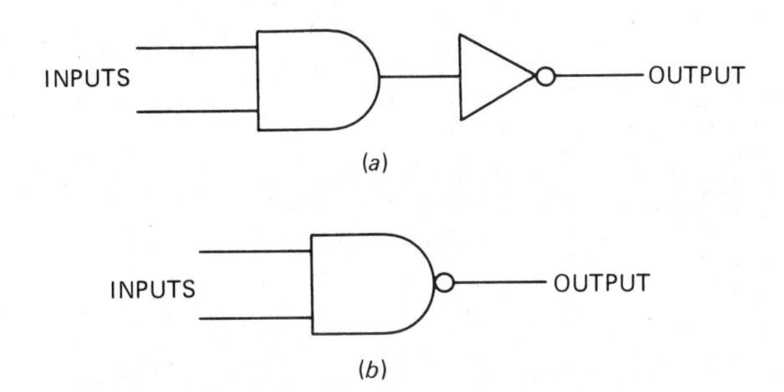

Fig. 48.2-1. (*a*) AND gate and inverter. (*b*) Logic symbol for NAND gate.

It is possible to show that the equation

$$\bar{A} + \bar{B} = C \qquad (48.2\text{-}2)$$

also defines the logic of a NAND gate.

By use of NAND logic, it is possible to construct other logic blocks, such as a NOT circuit, an OR gate, a NOR gate, and an AND gate.

IC NAND Gates

Figure 48.2-2 is the block symbol for the IC 7400, a quad two-input NAND gate. It shows the inputs to and output from each gate. In these positive-logic gates +5 V (approximately) stands for a binary 1, 0 V a binary 0. The IC requires that +5 V be connected between terminals 14 ($+V_{CC}$) and 7 (ground).

SUMMARY

1. A NAND gate is an AND circuit whose output is inverted. It is a NOT AND or simply a NAND gate.
2. The truth table for a NAND gate is that of an AND gate, but with the output inverted (see Table 48.2-1).
3. The logic symbol for a two-input NAND gate is shown in Fig. 48.2-1*b*.
4. NAND gates may have two, three, or more inputs. The IC used in this experiment, the 7400, is a quad two-input NAND gate.
5. The logic equation for a two-input NAND gate is

$$\overline{A \cdot B} = C$$

TABLE 48.2-1. NAND Gate Truth Table

Inputs			*Output*
A	*B*	*C*	*D*
0	0	0	1
0	0	1	1
0	1	0	1
0	1	1	1
1	0	0	1
1	0	1	1
1	1	0	1
1	1	1	0

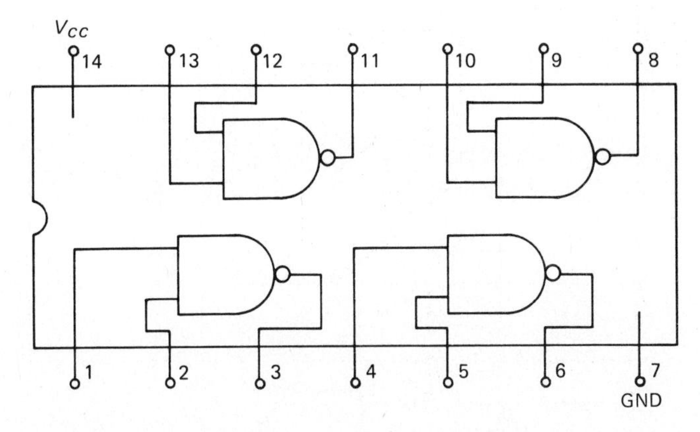

Fig. 48.2-2. Block diagram of IC 7400, quad two-input NAND gate.

SELF-TEST

Check your understanding by answering these questions.

1. If one or more inputs of a three-input NAND gate is DOWN, the output is ________.
2. The output of a NAND gate is DOWN when and only when all its inputs are ________.
3. The logic expression $\overline{1 \cdot 1}$ represents a(n) ________ gate, both of whose inputs are ________.
4. It is possible to construct a NOR gate from a combination of NAND gates. ________ (true/false)

MATERIALS REQUIRED

- Power supply: Variable, regulated low-voltage dc source
- Equipment: EVM or VOM
- Integrated circuit: 7400 (NAND) or equivalent
- Miscellaneous: Two SPDT, one SPST switch

PROCEDURE

CAUTION: Do not exceed the +5 V required.

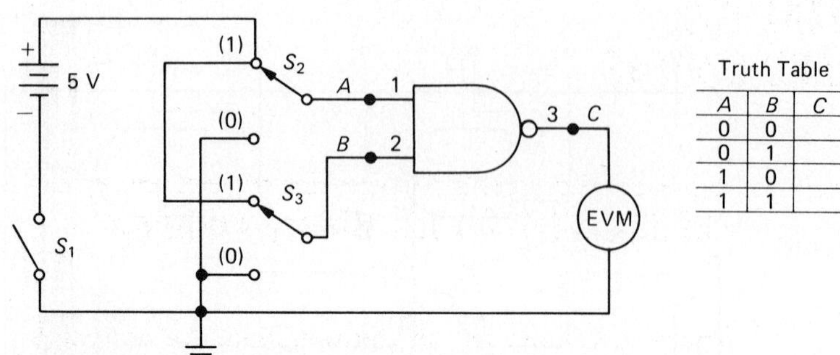

Truth Table

A	B	C
0	0	
0	1	
1	0	
1	1	

Fig. 48.2-3. Experimental NAND gate and truth table.

Truth Table

A	B
0	
1	

Fig. 48.2-4. Logic circuit constructed from NAND gate.

1. Construct the circuit of Fig. 48.2-3. Pin 14 receives +5 V; pin 7 is connected to the ground return.
2. Perform the logic and complete the truth table.
3. Connect the circuit of Fig. 48.2-4. Perform the logic and complete the truth table.
4. Connect the circuit of Fig. 48.2-5. Perform the logic and complete the truth table. Use the same type of input circuit as in Fig. 48.2-3.
5. Connect the circuit of Fig. 48.2-6. Perform the logic and complete the truth table. Use the same type of input circuit as in Fig. 48.2-3.

QUESTIONS

1. How does a NAND gate differ from an AND gate?
2. How would you convert a three-input NAND gate into a two-input NAND gate?
3. Draw the circuit which will satisfy this logic equation: $\bar{A} + \bar{B} = C$.
4. Complete the truth table for the logic block in question 3. What kind of gate is it?
5. Identify by name each of the logic circuits in Figs. 48.2-3 to 48.2-6.
6. How would you convert a quad two-input NAND gate into a NOR gate? Draw the circuit.

Answers to Self-Test

1. UP
2. UP
3. NAND; UP
4. true

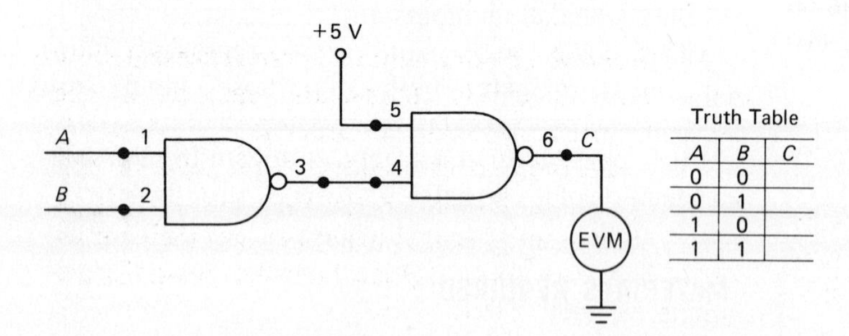

Truth Table

A	B	C
0	0	
0	1	
1	0	
1	1	

Fig. 48.2-5. Combination of NAND gates connected to form logic block.

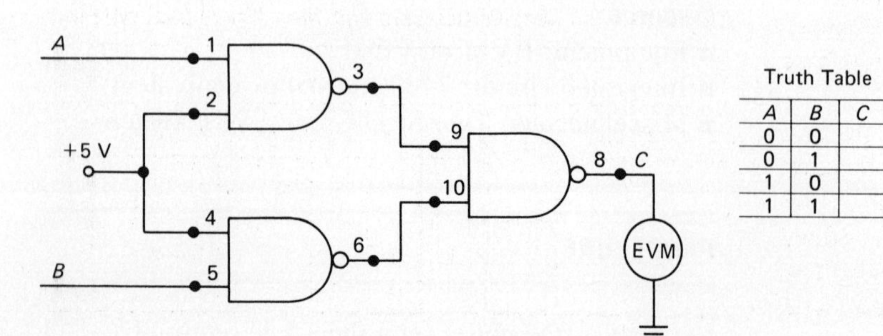

Truth Table

A	B	C
0	0	
0	1	
1	0	
1	1	

Fig. 48.2-6. Combination of NAND gates connected to form another logic block.

CHAPTER 49 DIGITAL ICs: BINARY ADDERS

EXPERIMENT 49.1. Binary Addition

OBJECTIVES

1. To convert a decimal into a binary number and a binary number into a decimal
2. To learn the rules of binary addition

INTRODUCTORY INFORMATION

Two-State Nature of Computer Components

As explained in a previous experiment, computer components generally have two operating states. A hole is either present or absent in a given location in a card; an electromagnetic relay holds its contacts either opened or closed; a piece of magnetic material may be magnetized in one direction to represent information or magnetized in the opposite direction to represent lack of information. Presence or absence of information can also be indicated by a dc level on a signal line.

The binary system of arithmetic uses only two symbols (0 and 1) to represent all quantities. This system is widely used in computers because the 0 and 1 are easily represented by the two-state digital circuits.

A relationship between the binary digits 0 and 1 and computer components may be made. For example, a hole in a card can represent a binary 1; absence of a hole can represent a binary 0. In the case of a magnetic material, magnetic flux in one direction can represent 1; flux in the opposite direction can represent 0.

In transmitting digits from one part of a computer to another, a binary 1 can be represented by a positive level on a line and a binary 0 can be represented by a negative level, and vice versa.

A computer must do more than store digits and transmit them from one place to another; it must calculate by proper manipulation of the digits. In general, the calculating operations can be reduced to adding, subtracting, multiplying, and dividing. The purpose of this experiment is to become familiar with binary addition, as an introduction to computer adding circuits.

Binary Number System

Counting is started in the binary system in the same way as in the decimal system with 0 for zero and 1 for one. But at 2 in the binary system there are no more symbols. Therefore, the same move must be taken at 2 in the binary system that is taken at 10 in the decimal system: it is necessary to place a 1 in the position to the left and start again with a 0 in the original position. Table 49.1-1 is a list of numbers shown in both decimal and binary form:

TABLE 49.1-1. Decimal and Binary Numbers

Decimal	*Binary*	*Decimal*	*Binary*
0	0	6	110
1	1	7	111
2	10	8	1000
3	11	9	1001
4	100	10	1010
5	101	11	1011

The order of a binary number is not represented by units, tens, hundreds, thousands, etc., as in the decimal system. Instead, the order is 1, 2, 4, 8, 16, 32, 64, 128, and so on, reading from right to left, with the position farthest to the right being 1. Table 49.1-2 shows more decimal quantities and their equivalents in binary form. Notice how the positions are numbered right to left.

TABLE 49.1-2. Decimal Numbers and Their Binary Equivalents

	Binary								
Decimal	*256*	*128*	*64*	*32*	*16*	*8*	*4*	*2*	*1*
34				1	0	0	0	1	0
15						1	1	1	1
225		1	1	1	0	0	0	0	1
75			1	0	0	1	0	1	1

To understand how a decimal quantity is changed to a binary quantity, consider the number 75. The first step is to find the largest number in the binary order that can be subtracted from 75. In this case the number is 64, so a binary 1 is placed in the column under 64 in the table. The first step leaves a remainder of 11. The next step is to find the largest number in the binary order that can be subtracted from 11. Because 8 is the largest number meeting the requirement, a 1 is placed under 8 in the table. The remainder is now 3. The largest number in the binary order that can be subtracted is 2, so a 1 is placed under position 2 in the table. Next, 1 is subtracted from the remainder of 1, leaving zero, and a 1 is placed under the binary order 1 in the table. The quantity 75 can therefore be written in binary form as 1001011. This actually represents the quantity $64 + 8 + 2 + 1$.

Binary Addition

Addition of binary quantities is very simple and is based on the following three rules:

$0 + 0 = 0$
$0 + 1 = 1$
$1 + 1 = 0$ with a 1 carry to the left

Table 49.1-3 is an example of binary addition using the rules stated. The factors to be added are 75 and 225. Refer to Table 49.1-3 for their binary equivalents and for the procedure which follows. Starting at the right, we have $1 + 1 = 0$ with a carry (rule 3). The next position to the left is added: $0 + 1 = 1$. However, when we add the 1 carry, the sum becomes 0 with 1 carried to the third position. The third position consists of $0 + 0 = 0 + 1$ (carry) $= 1$. This procedure is followed until all positions are added. The sum is given in binary form as 100101100, which is equal to $256 + 32 + 8 + 4 = 300$. This sum is exactly what we would get by adding the decimal quantities 225 and 75.

Binary quantities can also be subtracted, multiplied, and divided, using rules similar to those for addition.

TABLE 49.1-3. Adding Binary Numbers

	Binary value								
Decimal Value	*256*	*128*	*64*	*32*	*16*	*8*	*4*	*2*	*1*
	1	1					1	1	
225	0	1	1	1	0	0	0	0	1
75	0	0	1	0	0	1	0	1	1
Sum	1	0	0	1	0	1	1	0	0

SUMMARY

1. The binary number system, employed in digital computers, uses only two symbols, 1 and 0. These have the same meaning as 1 and 0 in the decimal number system with which you are so familiar.
2. In the decimal or base 10 system the value of each of the digits in a number is some power of 10 and depends on its position in the number. For example, in the number 527, the 7 is in the units (10^0) column and counts for 7, the 2 is in the tens column and counts for 2×10, or 20. The 5 is in the hundreds column and counts for 5×10^2, or 500.
3. Numbers in the binary system are formed as they are in the decimal, except that the value of a column is a power of 2 rather than of 10. The extreme righthand column has the value 2^0 or 1; the next column on the left has the value 2^1 or 2; the next is 2^2 or 4; the next is 2^3 or 8, etc. The values of the first seven binary columns, reading from right to left, are:

etc.	64	32	16	8	4	2	1
			1	0	1	1	0

4. To change a binary number into its base 10 (decimal) equivalent, place each of the binary digits in its proper binary column, just as the number 10110 is placed above. Now multiply each binary digit (1 or 0) by its value in the column and add the result. Thus the binary number $10110 = 1 \times 16 + 0 \times 8 + 1 \times 4 + 1 \times 2 + 0 \times 1 = 22$, the base 10 equivalent.
5. Addition of binary numbers is based on the following rules:

   $0 + 0 = 0$
   $1 + 0 = 1$
   $1 + 1 = 0$ with a 1 carry to the left

6. The reason digital computers use the binary number system is that digital circuits are two-state circuits; they are either ON (1) or OFF (0).

SELF-TEST

Check your understanding by answering these questions.

1. Binary numbers use how many symbols? ________
2. A number written in decimal form may be written in one and only one binary form. ________ (true/false)
3. The decimal value of each digit in the binary number 10111, reading from left to right, is: ________ ________ ________ ________ ________.

4. The decimal value of the number in question 3 is ________.
5. The sum of the binary numbers 111101 and 101101, in binary form, is ________.
6. The decimal value of the binary sum in question 5 is ________.

MATERIALS REQUIRED

- Miscellaneous: Pen or pencil and paper

PROCEDURE

1. Change the decimal number 239 into binary form and enter the result in Table 49.1-4. Show your computations.
2. Change the decimal number 60 into binary form and enter the result in Table 49.1-4.
3. Add the two binary numbers in steps 1 and 2 and enter the sum in binary form in Table 49.1-4.
4. Change the binary number in step 3 into decimal form and enter the result in Table 49.1-4. Show your computations.

TABLE 49.1-4. Adding Binary Numbers

	Decimal Number	*Binary Equivalent*
	239 60	
Sum		

QUESTIONS

1. What are the digits in the binary number system?
2. For the digits in question 1, what is their equivalent value in the decimal number system?
3. Why was the binary number system chosen for computer arithmetic?

Answers to Self-Test

1. Two
2. true
3. 16, 8, 4, 2, 1
4. 23
5. 1101010
6. 106

EXPERIMENT 49.2. Binary Full Adder

OBJECTIVE

To construct a binary full adder and experimentally determine its truth table

INTRODUCTORY INFORMATION

Full Adder

When a carry and the two quantities to be added are considered as inputs, there are eight input combinations, as shown in Table 49.2-1. An adder capable of producing the required outputs for the eight input combinations is called a *full adder*. The full adder is shown in the block diagram of Fig. 49.2-1.

Using the rules of AND-OR logic, it can be shown that any input combination to the full adder of Fig. 49.2-1 produces an output according to the rules of binary addition.

Consider the condition where a binary 1 is present on each of the three inputs *A*, *B*, and *C*. From the truth table we find that the sum and carry output

TABLE 49.2-1. Truth Table for a Full Adder

Inputs			*Outputs*	
A	*B*	*C*	*Sum*	*Carry*
0	0	0	0	0
1	0	0	1	0
0	1	0	1	0
0	0	1	1	0
1	1	0	0	1
1	0	1	0	1
0	1	1	0	1
1	1	1	1	1

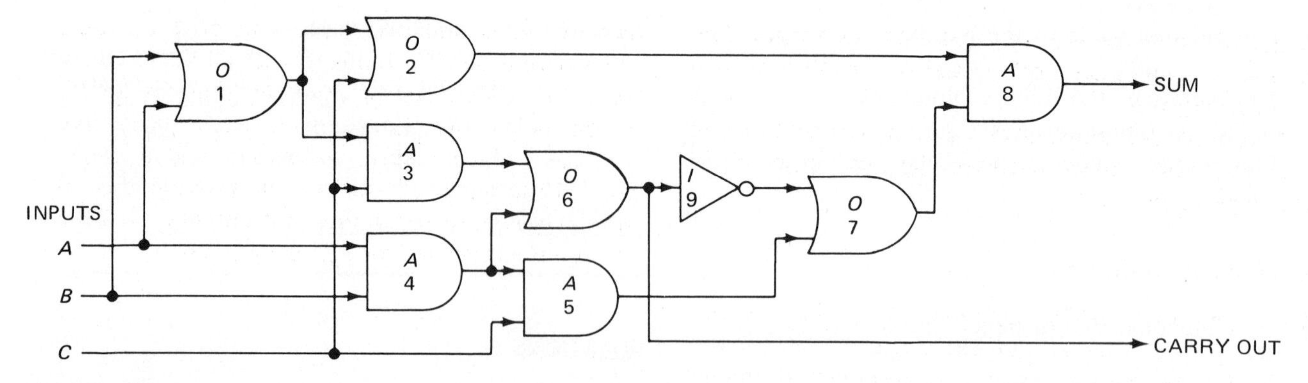

Fig. 49.2-1. Binary full adder.

lines should each contain a binary 1. With the input conditions stated, AND block 4 has an UP level on its output. This UP level is combined with the UP level on the input carry line in AND block 5 to produce an UP level on the lower input to OR block 7. The output of OR block 7 is now UP, producing an UP level on the lower leg of AND block 8.

Going back to the input, we find that OR block 1 has an UP level on its output, which in turn produces an UP level on the output of OR block 2. The output of OR block 2 and OR block 7 (both outputs being UP at this time) produces an UP level (binary 1) on the sum line at the output of AND block 8. At this point, one of the conditions in the truth table has been satisfied.

We must now obtain an UP level on the output carry line. Because the output of AND block 4 is UP at this time, the output of OR block 6 also has an UP level, which also appears on the output carry line. By the same analysis, the indicated output can be obtained for any of the other input combinations shown in the truth table.

The full adder shown represents a single position in a binary-adder system. Because many such adders are combined in a large computer, each full adder is represented as a block in the computer logic diagram. An example of a five-position binary adder is shown in Fig. 49.2-2. The actual number of positions in such an adder depends on the size of the computer and the type of calculations the computer is designed for.

Experimental Full Adder

The AND and OR gates of the binary full adder may consist of individual gates in ICs, such as those used in previous experiments, or they may be designed into a single IC full-adder block consisting of three inputs and two outputs—the sum and carry lines. If individual gates are used, 4 two-input AND gates, 4 two-input OR gates, and an inverter will be needed to build a full adder.

In this experiment the OR gates will come from a quad two-input OR gate IC (7432). We shall convert the triple three-input AND gate IC (7411) into 3 two-input AND gates. The fourth AND gate will be made from two gates in the quad two-input NAND gate IC (7400). A third NAND gate will be modified to act as an inverter.

Figure 49.2-3 shows how to connect 2 two-input NAND gates in the IC 7400 to form a two-input AND gate. NAND gate 2 is connected as an inverter, which NOTs the logic of NAND gate 1. The result is NOT NAND

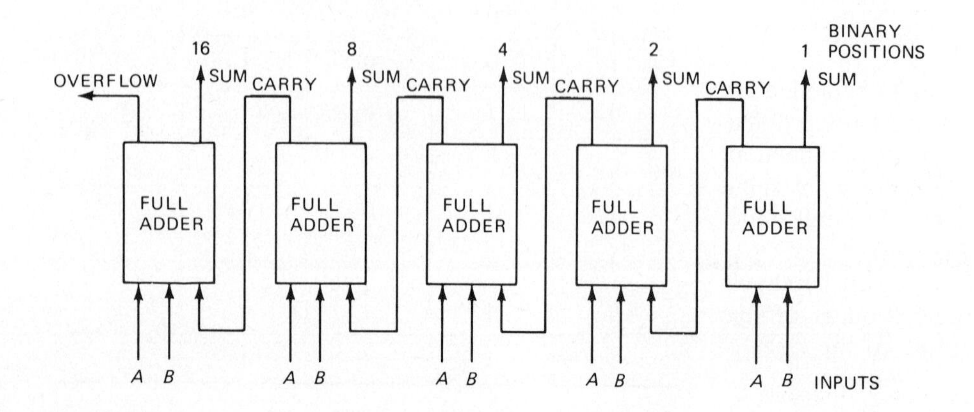

Fig. 49.2-2. Block diagram of a five-position binary adder.

logic, which is the same as NOT NOT AND logic. The two NOTs cancel each other, and the result is a two-input AND gate.

Figure 49.2-4 shows NAND gate 3 connected as an inverter. This will be used for the inverter block in the experimental full adder.

SUMMARY

1. A binary full adder combines *three* binary digits and provides an output and a carry. One of the inputs may be a carry from a previous arithmetic operation.
2. A binary full adder obeys all the laws of binary addition.
3. A full adder has eight possible input combinations.
4. A full adder may be constructed of a combination of four AND blocks, four OR blocks, and an inverter.
5. Though a full adder may be made from IC AND and OR gates and inverters, individual IC logic blocks are available complete with all the circuitry for a full adder. As an example, there is an IC which holds *two* full adders, each using five terminals—three inputs, an output, and a carry. So a 12-terminal IC block can provide the 10 terminals needed for inputs and outputs, plus the two terminals needed for V_{CC} and ground.

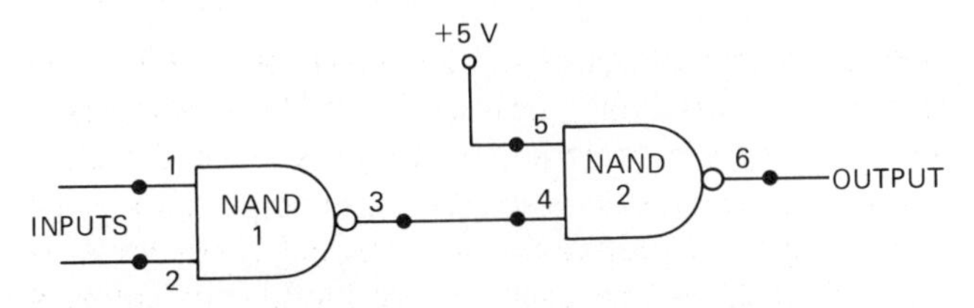

Fig. 49.2-3. Connecting 2 two-input NAND gates to operate as a two-input AND gate, using a 7400 IC.

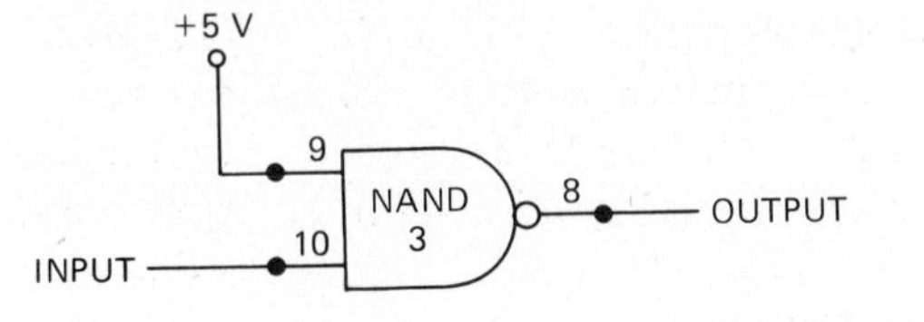

Fig. 49.2-4. NAND gate 3 in 7400 IC connected as an inverter.

SELF-TEST

Check your understanding by answering these questions.

1. In the full adder of Fig. 49.2-1 inputs A and B are 1, and C is 0. The sum line will be ________ (UP/DOWN); the carry line will be ________ (UP/DOWN).
2. How many other combinations will give the same result as in question 1? ________. These combinations are ________.
3. How many binary numbers can the adder in Fig. 49.2-2 combine? ________
4. The number of digits in each number that the adder in Fig. 49.2-2 can handle is ________.
5. If the inputs to the circuit in Fig. 49.2-3 are each 1, the output is ________.
6. If the input to the circuit in Fig. 49.2-4 is 1, the output is ________. If the input is 0, the output is ________. This circuit is a(n) ________.

MATERIALS REQUIRED

- Power supply: Variable, regulated low-voltage dc source
- Equipment: EVM or VOM

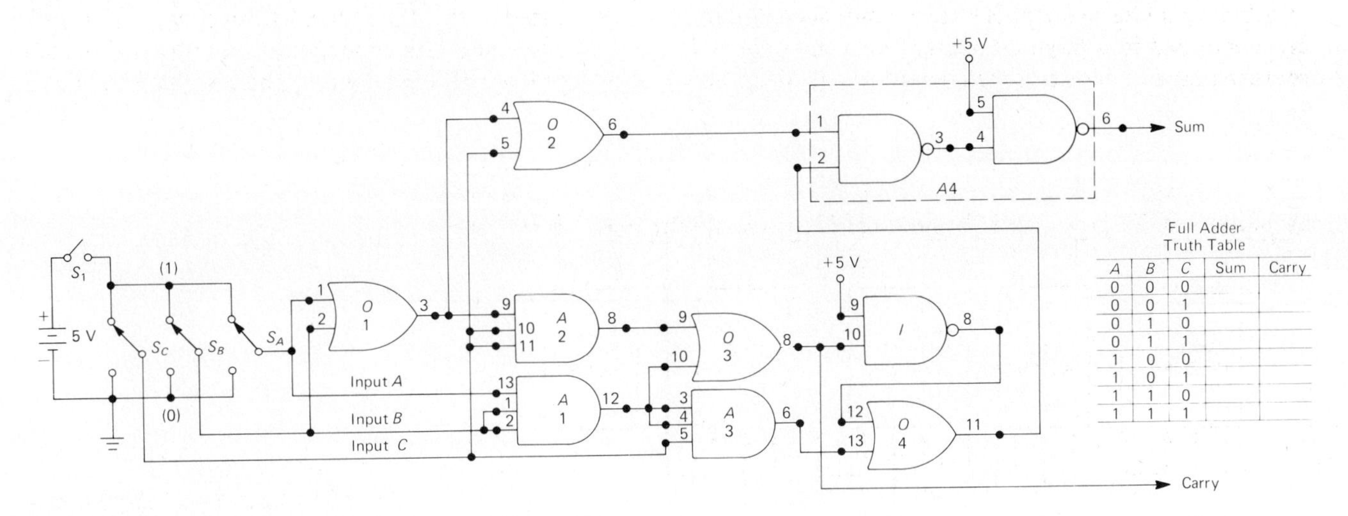

A	B	C	Sum	Carry
0	0	0		
0	0	1		
0	1	0		
0	1	1		
1	0	0		
1	0	1		
1	1	0		
1	1	1		

Fig. 49.2-5. Full-adder experimental circuit and truth table.

- Integrated circuits: 7432 (OR), 7411 (AND), 7400 (NAND), or the equivalents
- Miscellaneous: Three SPDT switches; SPST switch

PROCEDURE

1. Connect the full-adder circuit shown in Fig. 49.2-5. The four OR gates are contained in the 7432. AND gates 1, 2, and 3 are in the 7411, modified to two-input gates, as shown. AND gate 4 and the inverter are constructed from the NAND gates in the 7400. Connect pin 14 of each IC to +5 V and pin 7 to ground.
2. Perform the logic and enter the results in the truth table.

QUESTIONS

1. How does a full adder differ from a half adder?
2. In the circuit of Fig. 49.2-5, if block 3 were an AND block, what would be the sum and carry for an input $A = 1$, $B = 1$, $C = 0$? Would this circuit obey the rules of binary addition?

Answers to Self-Test

1. DOWN; UP
2. two; ($A = 1$, $B = 0$, $C = 1$); ($A = 0$, $B = 1$, $C = 1$)
3. Two
4. 5
5. 1
6. 0; 1; inverter

CHAPTER 50 LIGHT-EMITTING DIODE (LED)

EXPERIMENT 50.1. Light-Emitting Diode (LED)

OBJECTIVE

To determine experimentally the characteristics of an LED

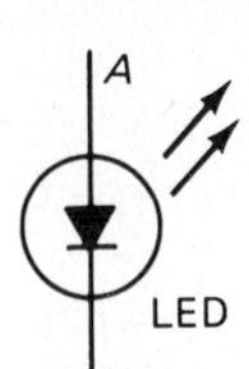

Fig. 50.1-1. LED symbol.

INTRODUCTORY INFORMATION

Light-Emitting Diodes (LEDs)

LEDs are specially constructed semiconductor diodes which emit light when they are properly forward-biased and conducting current. They can be used to replace filamentary-type pilot lights. Their *two-terminal* construction eliminates the need for special lamp sockets, since they can be wired directly into a circuit. Because they are solid-state devices, LEDs are more rugged than filamentary-type bulbs.

LEDs have many other applications; for example, they serve as the readout in alphanumeric displays used in digital computers and calculators. In this application they are rapidly replacing the gas-filled NIXIE tubes. LEDs are also being used as the number readouts in calculators. Together with light-sensing devices, LEDs serve as elements of other optoelectronic systems. These and other applications make the LED a most important electronic device.

LED Operation

Recombinations of holes and electrons occur in every semiconductor diode. Heat energy is released as a result of these recombinations when silicon or germanium serves as the semiconductor material. However, with certain other semiconductors, such as gallium arsenide, *light* energy is released when hole and electron recombinations occur at the semiconductor junction. The color of the emitted light depends on the material from which the LED is constructed and the concentration of the dopants or impurities in it. Gallium phosphide can be used to produce either green or red light in an LED. Fig. 50.1-1 is the symbol for an LED.

LEDs are characterized by the amount of current they must pass to emit light and by the amount of power they can dissipate. They must not be operated beyond their rated specifications. The LED used in this experiment operates well when it is passing about 20 mA of current and when the voltage across its terminals does not exceed 1.6 V.

In this experiment the LED will be connected as a logic "status" indicator, that is, as a device which will show whether the output of a logic circuit is a binary 1 or a binary 0. For this purpose the LED will be connected as a load in the collector circuit of a driver transistor, as in Fig. 50.1-2. A resistor R in series with the LED will limit its current to the desired value. The output of a positive logic NOR gate, connected as an inverter, will serve as the input to the

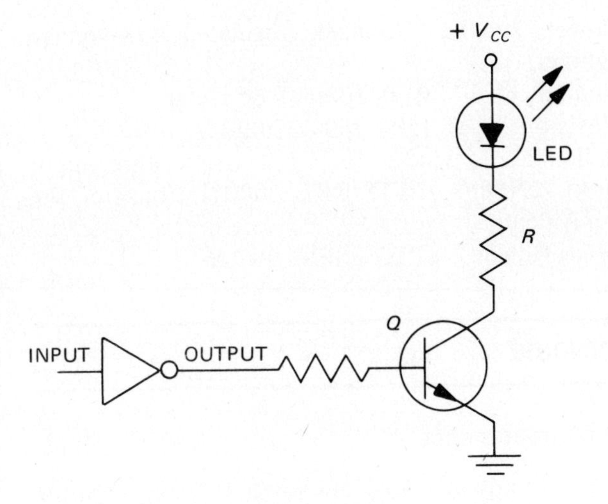

Fig. 50.1-2. LED connected as a logic status indicator.

driver. When the inverter holds a binary 1 in its output, that is, when its output is UP (positive), the driver transistor is forward-biased and the LED will light. When the output of the inverter is a binary 0 (DOWN), the driver transistor will not conduct, and the LED will remain dark.

SUMMARY

1. LEDs are semiconductor diodes which emit light when they are forward-biased and are passing current.
2. Because of their low power consumption they are replacing pilot lights as indicators in electronic circuits.
3. The LED in this experiment shines brightly when it is passing 20 mA of current and there is 1.6 V across it. It should not be operated beyond these values.
4. LEDs can be used as logic status indicators, that is, as devices which show whether a logic circuit has a binary 1 or 0 in its output.

SELF-TEST

Check your understanding by answering these questions.

1. An LED must be ________-biased in order to emit light.
2. LEDs should not be operated beyond the manufacturer's ratings. ________ (true/false)
3. When the LED is reversed in the circuit of Fig. 50.1-2, it ________ (will/will not) permit the transistor to pass collector current when the base-to-emitter junction of the transistor is forward-biased.

MATERIALS REQUIRED

- Power supply: Variable, regulated low-voltage source
- Equipment: EVM; 0–100-mA dc meter
- Resistors: ½-W 150-, 560-, 27,000-Ω
- Solid-state devices: IC 7427 (triple three-input NOR), 2N6004; LED #RL 2000 Litronix or the equivalents
- Miscellaneous: SPDT, SPST switch

PROCEDURE

LED Characteristics

NOTE: The longer lead of the RL2000 LED is the anode, the shorter lead, the cathode. An orange dot on the housing next to the short lead will also identify the cathode.

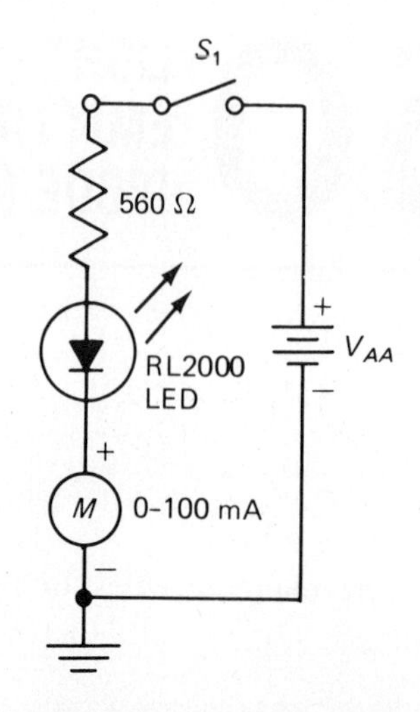

Fig. 50.1-3. Experimental circuit to determine LED characteristics.

CAUTION: Do *not* exceed 20 mA of LED current.

1. Connect the circuit of Fig. 50.1-3. Switch S_1 is open. Set the output of the V_{AA} at 0 V.
2. Close S_1. **Power on.** Gradually increase the output of the power supply, while observing the LED, until the LED *just glows*. Measure and record in Table 50.1-1 the current in the LED.
3. Adjust the output of the power supply until there is 10 mA current in the LED. Indicate in Table 50.1-1 whether the light from the LED is brighter or less bright than in step 2.
4. Repeat step 3 for LED current of 20 mA.
5. Open S_1. Reverse the leads of the LED in the circuit of Fig. 50.1-3. Set the output of the power supply at 10 V.
6. Close S_1. Measure and record in Table 50.1-1 the current, if any, in the LED.

LED as Logic Status Indicator

7. Connect the circuit of Fig. 50.1-4. One of the NOR gates in the 7427 is connected as an inverter.

TABLE 50.1-1. LED Characteristics

Step	*Intensity of LED*	I_{LED}, mA
2	Just glows	
3		10
4		20
6		

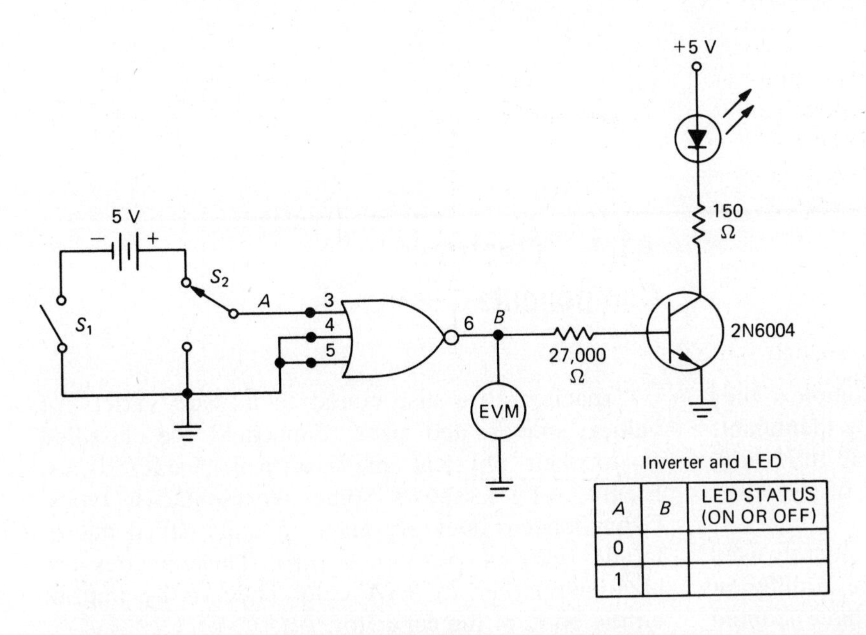

Inverter and LED

A	*B*	LED STATUS (ON OR OFF)
0		
1		

Fig. 50.1-4. Experimental circuit using LED as a logic status indicator, and truth table.

Terminal 14 receives +5 V. Terminal 7 is ground.

8. Perform the logic and record your results in the truth table. Indicate when the LED lights (is ON).

QUESTIONS

1. How does an LED differ from an ordinary semiconductor diode?
2. How does the amount of current an LED passes affect the intensity of light it emits?
3. What is meant by a *logic status indicator?*

Answers to Self-Test

1. forward
2. true
3. will not

APPENDIX A

A1.1. Electronic Components

Though electronic devices are fairly complex, they are made up of a small number of easily identifiable components. The technician must be able to identify these components readily. Photographs of a number of these parts are shown.

Resistors are the components most frequently used in electronic circuits. They are available in different values, shapes, and sizes. Manufacturers have adopted a standard Electronic Industries Association (EIA) color-coding system for determining resistance or ohmic values of low-power resistors. You will learn this coding system as you progress. The high-power resistors usually have the resistance value imprinted on their bodies.

There are fixed-value resistors, such as those shown in Fig. A1.1-1, and variable resistors, such as those shown in Fig. A1.1-2.

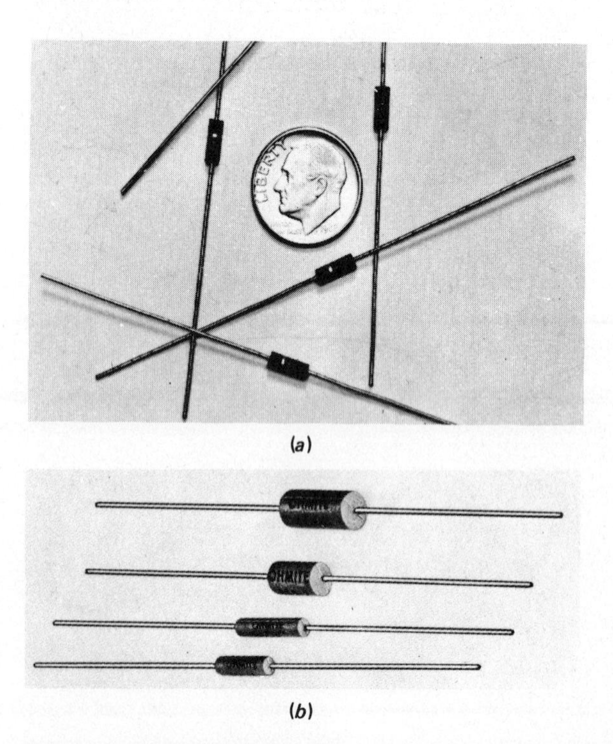

Fig. A1.1-1. Fixed-value resistors. (*a*) Quarter-watt resistors (*Speer Carbon Co.*); (*b*) molded silicon-ceramic wire-wound resistors. (*Ohmite Mfg. Co.*)

Capacitors are also found in a wide variety of values, shapes, and sizes. Capacitors are classified as to their physical and electrical characteristics. Figure A1.1-3 shows some representative types. Capacitance values are given in units called microfarads (μF) or picofarads (pF). These values are identified either by EIA color code or by imprint on the body of the capacitor.

Inductors and transformers make up another classification of components. Wire-wound coils and

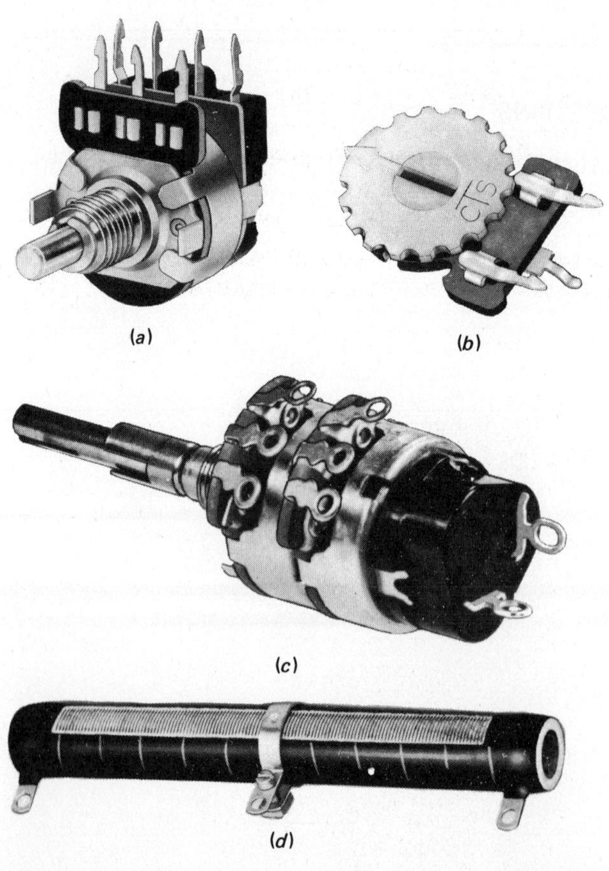

Fig. A1.1-2. Variable resistors. (*a*) Miniature composition variable resistor (*Chicago Telephone Supply Corp.*); (*b*) knob-operated composition trimmer (*Chicago Telephone Supply Corp.*); (*c*) dual-section composition variable resistor with switch (*Chicago Telephone Supply Corp.*); (*d*) vitreous enamel adjustable resistor. (*Ohmite Mfg. Co.*)

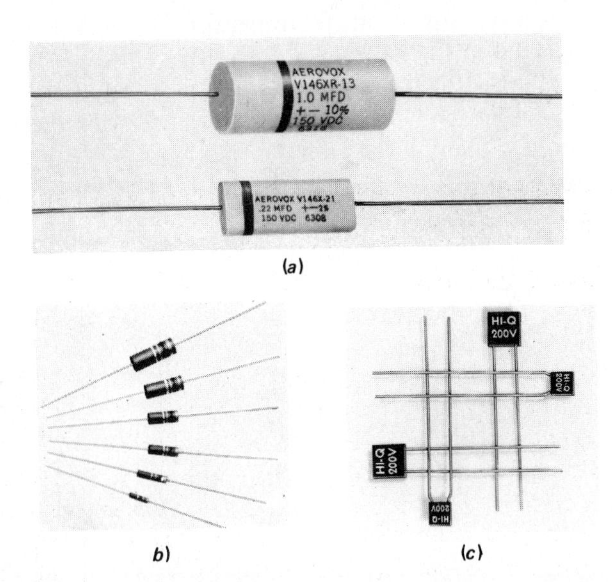

Fig. A1.1-3. Fixed-value capacitors. (*a*) Mylar capacitors; (*b*) molded cerafil capacitors; (*c*) molded ceramic capacitors for transistor circuits. (*Aerovox Corp.*)

chokes are classified as inductors. These are wound on different forms and cores. There are air-core coils, iron-core, powdered iron-core, etc. (Fig. A1.1-4). Transformers are coupled inductors. They serve many purposes. There are power transformers, filament transformers, radio-frequency (rf) transformers (Fig. A1.1-5*a*), television horizontal-sweep transformers (Fig. A1.1-5*b*), and many others.

Almost everyone familiar with tubes knows they

Fig. A1.1-4. Typical oscillator coil for broadcast band. (*F. W. Sickles, Div. of General Instrument Corp.*)

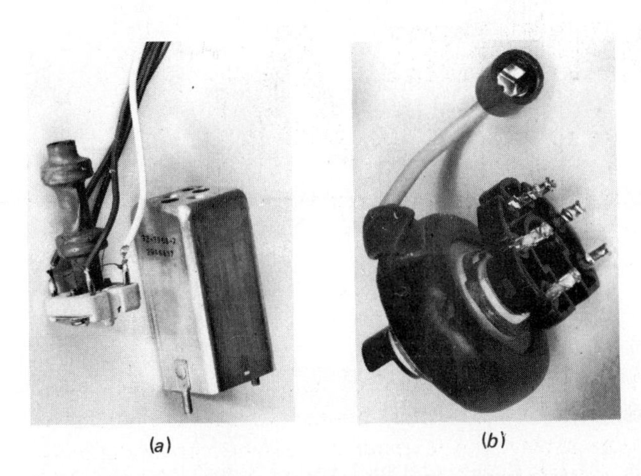

Fig. A1.1-5. Transformers: (*a*) Typical rf transformer (*F. W. Sickles, Div. of General Instrument Corp.*); (*b*) horizontal-sweep transformer. (*General Electric Company*)

were once the sinews of the electronics industry, but in new equipment they have been mainly replaced by transistors. Now, the transistor is rapidly being supplemented by the integrated circuit (IC). Tubes vary in size from tremendously large transmitter tubes to subminiature hearing-aid tubes. Tubes are described by their physical size, base arrangements, and electrical characteristics. Figure A1.1-6 shows a 9-pin miniature tube and a nuvistor.

Transistors (Fig. A1.1-7) are in the category of semiconductors. Despite their tiny size, they do a giant's job. Transistors have made possible present-day large-scale computers, guidance systems for space vehicles, space communications systems, etc. Now the IC is not only replacing the transistor but also placing entire circuits within the space normally

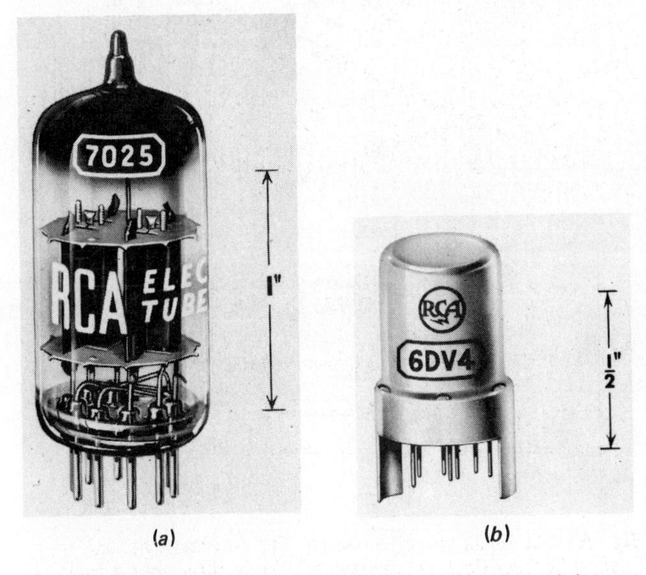

Fig. A1.1-6. Representative types of tubes: (*a*) 9-pin miniature; (*b*) nuvistor. (*RCA*)

necessary for a single transistor. Typical ICs are pictured in Fig. A1.1-8.

Many other components are also used in electronics. You will become acquainted with these as you use them in the course of your work.

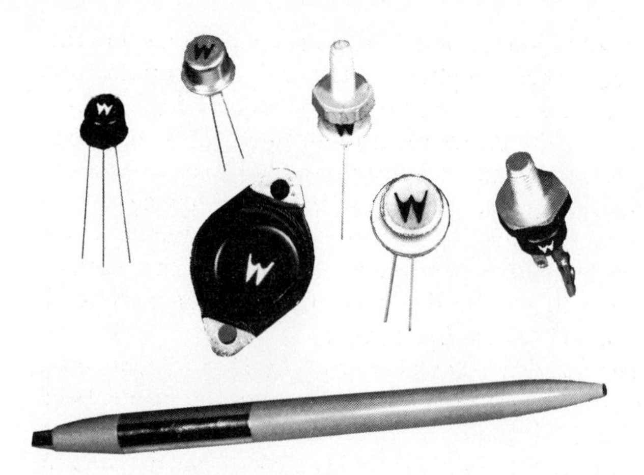

Fig. A1.1-7. Representative types of transistors. (*Motorola Semi-conductor Products, Inc.*)

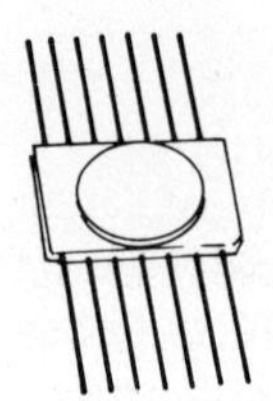

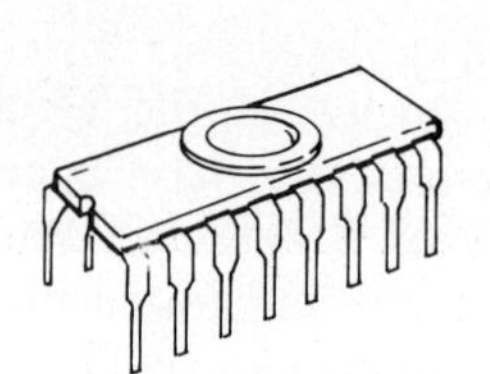

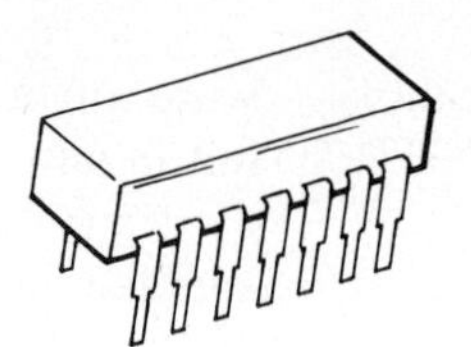

Fig. A1.1-8. Representative types of integrated circuits (*RCA*)

A1.2. Electronic Circuit Symbols

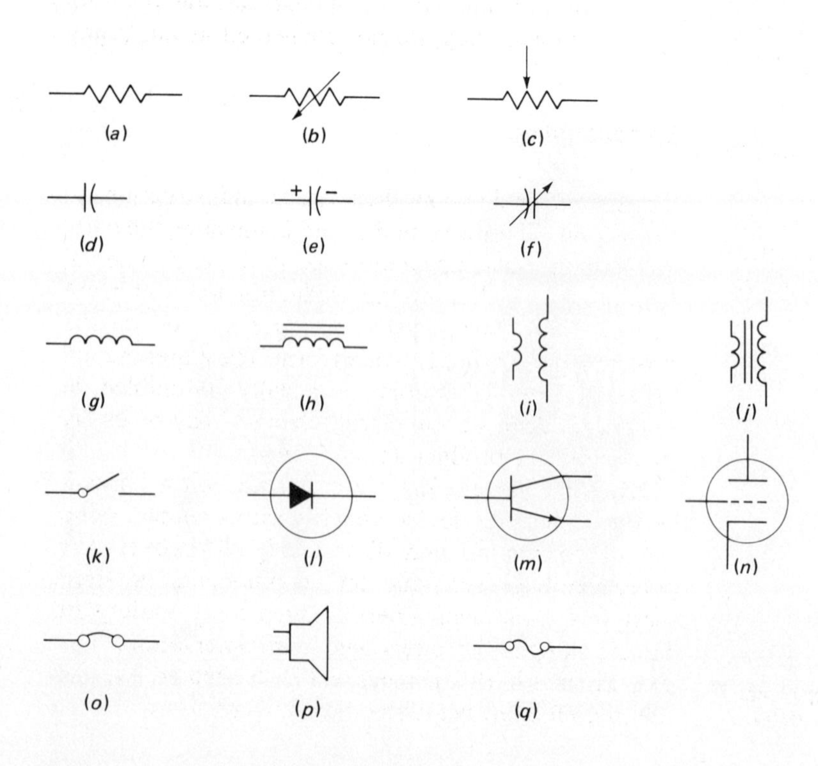

Fig. A1.2-1. Standard electronic component circuit symbols: (*a*) fixed resistor; (*b*) variable resistor; (*c*) potentiometer (resistor with adjustable tap); (*d*) capacitor (fixed-value); (*e*) polarized (electrolytic) capacitor; (*f*) variable capacitor; (*g*) inductor (air core); (*h*) inductor (iron core); (*i*) transformer (air core); (*j*) transformer (iron core); (*k*) switch; (*l*) diode (solid state or semiconductor); (*m*) transistor; (*n*) vacuum tube (triode); (*o*) earphones; (*p*) speaker (loudspeaker); (*q*) fuse.

In Appendix 1.1 you became acquainted with some of the more familiar electronic components. Here you will become acquainted with their circuit symbols.

Some standard electronic component symbols are shown in Fig. A1.2-1. They are for components which are used in electronic devices. The manner in which these components are connected in a device is shown in a drawing called a *schematic diagram.* In this diagram each component is represented by its symbol. In the next section you will learn how to interpret circuit or schematic diagrams.

Each symbol has two parts. One part actually identifies the component which the symbol represents. This part of the symbol may look like the drawings in Fig. A1.2-2.

The second part of each schematic symbol is the representation of the wires or connections on the device. These wires (leads) and connections are shown as an added part of the symbols in Fig. A1.2-3. Compare the drawings in these two figures. By looking closely at the schematic symbol, one can determine the number of leads on the actual device.

Most schematic symbols have been standardized throughout the electronics industry in the United States and will be the same regardless of the circuit diagram in which they appear.

The service technician must be able to recognize component symbols because they are the elements of a circuit diagram. The circuit diagram is used to show how the components in an electronic device are interconnected.

Fig. A1.2-2. Component identification symbols without leads.

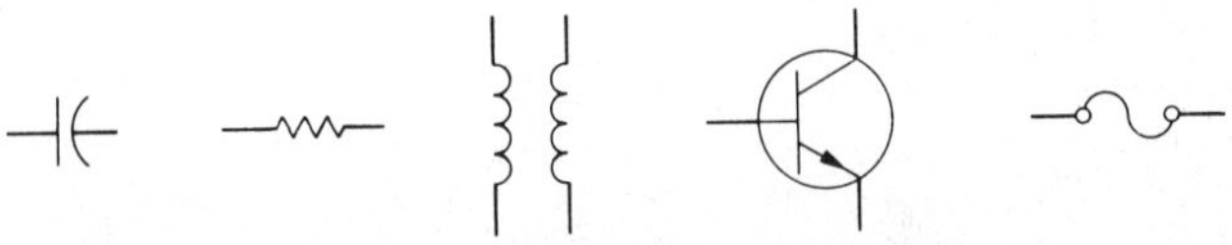

Fig. A1.2-3. Component identification symbols showing leads.

A2.1. The Schematic Diagram

The Schematic Diagram

The schematic diagram is an electronic blueprint. It tells the technician where and how the electrical parts are connected in a specific circuit, such as a radio circuit. There are a few simple rules to learn:

1. Parts are shown by their electronic symbols.
2. The leads which extend from a part are shown by straight lines.
3. (*a*) If two or more parts are joined physically at a common point (terminal), this point is indicated by a heavy dot on the schematic. For three resistors which have a lead of one connected to the leads of the others, we draw a simple schematic (Fig. A2.1-1). We can, if we wish, identify the junction point by some letter, such as *A* (Fig. A2.1-2).

   (*b*) If two lines in a schematic diagram cross one another and no dot appears at the point of crossing, they are not connected at this point.

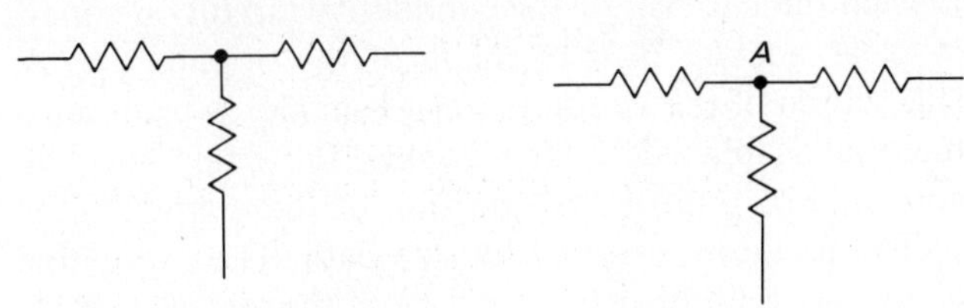

Fig. A2.1-1. Three resistors connected at a common point.

Fig. A2.1-2. Common point identified by the letter *A*.

Breadboarding

The schematic diagram, then, is a set of instructions to the technician for assembling and connecting the parts of a circuit. Schematics are used in the experimental laboratory as well as on the production line and in the service shop. The physical product created on the production line is in permanent form. The experimental circuit in the laboratory is usually assembled in temporary form so that circuit changes may be easily made until the product design is final.

Experimental circuits are assembled on a "breadboard." This is a device which ensures simple, rapid circuit connection and disassembly. Moreover, the breadboarding technique lets components be used over and over again, permitting a wide variety of circuit designs for study and experimentation. The experiments in this manual will be facilitated by, and are intended for, breadboarding devices.

A2.2. Circuit Tracing in Hand-Wired Circuits

Components in electronic circuits may be connected by handwiring or printed techniques. In this manual we shall work with hand-wired circuits only. Appendix A2.3 will be concerned with printed circuits.

In the handwiring process, component leads are connected to terminal posts and strips (Fig. A2.2-1). The pigtails (leads) of parts which must be electrically connected are secured at a common terminal point. If the parts are too far apart, their leads may not be long enough to reach a common point. In that case, the pigtails of these components (which must be joined electrically) are secured to *separate* terminal lugs. A wire is then used to connect these terminal lugs (Fig. A2.2-2). This is the handwiring technique for interconnecting components in a chassis or on a breadboard.

Most older equipment and even some more modern equipment use hand-wired circuits. It is, therefore, necessary for the technician to be able to identify component placement and common points of connection of components, in order to trace the electrical or signal flow in the equipment being repaired.

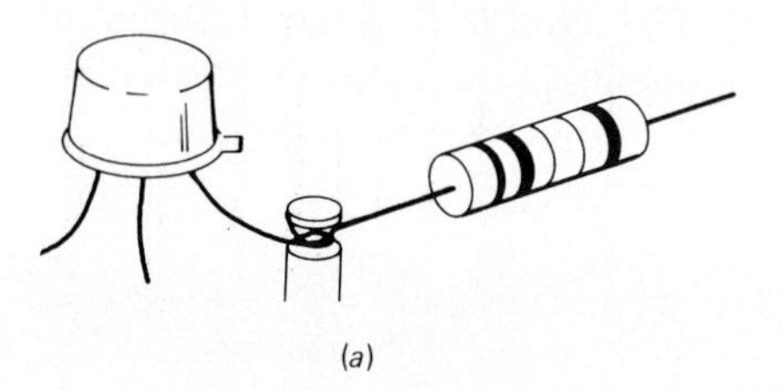

(*a*)

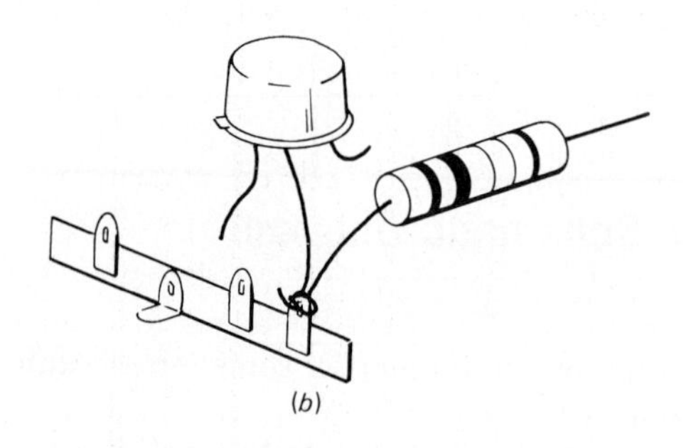

(*b*)

Fig. A2.2-1. (*a*) Components connected to a terminal post. (*b*) Components connected to a lug on a terminal strip. The component leads are fed through the lug hole. The leads are *not* wrapped around the lug, to provide ease of component replacement.

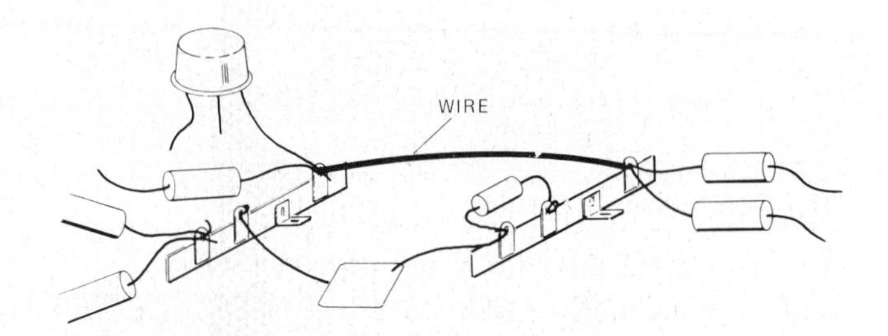

Fig. A2.2-2. A wire is used to connect two terminal lugs electrically.

A2.3. Circuit-Tracing Printed-Wired Circuits

An advanced means of interconnecting parts in mass-produced electronic devices uses etched (commonly called "printed") wiring connections. This method requires the use of a plastic, ceramic, glass, or phenolic board on which the components are mounted. Interconnections are made by means of conductive copper surfaces which remain, after a chemical bath has removed the unwanted copper.

One technique for making etched boards is as follows: Start with a piece of copper-clad plastic, phenolic, glass, etc. The copper-clad board may be coated on either one or both sides. For most uses only one side of the board is coated with copper. Design the pattern for the conductive circuit on paper. A clear acetate is then made of the original art. A silk screen of the circuit pattern is made. A protective material is placed on the copper foil using the silk screen. The unprotected portion of the copper foil is etched away in a ferric chloride bath. The protective coating is then also removed from the circuit pattern remaining on the board.

By another silk-screen process, solder resist is

applied to the pattern except in the areas where solder is to be applied. In some instances, silk-screen processes apply a "road map" to the top and bottom of the board. The holes are punched into the boards. This requires tooling, which is more accurate than drilling. A precise hole size which will produce a good solder fillet around the lead is mandatory.

Figure A2.3-1 shows an etched board ready for components. The components are mounted on the board by inserting their leads through the prepunched holes. The leads are then soldered to the conductive surfaces which "form" the pattern of interconnecting "printed" wiring between components. Figure A2.3-2 is a top view of a finished etched board. The circular blobs are points at which component leads have been soldered. The components are on the other side of the board (Fig. A2.3-3).

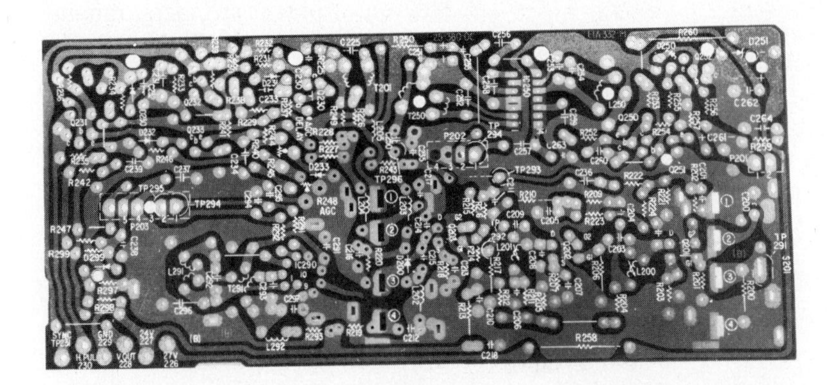

Fig. A2.3-1. Etched board ready for component insertion.

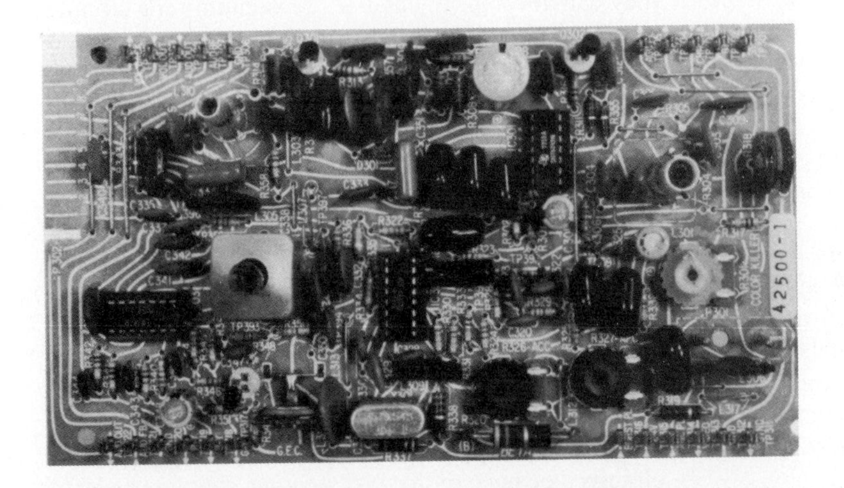

Fig. A2.3-2. Top view of a finished printed-circuit board (PCB).

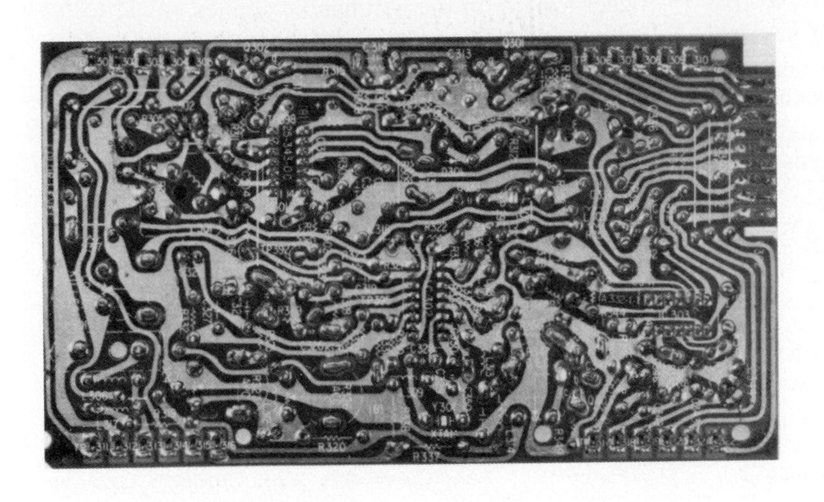

Fig. A2.3-3. Bottom view of a finished PCB.

A3.1. Familiarization with Hand Tools Used in Electronics

After an experimental circuit has been breadboarded, it is tested by a technician. Changes may be needed and are made as required. When the technician and engineer are satisfied that the circuit is performing as it should, it is ready for prototype assembly.

The proper chassis layout of circuit components is determined, the chassis is prepared, and the technician mounts the major components. In this phase of the chassis preparation and component assembly, the technician may use tools to form, drill, and cut the chassis. These tools may be power tools or hand tools such as scribe, punch, hammer, screwdriver, wrench, hacksaw, and file.

Hand Tools Used in Electronics

Electrical assembly follows the preparation of the chassis and the mounting of parts. This is the stage which requires electrical interconnection of components. It involves the preparation and soldering of conductive wiring between parts. The common hand tools used in electrical assembly include diagonal pliers, long-nose pliers, soldering aid, wire stripper, soldering iron, soldering pencil, soldering gun, knife, and heat sink.

Diagonal pliers, commonly called dykes, cutters, or diagonals, are used for cutting soft wire and component leads. They should not be used for cutting hard metals such as iron or steel. Some diagonals have a small, notched cutting surface for stripping the insulation from a wire. This stripping hole will accommodate #22 wire, which is the standard hookup wire used in electronics.

Long-nose pliers are used for holding wire so that the stripped end may be twisted around a terminal post or pushed through a terminal eye. Long-nose pliers sometimes include cutting edges so that the pliers can serve for both holding and cutting wires. The 5-in long-nose pliers is a popular size, as is the 5-in diagonal.

The needle-nose pliers is a variation of the long-nose. Its long, thin jaws can get at difficult-to-reach spots.

The soldering aid is a useful tool which simplifies soldering jobs. A standard aid has a sharp metal pointer at one end and a slotted V-type grip at the other. One function of the pointer end is to help clear solder out of terminal eyes on solder lugs. The gripping end is useful in unwrapping wire and component leads, when they are being unsoldered from terminal posts.

The wire stripper removes insulation from hookup wires. There are different types of strippers, ranging from the simple type found on diagonal pliers to the automatic strippers which can handle wires of different sizes. The automatic trip-type stripper is popular with electronics technicians. In addition to mechanical, there are also thermal strippers.

The soldering iron is still a standard electrical hand tool in the industry. A heating element inside the iron uses power from the power line. The heat is channeled to the tip. The tip is applied to and heats the area to be soldered.

Soldering irons are rated by the amount of power they dissipate, and so indirectly by the amount of heat they can develop. In hand-wired vacuum-tube circuits, where wires and leads are soldered to terminals, 100-W irons are standard. Heavier irons (250 W) may be used to solder connections to large metallic surfaces (such as a metal chassis) requiring more intense heat, in older vacuum-tube circuits.

The low-wattage soldering pencil is a soldering iron with a smaller heating capacity. It is used for soldering or unsoldering components from a printed wiring board, or in delicate soldering applications requiring low heat levels. Interchangeable tips in a wide variety of shapes and sizes add flexibility to the soldering pencil. In transistor circuits a 30-W soldering pencil is used.

The soldering gun was widely used in vacuum-tube circuits because of its fast heating characteristic. A trigger switch on the gun handle applies power to the heating and soldering tip and heats it in 10 to 30 s. Unlike the soldering iron, which is slow heating and must be left ON as long as it is in use, the soldering gun is heated only at the moment it is to be used. Between soldering operations it is left OFF. Popular sizes of guns are 100 and 125 W.

A pocket knife is useful in many small tasks requiring cutting and scraping. It can be used for scraping and cleaning terminal ends of wires and components, in preparation for soldering. Care must be exercised not to nick wires or component leads.

The heat sink is a small metal clip used to prevent overheating during soldering or unsoldering of heat-sensitive electronic parts. The heat sink is clipped onto the lead between the body of the component and the point at which heat from a soldering iron or gun is applied. The heat sink absorbs heat and reduces the amount of heat conducted to the component.

Desoldering tools simplify the job of cleaning etched (PC) board solder holes of solder while component leads are being removed from the holes. The holes must be free of solder before the terminals of a new component may be inserted. Desoldering tools used by the technician take several different forms. The most popular desoldering tool makes use of a vacuum to suck the solder from the PCB hole. One of these devices uses a spring-loaded suction tool, the second a suction device in the form of a hollow rubber ball, attached to a stainless steel or plastic tube. In operation, the open end of the tube is placed on a heated solder joint or solder hole. With the spring-loaded device, the spring is released and the molten solder is sucked into the tube, clearing the joint or hole of solder. With the ball-type tool, the ball is squeezed, and as it is released, the solder is sucked into the tube.

A third type of desoldering device is a wire braid. The braid is laid on the solder joint or hole to be cleared, and the soldering iron is placed on top of the braid. When the heat is sufficient to melt the solder, it is absorbed into the braid, leaving the hole or joint clear of solder.

CAUTION: A word of caution about the use of hand tools. They should never be used in a live circuit, that is, in a circuit where electric power is applied. These tools are made of metal, and metal is a conductor of electricity. Failure to observe this safety precaution may result in electrical shock or damage to the circuitry.

A3.2. Preparing Wire for Soldering

Wires

Hookup wire may come as a solid or stranded conductor. Stranded wire is the accepted standard as hookup wire. However, solid hookup wire is often used in circuits which are not subjected to movement. Solid wire is used for house wiring, bell wiring, transformer windings, etc. Component terminal leads are made of solid wire.

Stranded wire is made of wire threads twisted together. The number and diameter of the threads depend on the application and on the quantity of current the wire will carry. The larger the current, the larger must be the diameter of the wire. Stranded wire is flexible and may be bent and twisted without danger of breaking.

Wires conduct electricity. To prevent adjacent wires from making electrical contact, or from touching metallic or conductive surfaces, wires are covered with an insulated coating. The industry now makes extensive use of fire retardant insulation as wire covering. Older methods of insulating wires included solid-wire insulation, which used shellac or enamel into which the wire was dipped. Stranded wire was insulated by a plastic, rubber, or cloth coating.

Multiwire cables are frequently used in interconnecting electronic units or devices. For example, the several units of a radar set are interconnected by multiwire cables. A cable is made up of two or more insulated wires inside a common housing. The housing is usually a plastic or rubber sleeve, but the sleeve may also be a wire braid. Cables are designed and constructed for special applications requiring two or more separate insulated electrical conductors.

Coaxial cable is a very special type of shielded cable. It consists of an inner and outer conductor. The inner conductor may be solid or stranded. Around the inner conductor is a plastic or phenolic insulating sleeve. A braided shield is fitted around the insulating sleeve of the inner conductor. The braid acts as a second conductor around the inner conductor and is insulated from it. A rubber or plastic sleeve covers the braid and provides insulation for the coaxial assembly. Shielded cable is also manufactured without any external insulating cover.

Wire Preparation

Two or more wires that provide a conductive path for electricity must be electrically connected. This means that an uninsulated surface on one wire must be mechanically connected to an uninsulated surface on the other wire. To ensure that the wires will not separate, or the connection corrode, they are soldered at the junction.

Before wires may be connected and soldered, they must be properly prepared. This involves stripping away the insulation at the ends of the wire providing terminal leads which may be connected to each other or to a terminal lug. It also involves cleaning insulation or oxidation from the bare ends of the wire.

Mechanical wire strippers are generally used by the technician, though thermal strippers may also be employed. The technician selects the stripping hole which will fit the wire, inserts the wire into the tool, and squeezes the handles. The insulation is automatically removed, and the stripper resets in one operation.

After the insulation is removed, the technician examines the wire. If it has a shiny look, no further preparation is needed. However, a dull- or dark-appearing wire must be cleaned before it is connected and soldered. The wire may be cleaned by scraping it gently with a knife or with emery cloth. This same process is employed in removing the insulation from shellac- or enamel-covered solid wires. The technician must be careful not to "nick" the wire.

The technician may also be required to prepare coaxial or shielded wire. This type of wire is used in circuit connections where a shielded conductor is required. Coaxial cable is used for intercabling two assemblies of an electronic system and for instrument leads. The braid around the coaxial wire acts as the grounded shield for the inner conductor, which is the "hot" wire.

In preparing an uncovered shielded wire for use, the braid is combed away with a soldering aid, awl, or other pointed tool. The braid is then twisted together to form the ground lead. The insulated covering is stripped from the inner conductor. The inner and outer conductors are scraped, if necessary, before connection to the circuit.

The processes for preparing a coaxial cable for use as an instrument lead are illustrated in Fig. A3.2-1. The two ends of the cable require individual treatment because one end must be terminated in a microphone connector and the other in a probe or in two leads, depending on the use for which the cable is intended.

The preparation for mounting a microphone connector is shown in Fig. A3.2-1. One inch of cable

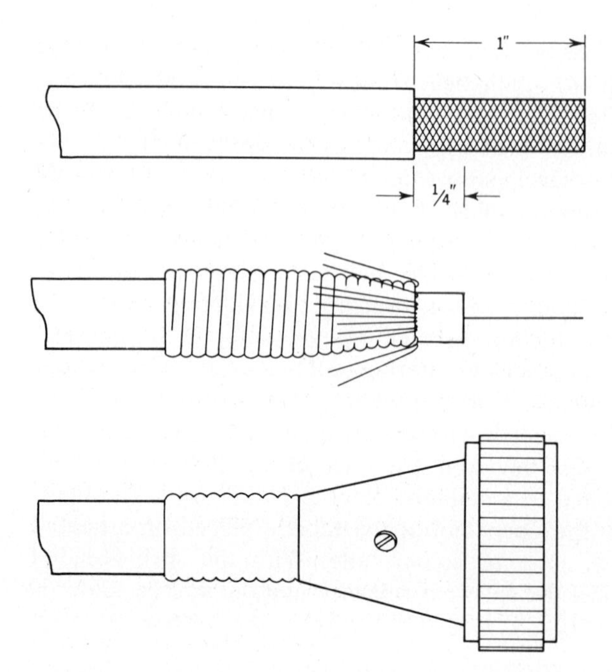

Fig. A3.2-1. Preparation of coaxial cable for attaching microphone connector.

insulation is first stripped away, exposing the braid. The spring shield of the connector is then placed over the braid. The braid is folded back and cut to fit the narrow neck of the spring. The insulation is stripped from the inner conductor. The knurled holding ring is placed over the spring so that the inner conductor protrudes through the eyelet. The ring is pushed back as far as it will go. The setscrew is tightened. Care must be taken not to "short" the inner conductor to the braid or the wire shield.

The center conductor protruding through the eyelet is cut flush with the eyelet. This completes preparation of the cable, and the inner conductor is ready to be soldered to the eyelet.

The insulation from the other end of the coaxial cable is stripped and the braid trimmed back. The insulation is stripped from the inner conductor. The braid is now combed and used directly, or a short piece of flexible ground wire may be soldered to it.

A4.1. Soldering Techniques I

Soldering is generally required to ensure permanent electric connections. Wires, or wires and terminals, are wrapped or twisted together, then solder is melted into the heated joint. When the heat is removed, the solder and wire cool, making the soldered joint look like a solid piece of metal. It is not possible, after proper soldering, to separate the wires at a joint except by breaking them or by unsoldering them.

Solder is an alloy of lead and tin. It has a low melting point and comes in wire form for electronics use. Electronics solder is made up of 60 percent tin and 40 percent lead, though the composition may vary for certain applications. Rosin-core and resin-core solder are used for soldering electronic components. Rosin and resin are fluxes which flow onto the circuit being soldered. A flux is a cleaning agent which helps to clean the surface to be soldered, ensuring a more perfect union. Acid and paste soldering flux should *not* be used in electronics.

Proper soldering requires the following:

- Clean metallic surfaces
- Sufficient heat applied to the joint to melt solder when solder is applied to the heated wire surface

The solder must not come in direct contact with the tip of the soldering iron or gun because it will melt rapidly. If the wires or terminals to be soldered have not been preheated sufficiently, the molten solder will *not* stick to their surface. The joint may look well soldered, but the chances are that it is not. *Cold-solder joints* provide poor electric contact and are difficult to discover when troubleshooting.

Soldering is not always used in the manufacturing process. There are many instances of solderless wire-wrap connections in use today.

Tinning

To ensure maximum transfer of heat from the iron to the surfaces at the joint, the tip of the iron or gun must be tinned. A new soldering tip or a tip that has been used for a long period of time must be cleaned by scraping it with a knife or emery paper, steel wool, a wire brush, or fine sandpaper. If the tip is badly pitted, it may be necessary to file it. This technique applies to *copper* tips.

Many modern tips are gold-plated or iron-bearing. A gold-plated tip should be cleaned by wiping it against a wet sponge. Iron-bearing tips can be cleaned with a wire brush. *These tips should never be filed or cleaned with sandpaper or emery cloth.*

After the tip is cleaned, it is heated. Solder is permitted to melt onto the tip, tinning it. If the iron is overheated and permitted to discolor before the solder is applied, it will be difficult to tin. Once an iron-bearing tip is tinned, it can also be cleaned with a wet sponge.

It is not only necessary to tin the iron; the surfaces to be soldered should also be cleaned and tinned. Tinned surfaces ensure good electric connections and proper bonding when soldered. Wire may be tinned by placing it on the tip of the iron and heating it sufficiently for the wire will melt solder. Stranded wire should be twisted together before it is tinned. Terminals should also be tinned. Most wire, component leads, and solder lugs used in electronics are pretinned at the factory and require no further tinning.

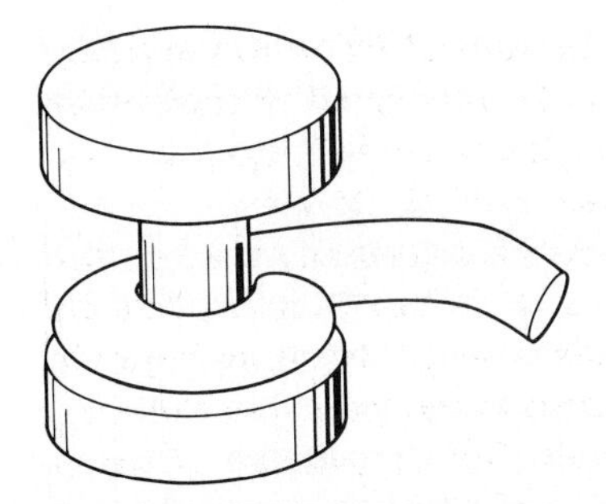

Fig. A4.1-1. Component lead secured to a terminal post before soldering.

The tip of a soldering iron tends to become dark and dirty when in use. To keep the iron at maximum heat-transfer efficiency, the tip should be cleaned periodically by wiping it on a damp sponge. A wire brush will also clean the tip.

Mechanical Connections

Wire and component leads should be secured by a three-quarter to a full turn around a terminal post, as in Fig. A4.1-1. The tendency of the novice is either to wind several turns around the post or to fail to connect the lead physically. Using too many turns around the terminal post is wasteful of terminal area and creates difficulties if it is ever necessary to unsolder and remove the lead. If the lead or wire is not physically secured to the terminal, a good mechanical connection will not be made and the joint will be weak. Also, a poor solder joint may be achieved because the lack of a physical connection may let the wires move while the solder is cooling, resulting in a cold solder joint. In Fig. A4.1-2 wire is secured to a terminal strip.

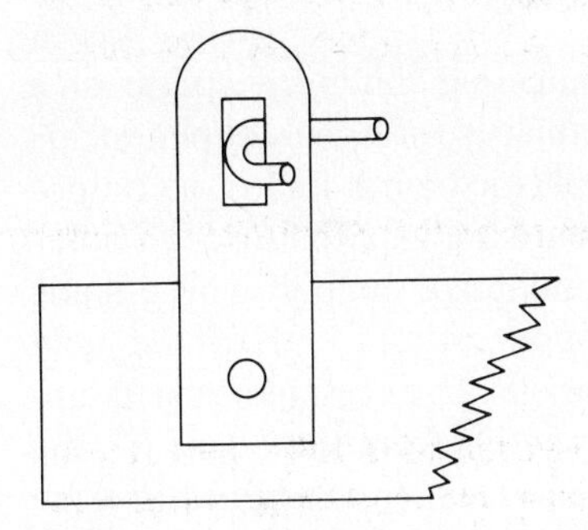

Fig. A4.1-2. Component lead secured to a terminal strip before soldering.

If two wires are to be joined together, a hook splice is used (Fig. A4.1-3). It is unnecessary to twist the wires together before soldering.

Heat Sinks

Heat sinks were mentioned in section A3.2, and you should now be able to recognize them. Heat sinks are used to avoid heat damage to heat-sensitive components during soldering. Solid-state devices such as transistors, ICs, and diodes are extremely sensitive to heat and can be permanently damaged if they are overheated. Any small metal clip, an alligator clip for example, may be used as a heat sink. Figure A4.1-4 illustrates the use of this device. It is clipped onto the lead between the component and the point at which the soldering iron is applied. The clip acts as a heat load and reduces heat transfer to the component. Low-wattage irons should be used, as too large an iron can overload the heat sink.

After the iron is removed, the heat sink should be kept in place until the joint is cool.

Wherever possible, the use of heat sinks is a good practice. However, they are not practical in many physical applications where solid-state devices have been soldered into circuit boards in such a manner that they do not allow space for heat sinks. The present trend in the manufacture of solid-state devices is toward plug-in components, for example, plug-in transistors and plug-in integrated circuits. Yet, most electric connections are soldered, and it is necessary for the technician to be able to make neat and efficient solder connections.

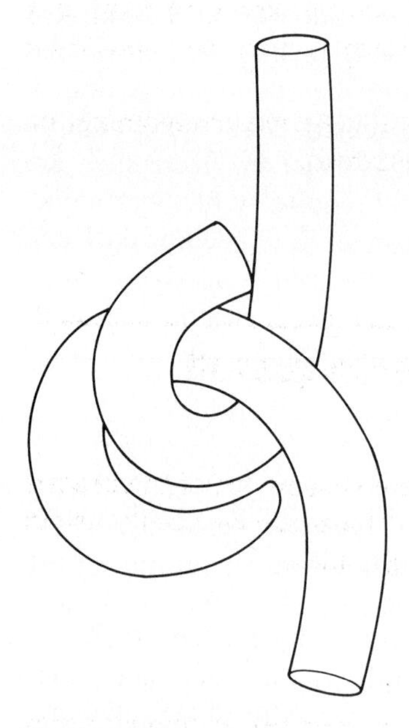

Fig. A4.1-3. Hook splice.

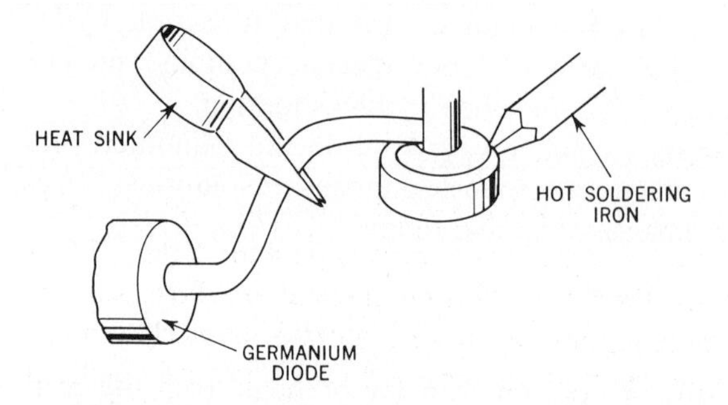

Fig. A4.1-4. Heat sink protecting germanium diode.

A4.2. Soldering Techniques II

Soldering in Printed Circuits

The network of interconnecting conductive paths on a printed-circuit board consists of thin copper strips and pads bonded to the plastic board. Leads of components mounted on the board are inserted through holes in the board and the conductive copper. These leads are soldered to the copper at the end of the hole through which they emerge. If excessive heat is applied to the copper, it may lift from the board, or the miniature components mounted on the board may be damaged.

A 30-W (approximately) soldering pencil is therefore used to heat the junction. This low-wattage iron provides an effective means of controlling heat. As in hand-wired circuits, component leads should be cleaned and tinned prior to soldering. Sixty-forty rosin-core solder or eutectic solder should be used. The surface of the copper bonded to the board should also be properly prepared prior to soldering. The copper can be cleaned by scraping it gently in the vicinity of the terminal hole, or an ink eraser may be used to clean the surface at the solder joint.

Another factor which must be considered in working on printed boards is the closeness of the conductive strips and pads to each other. Avoid excess solder to prevent bridging the gap between two copper paths. Excess solder should be avoided in hand-wired circuits also. Just enough solder should be melted into a junction to make a proper bond. If tiny solder globules form in the junction area, remove them by cleaning the soldering tip on a sponge or cloth, then apply the cleaned tip to the solder globules at the junction.

Rules for Soldering

Good soldering is part of a technician's skills. Solder connections must be mechanically strong, joints mechanically secure, so that they will not shake loose and cause loss of signal and possible damage to parts. Electrically, solder contacts must have low resistance to current flow for proper signal transfer. Some basic soldering rules are:

1. Soldering tip must be tinned and clean.
2. Metals to be soldered must be clean.
3. Support the joint mechanically where possible.
4. Pretin large surfaces before soldering them together.
5. Apply solder to joint, not to gun or iron tip. Solder must flow freely and have a shiny, smooth appearance.
6. Use only enough solder to make a solid connection.
7. Where additional flux is used, apply to joint, not to soldering tip. Only rosin or resin flux should be used.
8. Solder rapidly and do not permit components or insulation to burn or overheat.

A4.3. Soldering Techniques III

Unsoldering

Unsoldering is the reverse of the soldering process. To remove soldered leads or components from a PCB or chassis, they must first be unsoldered. To unsolder, the terminal point is heated with a properly tinned iron. The lead to be unsoldered is grasped with a soldering aid or with a pair of long-nose pliers. When the solder melts, the lead is gently unwrapped and removed from the post. If the lead was inserted through a terminal eye or a PCB hole, the eye or hole is cleaned so that it will be ready to accept a new lead. Use a heat sink if the part being unsoldered is heat-sensitive.

A terminal eye, or circuit board terminal "hole," may be cleaned by manipulating a soldering aid in the heated terminal hole, or by use of a desoldering suction tool as described in Appendix A3.1.

Excess solder or solder globules which may have separated from the post are shaken out of the chassis or board. Solder "sucker" tools are especially useful in removing these solder globules.

Safety Precautions

The soldering gun or iron operates at temperatures high enough to cause serious burns. Observe these safety precautions:

1. Do not permit hot solder to be sprayed into the air by shaking a hot gun or iron or a hot-soldered joint.
2. Always grasp a soldering gun or iron by its insulated handle. Do *not* grasp the bare metal part.
3. Do not permit the metal part of a soldering gun or iron to rest on combustible material. A soldering iron should always rest on a soldering stand.

A4.4. Soldering Techniques IV

Coaxial Cable

Coaxial cable is a special type of shielded cable. It consists of an inner and outer conductor. The inner conductor may be solid or stranded. Around the inner conductor is a plastic insulating sleeve. A shield is fitted around the insulating sleeve of the inner conductor. This shield is usually a braid of thin copper wire; however, for some uses it may be a thin sheet of foil spirally wound on top of the insulating sleeve of the inner conductor. The braid or foil acts as a second conductor around the inner conductor and is insulated from it. A rubber or plastic sleeve covers the braid and provides insulation for the coaxial assembly. Shielded cable is also manufactured without an external insulating cover.

Coaxial Cable Applications

Many test instruments use shielded cable for test leads. The inner conductor carries the "signal" to be measured by the test equipment. The shield is grounded so that interfering signals which might cause incorrect measurements are shielded from the test instrument and sent to ground. The technician must be able to make and repair test leads made of coaxial cable. Microphone connectors are frequently used to attach coaxial cable leads to test instruments.

Television signals are often carried on coaxial cable from the antenna to the set. This is generally true of large installations where the television signal is distributed throughout a motel, apartment building, etc. Many homes also have antenna distribution systems which carry the television signal to different rooms. To do this, the wire carrying the signal must go through the walls and floors, passing near electrical power lines in the home. If the television system wire were not shielded, electrical interference would result, affecting reception.

Television technicians are often called upon to install or repair antenna distribution systems and therefore must be able to work with coaxial cable and the standard television system coaxial connector, the F connector.

A5.1. Capacitor Color Code

Color Code

Presently, capacitance values of molded paper, mica, and disk ceramic capacitors are usually printed on the unit by the manufacturer. However, some equipment in use today may contain components marked according to several obsolete systems. These obsolete codes are shown here along with the codes which are presently being used for ceramic tubular-style capacitors.

The two types of molded capacitors, tubular and flat, are shown in Fig. A5.1-1. Color bands are used on the tubular type and color dots on the flat type in accordance with the code shown in Fig. A5.1-2. Figures A5.1-1 and A5.1-2 explain how the code is applied. All values are in picofarads (pF). It should be noted that the MIL code markings are somewhat different from the EIA markings. There are two types of molded mica, the flat postage-stamp type and the button silver mica, shown in Fig. A5.1-3. The accompanying obsolete color code is shown in Figs. A5.1-3 and A5.1-4. Figures A5.1-5 and A5.1-6 show an obsolete system for color-coding disk ceramic capacitors.

Figures A5.1-7 and A5.1-8 show the EIA color codes which are presently used for ceramic tubular-style capacitors.

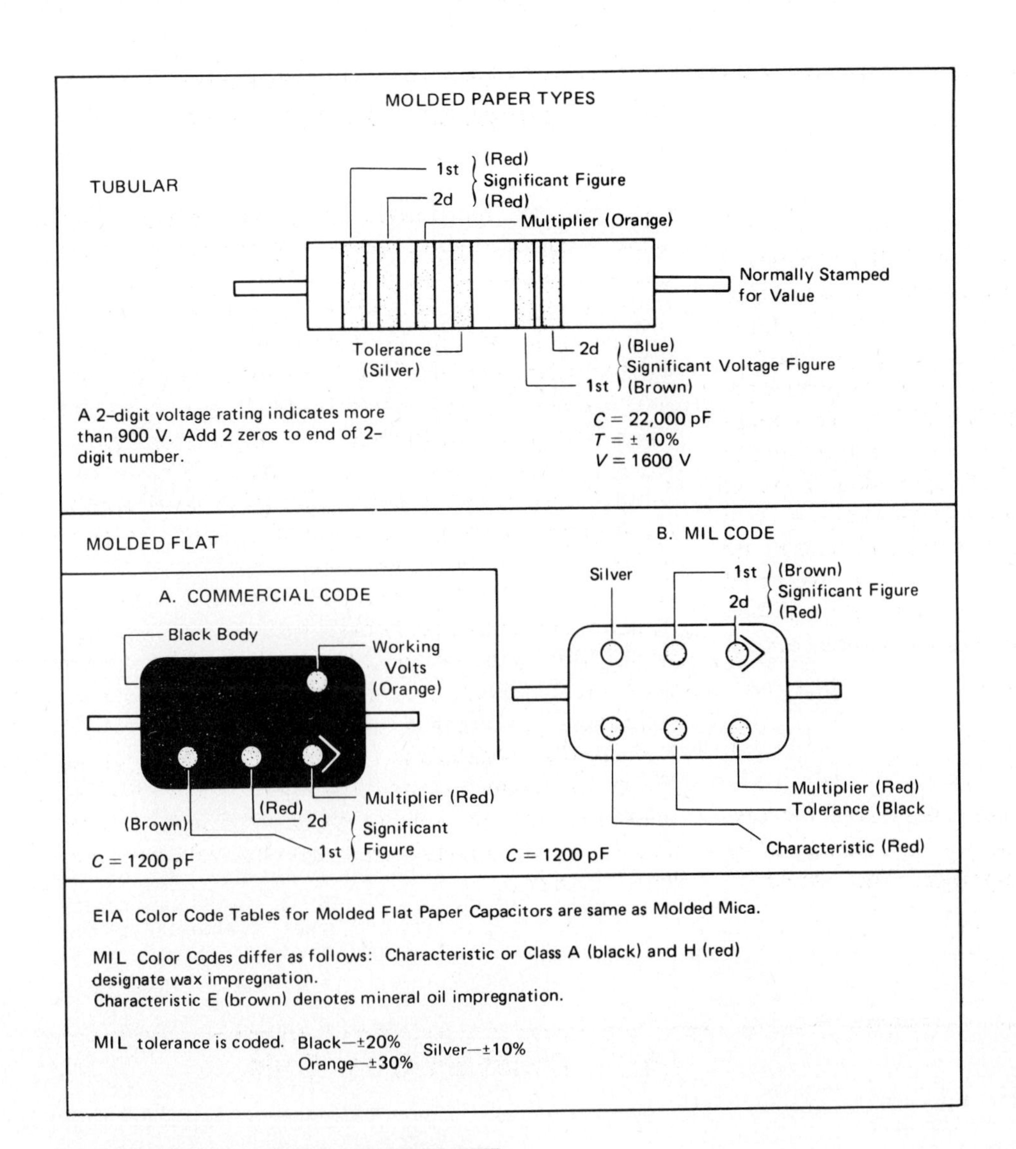

Fig. A5.1-1. Molded capacitor types. (*Centralab, a Div. of Globe-Union, Inc.*)

Color	Significant Figures	MOLDED PAPER TUBULAR CAPACITORS		
		Decimal Multiplier	Tolerance ±%	Voltage Volts
Black	0	1	20	—
Brown	1	10	—	100
Red	2	100	—	200
Orange	3	1000	30	300
Yellow	4	10000	40	400
Green	5	10^5	5	500
Blue	6	10^6	—	600
Violet	7	—	—	700
Gray	8	—	—	800
White	9	—	10_{EIA}	900
Gold	—	0.1	—	—
Silver	—	—	10_{EIA}	—
No Color	—	—	—	—

Fig. A5.1-2. Color code for molded capacitors. (*Centralab, a Div. of Globe-Union, Inc.*)

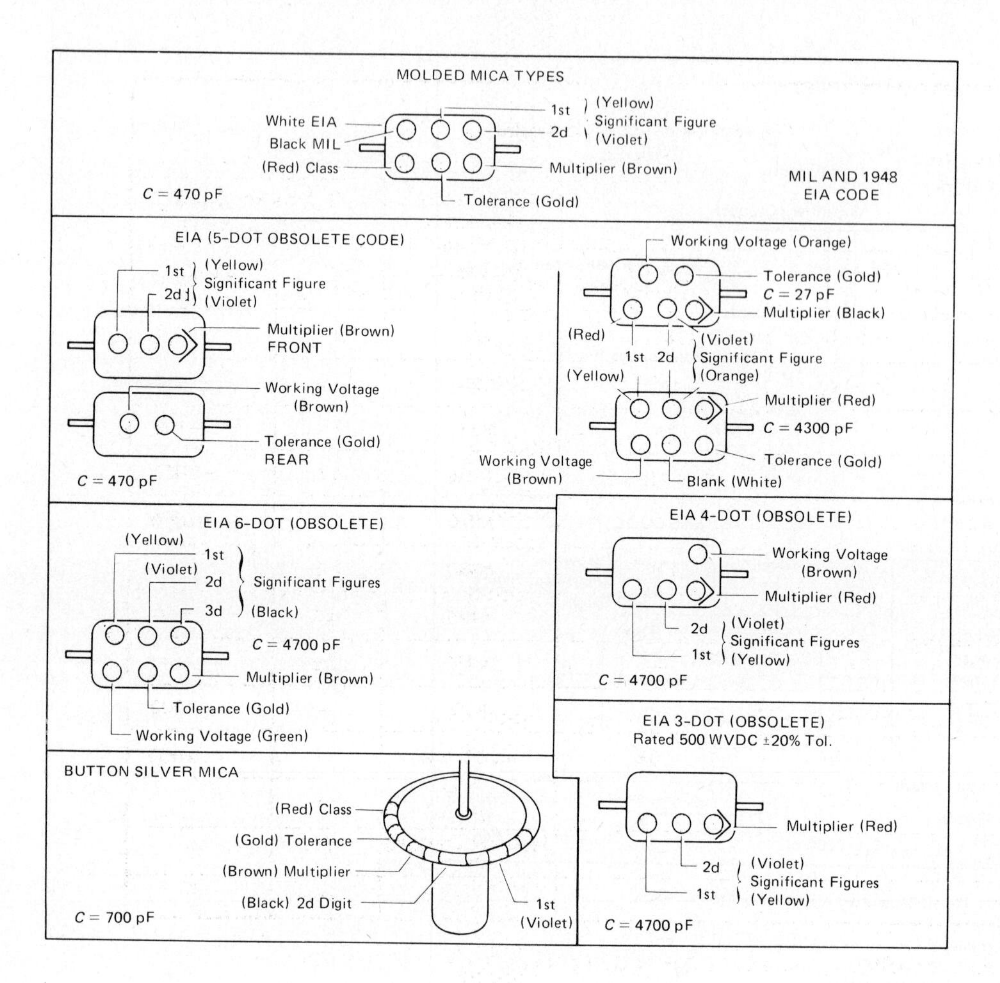

Fig. A5.1-3. Molded mica capacitor types. (*Centralab, a Div. of Globe-Union, Inc.*)

Color	Significant Figures	MOLDED MICA EIA Standard—MIL Specification Capacitors			
		Decimal Multiplier	Tolerance ±%	RMA Voltage Rating (All Capacitors)	Class
Black	0	1	20 (JAN RMA 1948)	—	A
Brown	1	10	—	100	B
Red	2	100	2	200	C
Orange	3	1000	3 (EIA)	300	D
Yellow	4	10000	—	400	E
Green	5	—	5 (EIA)	500	F (MIL)
Blue	6	—	—	600	G (MIL)
Violet	7	—	—	700	—
Gray	8	—	—	800	I (EIA)
White	9	—	—	900	J (EIA)
Gold	—	0.1	5 (MIL)	1000	—
Silver	—	0.01	10	2000	—
No Color	—	—	20 (old RMA)	500 (old RMA)	—

Fig. A5.1-4. Color code for molded mica capacitors. (*Centralab, a Div. of Globe-Union, Inc.*)

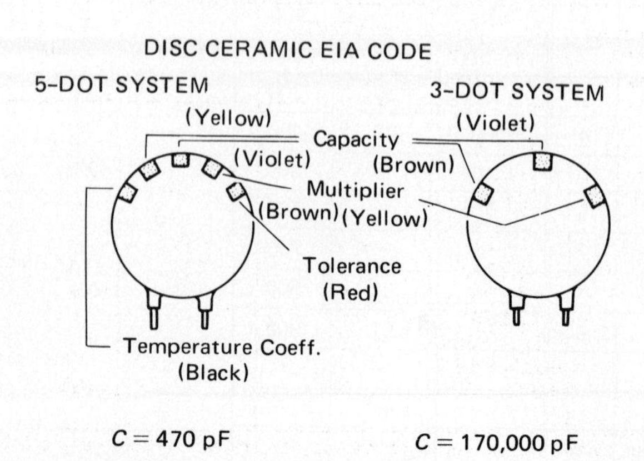

Fig. A5.1-5. Disk ceramic capacitor types. (*Centralab, a Div. of Globe-Union, Inc.*)

Color	Significant Figures	Decimal Multiplier	CERAMICS EIA Standard—MIL Specification Capacitors				
			Tolerance		Temperature Coefficient Parts Per Million Per Degree Celsius	T.C. for Extended Range TC HiCap	
			Capacity 10 pF or Less	Capacity More Than 10 pF		Significant Fig.	Multiplier
Black	0	1	±2.0 pF	±20%	NPO	0	−1
Brown	1	10	±0.1 pF	±1%	N33	—	−10
Red	2	100	—	±2%	N75	1	−100
Orange	3	1000	—	±2.5% (EIA)	N150	1.5	−1000
Yellow	4	10,000 (EIA)	—	—	N220	2.2	−10,000
Green	5	—	±0.5 pF	±5%	N330	3.3	+1
Blue	6	—	—	—	N470	4.7	+10
Violet	7	—	—	—	N750	7.5	+100
Gray	8	0.01	±0.25 pF	—	+30	—	+1000
White	9	0.1	±1.0 pF	±10%	N330±500	—	+10,000
Gold	—	—	—	—	+100 (MIL)	—	—
Silver	—	—	—	—	Bypass & Coupling (EIA)	—	—
No Color	—	—	—	—	—	—	—

Fig. A5.1-6. Color code for disk ceramic capacitors. (*Centralab, a Div. of Globe-Union, Inc.*)

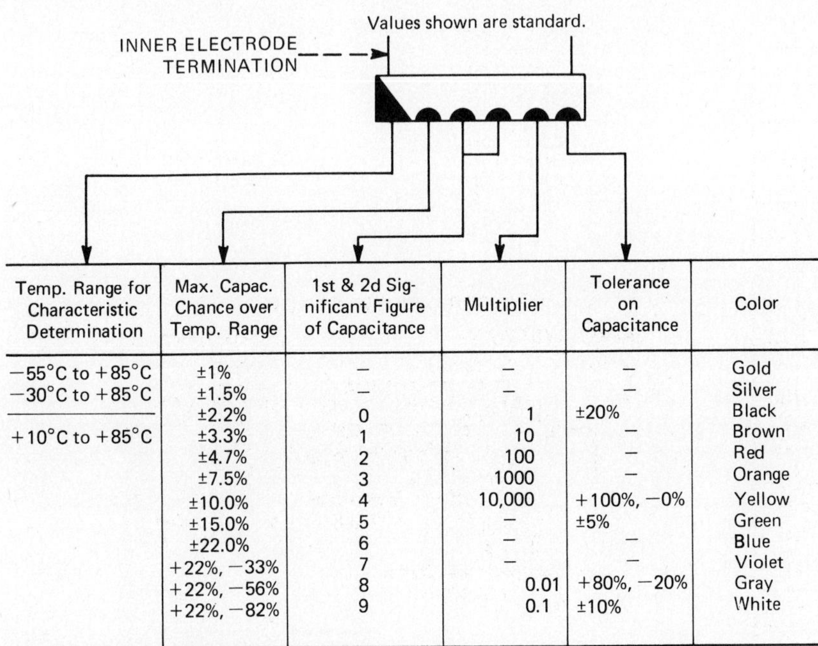

Temp. Range for Characteristic Determination	Max. Capac. Chance over Temp. Range	1st & 2d Significant Figure of Capacitance	Multiplier	Tolerance on Capacitance	Color
−55°C to +85°C	±1%	—	—	—	Gold
−30°C to +85°C	±1.5%	—	—	—	Silver
	±2.2%	0	1	±20%	Black
+10°C to +85°C	±3.3%	1	10	—	Brown
	±4.7%	2	100	—	Red
	±7.5%	3	1000	—	Orange
	±10.0%	4	10,000	+100%, −0%	Yellow
	±15.0%	5	—	±5%	Green
	±22.0%	6	—	—	Blue
	+22%, −33%	7	—	—	Violet
	+22%, −56%	8	0.01	+80%, −20%	Gray
	+22%, −82%	9	0.1	±10%	White

Note 1: Use lowest decimal multiplier to avoid alternate coding; for example, 2.0 pF should be red, black, white; not black, red, black.

Note 2: Listing of complete range of characteristics does not necessarily imply commercial availability of all values, but is for the purpose of providing a standard identification code for future development.

Fig. A5.1-7. Color coding for general-purpose tubular capacitors. (*Electronic Industries Association*)

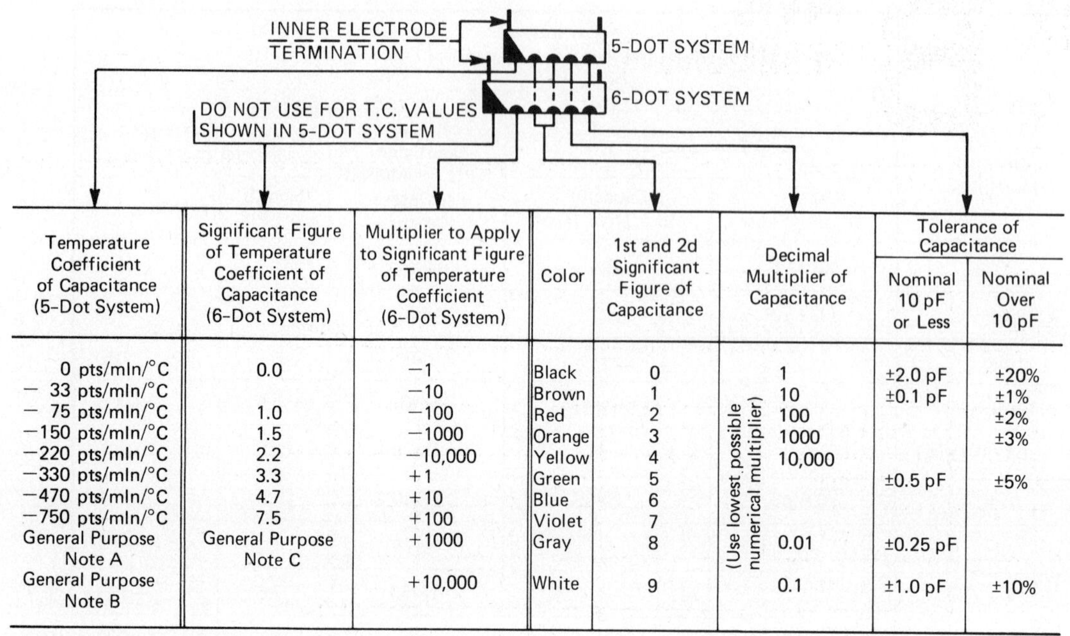

Temperature Coefficient of Capacitance (5-Dot System)	Significant Figure of Temperature Coefficient of Capacitance (6-Dot System)	Multiplier to Apply to Significant Figure of Temperature Coefficient (6-Dot System)	Color	1st and 2d Significant Figure of Capacitance	Decimal Multiplier of Capacitance	Tolerance of Capacitance	
						Nominal 10 pF or Less	Nominal Over 10 pF
0 pts/mln/°C	0.0	−1	Black	0	1	±2.0 pF	±20%
− 33 pts/mln/°C		−10	Brown	1	10	±0.1 pF	±1%
− 75 pts/mln/°C	1.0	−100	Red	2	100		±2%
−150 pts/mln/°C	1.5	−1000	Orange	3	1000		±3%
−220 pts/mln/°C	2.2	−10,000	Yellow	4	10,000		
−330 pts/mln/°C	3.3	+1	Green	5	(Use lowest possible numerical multiplier)	±0.5 pF	±5%
−470 pts/mln/°C	4.7	+10	Blue	6			
−750 pts/mln/°C	7.5	+100	Violet	7			
General Purpose Note A	General Purpose Note C	+1000	Gray	8	0.01	±0.25 pF	
General Purpose Note B		+10,000	White	9	0.1	±1.0 pF	±10%

Note A: This is a General Purpose Condenser having any nominal temperature coefficient between +150 and −1500 parts per million per degree Celsius, coefficient used to be at option of capacitor manufacturer.

Note B: This is a General Purpose Condenser having any nominal temperature coefficient between +100 and −750 parts per million per degree Celsius, coefficient used to be at option of capacitor manufacturer.

Note C: This is a General Purpose Condenser having any nominal temperature coefficient between −1000 and −5200 parts per million per degree Celsius, coefficient used to be at option of capacitor manufacturer. Use with multiplier color of black for −1.

Fig. A5.1-8. Color coding for temperature-compensating ceramic tubular capacitors. (*Electronic Industries Association*)

APPENDIX B PARTS REQUIREMENTS

Resistors, ½W

Resistance, Ω	*Quantity*	*Resistance,* Ω	*Quantity*	*Resistance,* Ω	*Quantity*
0.68*	1	2.7 kΩ	1	33 kΩ	1
68	1	3.3 kΩ	1	56 kΩ	1
150	1	4.7 kΩ	1	100 kΩ	1
330	1	5.1 kΩ	1	120 kΩ	1
390	1	5.6 kΩ	1	220 kΩ	1
470	1	6.8 kΩ	1	270 kΩ	2
560	1	8.2 kΩ	1	470 kΩ	1
680	1	10 kΩ	1	560 kΩ*	1
820	1	15 kΩ	1	1 MΩ	1
1 kΩ	4	18 kΩ	2	1.5 MΩ	1
1.2 kΩ	2	22 kΩ	1	2.2 MΩ	1
2.2 kΩ	1	27 kΩ	1	4.7 MΩ	1

* Do not mount.

Resistors, Higher Wattage

Resistance, Ω	W	*Quantity*	*Resistance,* Ω	W	*Quantity*
5	1	1	250	5	1
15	25	1	500	5	1
50	5	1	5000	5	1
100	1	2			

Potentiometers, 2 W

One each. 500-, 2500-, 5000-, 10,000-, 500,000-Ω

NOTE: 100-Ω pot will be required if there is no resistance decade box.

Tubes

One each. 6AL5, 6AV6, 12AU7A

Tube Sockets

One each. Miniature 7-pin, 9-pin

Transistors

One each unless otherwise specified. 3N187 (MOSFET), 2N2102 (two) with heat sinks, 2N4036 with heat sink, 2N6004, 2N6005

Capacitors

Capacitance, μF	*Quantity*	V	*Capacitance, μF*	*Quantity*	V
47 pF	1	35	0.05	2	35
250 pF	1	35	0.1	1	200
0.001	1	200	*0.1 miniature	1	12 or higher
0.002	1	200	0.5	1	200
0.005	1	200	1.0	1	200
0.01	2	200	1.0	2	12
*0.01 miniature	1	12 or higher	25.0	2	50
0.02	1	35	100.0	4	50

* Do not mount.

Diodes

One each unless otherwise specified. 1N34A, 1N5625 (two)

LED

RL 2000 (one) (Litronix) or equivalent

ICs and Their Mounts

One each.
CA3020, 7400, 7404, 7411, 7427, 7432

Varistor

Carborundum #333BNR-4 or equivalent (one)

LDR

CLAIREX CL5M2 or equivalent (one)

Transformers, Chokes, and Coils

- RF choke: 30 mH
- Oscillator coil: Hartley, broadcast band, Miller 2065 or equivalent
- Choke: 8 H @ 50 mA dc
- Filament transformer: 120-V primary, 26-V center-tapped secondary @ 1 A (Triad F40X or equivalent)
- Audio output transformer: Argonne type 174 or equivalent
- Audio output transformer: Argonne type AR119Z or equivalent
- Filament transformer: 120-V primary, 6.3-V secondary, required only if 6.3-V 60-Hz source is not otherwise available
- Audio output transformer: Argonne type 109 or equivalent

Miscellaneous

- Hook-up, 20 ft, #18, varnish-insulated copper
- Solenoid: 100 turns of #18 varnish-insulated copper in three layers on a 3-in-long × 1-in-diameter hollow cardboard or plastic cylindrical form
- Coil form: 2-in-long × ½-in diameter hollow cardboard or plastic cylinder, or copper tubing; 2-in soft-iron core to fit inside coil form
- Dry cells and holders: four 1½ V
- Switches: SPDT (four), SPST (three), DPDT (one)
- Fused 1-A line cord
- Loudspeaker: PM 3.2 Ω
- Magnets: Two bar, one horseshoe
- Iron filings, in salt-shaker-type dispenser
- Defective resistors: For Experiments 1.1, 9.1, 9.2, 9.3
- Defective components: For troubleshooting experiments
- Meter movement: 0–1 mA, 50 Ω, 50 mV
- Cardboard or glass plane, 8½ × 11-in
- Resistance decade box (2 W): Adjustable in 1-Ω steps from 1 to 99,999 Ω
- Magnetic compass
- Pilot light and socket: #49
- 25-W light bulb and socket
- 60-W light bulb and reflector
- Unmarked diode, for Experiment 23.3
- Unmarked transistor, for Experiment 29.1
- Audio source, such as microphone or loudspeaker
- Mounting board CA3020